LONDON MATHEMATICAL SOCIETY

Managing Editor Professor N. J. Hitchin, Mathematical Ins
Oxford OX1 3LB, United Kingdom

The titles below are available from booksellers, or from Cambridge University Press at www.cambridge/mathematics.org/

240 Stable groups, F.O. WAGNER
241 Surveys in combinatorics, 1997, R.A. BAILEY (eds)
242 Geometric Galois actions I, L. SCHNEPS & P. LOCHAK (eds)
243 Geometric Galois actions II, L. SCHNEPS & P. LOCHAK (eds)
244 Model theory of groups and automorphism groups, D. EVANS (ed)
245 Geometry, combinatorial designs and related structures, J.W.P. HIRSCHFELD *et al*
246 p-Automorphisms of finite p-groups, E.I. KHUKHRO
247 Analytic number theory, Y. MOTOHASHI (ed)
248 Tame topology and o-minimal structures, L. VAN DEN DRIES
249 The atlas of finite groups ten years on, R. CURTIS & R. WILSON (eds)
250 Characters and blocks of finite groups, G. NAVARRO
251 Gröbner bases and applications, B. BUCHBERGER & F. WINKLER (eds)
252 Geometry and cohomology in group theory, P. KROPHOLLER, G. NIBLO & R. STÖHR (eds)
253 The q-Schur algebra, S. D-ONKIN
254 Galois representations in arithmetic algebraic geometry, A.J. SCHOLL & R.L. TAYLOR (eds)
255 Symmetries and integrubility of difference equations, P.A. CLARKSON & F.W. NIJHOFF (eds)
256 Aspects of Gúlois theory, H. VŌLKLEIN *et al*
257 An introduction to noncommutative differential geometry and its physical applications, 2ed. J. MADORE
258 Sets and proofs, S.B. COOPER & J. TRUSS (eds)
259 Models and computability, S.B. COOPER & J. TRUSS (eds)
260 Groups St. Andrews 1997 in Bath, I, C.M. CAMPBELL
261 Groups St Andrews 1997 in Bath, II, C.M. CAMP
262 Analysis and logic, C.W. HENSON
263 Singularity theory, B. BRUCE &
264 New trends in algebraic geometry ... REID (eds)
265 Elliptic curves in cryptography, I
267 Surveys in combinatorics, 1999,
268 Spectral asymptotics in the semi
269 Ergodic theory and topological
270 Analysis on Lie groups, N.T. VA
271 Singular perturbations of differen
272 Character theory for the odd order theorem, T. PETERFALVI
273 Spectral theory and geometry, E.B. DAVIES & Y. SAFAROV (eds)
274 The Mandlebrot set, theme and variations, TAN LEI (ed)
275 Descriptive set theory and dynamical systems, M. FOREMAN *et al*
276 Singularities of plane curves, E. CASAS-ALVERO
277 Computational and geometric aspects of modern algebra, M.D. ATKINSON *et al*
278 Global attractors in abstract parabolic problems, J.W. CHOLEWA & T. DLOTKO
279 Topics in symbolic dynamics and applicatonis, F. BLANCHARD, A. MAASS & NOGUEIRA (eds)
280 Characters and automorphism groups of compact Riemann surfaces, T. BREUER
281 Explicit birational geometry of 3-folds, A. CORTI & M. REID (eds)
282 Auslander-Buchweitz approximations of equivariant modules, M. HASHIMOTO
283 Nonlinear elasticity, Y. FU & R.W. OGDEN (eds)
284 Foundations of computational mathemtics, R. DEVORE, A. ISERLES & E. SULI (eds)
285 Rational points on curves over finite, fields, H. NIEDERREITER & C. HING
286 Clifford algebras and spinors 2ed, P. LOUNESTO
287 Topics on Riemann surfaces and Fuchsian groups, E. BUJALANCE *et al*
288 Surveys in combinatorics, 2001. J. HIRSCHFELD (ed)
289 Aspects of Sobolev-type inequalities, I. SALOFF-COSTE
290 Quantum groups and Lie theory, A. PRESSLEY (ed)
291 Tins buildings and the model theory of groups, K. TENT (ed)
292 A qunatum groups primer, S. MAJID
293 Second order partial differential equations in Hilbert spaces, G. DA PRATO & ZABCZYK
294 Introduction to the theory of operator spaces, G. PISIER
295 Geometry and Integrability, L. MASON & YAVUZ NUTKU (eds)
296 Lectures on invariant theory, I. DOLGACHEV
297 The homotopy category of simply connected 4-manifolds, H.-J. BAIJES

298 Higher opernds, higher categories, T. LEINSTER
299 Kleinian Groups and Hyperbolic 3-Manifolds, Y. KOMORI, V. KARKOVIC & C. SERIES (eds)
300 Introduction to Mobius Differential Geometry, U. HERTRICH-JEROMIN
301 Stable Modules and the D(2)-Problem, F.E.A. JOHNSON
302 Discrete and Continuous Nonlinear Schrodinger Systems, M. J. ABLORWITZ, B. PRINARI & A. D. TRUBATCH
303 Number Theory and Algebraic Geometry, M. REID & A. SKOROBOGATOV (eds)
304 Groups St Andrews 2001 in Oxford Vol. 1, C.M. CAMPBELL, E.F. ROBERTSON & G.C. SMITH (eds)
305 Groups St Andrews 2001 in Oxford Vol. 2, C.M. CAMPBELL, E.F. ROBERTSON & G.C. SMITH (eds)
306 Peyresq lectures on geometric mechanics and symmetry, J. MONTALDI & T. RATIU (eds)
307 Surveys in Combinatorics 2003, C.D. WENSLEY (ed.)
308 Topology, geometry and quantum field theory, U.L. TILLMANN (ed)
309 Corings and Comodules, T. BRZEZINSKI & R. WISBAUER
310 Topics in Dynamics and Ergodic Theory, S. BEZUGLYI & S. KOLYADA (eds)
311 Groups topological combinatorial and arithmetic aspects, T.W. MÜLLER (ed)
312 Foundations of Computational Mathematics. Minneapolis 2002, FELIPE CUCKER *et al* (eds)
313 Transcendental aspects of algebraic cycles, S. MÜLLER-STACH & C. PETERS (eds)
314 Spectral generalizatoins of line graphis, D. CVETKOVIC. P. ROWLINSON & S. SIMIC
315 Structured ring spectra, A. BAKER & B. RICHTER (eds)
316 Linear Logic in Compuyter Science, T. EHRHARD *et al* (eds)
317 Advances in elliptic curve cryptography, I.F. BLAKE, G. SEROUSSI & N. SMART
318 Perturbation of the boundary in boundary-value problems of Partial Differential Equations, D. HENRY
319 Double Affine Hecke Algebras, I. CHEREDNIK
320 L-Functions and Galois Representations, D. BURNS, K. BUZZARD & J. NEKOVÁŘ (eds)
321 Surveys in Modern Mathematics, V. PRASOLOV & Y. ILYASHENKO (eds)
322 Recent perspectives in random matrix theory and number theory, F. MEZZADRI & N. C. SNAITH (eds)
323 Poisson geometry, deformation quantisation and group representations, S. GUTT *et al* (eds)
324 Singularities and Computer Algebra, C. LOSSEN & G. PFISTER (eds)
325 Lectures on the Ricci Flow, P. TOPPING
326 Modular Representations of Finite Groups of Lie Type, J.E. HUMPHREYS
328 Fundamentals of Hyperbolic Manifolds, R.D. CANARY, A. MARDEN & D.R.A. EPSTEIN (eds)
329 Spaces of Kleinian Groups, Y. MINSKY, M. SAKUMA & C. SERIES (eds)
330 Noncommutative Localization in Algebra and Topology, A. RANICKI (ed)
331 Foundations of Computational Mathematics, Santander 2005, L. PARDO, A. PINKUS, E. SULI & M. TODD (eds)
332 Handbook of Tilting Theory, L. ANGELERI HŪGEL, D. HAPPEL & H. KRAUSSE (eds)
333 Synthetic Differential Geometry 2ed. A. KOCK
334 The Navier-Stokes Equations, P.G. DRAZIN & N RILEY
335 Lectures on the Combinatorics of Free Probability, A. NICA & R. SPEICHER
336 Integral Closure of Ideals. Rings, and Modules, I. SWANSON & C. HUNEKE
337 Methods in Banach Space Theory, J.M.F. CASTILLO & W.H. JOHNSON (eds)
338 Surveys in Geometry and Number Theory, N. YOUNG (ed)
339 Groups St Andrews 2005 Vol. 1, C.M. CAMPBELL, M. R. QUICK, E.F. ROBERTSON & G.C. SMITH (eds)
340 Groups St Andrews 2005 Vol. 2, C.M. CAMPBELL, M. R. QUICK, E.F. ROBERTSON & G.C. SMITH (eds)
341 Ranks of Elliptic Curves and Random Matrix Theory, J. B. CONREY, D. W. FARMER, F. MEZZADRI & N. C. SNAITH (eds)
342 Elliptic Cohomology, H.R. MILLER & D.C. RAVENEL (eds)
343 Algebraic Cycles and Motives Vol. 1, J. NAGEL & C. PETERS (eds)
344 Algebraic Cycles and Motives Vol. 2, J. NAGEL & C. PETERS (eds)
345 Algebraic and Analytic Geometry, A. NEEMAN
346 Surveys in Combinatorics, 2007, A. HILTON & J. TALBOT (eds)
347 Surveys in Contemporary Mathematics, N. YOUNG & Y. CHOI (eds)
348 Transcendental Dynamics and Complex Analysis, P. RIPPON & G. STALLARD (eds)
349 Model Theory with Applications to Algebra and Analysis Vol. 1, Z. CHATZIDAKIS, D. MACPHERSON, A. PILLAY & A. WILKIE (eds)
350 Model Theory with Applications to Algebra and Analysis Vol. 2, Z. CHATZIDAKIS, D. MACPHERSON, A. PILLAY & A. WILKIE (eds)
351 Finite von Nuemann Algebras and Mases, A. SINCLAIR & R. SMITH
352 Number Theory and Polynomials, J. MCKEE & C. SMYTH (eds)
353 Trends in Stochastic Analysis, J. BLATH, P. MÖRTERS & M. SCHEUTZOW (eds)
354 Groups and Analysis, K. TENT (ed)
355 Non-equilibrium Statistical Mechanics and Turbulence, J. CARDY, G. FALKOVICH & K. GAWEDZKI
356 Elliptic Curves and Big Galois Representations, D. DELBOURGO

ELLIPTIC CURVES AND BIG GALOIS REPRESENTATIONS

DANIEL DELBOURGO

CAMBRIDGE
UNIVERSITY PRESS

CAMBRIDGE UNIVERSITY PRESS
Cambridge, New York, Melbourne, Madrid, Cape Town, Singapore, São Paulo, Delhi

Cambridge University Press
The Edinburgh Building, Cambridge CB2 8RU, UK

Published in the United States of America by Cambridge University Press, New York

www.cambridge.org
Information on this title: www.cambridge.org/9780521728669

First published 2008

Printed in the United Kingdom at the University Press, Cambridge

A catalogue record for this publication is available from the British Library

Library of Congress Cataloguing in Publication Data

Delbourgo, Daniel.
Elliptic curves and big Galois representations/Daniel Delbourgo.
p. cm.
Includes bibliographical references and index.
ISBN 978-0-521-72866-9 (pbk.)
1. Curves, Elliptic. 2. Galois theory. I. Title.

QA567.2.E44D36 2008
516.3′52–dc22 2008021192

ISBN 978-0-521-72866-9 paperback

For Mum, Dad, Tino, and Aunty Boo ...

Contents

Introduction

The connection between L-functions and arithmetic must surely be one of the most profound in mathematics. From Dirichlet's discovery that infinitely many primes occur in an arithmetic progression, right through to Wiles' celebrated proof of Fermat's Last Theorem, the applications of L-series to number theory seem to be limitless.

This book is concerned with the special values of L-functions of modular forms. The twentieth century saw many deep conjectures made about the interrelation between L-values and associated arithmetic invariants. Moreover, the last few years have seen a lot of these predictions proved correct, though much is still shrouded in mystery. Very frequently modular forms can be grouped together into families parametrised by a single analytic variable, and it is their properties which we intend to study here. Whilst we shall be primarily interested in the arithmetic of the whole family itself, the control theory often tells us something valuable about each individual member.

What then do we mean by a modular form? Let k and N be positive integers. An analytic function $f_k : \mathfrak{H} \cup \{\infty\} \longrightarrow \mathbb{C}$ is *modular of weight k and level N* if

$$f_k\left(\frac{az+b}{cNz+d}\right) = (cNz+d)^k f_k(z)$$

at all integers $a, b, c, d \in \mathbb{Z}$ such that $1 + bcN = ad$. In particular, the quadruple $(a, b, c, d) = (1, 1, 0, 1)$ clearly satisfies this condition, so we must have the identity $f_k(z+1) = f_k(z)$ for all $z \in \mathfrak{H}$. Moreover, if $f_k(z)$ is appropriately bounded as z approaches the cusps, we call f_k a modular form.

It follows that f_k is a periodic function of z, with a Fourier expansion

$$f_k(z) = \sum_{n=0}^{\infty} a_n(f_k) q^n \quad \text{where } q = \exp(2\pi i z).$$

Hecke proved that for a fixed level N and weight k, the space of modular forms is finitely-generated over $\mathbb{C}$. He introduced a system of operators 'the Hecke algebra', under whose action a basis of eigenforms can always be found.

N.B. It is very far from being true that every modular form occurring in nature hides deep secrets. For example, the Eisenstein series

$$\mathrm{Eis}_k(z) := \sum_{(0,0) \neq (m,n) \in \mathbb{Z} \times \mathbb{Z}} \frac{1}{(mz+n)^k} \quad \text{for integers } k > 2$$

whilst indispensable tools in the analytic theory, tell us precious little about the arithmetic of Diophantine equations.

We'll concern ourselves exclusively with the study of *newforms* of weight $k \geq 2$, a precise definition of which can be found in the next chapter. For the moment, we just mention that a newform f_k vanishes at the cusp ∞, so the constant term $a_0(f_k)$ in its Fourier expansion must be zero. The complex L-function

$$L(f_k, s) \quad := \quad \sum_{n=1}^{\infty} a_n(f_k).n^{-s}$$

converges in the right-half plane $\mathrm{Re}(s) > \frac{k}{2} + 1$, and can be analytically continued to the whole of the complex numbers. In contrast to the rather crude nature of Eisenstein series, newforms encode a lot of useful arithmetic data.

Let us fix an integer $s_0 \in \{1, ..., k-1\}$, and assume that $L(f_k, s_0)$ is non-zero. The conjectures of Bloch and Kato relate the value of the L-function at $s = s_0$ with the order of a mysterious group $\text{Ш}_k = \text{Ш}(f_k; s_0)$, which can be defined cohomologically. For their conjecture to make any sense, it is essential that the quantity $\#\text{Ш}_k$ be finite. In the special case where the weight $k = 2$, the object Ш_2 is the Tate-Shafarevich group of a modular elliptic curve which has f_2 as its associated newform; the Bloch-Kato Conjecture then reduces to the famous conjectures of Birch and Swinnerton-Dyer (see Chapter I for details).

Beilinson, and then subsequently Kato, discovered that the critical values of the L-function are governed by certain cohomology classes, which we now refer to as *Kato-Beilinson zeta-elements.*

Our first main result is purely technical, but nonetheless vital:

Theorem 0.1. *The space of zeta-elements generates the algebraic modular symbol associated to the cuspidal eigenform f_k.*

Let's see what the implications of this theorem are in deformation theory.

So far the weight k of our newforms has remained fixed, but we can relax this. We shall allow k to vary over the whole of the p-adic integers $\mathbb{Z}_p$, although only at positive integers can one say anything meaningful about the behaviour of newforms. Let $p > 3$ denote a prime number, and write Λ for the power series ring $\mathbb{Z}_p[\![X]\!]$. Hida showed that whenever the pth-Fourier coefficient is a p-adic unit, these modular forms come in ordinary families

$$\mathbf{f} \;=\; \sum_{n=1}^{\infty} a_n(\mathbf{f}; X)\, q^n \;\in\; \Lambda[\![q]\!]$$

where at infinitely many weights $k \geq 2$, the expansion

$$\mathbf{f}_k \;=\; \sum_{n=1}^{\infty} a_n\Big(\mathbf{f}; (1+p)^{k-2} - 1\Big)\, q^n \;\in\; \mathbb{Z}_p[\![q]\!]$$

is a classical eigenform of weight k.

This means we are no longer dealing with just a single Bloch-Kato Conjecture, rather a continuum of statements relating the quantities

$$\text{the special value } L(\mathbf{f}_k, s_0) \quad \overset{\text{Bloch-Kato}}{\longleftrightarrow} \quad \text{the } p\text{-part of } \#\text{Ш}_k.$$

We consider only the p-primary part of Ш_k above, because the choice of prime p is fundamental to the original deformation.

Not surprisingly, this raises a whole host of arithmetic questions:

Q1. *Is there a p-adic analytic L-function of k interpolating these special values?*

Q2. *What is the underlying object governing these Tate-Shafarevich groups Ш_k?*

Q3. *Can the Bloch-Kato Conjecture be formulated for the whole p-ordinary family, so that each individual conjecture is simply a manifestation at weight k?*

We shall look for answers to all these questions.

As a first step, we find an analytic parametrisation for the Galois cohomology of the representations interpolating $\{\mathbf{f}_k\}_{k\geq 2}$. Let $\mathbf{F}$ denote an abelian extension of $\mathbb{Q}$.

Theorem 0.2. *The étale Coleman exact sequence over $\mathbf{F}\otimes\mathbb{Z}_p$ deforms along the universal p-ordinary representation $\rho_\infty^{\text{univ}}: G_\mathbb{Q} \to \text{GL}_2(\mathbb{Z}_p[\![X]\!])$.*

The proof is based on the following generalisation of Theorem 0.1:

Theorem 0.3. *The weight-deformation of the zeta-elements over $\mathbb{Z}_p[\![X]\!]$, generates the universal Λ-adic modular symbol associated to the family* $\mathbf{f}$.

Of course, there is no reason at all why the point $s = s_0$ should have remain fixed. If we allow it to vary in exactly the same manner as the weight k varied, one can consider special values at all points of the critical strip, simultaneously.

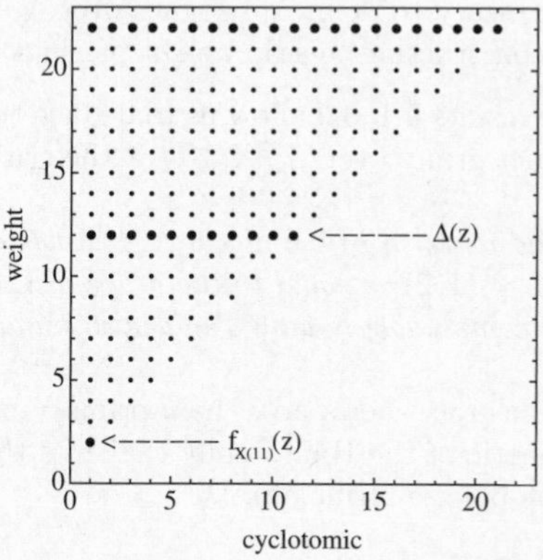

The critical region at $p = 11$ and tame level one

In terms of the deformation theory, this amounts to adding a second variable to the power series ring Λ, which means we are now working over $\mathbb{Z}_p[\![X,Y]\!]$ instead. The whole picture becomes clearer when we allow this extra cyclotomic variable Y because the full force of Iwasawa theory is at our disposal.

Our intention is to study Selmer groups associated to the following three lines: the vertical line $s=1$, the central line $s=k/2$, and the boundary line $s=k-1$. Let's consider $s=1$ and $s=k-1$ first.

Theorem 0.4. *The Selmer groups along both $s=1$ and $s=k-1$ are Λ-cotorsion, over all abelian extensions $\mathbf{F}$ of the rational numbers $\mathbb{Q}$.*

It is worthwhile remarking that the cotorsion of the cyclotomic Selmer group along the horizontal line $k=2$ was recently proven by Kato, and in the CM case Rubin. This remains one of the crown jewels in the Iwasawa theory of elliptic curves.

Unfortunately for the central line $s=k/2$, things are less clear cut.

Conjecture 0.5. *The Selmer group along $s=k/2$ should have Λ-corank equal to the generic order of vanishing of $L_p(\mathbf{f}_k, k/2)$ for even integers k.*

Greenberg predicted that the order of vanishing along $s=k/2$ was either almost always zero, or almost always one. Without knowing whether this statement holds true in general, alas 0.5 is destined to remain only a conjecture at best.

Granted we know something about the structure of these three Selmer groups, one can then compute the leading terms of their characteristic power series.

Theorem 0.6. *There are explicit formulae relating the Iwasawa invariants along $s=1$, $s=k/2$ and $s=k-1$ to the p-part of the BSD Conjecture.*

The author apologises profusely for stating the result in such a vacuous manner – for the full statements, we refer the reader to Theorems 9.18 and 10.1 in the text. To obtain these formulae is by no means trivial. If the function $L(\mathbf{f}_2, s)$ vanishes at the point $s=1$ it is necessary to define analogues over Λ of the elliptic regulator, which in turn involves constructing 'p-adic weight pairings' on an elliptic curve.

When combined, the results 0.1–0.6 allow us to deduce the arithmetic behaviour of the two-variable Selmer group over $\mathbb{Z}_p[\![X,Y]\!]$, at the critical point $(1,2)$:

Theorem 0.7. *The leading term of the algebraic two-variable p-adic L-function at $(s,k)=(1,2)$ is equal to the order of Ш_2, multiplied by some readily computable Λ-adic Tamagawa numbers.*

In particular, this last theorem shows how the p-primary part of Ш is completely controlled by the arithmetic of the Hida family that lifts the classical eigenform $\mathbf{f}_2$ (c.f. Section 10.3 for the precise formulae).

The organisation of this book is as follows. The first chapter is meant to be purely introductory, containing a very brief review of elliptic curves and modular forms. In Chapter II we recall the work of Perrin-Riou and Kato on the theory of Euler systems for modular forms. Then in Chapter III we describe a brand new method for constructing p-adic L-functions using these tools. The main advantage of our constructions is that each Euler system is assigned a modular symbol, and there is a particularly nice deformation theory for these symbols.

Once we have a working model in place for the cyclotomic variable Y, it is then time to introduce the weight variable X. In Chapter IV we provide a short description of Hida's ordinary deformation theory, which exerts strict control over the modular forms occurring in the family. The two chapters that follow contain the technical heart of the book. We develop a theory of two-variable Euler systems over $\mathbb{Z}_p[\![X, Y]\!]$, in terms of the Λ-deformation of the space of modular symbols. Since there are already ambiguities present in certain of the objects considered, we will give a construction compatible with the analytic theory of Greenberg-Stevens and Kitagawa.

The remainder of the book is completely devoted to a study of the arithmetic of p-ordinary families. In Chapter VII we explain how to associate Selmer groups over a one-variable deformation ring $\Lambda = \mathbb{Z}_p[\![X]\!]$, and hence compute their Λ-coranks. In the next two chapters we prove formulae for the p-part of the Tate-Shafarevich group of an elliptic curve (under the assumption that the number field is abelian). Finally, Chapter X ties everything together in what is rather grandly called the "Two-Variable Main Conjecture". This statement is now over the larger power series ring $\mathbb{Z}_p[\![X, Y]\!]$, and our previous Euler characteristic computations allow us to formulate the conjecture without error terms.

The reader who has done a graduate-level course in algebraic number theory, should have no trouble at all in understanding most of the material. A passing acquaintance with algebraic geometry could also be helpful. However, someone with a number theory background could easily skip the first couple of chapters, and the battle-hardened Iwasawa theorist could probably dive straight into Chapter IV.

Acknowledgements: Firstly, the author is greatly indebted to Adrian Iovita for explaining his generalisation of $B_{\max}$, and to Denis Benois for sharing his knowledge of the dual exponential map. He also thanks Paul Smith for writing Appendix C, and for both his and John Cremona's computer experiments.

For their moral support, he is particularly grateful to John Coates, David Burns, and his work colleagues at the University of Nottingham. He thanks D.P.M.M.S. for their very kind hospitality during a sabbatical leave period at the end of 2006. Lastly, many thanks to Nigel Hitchin and the team at Cambridge University Press, for their assistance and technical knowhow.

List of Notations

(a) For a field K we write $\overline{K}$ for its separable algebraic closure. At each prime number p, let μ_{p^n} denote the group of p^n-th roots of unity living inside of $\overline{K}$. If M is a $\mathbb{Z}_p[\mathrm{Gal}(\overline{K}/K)]$-module and the integer $j \geq 0$, then '$M(j)$' denotes the Tate twist $M \otimes_{\mathbb{Z}_p} \big(\varprojlim_n \mu_{p^n}\big)^{\otimes j}$. On the other hand, if $j < 0$ then it denotes the twist $M \otimes_{\mathbb{Z}_p} \mathrm{Hom}_{\mathbb{Z}_p}\big(\varprojlim_n \mu_{p^n}^{\otimes -j}, \mathbb{Z}_p\big)$.

(b) Throughout we shall fix embeddings $\overline{\mathbb{Q}} \hookrightarrow \mathbb{C}$ and $\overline{\mathbb{Q}} \hookrightarrow \overline{\mathbb{Q}}_p$ at each prime p. We write $\mathbb{C}_p$ for the completion of $\overline{\mathbb{Q}}_p$ with respect to the p-adic metric (it is an algebraically closed field). Thus we may consider all Dirichlet characters $\psi : (\mathbb{Z}/M\mathbb{Z})^\times \to \overline{\mathbb{Q}}^\times$ as taking values in both $\mathbb{C}^\times$ and $\mathbb{C}_p^\times$ under our embeddings.

(c) The maximal unramified extension $\mathbb{Q}_p^{\mathrm{nr}}$ of the p-adic numbers, has Galois group $\mathrm{Gal}\big(\mathbb{Q}_p^{\mathrm{nr}}/\mathbb{Q}_p\big) \cong \mathrm{Gal}\big(\overline{\mathbb{F}}_p/\mathbb{F}_p\big)$. The arithmetic Frobenius element $\mathrm{Frob}_p : x \mapsto x^p$ in the latter group can be considered as generating $\mathrm{Gal}\big(\mathbb{Q}_p^{\mathrm{nr}}/\mathbb{Q}_p\big)$ topologically. Moreover, we abuse notation and write 'Frob_p' for any of its lifts to $\mathrm{Gal}\big(\overline{\mathbb{Q}}_p/\mathbb{Q}_p\big)$, which are only well-defined modulo the inertia group I_p.

(d) For a ring R we denote the ith-étale cohomology group $H^i_{\text{ét}}\big(\mathrm{Spec}(R), -\big)$ by $H^i_{\text{ét}}(R, -)$, or sometimes just by $H^i(R, -)$. Assume further that R is an integral domain with field of fractions K, and write $j : \mathrm{Spec}(K) \to \mathrm{Spec}(R)$ for the inclusion morphism. Then for any sheaf $\mathcal{A}$ of abelian groups on $\mathrm{Spec}(K)$, we abbreviate $H^i\big(R, j_*(\mathcal{A})\big)$ simply by $H^i(R, \mathcal{A})$.

(e) Given an integer level $N \geq 1$, let $\Gamma_0(N)$ denote the group of unimodular matrices $\begin{pmatrix} a & b \\ c & d \end{pmatrix}$ satisfying the congruence $c \equiv 0 \pmod{N}$. Similarly, the subgroup $\Gamma_1(N)$ consists of matrices satisfying $c \equiv 0 \pmod{N}$ and $a \equiv d \equiv 1 \pmod{N}$. If $\Phi = \Gamma_0(N)$ or $\Gamma_1(N)$, then $\mathcal{S}_k(\Phi)$ is the space of cusp forms of weight k on Φ. Finally, for each primitive Dirichlet character $\epsilon : (\mathbb{Z}/N\mathbb{Z})^\times \to \mathbb{C}^\times$, we will write $\mathcal{S}_k\big(\Gamma_0(N), \epsilon\big)$ to indicate the ϵ-eigenspace

$$\left\{ f \in \mathcal{S}_k\big(\Gamma_1(N)\big) \text{ such that } f\Big|_k \begin{pmatrix} a & b \\ c & d \end{pmatrix} = \epsilon(d) f \text{ for all } \begin{pmatrix} a & b \\ c & d \end{pmatrix} \in \Gamma_0(N) \right\}.$$

CHAPTER I

Background

Although the study of elliptic curves can be traced back to the ancient Greeks, even today there remain surprisingly many unanswered questions in the subject. The most famous are surely the conjectures of Birch and Swinnerton-Dyer made almost half a century ago. Their predictions have motivated a significant portion of current number theory research, however they seem as elusive as they are elegant. Indeed the Clay Institute included them as one of the seven millenium problems in mathematics, and there is a million dollar financial reward for their resolution.

This book is devoted to studying the Birch, Swinnerton-Dyer (BSD) conjecture over the universal deformation ring of an elliptic curve. A natural place to begin is with a short exposition of the basic theory of elliptic curves, certainly enough to carry us through the remaining chapters. Our main motivation here will be to state the BSD conjecture in the most succinct form possible (for later reference). This seems a necessary approach, since the arithmetic portion of this work entails searching for their magic formula amongst all the detritus of Galois cohomology, i.e. we had better recognise the formula when it finally does appear!

After defining precisely what is meant by an elliptic curve E, we introduce its Tate module $\mathrm{Ta}_p(E)$ which is an example of a two-dimensional Galois representation. The image of the Galois group inside the automorphisms of $\mathrm{Ta}_p(E)$ was computed by Serre in the late 1960's. We next explain how to reduce elliptic curves modulo prime ideals, which then enables us to define the L-function of the elliptic curve E. This L-function is a pivotal component in the BSD formula in §1.4.

One of the highlights of the subject is the Mordell-Weil theorem, which asserts that the group of rational points on an elliptic curve is in fact finitely-generated. We shall sketch the proof of this famous result, primarily because it involves the application of 'height pairings' which will be invaluable tools in later chapters. Lastly, the connection between elliptic curves and modular forms is covered in §1.5. These important objects are introduced from a purely algebraic standpoint, because this gives us greater flexibility when visualising Beilinson's K_2-elements.

The excellent volumes of Silverman [Si1,Si2] cover just about everything you would want to know about the fundamental theory of elliptic curves, and about two thirds of this chapter is no more than a selective summary of his first tome. For the complex analytic theory there is the book of Knapp [Kn], which covers the connection with modular forms in some detail. A gentler introduction is the volume of Silverman-Tate [ST], and of course Cassels' text [Ca1] is a classic.

1.1 Elliptic curves

We say that E is an elliptic curve if it is a smooth projective curve of genus one, equipped with a specified base-point O_E. Furthermore, E is said to be defined over a field K if the underlying curve is, and in addition the base-point O_E has K-rational coordinates. Since every elliptic curve may be embedded as a cubic in projective space, we can equally well picture it in *Weierstrass form*

$$E : Y^2Z + a_1XYZ + a_3YZ^2 = X^3 + a_2X^2Z + a_4XZ^2 + a_6Z^3$$

where the a_1, a_2, a_3, a_4, a_6 all lie inside K. Under this identification, the origin O_E will be represented by the point at infinity $(X, Y, Z) = (0, 1, 0)$.

If the characteristic of K is neither 2 nor 3, one can always change coordinates to obtain (birationally) a simpler affine equation

$$E : y^2 = x^3 + Ax + B$$

where again A, B are elements of K. The non-singularity of our curve is then equivalent to the cubic $x^3 + Ax + B$ possessing three distinct roots, i.e. to the numerical condition

$$\Delta(E) = -16(4A^3 + 27B^2) \neq 0.$$

Remark: The above quantity is called *the discriminant of* E and depends on that particular choice of Weierstrass equation. On the other hand, the j-*invariant*

$$j(E) := 1728 \times \frac{4A^3}{4A^3 + 27B^2}$$

is independent of this choice, and classifies elliptic curves up to isomorphism.

The principal reason why the theory of elliptic curves is so rewarding is because the points on an elliptic curve are endowed with the structure of an abelian group. If P_1 and P_2 are two such points on E, then their sum P_3 = "$P_1 + P_2$" is the unique point satisfying

$$(P_3) - (O_E) \sim (P_1) + (P_2) - 2(O_E) \qquad \text{inside } \mathrm{Pic}^0(E),$$

the degree zero part of the divisor class group of E. In terms of the Weierstrass equation $y^2 = x^3 + Ax + B$, it can be shown that the x-coordinate of P_3 is

$$x(P_3) = x(P_1 + P_2) = \left(\frac{y_1 - y_2}{x_1 - x_2}\right)^2 - (x_1 + x_2)$$

when $P_1 = (x_1, y_1)$ differs from $P_2 = (x_2, y_2)$, and if they are the same point

$$x(P_3) = x(P_1 + P_1) = \frac{x_1^4 - 2Ax_1^2 - 8Bx_1 + A^2}{4x_1^3 + 4Ax_1 + 4B}.$$

Geometrically three points sum to zero if and only if they lie on the same line, so the additive inverse of $P_1 = (x_1, y_1)$ must be $-P_1 = (x_1, -y_1)$.

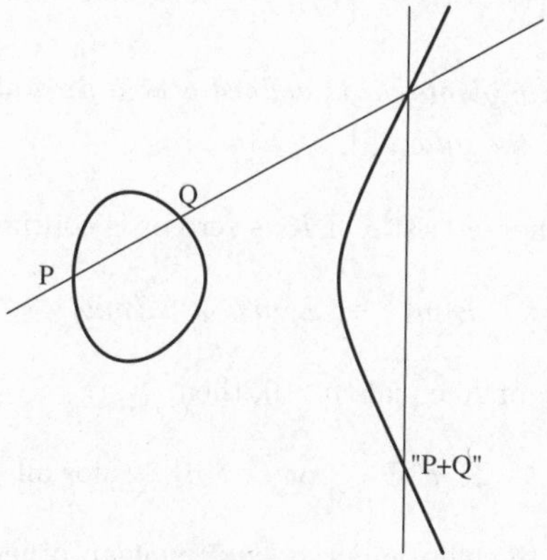

Figure 1.1

Adding the same point repeatedly to itself a fixed number of times gives rise to an endomorphism of E, defined over K. More precisely, for an integer $m \in \mathbb{Z}$ we denote by $[\times m] \in \mathrm{End}(E)$ the map for which

$$[\times m]P \quad = \quad \begin{cases} P + \cdots + P & \text{if } m > 0 \\ O_E & \text{if } m = 0 \\ -P - \cdots - P & \text{if } m < 0 \end{cases} \qquad \text{“ } |m| \text{-times ”}$$

at all points $P \in E$.

Actually there are not that many possibilities for the endomorphism ring of E. If the field K has characteristic zero then $\mathrm{End}(E)$ is either $\mathbb{Z}$, or an order in an imaginary quadratic field in which case we say that E has *complex multiplication.* Note that if K has positive characteristic then $\mathrm{End}(E)$ could also be a maximal order in a quaternion algebra.

Isogenies and the Tate module.
Suppose now that E' is another elliptic curve defined over K. An isogeny between E and E' is a non-constant morphism $\phi : E \to E'$ of curves such that $\phi(O_E) = O_{E'}$. In particular, ϕ is a group homomorphism whence $\phi(P_1 + P_2) = \phi(P_1) + \phi(P_2)$. The kernel of ϕ will be a finite subgroup of E, and the degree of ϕ is its degree as a finite map of curves.

The dual isogeny $\widehat{\phi} : E' \to E$ is then characterized by the property that

$$\widehat{\phi} \circ \phi \;=\; [\times n]_E \quad \text{and} \quad \phi \circ \widehat{\phi} \;=\; [\times n]_{E'} \quad \text{where } n = \deg(\phi).$$

If $\lambda : E \to E'$ and $\theta : E' \to E$ are two further isogenies, their duals satisfy

$$\widehat{\widehat{\phi}} = \phi, \qquad \widehat{\phi + \lambda} \;=\; \widehat{\phi} + \widehat{\lambda} \qquad \text{and} \qquad \widehat{\phi \circ \theta} \;=\; \widehat{\theta} \circ \widehat{\phi}.$$

Definition 1.1. *The kernel of the isogeny $[\times m] : E \to E$ is denoted by*

$$E[m] \;:=\; \mathrm{Ker}\big([\times m]\big) \;=\; \Big\{P \in E \quad \textit{such that} \quad [\times m]P \;=\; O_E\Big\},$$

and consists of geometric points on E defined over a fixed algebraic closure $\overline{K}$. We also write E_{tors} for the union $\bigcup_{m\geq 1} E[m]$.

Remarks: (a) If the characteristic of K is zero or is coprime to $m \geq 1$, then

$$E[m] \;\cong\; \mathbb{Z}/m\mathbb{Z} \times \mathbb{Z}/m\mathbb{Z};$$

(b) If the characteristic of K equals $p > 0$, then

$$E[p^n] \;\cong\; \mathbb{Z}/p^n\mathbb{Z} \qquad \text{or} \qquad \{0\} \qquad \text{for all } \; n \geq 1;$$

if $E[p^n]$ is zero we call the elliptic curve *supersingular*, otherwise it is *ordinary.*

Let us fix a prime number p. The multiplication-by-p endomorphism induces a transition map $[\times p] : E[p^{n+1}] \longrightarrow E[p^n]$ of finite p-groups, for all integers $n \geq 1$. The projective limit is called the p-adic Tate module of E, and is written as

$$\mathrm{Ta}_p(E) \;=\; \varprojlim_n E[p^n].$$

Whenever the characteristic of K is coprime to p, we see from part (a) of the above remark that there is a naive decomposition

$$\mathrm{Ta}_p(E) \;\cong\; \mathbb{Z}_p \oplus \mathbb{Z}_p,$$

or in terms of $\mathbb{Q}_p$-vector spaces

$$V_p(E) \;:=\; \mathrm{Ta}_p(E) \otimes_{\mathbb{Z}_p} \mathbb{Q}_p \;\cong\; \mathbb{Q}_p \oplus \mathbb{Q}_p.$$

The advantage of studying torsion points on elliptic curves is that they provide us with many examples of Galois representations, which we describe below.

Recall that E is an elliptic curve defined over K. In addition, we shall now suppose our field K to be perfect. The action of the Galois group $G_K = \mathrm{Gal}(\overline{K}/K)$ commutes with the group law on E, so leaves the finite subgroup $E[p^n]$ stable. Provided the characteristic of K is coprime to p, one obtains a two-dimensional representation

$$\rho_{E,p}^{(n)} : G_K \longrightarrow \mathrm{Aut}\big(E[p^n]\big) \;\cong\; \mathrm{GL}_2\big(\mathbb{Z}/p^n\mathbb{Z}\big)$$

for all integers $n \geq 1$. Passing to the limit over n yields

$$\rho_{E,p} : G_K \longrightarrow \mathrm{Aut}\big(\mathrm{Ta}_p(E)\big) \;\cong\; \mathrm{GL}_2(\mathbb{Z}_p),$$

and we shall also write $\rho_{E,p} : G_K \longrightarrow \mathrm{Aut}\big(V_p(E)\big) \;\cong\; \mathrm{GL}_2(\mathbb{Q}_p)$ for the associated vector space Galois representation.

It is only natural to ask how large the image of Galois in the automorphism group of the Tate module can be. If our elliptic curve E admits complex multiplication and K is a number field, then $\rho_{E,p}$ is neatly described using global class field theory, in fact the image of the Galois group G_K is "almost abelian" (see Theorem 1.9 for an explanation).

In the non-CM case we have the following description.

Theorem 1.2. *(Serre) Assume that K is a number field, and the elliptic curve E is without complex multiplication. Then*
(a) The image of $\rho_{E,p}$ is of finite index in $\mathrm{GL}_2(\mathbb{Z}_p)$ for all primes p;
(b) The image of $\rho_{E,p}$ is equal to $\mathrm{GL}_2(\mathbb{Z}_p)$ for all but finitely many primes p.

Let us focus attention for the moment on the relationship between $\mathrm{Ta}_p(E)$ and its dual module $\mathrm{Ta}_p(E)^* = \mathrm{Hom}_{\mathbb{Z}_p}\big(\mathrm{Ta}_p(E), \mathbb{Z}_p\big)$. Again fix an integer $m \geq 2$ which is coprime to the characteristic of K. The *Weil pairing* is a pairing

$$[,]_E : E[m] \times E[m] \longrightarrow \mu_m(\overline{K})$$

satisfying the following properties:

(i) *Bilinearity:* $$[P_1 + P_2, Q]_E = [P_1, Q]_E \times [P_2, Q]_E, \qquad [P, Q_1 + Q_2]_E = [P, Q_1]_E \times [P, Q_2]_E;$$

(ii) *Alternation:* $[P, P]_E = 1$;

(iii) *Non-degeneracy:* $[P, Q]_E = 1$ for all $Q \in E[m]$ if and only if $P = O_E$;

(iv) *Equivariance:* $[P^\sigma, Q^\sigma]_E = [P, Q]_E^\sigma$ for all $\sigma \in G_K$.

N.B. Here $\mu_m(\overline{K})$ denotes the group of mth-roots of unity inside the multiplicative group $\mathbb{G}_m$. For a more explicit description of this pairing, choose some element g in the function field $K(E)$ whose divisor satisfies

$$\mathrm{div}(g) = \sum_{P' \in E[m]} (P + P') - (P').$$

Then

$$[P, Q]_E = \frac{g(R + Q)}{g(R)}$$

at any point $R \in E$ such that g does not vanish at R nor $R + Q$.

If $m = p^n$ clearly the Weil pairing identifies the second exterior power of $E[p^n]$ with the group μ_{p^n} for all integers $n \geq 1$. Further, the determinant of $\rho_{E,p}$ must equal the pth-cyclotomic character

$$\chi_{\mathrm{cy}} : G_K \longrightarrow \mathbb{Z}_p^\times \quad \text{where} \quad \sigma(\zeta_{p^n}) = \zeta_{p^n}^{\chi_{\mathrm{cy}}(\sigma)} \quad \text{for all } \sigma \in G_K \text{ and } \zeta_{p^n} \in \mu_{p^n}.$$

In terms of the Tate module of E, we obtain a bilinear, alternating pairing

$$[\,,\,]_E : \mathrm{Ta}_p(E) \times \mathrm{Ta}_p(E) \longrightarrow \mathbb{Z}_p(1)$$

which is both perfect and Galois equivariant. The notation $\mathbb{Z}_p(1)$ denotes the projective limit $\varprojlim_n \mu_{p^n}$, viewed as the Tate module of the multiplicative group. In fact the module $\mathrm{Ta}_p(E)$ is dual to the étale cohomology group $H^1_{\text{ét}}(E \otimes_K \overline{K}, \mathbb{Z}_p)$, which is sometimes a good way to think of it.

Finally, it would be inappropriate not to mention the following deep result, which was first conjectured by Tate in the 1960's.

Theorem 1.3. *(Faltings) Assume that K is any number field, and p is a prime. If E and E' are two elliptic curves defined over K, then the natural map*

$$\mathrm{Hom}_K(E, E') \otimes \mathbb{Z}_p \longrightarrow \mathrm{Hom}_{\mathbb{Z}_p}\big(\mathrm{Ta}_p(E), \mathrm{Ta}_p(E')\big)^{G_K}$$

is an isomorphism.

Reducing elliptic curves over finite fields.
Consider now a field F containing the field of definition K of our elliptic curve E. We write

$$E(F) = \Big\{P = (x, y) \in E \quad \text{such that} \quad x, y \in F\Big\} \cup \{O_E\}$$

for the group of F-rational points on E. In this section we will discuss the structure of $E(F)$ when F is a local field. However before we do this, we say a word or two about the behaviour of elliptic curves over finite fields.

Assume that $K = \mathbb{F}_q$ is a finite field consisting of $q = p^d$ elements. The qth-power Frobenius map $\mathfrak{f} : E \to E$ is given on Weierstrass coordinates by $\mathfrak{f}(x, y) = (x^q, y^q)$. Clearly $E(\mathbb{F}_q) = \mathrm{Ker}(1 - \mathfrak{f})$, thus

$$\#E(\mathbb{F}_q) = \#\mathrm{Ker}(1 - \mathfrak{f}) = \deg(1 - \mathfrak{f})$$

as the morphism $1 - \mathfrak{f}$ is separable.

Furthermore, if we put $a = \mathfrak{f} + \widehat{\mathfrak{f}} \in \mathbb{Z} \subset \mathrm{End}(E)$, and in addition observe that $\mathfrak{f} \circ \widehat{\mathfrak{f}} = \deg(\mathfrak{f}) = q \in \mathbb{Z} \subset \mathrm{End}(E)$, then

$$m^2 + amn + qn^2 = \Big(m + n\mathfrak{f}\Big) \circ \Big(m + n\widehat{\mathfrak{f}}\Big) \geq 0$$

for any $(m, n) \in \mathbb{Z} \times \mathbb{Z}$.

As a quadratic form it is positive semi-definite, so its discriminant must satisfy $a^2 - 4q \leq 0$. For the particular value $(m, n) = (1, -1)$ one finds that

$$\#E(\mathbb{F}_q) = \deg(1 - \mathfrak{f}) = 1 - a + q$$

and since $|a| \leq 2\sqrt{q}$, we have shown the following estimate.

Theorem 1.4. *(Hasse) There is an upper bound* $\left|q+1-\#E(\mathbb{F}_q)\right| \leq 2\sqrt{q}$.

Even more is true. Weil proved that the zeta-function

$$Z(E/\mathbb{F}_q, T) \;=\; \exp\left(\sum_{n=1}^{\infty} \#E(\mathbb{F}_{q^n})\frac{T^n}{n}\right)$$

is none other than the rational function $\left(1-aT+qT\right)\left(1-T\right)^{-1}\left(1-qT\right)^{-1}$ of T. Moreover, the numerator has roots of weight $\frac{1}{2}$, i.e.

$$1-aT+qT^2 \;=\; (1-\alpha T)(1-\beta T) \qquad \text{with } |\alpha| = |\beta| = \sqrt{q}.$$

See Silverman's book [Si1] for a proof of this statement.

Suppose that K is a finite algebraic extension of $\mathbb{Q}_p$ with ring of integers $\mathcal{O}_K$, maximal ideal $\mathfrak{p}$ and residue field $\mathbb{F}_q = \mathcal{O}_K/\mathfrak{p}$. Let E be an elliptic curve over K. By making a suitable change of coordinates, one may assume that our elliptic curve is given by a Weierstrass equation

$$E \,:\, y^2 + a_1xy + a_3y \;=\; x^3 + a_2x^2 + a_4x + a_6$$

for which $a_1, a_2, a_3, a_4, a_6 \in \mathcal{O}_K$ and $\mathrm{ord}_{\mathfrak{p}}\big(\Delta(E)\big)$ is *minimal.*

Definition 1.5. *The reduction of* E *modulo* $\mathfrak{p}$ *is the equation*

$$\widetilde{E} \,:\, y^2 + \widetilde{a_1}xy + \widetilde{a_3}y \;=\; x^3 + \widetilde{a_2}x^2 + \widetilde{a_4}x + \widetilde{a_6}$$

where the superscript $\widetilde{\ }$ *denotes the image inside the residue field* $\mathbb{F}_q$.

Warning: Be careful, the curve $\widetilde{E}$ might very well be singular! If $\widetilde{E}$ is itself an elliptic curve, we say that E has *good reduction* over K. Otherwise if $\widetilde{E}$ has a singularity, then E has *bad reduction* over K.

Let us write $\widetilde{E}^{\mathrm{ns}}$ for the group of non-singular points on the reduction of E modulo $\mathfrak{p}$. The following three possibilities can occur:

(i) $\widetilde{E} = \widetilde{E}^{\mathrm{ns}}$ is an elliptic curve;

(ii) $\widetilde{E}$ is a nodal cubic and $\widetilde{E}^{\mathrm{ns}} \cong \mathbb{G}_{\mathrm{mult}}$, the multiplicative group over $\mathbb{F}_q$;

(iii) $\widetilde{E}$ is a cuspidal cubic and $\widetilde{E}^{\mathrm{ns}} \cong \mathbb{G}_{\mathrm{add}}$, the additive group over $\mathbb{F}_q$.

In case (iii) we say that E has *additive* reduction at $\mathfrak{p}$ – it turns out that one can replace K by a finite algebraic field extension, so that the reduction type becomes either (i) or (ii).

In case (ii) there are yet two further possibilities. If the tangent directions at the node of $\widetilde{E}$ are defined over $\mathbb{F}_q$ we say that E has *split multiplicative* reduction at $\mathfrak{p}$, otherwise it has *non-split multiplicative* reduction. The reduction type can always be made split by passing to an appropriate unramified quadratic extension of K.

Remark: Tate [Ta1] proved that every elliptic curve E which is split multiplicative over the local field K, has a p-adic uniformization

$$E(K) \xrightarrow{\sim} K^\times/\mathbf{q}_E^{\mathbb{Z}}.$$

The Tate period $\mathbf{q}_E \in \mathfrak{p}$ itself is related to the j-invariant via the transcendental equation $j(E) = J(\mathbf{q}_E)$, where $J(q) = q^{-1} + 744 + 196884q + 21493760q^2 + \cdots$ denotes the modular J-function.

We now define subgroups $E_1 \subset E_0$ by

$$E_0(K) \ := \ \left\{ P \in E(K) \ \text{ such that } \ \widetilde{P} \in \widetilde{E}^{\mathrm{ns}}(\mathbb{F}_q) \right\}$$

and likewise

$$E_1(K) \ := \ \left\{ P \in E(K) \ \text{ such that } \ \widetilde{P} = \widetilde{O}_E \right\}.$$

For example, if E has good reduction over K then $\widetilde{E} = \widetilde{E}^{\mathrm{ns}}$ is an elliptic curve, hence $E_0(K)$ coincides with $E(K)$.

In general, the two groups are connected by the short exact sequence

$$0 \longrightarrow E_1(K) \longrightarrow E_0(K) \longrightarrow \widetilde{E}^{\mathrm{ns}}(\mathbb{F}_q) \rightarrow 0.$$

In other words, the quotient group $E_0(K)/E_1(K)$ is the group of non-singular points on the reduced curve. It remains to discuss the structure of $E_1(K)$ and $E(K)/E_0(K)$, in order to completely describe the group of local points on E.

Proposition 1.6. *(a) The group $E(K)/E_0(K)$ is cyclic, and the index*

$$[E(K) : E_0(K)] = \begin{cases} 1 & \textit{if } E \textit{ has good reduction over } K \\ 2, 3 \textit{ or } 4 & \textit{if } E \textit{ has additive reduction over } K \\ 2 & \textit{if } E \textit{ has non-split multiplicative reduction over } K \\ \mathrm{ord}_{\mathfrak{p}}(\mathbf{q}_E) & \textit{if } E \textit{ has split multiplicative reduction over } K \end{cases}$$

is called the Tamagawa factor of E over K.

(b) There is an isomorphism $E_1(K) \cong \widehat{E}(\mathfrak{p})$ where $\widehat{E}$ is the formal group associated to E over $\mathcal{O}_K$. If the reduction $\widetilde{E}$ at $\mathfrak{p}$ is a supersingular elliptic curve, then the height of $\widehat{E}$ is 2. If $\widetilde{E}$ is an ordinary elliptic curve or E has bad multiplicative reduction at $\mathfrak{p}$, then the height of $\widehat{E}$ equals 1.

(c) The group $E_1(K)$ has no m-torsion if $m \notin \mathfrak{p}$. Furthermore, if $K/\mathbb{Q}_p$ is an unramified extension then $E_1(K)$ is torsion-free, unless $p = 2$ in which case it may possess elements of order 2.

The proof of statement 1.6(a) exploits the fact that the group $E(K)/E_0(K)$ is isomorphic to the group of connected components of the Néron model of the curve, whose structure was found by Kodaira and Néron. In characteristic 2 or 3 there is a lengthy algorithm of Tate's which can compute the special fiber.

The proof of statement 1.6(c) is based on the following formula:

$$\mathrm{ord}_{\mathfrak{p}}(x) \;\leq\; \frac{\mathrm{ord}_{\mathfrak{p}}(p)}{p^n - p^{n-1}}$$

giving an upper bound on the valuation of an element $x \in \widehat{E}(\mathfrak{p})$ of exact order p^n. If $K/\mathbb{Q}_p$ is unramified then $\mathrm{ord}_{\mathfrak{p}}(p) = 1$, so x must be zero unless $p = 2$.

It is wise to avoid the prime 2 at all costs.

1.2 Hasse-Weil *L*-functions

We now introduce an analytic invariant of an elliptic curve called its L-function, whose mysterious properties motivate much current research into number theory. Let $\mathbf{F}$ be a number field, and write $\mathcal{O}_F$ for its ring of integers. For a prime $\mathfrak{p}$ of $\mathbf{F}$ we shall write $\mathbf{F}_{\mathfrak{p}}$ for its completion at $\mathfrak{p}$. If $\mathfrak{p}$ is archimedean then $\mathbf{F}_{\mathfrak{p}} = \mathbb{R}$ or $\mathbb{C}$, otherwise it will be a finite extension of $\mathbb{Q}_p$ where $p > 0$ is the characteristic of the residue field at $\mathfrak{p}$.

Assume that E is an elliptic curve defined over $\mathbf{F}$. If $\mathbf{F}$ has class number one, then it is possible to find a Weierstrass equation

$$E \,:\, y^2 + a_1xy + a_3y \;=\; x^3 + a_2x^2 + a_4x + a_6$$

which is simultaneously minimal at all non-archimedean primes $\mathfrak{p}$ of $\mathbf{F}$. However if the class number is strictly greater than one, we need to work with $\mathfrak{p}$-minimal Weierstrass equations separately at each prime $\mathfrak{p}$.

In any case, if E has good reduction over $\mathbf{F}_{\mathfrak{p}}$ then we assign an integer

$$a_{\mathfrak{p}}(E) \;:=\; q_{\mathfrak{p}} \;+\; 1 \;-\; \#\widetilde{E}(\mathbb{F}_{q_{\mathfrak{p}}})$$

where the finite residue field $\mathbb{F}_{q_{\mathfrak{p}}} = \mathcal{O}_{\mathbf{F}}/\mathfrak{p}$ contains $q_{\mathfrak{p}}$ elements.

Definition 1.7. *(a) The local L-factor of E at $\mathfrak{p}$ is the polynomial*

$$L_{\mathfrak{p}}(E,T) \;=\; \begin{cases} 1 - a_{\mathfrak{p}}(E)T + q_{\mathfrak{p}}T^2 & \textit{if } E \textit{ has good reduction at } \mathfrak{p} \\ 1 - T & \textit{if } E \textit{ is split multiplicative at } \mathfrak{p} \\ 1 + T & \textit{if } E \textit{ is non-split multiplicative at } \mathfrak{p} \\ 1 & \textit{if } E \textit{ has additive reduction at } \mathfrak{p}. \end{cases}$$

(b) The Hasse-Weil L-function of E over $\mathbf{F}$ has the Euler product expansion

$$L(E/\mathbf{F},s) \;=\; \prod_{\text{primes } \mathfrak{p}} L_{\mathfrak{p}}(E,\, q_{\mathfrak{p}}^{-s})^{-1} \qquad \textit{for } \mathrm{Re}(s) \gg 0$$

where the product ranges over all non-archimedean primes of $\mathbf{F}$.

Hasse's Theorem 1.4 tells us that the modulus of $a_{\mathfrak{p}}(E)$ is bounded above by $2\sqrt{q_{\mathfrak{p}}}$, hence the Euler product will certainly converge in the right half-plane $\mathrm{Re}(s) > \frac{3}{2}$. It is fairly straightforward to prove that if E and E' are two $\mathbf{F}$-isogenous elliptic curves then their two L-series coincide, in other words $L(E/\mathbf{F}, s) = L(E'/\mathbf{F}, s)$. The converse statement is harder to deduce, relying heavily on Theorem 1.3.

Conjecture 1.8. *The L-function $L(E/\mathbf{F}, s)$ has an analytic continuation to the whole complex plane, and the completed L-function*

$$\Lambda_\infty(E/\mathbf{F}, s) \quad := \quad N_{E/\mathbf{F}}^{s/2}\,(2\pi)^{-s}\,\Gamma(s)\,L(E/\mathbf{F}, s)$$

satisfies the functional equation

$$\Lambda_\infty(E/\mathbf{F}, s) \quad = \quad w_\infty\,\Lambda_\infty(E/\mathbf{F}, 2-s) \quad \textit{with the root number } w_\infty = +1 \textit{ or } -1.$$

The conductor $N_{E/\mathbf{F}}$ above is a positive integer of the form

$$N_{E/\mathbf{F}} \;=\; \prod_{\text{primes } \mathfrak{p}} \mathfrak{p}^{f_{\mathfrak{p}}(E/\mathbf{F})}$$

where the exponents

$$f_{\mathfrak{p}}(E/\mathbf{F}) \;=\; \begin{cases} 0 & \text{if } E \text{ has good reduction at } \mathfrak{p} \\ 1 & \text{if } E \text{ is bad multiplicative at } \mathfrak{p} \\ 2 & \text{if } E \text{ has additive reduction at } \mathfrak{p} \nmid 6. \end{cases}$$

If the place $\mathfrak{p}$ lies above 2 or 3, then the exponent can be readily computed using a formula of Ogg and Saito.

Remark: Observe that the conjectured functional equation for $\Lambda_\infty(E/\mathbf{F}, s)$ relates the value at s with the value at $2-s$. This indicates that $s = 1$ is a central point of symmetry for the L-function. For example, if the root number $w_\infty = -1$ then

$$\Lambda_\infty(E/\mathbf{F}, 1) \;=\; -\,\Lambda_\infty(E/\mathbf{F}, 2-1)$$

so $L(E/\mathbf{F}, s)$ is forced to vanish at $s = 1$.

It is worthwhile to mention some known cases of Conjecture 1.8. If E is *modular*, i.e. its L-function is the Mellin transform of a newform of weight two, both the analytic continuation and functional equation are known. Wiles et al [Wi1,BCDT] proved that every elliptic curve over $\mathbb{Q}$ is modular, so this already yields a large class of curves to work with.

On the other hand, the theory of complex multiplication supplies us with yet more examples where the analytic properties are known. For the remainder of this section, assume that our curve E has CM by an order in an imaginary quadratic field $\mathbf{K} = \mathbb{Q}(\sqrt{-D})$. As the L-function is isogeny-invariant, without loss of generality we can replace E by an isogenous elliptic curve to ensure that $\mathrm{End}(E) = \mathcal{O}_{\mathbf{K}}$.

For a fixed choice of Weierstrass equation

$$E \,:\, y^2 \;=\; x^3 + Ax + B \quad \text{of discriminant } \Delta(E) = -16(4A^3 + 27B^2) \neq 0,$$

define the Weber function at points P of E by

$$\theta_E(P) \;:=\; \begin{cases} \frac{AB}{\Delta(E)} \times x(P) & \text{if } j(E) \neq 0,\, 1728 \\ \frac{A^2}{\Delta(E)} \times x(P)^2 & \text{if } j(E) = 1728 \\ \frac{B}{\Delta(E)} \times x(P)^3 & \text{if } j(E) = 0. \end{cases}$$

Theorem 1.9. *(Kronecker, Weber)*

(a) The Hilbert class field $\mathbf{H}$ *of* $\mathbf{K}$ *is obtained by adjoining the* j*-invariant of* E*;*

(b) The maximal abelian extension of $\mathbf{K} = \mathbb{Q}(\sqrt{-D})$ *is precisely the field*

$$\mathbf{K}^{\mathrm{ab}} \;=\; \mathbf{H}\Big(\theta_E(P),\; P \in E_{\mathrm{tors}}\Big).$$

This justifies an earlier statement asserting the image of the Galois representation

$$\rho_{E,p} \,:\, G_{\mathbf{F}} \longrightarrow \mathrm{Aut}\big(\mathrm{Ta}_p(E)\big) \qquad \text{was "almost abelian"},$$

since adjoining $\sqrt{-D}$ to the field of definition for our elliptic curve E forces the representation to factor through the infinite abelian group $\mathrm{Gal}\Big(\mathbf{K}^{\mathrm{ab}}/\mathbf{F}\big(\sqrt{-D}\big)\Big)$.

Let us return to the question of analycity for the Hasse-Weil L-function. Recall that we are permitted to assume that E has CM by the full ring of integers $\mathcal{O}_{\mathbf{K}}$. We shall write $\mathbb{A}_{\mathbf{F}}$ for the topological ring of adeles over $\mathbf{F}$. To each elliptic curve E as above, there is an associated Grössencharacter

$$\psi_{E/\mathbf{F}} \,:\, \mathbb{A}_{\mathbf{F}}^{\times} \;\longrightarrow\; \mathbf{K}^{\times}$$

such that at all primes $\mathfrak{p}$ of good reduction, the element $\psi_{E/\mathbf{F}}(\mathfrak{p}) \in \mathcal{O}_{\mathbf{K}} = \mathrm{End}(E)$ reduces modulo $\mathfrak{p}$ to the Frobenius endomorphism on $\widetilde{E}$ (N.B. this is not quite enough to determine $\psi_{E/\mathbf{F}}$ uniquely, but we shall skip over this point).

Attached to any Grössencharacter $\psi : \mathbb{A}_{\mathbf{F}}^{\times} \;\longrightarrow\; \mathbf{K}^{\times}$ is a Hecke L-series

$$L(s,\psi) \;=\; \prod_{\text{primes } \mathfrak{p}} \Big(1 - \psi(\mathfrak{p}).\big(\mathrm{Norm}_{\mathbf{F}/\mathbb{Q}}\, \mathfrak{p}\big)^{-s}\Big)^{-1}$$

and these are well-known to be analytic over $\mathbb{C}$, possessing nice functional equations.

Theorem 1.10. *(Deuring) (a) If* $\mathbf{K}$ *is contained in* $\mathbf{F}$*, then*

$$L(E/\mathbf{F},s) \;=\; L(s,\psi_{E/\mathbf{F}}) \;\times\; L(s,\overline{\psi_{E/\mathbf{F}}});$$

(b) If $\mathbf{K}$ *is not contained in* $\mathbf{F}$*, then*

$$L(E/\mathbf{F},s) \;=\; L(s,\psi_{E/\mathbf{F}'}) \qquad \textit{where} \quad \mathbf{F}' = \mathbf{F}.\mathbf{K}.$$

As a corollary, the analycity and functional equation for $L(E/\mathbf{F},s)$ then follow immediately from this description in terms of Hecke L-series.

1.3 Structure of the Mordell-Weil group

For the rest of this book, we concern ourselves exclusively with elliptic curves over number fields, although there are analogous statements over function fields. Assume that E is an elliptic curve defined over a number field $\mathbf{F}$. Its subgroup of $\mathbf{F}$-rational points, $E(\mathbf{F})$, is named in honour of the two mathematicians who proved the following fundamental result.

Theorem 1.11. *(Mordell, Weil) The abelian group $E(\mathbf{F})$ is finitely-generated.*

Thus to find every $\mathbf{F}$-rational point on E, all we require is a finite generating set of points in $E(\mathbf{F})$, together with an explicit description of the group law on E.

It was Mordell who originally proved the result over $\mathbb{Q}$, and Weil who extended the argument to number fields. We spend the remainder of this section outlining their proof, primarily because it introduces the notion of a height function which we will need later on. Full details of the proof itself can be found in [Si1,§10].

Remark: For a field K and a discrete G_K-module M, we write $H^i(K,M)$ for the ith-cohomology group $H^i\big(\mathrm{Gal}(\overline{K}/K), M\big)$. For example, if $i=0$ then

$$H^0(K,M) \;=\; M^{G_K} \;=\; \Big\{m\in M \quad \text{such that } m^\sigma - m = 0 \text{ for all } \sigma\in G_K\Big\}.$$

Similarly, if $i=1$ then $H^1(K,M)$ equals the quotient

$$\frac{\Big\{\xi\in \mathrm{Cont}(G_K,M) \;\text{ such that } \xi(\sigma\tau)=\xi(\sigma)^\tau+\xi(\tau) \text{ for all } \sigma,\tau\in G_K\Big\}}{\Big\{\xi\in \mathrm{Cont}(G_K,M) \;\text{ such that } \xi(\sigma)=m_\xi^\sigma - m_\xi \text{ for some } m_\xi\in M\Big\}}.$$

The numerator is called the group of 1-cocycles, and the denominator is the group of 1-coboundaries.

In particular, were G_K to act trivially on M it would follow that $m^\sigma - m = 0$ for all $\sigma\in G_K$ and $m\in M$, whence

$$H^0(K,M) \;=\; M \qquad \text{and} \qquad H^1(K,M) \;=\; \mathrm{Hom}_{\mathrm{cont}}(G_K,M).$$

Fix an integer $m\geq 2$. Taking G_K-invariants of the tautological exact sequence

$$0 \longrightarrow E[m] \longrightarrow E \overset{\times m}{\longrightarrow} E \longrightarrow 0$$

one obtains the long exact sequence in G_K-cohomology

$$\ldots \longrightarrow E(K) \overset{\times m}{\longrightarrow} E(K) \overset{\partial}{\longrightarrow} H^1\big(K,E[m]\big) \longrightarrow H^1(K,E) \overset{\times m}{\longrightarrow} H^1(K,E).$$

Truncating both the left and the right-hand sides, gives rise to the well-known Kummer sequence

$$0 \longrightarrow E(K)/mE(K) \overset{\partial}{\longrightarrow} H^1\big(K,E[m]\big) \longrightarrow H^1\big(K,E\big)[m] \longrightarrow 0.$$

Note that the group $H^1\big(K,E[m]\big)$ is not necessarily finite even though $E[m]$ is.

Viewing this sequence firstly over the number field $\mathbf{F}$, and secondly over *all* its completions $\mathbf{F}_\mathfrak{p}$, yields a commutative diagram

$$\begin{array}{ccccccccc} 0 & \longrightarrow & \dfrac{E(\mathbf{F})}{mE(\mathbf{F})} & \longrightarrow & H^1(\mathbf{F}, E[m]) & \longrightarrow & H^1(\mathbf{F}, E(\overline{\mathbf{F}}))[m] & \longrightarrow & 0 \\ & & \downarrow & & \downarrow & & \downarrow & & \\ 0 & \longrightarrow & \prod_\mathfrak{p} \dfrac{E(\mathbf{F}_\mathfrak{p})}{mE(\mathbf{F}_\mathfrak{p})} & \longrightarrow & \prod_\mathfrak{p} H^1(\mathbf{F}_\mathfrak{p}, E[m]) & \longrightarrow & \prod_\mathfrak{p} H^1(\mathbf{F}_\mathfrak{p}, E(\overline{\mathbf{F}}_\mathfrak{p}))[m] & \longrightarrow & 0 \end{array}$$

with exact rows, and whose columns are composed of restriction maps.

Definition 1.12. *(a) The Selmer group of E over $\mathbf{F}$ is given by*

$$\mathrm{Sel}_\mathbf{F}(E) \quad := \quad \mathrm{Ker}\left(H^1(\mathbf{F}, E_{\mathrm{tors}}) \overset{\prod \mathrm{res}_\mathfrak{p}}{\longrightarrow} \prod_{\text{all } \mathfrak{p}} H^1(\mathbf{F}_\mathfrak{p}, E(\overline{\mathbf{F}}_\mathfrak{p})) \right).$$

(b) The Tate-Shafarevich group of E over $\mathbf{F}$ is the group

$$\text{Ш}_\mathbf{F}(E) \quad := \quad \mathrm{Ker}\left(H^1(\mathbf{F}, E) \overset{\prod \mathrm{res}_\mathfrak{p}}{\longrightarrow} \prod_{\text{all } \mathfrak{p}} H^1(\mathbf{F}_\mathfrak{p}, E(\overline{\mathbf{F}}_\mathfrak{p})) \right).$$

It is an easy exercise to show that the sequence

$$0 \longrightarrow E(\mathbf{F})/mE(\mathbf{F}) \longrightarrow \mathrm{Sel}_\mathbf{F}(E)[m] \longrightarrow \text{Ш}_\mathbf{F}(E)[m] \longrightarrow 0$$

is exact, just apply the Snake lemma to our commutative diagram above. We can then pass to the direct limit over m to obtain the isomorphism

$$\text{Ш}_\mathbf{F}(E) \;\cong\; \frac{\mathrm{Sel}(E/\mathbf{F})}{E(\mathbf{F}) \otimes \mathbb{Q}/\mathbb{Z}}.$$

The Selmer group is frequently infinite, e.g. whenever $E(\mathbf{F})$ contains a point of infinite order.

We now give a proof of a weak version of the Mordell-Weil Theorem.

Theorem 1.11'. *For any integer $m \geq 2$ the m-Selmer group $\mathrm{Sel}_\mathbf{F}(E)[m]$ is finite, hence both $E(\mathbf{F})/mE(\mathbf{F})$ and $\text{Ш}_\mathbf{F}(E)[m]$ are finite too.*

Proof: Let S be a finite set of places of $\mathbf{F}$ containing primes lying above m, and also the primes where E has bad reduction. Write $\mathbf{F}_S$ for the maximal algebraic extension of $\mathbf{F}$ unramified outside S and the infinite places.

If $\mathfrak{p} \notin S$ then E has good reduction over $\mathbf{F}_\mathfrak{p}$, the sequence

$$0 \longrightarrow E_1(\overline{\mathbf{F}}_\mathfrak{p}) \longrightarrow E(\overline{\mathbf{F}}_\mathfrak{p}) \longrightarrow \widetilde{E}(\overline{\mathbb{F}}_{q_\mathfrak{p}}) \to 0$$

is exact, and the group $E_1(\overline{\mathbf{F}}_\mathfrak{p})[m]$ is trivial by Proposition 1.6(c). As a consequence the finite group $E[m]$ injects into $\widetilde{E}(\overline{\mathbb{F}}_{q_\mathfrak{p}})$ under the reduction map; in particular, the $G_\mathbf{F}$-module $E[m]$ is unramified at $\mathfrak{p}$.

It follows that the m-Selmer group consists of 1-cocycles that can ramify only at S and infinity, hence we have an inclusion

$$\mathrm{Sel}_{\mathbf{F}}(E)[m] \hookrightarrow H^1\Big(\mathrm{Gal}(\mathbf{F}_S/\mathbf{F}),\, E[m]\Big).$$

We claim that the latter group is finite.

Choose an open normal subgroup H of $\mathrm{Gal}(\mathbf{F}_S/\mathbf{F})$ such that H acts trivially on $E[m]$. Let $\mathbf{F}'$ be the fixed field of H. By inflation-restriction one knows that

$$0 \longrightarrow H^1(\mathrm{Gal}(\mathbf{F}'/\mathbf{F}), E[m]) \stackrel{\mathrm{infl}}{\longrightarrow} H^1\Big(\mathrm{Gal}(\mathbf{F}_S/\mathbf{F}),\, E[m]\Big) \stackrel{\mathrm{rest}}{\longrightarrow} H^1(H,\, E[m])$$

is exact, and the left-most group is clearly finite because $\#\mathrm{Gal}(\mathbf{F}'/\mathbf{F}) < \infty$.

On the other hand, the right-most group $H^1(H, E[m]) = \mathrm{Hom}_{\mathrm{cont}}(H, E[m])$ classifies Galois extensions of $\mathbf{F}'$ unramified outside S and infinity, whose Galois group is isomorphic to a subgroup of $E[m]$. However, a theorem of Hermite and Minkowski tells us that there are only finitely many of these extensions. □

To complete the proof of the Mordell-Weil Theorem, we need to discuss the existence of height functions on elliptic curves, whose p-adic avatars crop up in later chapters. For projective n-space $\mathbb{P}^n$, the absolute logarithmic height $H : \mathbb{P}^n(\overline{\mathbf{F}}) \to [0,\infty)$ is the function

$$H\big(\{x_0, ..., x_n\}\big) = \sum_{\text{all places } \mathfrak{p}} \log\left(\max_{0\le i\le n} |x_i|_{\mathfrak{p}}\right)$$

where the absolute values $|\;|_{\mathfrak{p}}$ are normalized so that they are compatible with the product formula.

If our elliptic curve E is given by a Weierstrass equation in coordinates (x, y), the naive height is the function

$$h : E(\overline{\mathbf{F}}) \longrightarrow [0,\infty), \qquad h(P) = H\big(\{x(P), 1\}\big).$$

For instance, one useful property of the naive height is that the set

$$\Big\{P \in E(\mathbf{F}) \quad \text{such that } h(P) \le B\Big\}$$

is finite for any positive bound $B \in \mathbb{R}$.

Remark: The absolute logarithmic height is almost quadratic. To cook up a function that is truly quadratic, define the *canonical height* $\hat{h} : E(\overline{\mathbf{F}}) \longrightarrow [0,\infty)$ to be the limit

$$\hat{h}(P) := \lim_{n\to\infty} 4^{-n} h\Big([\times 2]^n P\Big)$$

which fortunately exists and is well-defined.

Proposition 1.13. *(Néron, Tate) The canonical height is a positive semi-definite quadratic form on* $E(\mathbf{F})$ *satisfying:*

(a) $\hat{h}(P) = h(P) + O(1)$ *for all* $P \in E(\overline{\mathbf{F}})$;

(b) $\hat{h}(P) = 0$ *if and only if* P *is a torsion point;*

(c) $\hat{h}$ *extends* $\mathbb{R}$*-linearly to give a positive definite quadratic form on* $E(\mathbf{F}) \otimes \mathbb{R}$.

We have skipped quite a bit here, proving this proposition is by no means trivial. Nevertheless, taking these properties of the canonical height for granted, one can polish off the demonstration of the Mordell-Weil Theorem.

Proof of 1.11: Courtesy of the weak version Theorem 1.11', we know that the group $E(\mathbf{F})/mE(\mathbf{F})$ is finite. Let $P_1, ..., P_n \in E(\mathbf{F})$ be a set of coset representatives, and put $B = \max \hat{h}(P_i)$. The set

$$S(B) \;=\; \Big\{ P \in E(\mathbf{F}) \quad \text{such that } \hat{h}(P) \leq B \Big\}$$

is certainly finite, since $\hat{h} = h + O(1)$ and we know the statement to be true for the absolute logarithmic height h.

We claim that the points in $S(B)$ generate the group $E(\mathbf{F})$. Suppose instead that they do not generate it. Then we could find a point $Q \in E(\mathbf{F})$ of minimal canonical height which does not lie in the span of $S(B)$. It would have to be of the form $Q = P_i + mR$ for some $i \in \{1, \ldots, n\}$ and some $R \in E(\mathbf{F})$ not lying in the span of $S(B)$. As a consequence

$$\hat{h}(Q) \leq \hat{h}(R) = \frac{1}{m^2}\hat{h}(mR) = \frac{1}{m^2}\hat{h}(Q - P_i) \leq \frac{2}{m^2}\Big(\hat{h}(Q) + \hat{h}(P_i)\Big)$$

since $\hat{h}(Q - P_i) + \hat{h}(Q + P_i) = 2\hat{h}(Q) + 2\hat{h}(P_i)$.

It would then follow that

$$\hat{h}(Q) \leq \frac{2}{m^2}\Big(\hat{h}(Q) + B\Big), \qquad \text{that is} \quad \hat{h}(Q) \leq \frac{2}{m^2 - 2}\, B \leq B$$

implying that $Q \in S(B)$, which supplies the contradiction.

□

Corollary 1.14. *The Mordell-Weil group of* E *over* $\mathbf{F}$ *has the form*

$$E(\mathbf{F}) \;\cong\; \mathbb{Z}^{r_{\mathbf{F}}} \,\oplus\, E(\mathbf{F})_{\text{tors}}$$

where $E(\mathbf{F})_{\text{tors}}$ *is a finite group, and the integer* $r_{\mathbf{F}} \geq 0$.

Quite a lot is known about the structure of the torsion subgroup.

If $\mathbf{F} = \mathbb{Q}$ then Mazur proved that $E(\mathbb{Q})_{\text{tors}}$ is one of the 15 groups

$$\begin{array}{ll} \mathbb{Z}/n\mathbb{Z} & \text{with } 1 \leq n \leq 12,\ n \neq 11 \qquad \text{or} \\ \mathbb{Z}/2\mathbb{Z} \times \mathbb{Z}/2n\mathbb{Z} & \text{with } 1 \leq n \leq 4 \end{array}$$

and all these possibilities occur in nature.

For a general number field, Kamienny and Merel showed there exists a constant $C = C(d)$ depending only on the degree $d = [\mathbf{F} : \mathbb{Q}]$, such that

$$\#E(\mathbf{F})_{\mathrm{tors}} \;\leq\; C \quad \text{for any elliptic curve } E \text{ defined over } \mathbf{F}.$$

Thus to understand completely the structure of the Mordell-Weil group, one is left with the seemingly innocuous task of determining the rank $r_{\mathbf{F}}$ of its free part. Before discussing this problem in the next section, let's introduce a further invariant arising from the theory of heights.

Definition 1.15. *(a) The Néron-Tate height pairing is the bilinear form*

$$\langle P, Q \rangle_{\mathbf{F},\infty}^{\mathrm{NT}} \;=\; \frac{1}{2}\Big(\hat{h}(P+Q) \;-\; \hat{h}(P) \;-\; \hat{h}(Q)\Big) \quad \textit{for all } P, Q \in E(\mathbf{F}).$$

(b) The elliptic regulator of E over $\mathbf{F}$ is the positive real number defined by

$$\mathrm{Reg}_{\infty,\mathbf{F}}(E) \;:=\; \det\Big(\langle P_i, P_j \rangle_{\mathbf{F},\infty}^{\mathrm{NT}}\Big)_{1 \leq i,j \leq r_{\mathbf{F}}}$$

where $P_1, \ldots, P_{r_{\mathbf{F}}}$ is any basis for $E(\mathbf{F})/E(\mathbf{F})_{\mathrm{tors}}$.

1.4 The conjectures of Birch and Swinnerton-Dyer

In the early 1960's, two Cambridge number theorists Birch and Swinnerton-Dyer conducted numerical experiments on the CM elliptic curves

$$y^2 \;=\; x^3 - Dx \qquad \text{and} \qquad y^2 \;=\; x^3 + 4Dx$$

for D a rational integer. Based on their extensive computational evidence, they predicted a startling link between the rank of the Mordell-Weil group, and the behaviour of the Hasse-Weil L-function at its central point.

Again we suppose that E is an elliptic curve that is defined over a number field $\mathbf{F}$. Let $\sigma_1, \ldots, \sigma_s : \mathbf{F} \hookrightarrow \mathbb{R}$ be the real embeddings of $\mathbf{F}$, and $\tau_1, \ldots, \tau_t, \overline{\tau_1}, \ldots, \overline{\tau_t} : \mathbf{F} \hookrightarrow \mathbb{C}$ its complex embeddings; in particular $s + 2t = [\mathbf{F} : \mathbb{Q}]$.

Remark: If our elliptic curve is given in Weierstrass form by the affine equation

$$E : y^2 \,+\, a_1xy \,+\, a_3y \;=\; x^3 \,+\, a_2x^2 \,+\, a_4x \,+\, a_6,$$

then the *archimedean period* of E over $\mathbf{F}$ is the non-zero complex number

$$\Omega_{E/\mathbf{F}} \;=\; \prod_{i=1}^{s} \int_{E(\mathbb{R})} |\omega_E^{\sigma_i}| \;\times\; \prod_{j=1}^{t} 2\int_{E(\mathbb{C})} \omega_E^{\tau_j} \wedge \overline{\omega_E^{\tau_j}}$$

where $\omega_E = \frac{dx}{2y+a_1x+a_3} \in H^1_{\mathrm{dR}}(E/\mathbf{F})$ indicates a Néron differential on E.

Conjecture 1.16. *(Birch, Swinnerton-Dyer)*

(a) $\text{order}_{s=1} L(E/\mathbf{F}, s) \stackrel{?}{=} r_{\mathbf{F}}$;

(b) $$\lim_{s\to 1} \frac{L(E/\mathbf{F}, s)}{(s-1)^{r_{\mathbf{F}}}} \stackrel{?}{=} \Omega_{E/\mathbf{F}} \times \text{Reg}_{\infty,\mathbf{F}}(E) \times \frac{\#\text{Ш}_{\mathbf{F}}(E) \prod_{\mathfrak{p}\le\infty} [E(\mathbf{F}_{\mathfrak{p}}) : E_0(\mathbf{F}_{\mathfrak{p}})]}{\sqrt{\text{disc}_{\mathbf{F}}} \times \#E(\mathbf{F})^2_{\text{tors}}}.$$

Some explanations are in order.

Firstly, we should bear in mind that the analycity of the L-function at the point $s = 1$ has yet to be established in general, so the left-hand sides of 1.16(a) and 1.16(b) might not have any meaning. This conjecture relates arithmetic invariants of E with an analytic invariant, i.e. the leading term of its Hasse-Weil L-function. The formulae are thus reminiscent of the analytic class number formula for the zeta-function of a number field.

Remarks: (i) The archimedean period $\Omega_{E/\mathbf{F}}$ a priori lies in $\mathbb{C}^\times$, but if our number field $\mathbf{F}$ is assumed to be totally real, the period itself will be a positive real number. The elliptic regulator $\text{Reg}_{\infty,\mathbf{F}}(E)$ is always positive, no matter what the number field of definition. In particular, Conjecture 1.16(b) predicts that the quantity

$$\lim_{s\to 1} \frac{L(E/\mathbf{F}, s)}{(s-1)^{r_{\mathbf{F}}}} \times \sqrt{\text{disc}_{\mathbf{F}}} \times \left(\Omega_{E/\mathbf{F}} \times \text{Reg}_{\infty,\mathbf{F}}(E)\right)^{-1}$$

is a non-zero rational number. In the special case where $\mathbf{F} = \mathbb{Q}$, it is a worthwhile exercise in homology to show that $L(E/\mathbb{Q}, 1) \times \left(\Omega_{E/\mathbb{Q}}\right)^{-1} \in \mathbb{Q}$.

(ii) At non-archimedean primes $\mathfrak{p}$, all possible Tamagawa numbers $[E(\mathbf{F}_{\mathfrak{p}}) : E_0(\mathbf{F}_{\mathfrak{p}})]$ are listed in Proposition 1.6(a). Whenever E has good reduction at $\mathfrak{p}$, the index $[E(\mathbf{F}_{\mathfrak{p}}) : E_0(\mathbf{F}_{\mathfrak{p}})]$ is equal to one. Since this is the true for all but finitely many primes, it follows that the infinite product

$$\prod_{\mathfrak{p}\le\infty} [E(\mathbf{F}_{\mathfrak{p}}) : E_0(\mathbf{F}_{\mathfrak{p}})]$$

occurring in the Birch and Swinnerton-Dyer formula, is really a finite product.

(iii) At archimedean primes $\mathfrak{p}$, there are essentially three cases to consider when computing the size of the group of connected components. If the completion $\mathbf{F}_{\mathfrak{p}} = \mathbb{C}$ then $[E(\mathbb{C}) : E_0(\mathbb{C})] = 1$ as E is geometrically connected. If $\mathbf{F}_{\mathfrak{p}} = \mathbb{R}$ and our elliptic curve E is connected over the real numbers, then again we must have $[E(\mathbb{R}) : E_0(\mathbb{R})]$ equal to one. Conversely, if E is not connected over $\mathbb{R}$ then it has precisely two connected components, corresponding to the intersection of the torus with a plane; in this case $[E(\mathbb{R}) : E_0(\mathbb{R})] = 2$.

Finally, what is this mysterious number $\#\text{Ш}_{\mathbf{F}}(E)$? Recall that Theorem 1.11' stated that for any integer $m \ge 2$, the group $\text{Ш}_{\mathbf{F}}(E)[m]$ is finite. Regrettably this does not imply that $\text{Ш}(E/\mathbf{F})$ itself is a finite group. Indeed there is no known proof of the finiteness of the Tate-Shafarevich group, except in certain specialized situations where the analytic rank is miniscule.

Conjecture 1.17. $\text{Ш}_{\mathbf{F}}(E)$ *is always a finite group.*

It was Tate who in the early 1970's commented on Birch and Swinnerton-Dyer's prediction that "this remarkable conjecture relates the behaviour of a function L at a point where it is not at present known to be defined, to the order of a group Ш which is not known to be finite!"

Browsing through the first few elliptic curves occurring in Cremona's tables [Cr], one might be led naively to believe that the size of $\text{Ш}_{\mathbb{Q}}(E)$ is bounded above. However, there are results which show that the order of the Tate-Shafarevich group can become arbitrarily large in general.

Theorem 1.18. *(Cassels) There is an alternating, bilinear pairing*

$$\text{Ш}_{\mathbf{F}}(E) \times \text{Ш}_{\mathbf{F}}(E) \longrightarrow \mathbb{Q}/\mathbb{Z}$$

whose left and right-kernels consist of the divisible elements in $\text{Ш}_{\mathbf{F}}(E)$.

As a corollary, let us remark that if the Tate-Shafarevich group is finite then the above pairing is perfect, therefore the quantity $\#\text{Ш}_{\mathbf{F}}(E)$ must be a perfect square. On the other hand, from the short exact sequence

$$0 \longrightarrow E(\mathbf{F}) \otimes \mathbb{Q}/\mathbb{Z} \longrightarrow \mathrm{Sel}_{\mathbf{F}}(E) \longrightarrow \text{Ш}_{\mathbf{F}}(E) \longrightarrow 0$$

we see the finiteness of $\text{Ш}_{\mathbf{F}}(E)$ is equivalent to the assertion that the maximal divisible subgroup of the Selmer group is precisely $E(\mathbf{F}) \otimes \mathbb{Q}/\mathbb{Z}$.

Assuming the finiteness of the Tate-Shafarevich group, one can legally define

$$\mathbf{BSD}_{\infty,\mathbf{F}}(E) \quad := \quad \frac{\#\text{Ш}_{\mathbf{F}}(E) \times \prod_{\mathfrak{p} \le \infty} \left[E(\mathbf{F}_{\mathfrak{p}}) : E_0(\mathbf{F}_{\mathfrak{p}})\right]}{\sqrt{\mathrm{disc}_{\mathbf{F}}} \times \#E(\mathbf{F})^2_{\mathrm{tors}}} \times \mathrm{Reg}_{\infty,\mathbf{F}}(E).$$

The following result is often referred to as the Isogeny Theorem.

Theorem 1.19. *(Cassels) Assume that E and E' are $\mathbf{F}$-isogenous elliptic curves. If $\text{Ш}_{\mathbf{F}}(E)$ is finite then so is $\text{Ш}_{\mathbf{F}}(E')$, and moreover*

$$\Omega_{E/\mathbf{F}} \times \mathbf{BSD}_{\infty,\mathbf{F}}(E) \quad = \quad \Omega_{E'/\mathbf{F}} \times \mathbf{BSD}_{\infty,\mathbf{F}}(E').$$

The L-function $L(E/\mathbf{F}, s)$ does not change if we swap E with E', and neither does the left-hand side of the formula in 1.16(b). Thus Cassels' result provides (weak) support for Conjecture 1.16, since the right-hand side of the formula is now seen to be an invariant of the $\mathbf{F}$-isogeny class too.

We shall conclude by describing situations where the Birch and Swinnerton-Dyer Conjecture has been partially proven. From now on let us put $\mathbf{F} = \mathbb{Q}$, so our elliptic curve E is defined over the rationals. By the work of Wiles and others, at least it makes good sense to speak of the behaviour of the Hasse-Weil L-function at the critical point $s = 1$.

Theorem 1.20. *(Coates, Wiles) Assume the curve E has complex multiplication. If $L(E/\mathbb{Q},1) \neq 0$ then $E(\mathbb{Q})$ is finite.*

The demonstration of this result involves the theory of elliptic units. These special units satisfy various norm-compatibility relations, and form examples of what is now commonly referred to as an 'Euler system'.

Euler systems crop up all the time in a subject called Iwasawa theory, and their cohomological properties are of great help in bounding the orders of Selmer groups. It is widely conjectured that any sensible arithmetic object, which possesses an associated L-function, should also have an Euler system attached to it.

Theorem 1.21. *(Kolyvagin, Rubin, Gross-Zagier)*
(a) If $L(E/\mathbb{Q},1) \neq 0$, then both $E(\mathbb{Q})$ and $\text{Ш}_{\mathbb{Q}}(E)$ are finite.
(b) If $\text{order}_{s=1} L(E/\mathbb{Q},s) = 1$, then $\text{Ш}_{\mathbb{Q}}(E)$ is finite and $E(\mathbb{Q})$ has rank one.

In fact quite a bit more is true. If E has complex multiplication, Rubin has shown that under these conditions Conjecture 1.16(b) is true (up to powers of 2 and 3). In the non-CM case, Kolyvagin has proven a similar type of result.

The basis of their arguments is the following very beautiful analytic formula. For an appropriate 'good choice' of imaginary quadratic field $\mathbf{K}$, Gross and Zagier showed that the derivative of the L-function over $\mathbf{K}$ satisfies

$$L'(E/\mathbf{K},1) \;=\; \Omega_{E/\mathbf{K}} \times \frac{\langle P_{\mathbf{K}}, P_{\mathbf{K}} \rangle^{\mathrm{NT}}_{\mathbf{F},\infty}}{c^2\, u_{\mathbf{K}}^2\, \sqrt{-\mathrm{disc}_{\mathbf{K}}}}$$

where $u_{\mathbf{K}}$ is the number of roots of unity inside $\mathbf{K}$, and $c \in \mathbb{Z}$ is Manin's constant.

The Heegner point $P_{\mathbf{K}}$ arises naturally in the theory of modular curves, and under the right circumstances it has infinite order inside $E(\mathbf{K})$. Kolyvagin exploited the fact that Heegner points themselves form an Euler system, in order to drive his descent machinery.

Theorem 1.22. *(Greenberg, Nekovář) At infinitely many primes $p > 3$ for which the elliptic curve E has ordinary reduction,*

$$\text{order}_{s=1} L(E/\mathbb{Q},s) \;\equiv\; r_{\mathbb{Q}} + \text{corank}_{\mathbb{Z}_p}\Big(\text{Ш}_{\mathbb{Q}}(E)[p^\infty]\Big) \pmod 2$$

subject to Greenberg's conjecture holding (for the two-variable p-adic L-function).

This result is of a rather different nature to the preceding two theorems, since there is no requirement for the order of vanishing of the L-function at $s = 1$ to be small. The $\mathbb{Z}_p$-corank of $\text{Ш}_{\mathbb{Q}}(E)[p^\infty]$ counts the number of copies of $\mathbb{Q}_p/\mathbb{Z}_p$ contained inside the Tate-Shafarevich group. If Conjecture 1.17 holds there will be no copies of $\mathbb{Q}_p/\mathbb{Z}_p$ occurring at all; we would then obtain the congruence

$$\text{order}_{s=1} L(E/\mathbb{Q},s) \;\equiv\; r_{\mathbb{Q}} \pmod 2$$

i.e. that the Birch and Swinnerton-Dyer Conjecture is true modulo 2.

1.5 Modular forms and Hecke algebras

In this last section, we introduce the notion of a modular form and its associated Hecke operators. The constructions that occur later on in the book require us to work over schemes rather than fields, hence we adopt a more algebraic viewpoint than some of the standard texts [Ko,My,Sh]. Let $S = \mathrm{Spec}(R)$ be an affine scheme. An elliptic curve E/S is an abelian scheme over S of dimension one.

Fix an integer $N \geq 5$, and assume that R has the structure of a $\mathbb{Z}[1/N]$-algebra. The modular curve $Y_1(N)$ is the open modular curve, parametrising elliptic curves E together with a point of exact order N on them. Its compactification $X_1(N)$ then classifies pairs $\big(E, \alpha : \mu_N \hookrightarrow E[N]\big)$ where E is a (generalised) elliptic curve, and $\mathrm{Im}(\alpha)$ meets every irreducible component in each geometric fiber of E.

Remark: Because $N \geq 5$, there exists a universal elliptic curve $\mathbb{E}_{\mathrm{univ}/R}$ which is the final object in the category of generalised elliptic curves $E/X_1(N)$, defined over $\mathrm{Spec}(R)$ and with group scheme embeddings α. For instance, if $R = \mathbb{C}$ then the complex points of $\mathbb{E}_{\mathrm{univ}}$ are described by

$$\Big(\mathbb{E}_{\mathrm{univ}}\Big)(\mathbb{C}) \quad = \quad \Gamma_1(N)\Bigg\backslash\Bigg(\bigcup_{\tau\in\mathfrak{H}} \mathbb{C}/\mathbb{Z}\oplus\mathbb{Z}.\tau\Bigg)$$

where $\mathfrak{H} := \big\{z \in \mathbb{C} \text{ such that } \mathrm{Im}(z) > 0\big\}$ denotes the upper half-plane.

Definition 1.23. *If $\delta : E \to S$ is a smooth morphism, set $\underline{\omega}_E := \delta_* \Omega^1_{E/S}$.*

(a) The R-module of modular forms of weight $k \geq 1$ on $\Gamma_1(N)$ over R, is defined to be the set of global sections

$$M_k\big(X_1(N)\big)_{/R} \quad := \quad H^0\Big(X_1(N)_{/R},\ \underline{\omega}^{\otimes k}\Big).$$

In other words, a modular form $f \in H^0\big(X_1(N)_{/R},\ \underline{\omega}^{\otimes k}\big)$ is a rule assigning an element $f(E,\alpha) \in \underline{\omega}_E^{\otimes k}$ to each R-point $\big(E, \alpha : \mu_N \hookrightarrow E[N]\big)$ on the modular curve. This definition is compatible with base-change.

The Tate curve $E^{\mathrm{Tate}} = \mathbb{G}_{\mathrm{mult}}/q^{\mathbb{Z}}$ is an elliptic curve over the ring $\mathbb{Z}[\![q]\!][1/q]$, which extends to a generalised elliptic curve over $\mathbb{Z}[\![q]\!]$. The natural inclusion map $i_N : \mu_N \hookrightarrow E^{\mathrm{Tate}}$ arising from the containment $\mu_N \subset \mathbb{G}_{\mathrm{mult}}$, allows us to evaluate modular forms on the Tate curve. If $f \in H^0\big(X_1(N)_{/R},\ \underline{\omega}^{\otimes k}\big)$ then its q-expansion is the power series $F(f)(q) \in R[\![q]\!]$ given by

$$f\Big(\mathbb{G}_{\mathrm{mult}}/q^{\mathbb{Z}}, i_N\Big) \quad = \quad F(f)(q)\left(\frac{dt}{t}\right)^{\otimes k}$$

where t is the coordinate on the multiplicative group $\mathbb{G}_{\mathrm{mult}}$. In particular, the q-development map $F : H^0\big(X_1(N)_{/R},\ \underline{\omega}^{\otimes k}\big) \longrightarrow R[\![q]\!]$ is injective.

Definition 1.23. *(b) The cusp forms $\mathcal{S}_k\big(X_1(N)\big)_{/R}$ of weight $k \geq 1$ on $\Gamma_1(N)$ is the R-submodule of $H^0\Big(X_1(N)_{/R},\, \underline{\omega}^{\otimes k}\Big)$ consisting of those modular forms whose q-expansions have zero as their constant coefficient.*

Equivalently, one can view the submodule of cusp forms as being modular forms which vanish on degenerate elliptic curves, i.e. at precisely the set of cusps of the modular curve $X_1(N)_{/R}$.

Hecke operators, involutions and degeneration maps.
There are two possible ways one can introduce these operators. The first way is geometric, and involves defining Hecke correspondences on the modular curves $X_1(N)$ and $X_1(N;l)$, then pulling them back via the Kodaira-Spencer isomorphism. Instead we'll adopt the second approach, which is just to write them down!

For a congruence class $a \in (\mathbb{Z}/N\mathbb{Z})^\times$ the diamond operator $< a >$ acts on $X_1(N)$ by $< a > (E, \alpha) = (E, a.\alpha)$, thence also on modular forms $f \in M_k\big(X_1(N)\big)$ via

$$(< a > f)(E,\alpha) \quad = \quad f(E, a.\alpha).$$

Definition 1.24. *Let $n \geq 1$ be an integer. The covariant (or Albanese) operator T_n has an action on f, given by the formula*

$$(T_n f)(E,\alpha) \quad = \quad \frac{1}{n} \sum_{\lambda: E \to E',\ \deg(\lambda)=n} \lambda^*\Big(f\big(E', \lambda \circ \alpha\big)\Big).$$

Alternatively, the contravariant (or Picard) operator $T_n^{\mathcal{D}}$ acts on f through

$$(T_n^{\mathcal{D}} f)(E,\alpha) \quad = \quad \frac{1}{n} \sum_{\mu: E' \to E,\ \deg(\mu)=n} \mu_*\Big(f\big(E', \mu^{-1} \circ \alpha\big)\Big).$$

By expanding both modular forms $T_l f$ and $T_l^{\mathcal{D}} f$ over the Tate curve, one easily checks that $T_l = < l > T_l^{\mathcal{D}}$ for all primes $l \nmid N$. We should point out if there exists a Dirichlet character ϵ modulo N such that $< a > f = \epsilon(a) f$ for all $a \in (\mathbb{Z}/N\mathbb{Z})^\times$, the modular form f is said to have type (N, k, ϵ).

The Fricke involution $w_N : X_1(N) \longrightarrow X_1(N)$ defined over $\mathrm{Spec}\Big(\mathbb{Z}\big[1/N, \mu_N\big]\Big)$ is explicitly described on the points of $Y_1(N)$ by

$$w_N\Big(E, \alpha : \mu_N \hookrightarrow E[N]\Big) \quad = \quad (E', \beta);$$

N.B. here $E' = E/\mathrm{Im}(\alpha)$ as elliptic curves over R, and $\beta : \mu_N \hookrightarrow E'[N]$ satisfies

$$\Big[\alpha(\zeta),\ \text{lift of } \beta(\zeta)\Big]_E \quad = \quad \zeta \qquad \text{under the Weil pairing on } E.$$

On the space of modular forms $H^0\big(X_1(N)_{/R}, \underline{\omega}^{\otimes k}\big)$ itself, this involution transforms into the matrix operator $W_N = \begin{pmatrix} 0 & -1 \\ N & 0 \end{pmatrix}$ for the standard action of $\mathrm{GL}_{2/R}$.

Remark: It can be shown that $T_l^{\mathcal{D}} = W_N \circ T_l \circ W_N$ for all prime numbers $l \nmid N$. Therefore on that portion of $M_k(X_1(N))$ which is fixed under the action of W_N, the Albanese and Picard actions are self-adjoint (with respect to the inner product). However on the remainder of $M_k(X_1(N))$, this is definitely not the case.

Finally let us explain how to change levels. Pick divisors $d, M \geq 1$ such that $dM | N$. The finite degeneration map $\pi_d : X_1(N) \longrightarrow X_1(M)$ is given on affine points by

$$\pi_d\Big(E,\ \alpha : \mu_N \hookrightarrow E[N]\Big) \quad = \quad \Big(E' = E/\alpha(\mu_d),\ \alpha' : \mu_M \hookrightarrow E'[M]\Big)$$

where the above composition $\alpha' : \mu_M \hookrightarrow \mu_{N/d} \cong \mu_N/\mu_d \overset{\alpha \bmod \mu_d}{\hookrightarrow} E/\alpha(\mu_d) = E'$. These level-changing maps exhibit an induced action on cohomology "$\pi_{d\,*}$" say, and in particular on the R-module $H^0\Big(X_1(N)_{/R},\ \underline{\omega}^{\otimes k}\Big)$ and submodule $\mathcal{S}_k\big(X_1(N)\big)_{/R}$. For example, given any prime $l|M$ we have relations

$$\pi_{1\,*} \circ T_l \quad = \quad T_l \circ \pi_{1\,*} \quad \text{and} \quad \pi_{1\,*} \circ T_l^{\mathcal{D}} \quad = \quad T_l^{\mathcal{D}} \circ \pi_{1\,*}.$$

In fact, the trace map $\pi_{1\,*}$ commutes with both the diamond and Hecke operators at all integers coprime to N.

The Galois representations attached to cusp forms.
For simplicity we now assume that $k \geq 2$, and work over either $R = \mathbb{Z}$ or $R = \mathbb{C}$. Let us denote by $h_k(N; R)$ the R-subalgebra of $\mathrm{End}_R\left(\mathcal{S}_k\big(X_1(N)\big)\right)$ generated by all the T_n's with $n \geq 1$, and the diamond operators $< a >$ with $a \in (\mathbb{Z}/N\mathbb{Z})^\times$.

Notation: By common convention, for any cusp form $f \in \mathcal{S}_k\big(X_1(N)\big)$ we write

$$F(f)(q) \quad = \quad \sum_{n=1}^{\infty} a_n(f) q^n$$

as its q-coordinate expansion on the curve $E^{\mathrm{Tate}} = \mathbb{G}_{\mathrm{mult}}/q^{\mathbb{Z}}$.

The *Hecke algebra* $h_k(N; R)$ is commutative, and is also free of finite rank over R. Moreover for $R = \mathbb{Z}$ or $\mathbb{C}$, there exists a perfect duality

$$\mathcal{S}_k\big(X_1(N)\big)_{/R} \times h_k(N; R) \longrightarrow R, \quad (f, T) \mapsto a_1(f|T).$$

The space of oldforms is defined to be the subspace of $\mathcal{S}_k\big(X_1(N)\big)$ spanned by the images of the maps $f \mapsto \pi_d^*(f)$, namely

$$\mathcal{S}_k^{\mathrm{old}}\big(X_1(N)\big) \;=\; \text{the span of} \bigcup_{M|N,\, M \neq N}\ \bigcup_{d|N/M} \Big\{\pi_d^*(f) \text{ where } f \in \mathcal{S}_k\big(X_1(M)\big)\Big\}.$$

Analogously, the space of newforms $\mathcal{S}_k^{\mathrm{new}}\big(X_1(N)\big)_{/\mathbb{C}}$ is the orthogonal complement of $\mathcal{S}_k^{\mathrm{old}}\big(X_1(N)\big)_{/\mathbb{C}}$ under the Petersson inner product, and must necessarily comprise cusp forms of exact level N but no smaller.

Proposition 1.25. *The space of newforms has a normalised basis, consisting of simultaneous eigenforms under the action of the whole Hecke algebra* $h_k(N, \mathbb{C})$.

It follows that to study cusp forms, first isolate the newforms at a certain level N, then diagonalise under $h_k(N, \mathbb{C})$ so as to consider each Hecke eigenform individually. The eigenvectors $f \in \mathcal{S}_k^{\text{new}}$ in Proposition 1.25 are normalised so that $a_1(f) = 1$; such vectors are called *primitive forms.*

The following result was first proved by Eichler-Shimura at weight $k = 2$, then generalised by Deligne to weight $k > 2$ as part of his proof of the Weil conjectures.

Theorem 1.26. [De,ES] *Let f be a normalised cuspidal eigenform of type (N, k, ϵ). Then for every prime number p, there exists an irreducible Galois representation*

$$\rho_{f,p} : \text{Gal}(\overline{\mathbb{Q}}/\mathbb{Q}) \longrightarrow \text{GL}_2(\mathcal{K}_f) \quad \textit{with } \mathcal{K}_f = \mathbb{Q}_p\Big(a_n(f) \mid n \in \mathbb{N}\Big),$$

unramified outside the support of Np and $\{\infty\}$, which is uniquely determined by the identities at every prime $l \nmid Np$:

$$\text{Tr}\Big(\rho_{f,p}(\text{Frob}_l^{-1})\Big) = a_l(f) \quad \textit{and} \quad \det\Big(\rho_{f,p}(\text{Frob}_l^{-1})\Big) = \epsilon(l) l^{k-1}.$$

L-functions and modularity.
In the same way every elliptic curve E has a complex L-function attached to it, a similar association works with modular forms (in fact, it works even better). Assume f is a cusp form of level N and weight $k \geq 2$, possessing the q-expansion $f(z) = \sum_{n=1}^{\infty} a_n(f) q^n$ with $q = \exp(2\pi i z)$.

Definition 1.27. *The L-function of f over $\mathbb{Q}$ is the Dirichlet series*

$$L(f, s) := \sum_{n=1}^{\infty} a_n(f).n^{-s} \qquad \textit{which converges for } \text{Re}(s) > \frac{k+2}{2}.$$

Provided f is a simultaneous eigenform for the whole Hecke algebra, its L-function further admits an Euler product expansion in the same right half-plane.

Theorem 1.28. *(Hecke) If f is a primitive form of level N and weight $k \geq 2$, then the completed L-function*

$$\Lambda(f, s) := N^{s/2}(2\pi)^{-s}\Gamma(s)\, L(f, s)$$

has an analytic continuation to all of $\mathbb{C}$, and satisfies the functional equation

$$\Lambda(f, s) = i^k\, \Lambda\Big(f|W_N,\, k - s\Big).$$

Note that $f|W_N = i^{-k} w_\infty f^*$ where the dual cusp form f^* has the q-expansion $f^*(z) = \sum_{n=1}^{\infty} \overline{a_n(f)} q^n$, and the complex sign $w_\infty \in \{\pm 1\}$.

Remark: In order to describe the link between modular forms and elliptic curves, it is first necessary to introduce the modular curves $X_0(N)$ for an integer $N \geq 1$. The R-points of $X_0(N)$ are pairs (E, C) where E is a (generalised) elliptic curve defined over $\mathrm{Spec}(R)$, and $C \subset E[N]$ is a cyclic subgroup scheme of exact order N. More specifically, if $R = \mathbb{C}$ then the complex points of $X_0(N)$ have the structure of a Riemann surface, isomorphic to the extended upper half-plane $\mathfrak{H}^* = \mathfrak{H} \cup \mathbb{P}^1(\mathbb{Q})$ quotiented by the Möbius action of $\Gamma_0(N)$.

One says that an elliptic curve E defined over $\mathbb{Q}$ having conductor N_E is *modular*, if one of the following equivalent conditions hold:

(a) There exists a newform f_E of type $(N_E, 2, \mathbf{1})$ such that $L(E, s) = L(f_E, s)$;

(b) There exists a non-constant $\mathbb{Q}$-rational morphism $X_0(N_E) \twoheadrightarrow E$;

(c) For every prime p, the Galois representation $\rho_{E,p} : G_{\mathbb{Q}} \longrightarrow \mathrm{Aut}_{\mathbb{Z}_p}(\mathrm{Ta}_p(E))$ is equivalent to the contragredient $\rho^{\vee}_{f,p}$ for some newform f of type $(N_E, 2, \mathbf{1})$.

The next result is arguably the most celebrated from last century in number theory. Before the theorem's final proof it was known as the *Shimura-Taniyama conjecture*, and was subsequently also attributed to Weil who was a great believer in it.

Theorem 1.29. [Wil,TW,BCDT] *Every elliptic curve E over $\mathbb{Q}$ is modular.*

One cannot emphasize enough the beauty and complexity encapsulated in this single statement. For an account of the principal ingredients in its proof – together with the famous connection with 'Fermat's Last Theorem' – the conference proceedings [CSS] are extremely readable.

That's more or less all the background we should require for the rest of the book. As was discussed at the end of §1.2, the theory of complex multiplication provides us with a whole bestiary of elliptic curves whose L-values one can study in detail. However, we shall instead work in the context of elliptic curves E over $\mathbb{Q}$, which by Theorem 1.29 are necessarily modular. The main reason is that their universal deformation rings at p have an extra structure as Hecke algebras; our ultimate goal is to understand the arithmetic of E over its p-ordinary deformations ...

CHAPTER II

p-Adic L-functions and Zeta Elements

One of the most fruitful developments in the last couple of decades, has been the use of K-theory in the study of special values of L-functions of modular forms. This link is manifested in several ways. The first is through the work of Beilinson, who constructed zeta-elements living in K_2 of a modular curve. He managed to prove that their image under the regulator map yielded the residue of the L-series at $s = 0$. In an orthogonal direction, Coates and Wiles found a non-archimedean approach to relating the L-function of a CM elliptic curve at $s = 1$, with a certain Euler system of elliptic units. They were then able to prove the first concrete theorems in the direction of the Birch and Swinnerton-Dyer conjecture.

It was Kato who realised the approach of Beilinson and that of Coates-Wiles could be combined, with spectacular results. He noticed that the zeta-elements considered by Beilinson satisfied norm-compatibility relations, reminiscent of those satisfied by elliptic units. Moreover, he devised a purely p-adic method whereby these elements could be used to study L-values. Underlying his discoveries was the ground-breaking work of Perrin-Riou in the early nineties, which extended the local Iwasawa theory used by Coates and Wiles to a very general framework (required for modular forms and non-CM elliptic curves). We shall review some of the highlights of their work, and in Chapters III, V and VI we will generalise it to modular symbols and Λ-adic families of modular forms.

It is natural to ask whether BSD conjectures have non-archimedean counterparts. In order to phrase such questions, we need to replace the complex L-function with a p-adic avatar. For newforms of weight $k \geq 2$, there are analytic constructions due to various authors [MSD,Mn,MTT]. Perhaps the most elegant way to introduce these p-adic L-functions is through the language of measure theory, and it is the viewpoint we adopt at the start of this chapter.

2.1 The p-adic Birch and Swinnerton-Dyer conjecture

Let R be a commutative ring with identity, and X some compact topological space. An R-valued distribution $\mu \in \mathrm{Dist}(X, R)$ is a finitely additive function on the compact open subsets of X, taking values in R. Most commonly, we will consider spaces X which are p-adic Lie groups of dimension ≤ 2, and rings R which are complete, Noetherian and local (CNL's for short).

For simplicity, assume that $X = \mathbb{Z}_p$ and R is the integer ring of the Tate field, although later on we shall consider measures in more general situations than this.

If $g \in \mathrm{Step}(X, R)$ is a locally constant function and providing the sums converge, we employ the integral notation

$$\int_{x\in\mathbb{Z}_p} g(x).d\mu(x) \quad := \quad \lim_{j\to\infty}\left(\sum_{X=\bigcup_{i=1}^{d_j} X_i^{(j)}} g\big|_{X_i^{(j)}} \times \mu\left(X_i^{(j)}\right)\right)$$

where the above limit ranges through ever finer disjoint covers $\left\{X_i^{(j)}\right\}_{1\leq i\leq d_j}$ of X. We say that μ is *supported on* $\mathbb{Z}_p^\times$, if $\mu(U) = 0$ for every compact open set $U \subset p\mathbb{Z}_p$.

Let r be a positive number.

Definition 2.1. *(a) One calls μ an 'r-admissible distribution' on $\mathbb{Z}_p^\times$ if for all integers $j \in [0, r-1]$, the moments*

$$\left|\int_{x\in a+p^n\mathbb{Z}_p} (x-a)^j.d\mu(x)\right|_p \quad \text{are of type } o\big(p^{n(r-j)}\big) \text{ where } (n,a) \in \mathbb{N}\times\mathbb{Z}_p^\times.$$

(b) Furthermore, μ is a 'bounded measure' if the values

$$\left|\int_{x\in a+p^n\mathbb{Z}_p} d\mu(x)\right|_p \quad \text{are bounded above, for every pair } (n,a) \in \mathbb{N}\times\mathbb{Z}_p^\times.$$

When μ is supported on $\mathbb{Z}_p^\times$, there are two sorts of function which we will integrate. The first of these arises by considering a Dirichlet character ψ of conductor p^n. Under our embedding $\iota_p : \overline{\mathbb{Q}} \hookrightarrow \mathbb{C}_p$, we obtain a homomorphism

$$\mathbb{Z}_p^\times \overset{\text{mod } p^n}{\twoheadrightarrow} (\mathbb{Z}_p/p^n\mathbb{Z}_p)^\times \xrightarrow{\sim} (\mathbb{Z}/p^n\mathbb{Z})^\times \xrightarrow{\psi} \overline{\mathbb{Q}}^\times \overset{\iota_p}{\hookrightarrow} \overline{\mathbb{C}}_p^\times$$

which is locally constant, denoted once more by $\psi(x)$. The second type of function is defined by a power series expansion

$$< x >^s \ := \ \exp_p\big(s\log_p(x)\big) \quad \text{which converges provided } |s|_p < p^{(p-2)/(p-1)}.$$

Bearing in mind these notions, we can now describe the method of Mazur, Tate and Teitelbaum, which associates a p-adic measure to a cusp form f of weight $k \geq 2$. The initial step is to set

$$h_p := \mathrm{ord}_p(\alpha_p) \quad \text{where } \alpha_p \text{ is a root of } X^2 - a_p(f)X + \epsilon(p)p^{k-1}$$

satisfying $\mathrm{ord}_p(\alpha_p) < k-1$. In particular, if $a_p(f)$ is a p-adic unit then so is α_p. Furthermore, we must fix a choice of periods $\Omega^\pm \in \mathbb{C}^\times$ so that the special values

$$\frac{L(f,\psi,j+1)}{(2\pi i)^j\,\Omega^{\mathrm{sign}\left(\psi(-1)(-1)^j\right)}} \qquad \text{with } j \in \{0,\ldots,k-2\}$$

are algebraic numbers, at all twists by Dirichlet characters ψ.

Theorem 2.2. *([MTT, Sect.1]) There exists a unique h_p-admissible distribution μ_{f,α_p} supported on $\mathbb{Z}_p^\times$, such that*

$$\int_{x\in\mathbb{Z}_p^\times} \psi^{-1}(x)x^j.d\mu_{f,\alpha_p}(x) = \frac{j!\,p^{nj}}{\alpha_p^n}\left(1-\psi^{-1}(p)\frac{p^j}{\alpha_p}\right)\left(1-\psi(p)\epsilon(p)\frac{p^{k-2-j}}{\alpha_p}\right) \times \left(\sum_{m=1}^{p^n}\psi^{-1}(m)e^{2\pi i m/p^n}\right) \times \frac{L(f,\psi,j+1)}{(2\pi i)^j\Omega^\pm}$$

for all $j \in \{0,\dots,k-2\}$ and characters ψ of conductor p^n, with $\pm = \psi(-1)(-1)^j$. Furthermore, if $a_p(f)$ is a p-adic unit then μ_{f,α_p} is a bounded measure.

To manufacture a p-adic L-function from an admissible distribution, we make a choice of root α_p. If $a_p(f)$ is a p-adic unit, this choice is forced upon us anyway. By integration, one can then define

$$\mathbf{L}_p(f,\psi,s) = \mathbf{L}_{p,\alpha_p,\Omega^\pm}(f,\psi,s) := \int_{x\in\mathbb{Z}_p^\times} \psi^{-1}(x) < x >^{s-1} .d\mu_{f,\alpha_p}(x)$$

which is a p-adic analytic function of $s \in \mathbb{Z}_p$. If $\mathrm{ord}_p(\alpha_p) = 0$ i.e. the ordinary case, then in fact $\mathbf{L}_p(f,\psi,s)$ is an Iwasawa function (a continuous function on the unit disk with bounded sup-norm).

A special case

For the time being, we concern ourselves exclusively with modular elliptic curves. Let E be an elliptic curve over $\mathbb{Q}$, and f_E its associated newform of weight two. In this situation there is only one critical point. Consequently, the p-adic L-function should interpolate Dirichlet twists of the algebraic part of the L-value at $s = 1$. Put $\mathbf{L}_p(E,\psi,s) = \mathbf{L}_{p,\alpha_p,\Omega_E^\pm}(f_E,\psi,s)$, so

$$\mathbf{L}_p(E,\psi,1) = \frac{1}{\alpha_p^n} \times \left(\sum_{m=1}^{p^n}\psi^{-1}(m)e^{2\pi i m/p^n}\right) \times \frac{L(E,\psi,j+1)}{\Omega_E^{\mathrm{sign}(\psi)}}$$

whilst at trivial $\psi = \mathbf{1}$,

$$\mathbf{L}_p(E,1) = \begin{cases} \left(1-\frac{1}{\alpha_p}\right)^2 \times \frac{L(E,1)}{\Omega_E^+} & \text{if } p \text{ does not divide } N_E \\ \left(1-\frac{1}{a_p(E)}\right) \times \frac{L(E,1)}{\Omega_E^+} & \text{if } p \text{ exactly divides } N_E. \end{cases}$$

Remark: The reader will notice that when $E_{/\mathbb{Q}_p}$ has split multiplicative reduction the coefficient $a_p(E) = +1$, hence $\left(1-\frac{1}{a_p(E)}\right) = 0$. Thus $\mathbf{L}_p(E,1) = 0$ regardless of whether the complex L-function vanishes at $s = 1$. This is commonly called the *exceptional zero phenomenon*, and can sometimes occur at weight $k > 2$ as well.

The following are p-adic analogues of Conjecture 1.16(a),(b) over the field $\mathbf{F} = \mathbb{Q}$.

Conjecture 2.3. *(i) In the non-exceptional zero situation,*

$$\mathrm{order}_{s=1}\mathbf{L}_p(E,s) \ \stackrel{?}{=}\ \mathrm{rank}_{\mathbb{Z}}E(\mathbb{Q});$$

(ii) In the exceptional zero situation i.e. $p\|N_E$ *and* $a_p(E)=+1$,

$$\mathrm{order}_{s=1}\mathbf{L}_p(E,s) \ \stackrel{?}{=}\ 1\ +\ \mathrm{rank}_{\mathbb{Z}}E(\mathbb{Q}).$$

Kato and Rubin [Ka1,Ru1] have proven that Conjecture 2.3 is true when the '=' is replaced with a '$\geq$' sign. In other words, the order of $\mathbf{L}_p(E,s)$ at $s=1$ gives an upper bound on the Mordell-Weil rank of E. The demonstration requires the theory of Euler systems, which we shall rely on heavily in due course.

Conjecture 2.4. *(i) The leading term of* $\mathbf{L}_p(E,s)$ *at* $s=1$, *should equal*

$$\frac{\det\Big(\langle -,-\rangle^{\mathrm{cy}}_{\mathbb{Q},p}\Big)\times\prod_{l\leq\infty}\left[E(\mathbb{Q}_l):E_0(\mathbb{Q}_l)\right]\times\#\mathrm{III}_{\mathbb{Q}}(E)[p^\infty]}{\#E(\mathbb{Q})[p^\infty]^2}\qquad \textit{modulo}\ \mathbb{Z}_p^\times,$$

where $\det\Big(\langle\cdots\rangle^{\mathrm{cy}}_{\mathbb{Q},p}\Big)$ *denotes the p-adic regulator of Schneider et al.* [Sn1,Sn2]*;*

(ii) In the exceptional zero situation, the above formula holds with the additional factor $\frac{\log_p(\mathbf{q}_E)}{\mathrm{ord}_p(\mathbf{q}_E)}$, *where* $\mathbf{q}_E$ *denotes the Tate period of the elliptic curve.*

In contrast to its archimedean cousin, the non-degeneracy or otherwise of the p-adic height pairing

$$\langle -,-\rangle^{\mathrm{cy}}_{\mathbf{K},p}: E(\mathbf{K})\ \times\ E(\mathbf{K})\ \longrightarrow\ \mathbb{Q}_p$$

is still an open question. We will return to this pairing in the final two chapters, where we examine its behaviour within an analytically varying family.

We should point out that both conjectures rely on the finiteness of the p-primary part of the Tate-Shafarevich group. For the heretics who disbelieve this, one should then replace the Mordell-Weil rank by the $\mathbb{Z}_p$-corank of the p^∞-Selmer group of E. The adjustments required in 2.3(ii) and 2.4(ii) are caused by the trivial zero in the p-adic multiplier term. The quantity $\frac{\log_p(\mathbf{q}_E)}{\mathrm{ord}_p(\mathbf{q}_E)}$ is usually called an $\mathcal{L}$-invariant, and it has a purely local description in this instance.

2.2 Perrin-Riou's local Iwasawa theory

In order to study these conjectures, the first step is to replace the analytic p-adic L-function with a more algebraic object. In the early part of the 90's, Perrin-Riou [PR1,PR2,PR3] developed a general machine for precisely this purpose, building on some earlier works of Coates, Wiles and Coleman. Her theory is phrased in the language of p-adic representations, and we will introduce it gradually. We start with some preliminaries on Fontaine's mysterious rings B_{dR} and B_{cris}.

Rings of p-adic periods

Let $\widetilde{\mathbf{E}}$ denote the set of sequences of the form $x = \left(x^{(0)}, x^{(1)}, \dots\right)$ where each $x^{(n)} \in \mathbb{C}_p$ satisfies $\left(x^{(n+1)}\right)^p = x^{(n)}$. If $x, y \in \widetilde{\mathbf{E}}$, one defines $x + y := \left(s^{(0)}, s^{(1)}, ...\right)$ where

$$s^{(n)} = \lim_{m\to\infty} \left(x^{(n+m)} + y^{(n+m)}\right)^{p^m},$$

the multiplication in $\widetilde{\mathbf{E}}$ being defined componentwise. This endows $\widetilde{\mathbf{E}}$ with the structure of a field of characteristic p, complete with respect to the valuation $\mathrm{val}_{\widetilde{\mathbf{E}}}(x) = \mathrm{val}_{\mathbb{C}_p}(x^{(0)})$. For example, if $\varepsilon = \left(1, \zeta_p, \zeta_{p^2}, ...\right)$ then $\mathrm{val}_{\widetilde{\mathbf{E}}}(\varepsilon - 1) = \frac{p}{p-1}$; in particular, $\varepsilon - 1$ lies in the maximal ideal of the ring of integers $\widetilde{\mathbf{E}}^+$ so $\mathbb{F}_p((\varepsilon - 1))$ is a subfield of $\widetilde{\mathbf{E}}$.

Remark: Let A_{inf} denote the Witt ring $W(\widetilde{\mathbf{E}}^+)$ of $\widetilde{\mathbf{E}}^+$, and $[\ ] : \widetilde{\mathbf{E}}^+ \to W(\widetilde{\mathbf{E}}^+)$ the Teichmüller lift to characteristic zero. The homomorphism $\theta : A_{\mathrm{inf}} \to \mathcal{O}_{\mathbb{C}_p}$ sending a Witt expansion $\sum_{n=0}^{\infty} p^n [x_n]$ to $\sum_{n=0}^{\infty} p^n x_n^{(0)}$ is surjective; moreover θ linearly extends to a map from $B_{\mathrm{inf}}^+ = A_{\mathrm{inf}}[\frac{1}{p}]$ onto $\mathbb{C}_p$.

One defines the topological ring B_{dR}^+ to be the adic completion $\varprojlim_n B_{\mathrm{inf}}^+/\mathrm{Ker}(\theta)^n$. The $\mathrm{Gal}(\overline{\mathbb{Q}}_p/\mathbb{Q}_p)$-action on B_{inf}^+ extends continuously to B_{dR}^+. Also, the power series

$$\log[\varepsilon] = \sum_{n=1}^{\infty} \frac{(-1)^{n-1}\left([\varepsilon] - 1\right)^n}{n} \qquad \text{converges in } B_{\mathrm{dR}}^+$$

to an element t which generates $\mathrm{Ker}(\theta)$. It is a straightforward exercise to check that the action of every $\sigma \in G_{\mathbb{Q}_p}$ on t, is described by the formula $t^\sigma = \chi_{\mathrm{cy}}(\sigma)t$. The filtration on the topological field $B_{\mathrm{dR}} = B_{\mathrm{dR}}^+[t^{-1}]$ is then taken to be $t^m B_{\mathrm{dR}}^+$, which is clearly Galois stable (many people regard t as the p-adic analogue of $2\pi i$).

Lastly, we write A_{cris} for the collection of elements $x \in B_{\mathrm{dR}}^+$ for which there exist sequences $\{a_n\}_{n\geq 0}$ inside A_{inf} tending to 0, with the property that $x = \sum_{n=0}^{\infty} a_n \frac{t^n}{n!}$. In fact $B_{\mathrm{cris}}^+ := A_{\mathrm{cris}}[\frac{1}{p}]$ is a $G_{\mathbb{Q}_p}$-stable subring of B_{dR}^+ containing the element t. Further, raising to p^{th}-powers in $\widetilde{\mathbf{E}}^+$ will induce a Frobenius operator φ on each of A_{inf}, B_{inf}^+, A_{cris} and B_{cris}^+ (however the Frobenius φ does **not** extend to B_{dR}^+). It follows immediately from the definition of t that $\varphi(t) = p\, t$, and henceforth we denote by B_{cris} the subring $B_{\mathrm{cris}}^+[t^{-1}]$ of B_{dR}.

The exponential map and its dual

Armed with these slightly bizarre topological constructions, one can study the local representation theory of modular forms. In particular, it is possible to define quite general analogues of the classical Lie group exponentials occurring in nature. Let K be a finite algebraic extension of $\mathbb{Q}_p$ and write $H^i(K, -)$ for the continuous Galois cohomology group $H^i_{\mathrm{cont}}(\mathrm{Gal}(\overline{K}/K), -)$.

Suppose that V is a finite dimensional p-adic representation of $\mathrm{Gal}(\overline{K}/K)$.

Fontaine and Messing [FM] devised a fundamental exact sequence

$$0 \longrightarrow \mathbb{Q}_p \longrightarrow B_{\mathrm{cris}}^{\varphi=1} \longrightarrow B_{\mathrm{dR}}/B_{\mathrm{dR}}^{+} \longrightarrow 0,$$

the right-hand map being reduction modulo B_{dR}^{+}. Tensoring this over $\mathbb{Q}_p$ by V yields a long exact sequence in Galois cohomology

$$\begin{aligned} 0 \longrightarrow V^{\mathrm{Gal}(\overline{K}/K)} \longrightarrow \Big(V \otimes_{\mathbb{Q}_p} B_{\mathrm{cris}}^{\varphi=1}\Big)^{\mathrm{Gal}(\overline{K}/K)} &\longrightarrow \Big(V \otimes_{\mathbb{Q}_p} B_{\mathrm{dR}}/B_{\mathrm{dR}}^{+}\Big)^{\mathrm{Gal}(\overline{K}/K)} \\ \xrightarrow{\ \partial\ } H^1(K,V) \longrightarrow H^1\Big(K, V \otimes_{\mathbb{Q}_p} B_{\mathrm{cris}}^{\varphi=1}\Big) &\longrightarrow \cdots . \end{aligned}$$

The boundary map ∂ is more commonly referred to as the 'exponential'.

Let us denote by $\mathbf{D}_{\mathrm{dR}}(V)$ the K-vector space $H^0(K, V \otimes_{\mathbb{Q}_p} B_{\mathrm{dR}})$, and write $\mathrm{Fil}^0\mathbf{D}_{\mathrm{dR}}(V)$ for the submodule $H^0(K, V \otimes_{\mathbb{Q}_p} B_{\mathrm{dR}}^{+})$. If V is a de Rham representation so that $\dim_K \mathbf{D}_{\mathrm{dR}}(V) = \dim_{\mathbb{Q}_p} V$, then

$$\exp_{K,V} : \mathrm{tang}(V/K) \cong \Big(V \otimes_{\mathbb{Q}_p} B_{\mathrm{dR}}/B_{\mathrm{dR}}^{+}\Big)^{\mathrm{Gal}(\overline{K}/K)} \xrightarrow{\ \partial\ } H^1(K,V)$$

N.B. the tangent space is the quotient of $\mathbf{D}_{\mathrm{dR}}(V)$ by the subspace $\mathrm{Fil}^0\mathbf{D}_{\mathrm{dR}}(V)$. The image of $\exp_{K,V}$ is usually abbreviated as $H^1_e(K,V)$.

Example 2.5 Consider the special case when A is an abelian variety over K, and $V = \mathrm{Ta}_p(A) \otimes \mathbb{Q}$ is the p-adic representation associated to the Tate module of A. Bloch and Kato [BK, Sect.3] proved the commutativity of the square

$$\begin{array}{ccc} \mathrm{tang}_K(A) & \xrightarrow{\text{Lie group exponential}} & A(K)\widehat{\otimes}\mathbb{Q}_p \\ \cong\big\downarrow \text{Hodge theory} & & \cong\big\downarrow \text{Kummer map} \\ \mathbf{D}_{\mathrm{dR}}(V)/\mathrm{Fil}^0 & \xrightarrow{\exp_{K,V}} & H^1_e(K,V). \end{array}$$

Here the Kummer map on $A(K)/p^n$ sends a point $P \pmod{p^n}$ to the class of the one-cochain $\sigma \mapsto Q^\sigma - Q$, where $Q \in A(\overline{K})$ is any local point satisfying $p^nQ = P$. On an integral level, this identifies the image of the completed points $A(K)\widehat{\otimes}\mathbb{Z}_p$ with the exact kernel of $H^1(K, \mathrm{Ta}_p(A)) \longrightarrow H^1\Big(K, \mathrm{Ta}_p(A) \otimes_{\mathbb{Z}_p} B_{\mathrm{cris}}^{\varphi=1}\Big)$.

Since the target of the exponentials is Galois cohomology, if we want to measure cohomology we should examine their duals instead. The invariant map of local class field theory relates the p^n-torsion in the Brauer group $H^2(K, \mu_{p^n})$ with $p^{-n}\mathbb{Z}_p/\mathbb{Z}_p$. Upon passage to the limit, one obtains an isomorphism $\mathrm{inv}_K : H^2(K, \mathbb{Z}_p(1)) \xrightarrow{\sim} \mathbb{Z}_p$. The dual exponential map $\exp^*_{K,V}$ is defined by the commutativity of the diagram

$$\begin{array}{ccccc} \mathrm{cotang}(V/K) \times \mathrm{tang}(V^*(1)/K) & \xrightarrow{\cup_{\mathrm{dR}}} & \mathbf{D}_{\mathrm{dR}}(\mathbb{Q}_p(1)) \cong K & \xrightarrow{\mathrm{Tr}_{K/\mathbb{Q}_p}} & \mathbb{Q}_p \\ \exp^*_{K,V}\big\uparrow \qquad\qquad \big\downarrow \exp_{K,V^*(1)} & & & & \big\| \\ H^1(K,V) \times H^1(K,V^*(1)) & \xrightarrow{\cup} & H^2(K, \mathbb{Q}_p(1)) & \xrightarrow{\mathrm{inv}_K} & \mathbb{Q}_p \end{array}$$

where $\mathrm{cotang}(V)$ is identified with $\mathrm{Fil}^0\mathbf{D}_{\mathrm{dR}}(V)$.

Equivalently,

$$\mathrm{Tr}_{K/\mathbb{Q}_p}\Big(\exp^*_{K,V}(x) \cup_{\mathrm{dR}} \mathbf{v} \mod \mathrm{Fil}^0\Big) \quad = \quad \mathrm{inv}_K\Big(x \cup \exp_{K,V^*(1)}(\mathbf{v})\Big)$$

which is a formula we shall make good use of in later chapters.

Behaviour of $\exp^*$ *in cyclotomic towers*

Essentially deforming the dual exponential maps in the cyclotomic direction yields the first of the two variables in our p-adic interpolation. Let μ_{p^∞} denote the group of p-power roots of unity inside $\overline{K}$. We set $G_{\infty,K}$ equal to $\mathrm{Gal}(K(\mu_{p^\infty})/K)$; in particular, the pth-cyclotomic character χ_{cy} exhibits an isomorphism between $G_{\infty,K}$ and an open subgroup of $\mathbb{Z}_p^\times$. As the prime $p > 2$, we have the decomposition $G_{\infty,K} \cong \Gamma_K \times \Delta_K$, where Γ_K is the Galois group of the cyclotomic $\mathbb{Z}_p$-extension of K, and Δ_K is a finite cyclic group of order dividing $p-1$.

To study the dual maps $\exp^*_{K(\mu_{p^n}),V}$ as $n \in \mathbb{N}$ varies, we begin with a short discussion of Iwasawa modules. Here the cyclotomic Iwasawa algebra is taken to be the completed group algebra

$$\Lambda := \mathbb{Z}_p\big[\big[G_{\infty,\mathbb{Q}_p}\big]\big] = \varprojlim_n \mathbb{Z}_p\big[\mathrm{Gal}(\mathbb{Q}_p(\mu_{p^n})/\mathbb{Q}_p)\big].$$

We say that a Λ-module has rank r if each eigenspace under the action of Δ has rank r over $\mathbb{Z}_p[\![\Gamma]\!]$. Note that $\mathbb{Z}_p[\![\Gamma]\!]$ can be identified with the power series ring $\mathbb{Z}_p[\![X]\!]$ by sending a topological generator γ_0 of Γ to $1+X$.

Definition 2.6. *(a) We define* $H^1_{\mathrm{Iw}}(K,V) := H^1_{\mathrm{cont}}(K, V \otimes_{\mathbb{Z}_p} \Lambda)$*, in fact for any Galois-stable lattice* $\mathbf{T}$ *in* V

$$H^1_{\mathrm{Iw}}(K,V) \cong \varprojlim_n H^1(K(\mu_{p^n}), \mathbf{T}) \otimes_{\mathbb{Z}_p} \mathbb{Q}_p$$

the isomorphism arising by application of Shapiro's lemma.

(b) We shall write $H^1_{\#}(K(\mu_{p^m}), \mathbf{T})$ *for the projection to the* m^{th}*-layer of the limit* $\varprojlim_n H^1(K(\mu_{p^n}), \mathbf{T})$*; analogously* $H^1_{\#}(K(\mu_{p^m}), V)$ *is the projection of* $H^1_{\mathrm{Iw}}(K,V)$.

Assume now that K is unramified over $\mathbb{Q}_p$. Consequently $H^1_{\mathrm{Iw}}(K,V)$ becomes a Λ-module, as the action of $G_{\infty,\mathbb{Q}_p} \cong \varprojlim_n \mathrm{Gal}\big(K(\mu_{p^n})/K\big)$ on $\varprojlim_n H^1\big(K(\mu_{p^n}), \mathbf{T}\big)$ extends linearly and continuously to the whole Iwasawa algebra.

For all $j \in \mathbb{Z}$, there is a natural Λ-isomorphism

$$\begin{aligned} H^1_{\mathrm{Iw}}(K,V) &\xrightarrow{\sim} H^1_{\mathrm{Iw}}(K,V(j)) \\ (x_{p^n})_n &\mapsto (x_{p^n})_n \otimes \varepsilon^{\otimes j} \end{aligned}$$

which depends upon fixing some generator $\varepsilon = (\zeta_{p^n})_n$ of the Tate module $\mathbb{Z}_p(1)$. To be totally consistent, we assume the exact same ζ_{p^n}'s are chosen which define the uniformiser $t = \log[\varepsilon]$ of B_{dR}.

Warning: Throughout the remainder of this book, we adopt the non-standard convention that

$$\text{“ } x_{p^m} \otimes \zeta_{p^m}^{\otimes j} \text{ ”} \quad \text{refers to the } m^{\text{th}}\text{-layer term} \quad \Big((x_{p^n})_n \otimes \varepsilon^{\otimes j} \Big)_m$$

whenever the elements x_{p^n} live in a norm-compatible family. The justification for this change is pretty shallow – by abbreviating the projection in this manner, our formulae shorten dramatically.

Perrin-Riou [PR1] showed that the Λ-rank of $\varprojlim_n H^1(K(\mu_{p^n}), \mathbf{T})$ is equal to the quantity $[K:\mathbb{Q}_p]\dim_{\mathbb{Q}_p} V$. For example, when $V = V_f^*$ is the (dual of the) Deligne representation attached to any p-stabilised eigenform $f \in \mathcal{S}_k(\Gamma_0(Np^r), \epsilon)$, we have

$$\text{rank}_{\mathbb{Q}\otimes\Lambda_f}\Big(H^1_{\text{Iw}}(K, V_f^*)\Big) \;=\; 2[K:\mathbb{Q}_p] \quad \text{where } \Lambda_f = \Lambda \otimes_{\mathbb{Z}_p} \mathcal{O}_f.$$

In general, one requires a larger ring than Λ to capture all the p-adic L-functions arising from Perrin-Riou's theory. Let $\mathcal{K}$ be a local field of residue characteristic p. The ring of tempered $\mathcal{K}$-valued functions on Γ, is defined to be

$$\mathcal{H}(\Gamma) := \left\{ \sum_{n\geq 0} h_n(\gamma_0 - 1)^n \text{ with } h_n \in \mathcal{K}, \lim_{n\to\infty} \frac{|h_n|_p}{n^r} = 0 \text{ for some } r > 0 \right\}$$

and we also set $\mathcal{H}(G_\infty) := \mathcal{H}(\Gamma) \otimes_{\mathbb{Z}_p[\![\Gamma]\!]} \Lambda$. The p-adic L-functions mentioned in §2.1 lie inside $\mathcal{H}(G_\infty)$ but not $\mathcal{O}_{\mathcal{K}}[\![G_\infty]\!]$, when the cusp form is non-ordinary at p.

For any local Galois character ψ modulo p^n, let us define its Gauss sum by

$$G(\psi, \zeta_{p^n}) \;=\; \sum_{\sigma\in \text{Gal}\left(K(\mu_{p^n})/K\right)} \psi(\sigma) \times \zeta_{p^n}^{\sigma}.$$

Assume V is a de Rham $\text{Gal}(\overline{K}/K)$-representation with underlying field $\mathcal{K}$. The following deep result was first proved by Perrin-Riou for crystalline representations, and extended by Kato et al [KKT,Cz] in the non-crystalline case. As before, K is assumed to be absolutely unramified.

Theorem 2.7. *(Perrin-Riou* [PR2]*) There is a unique $\mathcal{O}_{\mathcal{K}}[\![G_{\infty,K}]\!]$-homomorphism*

$$\text{PR} : H^1_{\text{Iw}}(K, V^*) \otimes_{\mathcal{K}} \mathbf{D}_{\text{cris}}(V) \longrightarrow \mathcal{H}(G_{\infty,K}) \otimes_{\mathcal{K}} \mathbf{D}_{\text{dR}}(\mathcal{K})$$

such that for all primitive characters $\psi : \text{Gal}\big(K(\mu_{p^n})/K\big) \to \overline{\mathbb{Q}}_p^\times$ and $j \geq 0$, we have

$$\psi\chi_{\text{cy}}^j\Big(\text{PR}(\underline{x}\otimes\mathbf{v})\Big) = j!\, p^{nj} G(\psi^{-1}, \zeta_{p^n}) \sum_{\sigma\in\text{Gal}\left(K(\mu_{p^n})/K\right)} \psi(\sigma)\Big(\exp^*_{V^*(-j)}\Big(x_{p^n} \otimes \zeta_{p^n}^{\otimes -j}\Big)^{\sigma} \cup \varphi^{-n}\mathbf{v}\Big)$$

whilst at trivial ψ

$$\chi_{\text{cy}}^j\Big(\text{PR}(\underline{x}\otimes\mathbf{v})\Big) = j!\exp^*_{V^*(-j)}(x_1) \cup (1 - p^j\varphi^{-1})(1 - p^{-j-1}\varphi)^{-1}\mathbf{v}.$$

2.3 Integrality and (φ,Γ)-modules

Whilst the preceding result works well with any pseudo-geometric representation, for our purposes it is just a starting point. Let's focus first on what this local theory means for modular forms on GL_2. As will shortly become evident, in the ordinary case one can refine Theorem 2.7 somewhat, so that the image of the interpolation lies in an Iwasawa algebra.

Fix a prime $p \neq 2$ and a positive integer N coprime to p. Let $f \in \mathcal{S}_k(\Gamma_0(Np^r),\epsilon)$ be a p-ordinary normalised eigenform (for the Hecke algebra) of positive weight k, nebentypus ϵ and level Np^r; in particular if $f(z) = \sum_{n\geq 1} a_n(f)q^n$ with $q = e^{2\pi i z}$ then $a_p(f)$ is a p-adic unit.

Notation: We say that f is a *p-stabilised ordinary newform* if either

(i) f is a newform of conductor Np^r; or

(ii) $f(z) = g(z) - \beta_p g(pz)$ where g is a newform of conductor N and β_p is the non-unit root of Frobenius for g.

As always, $\mathcal{K}_f$ is the finite extension of $\mathbb{Q}_p$ obtained by adjoining all the $a_n(f)$'s. Recall also from Chapter I that for weight $k \geq 2$, Deligne [De] attached a continuous representation

$$\rho_f : G_{\mathbb{Q}} \to \mathrm{Aut}(V_f) \quad \text{with} \quad \dim_{\mathcal{K}_f} V_f = 2$$

of the absolute Galois group $G_{\mathbb{Q}} := \mathrm{Gal}(\overline{\mathbb{Q}}/\mathbb{Q})$. This p-adic representation was unramified outside of Np, and characterised by the identity

$$\det\left(1 - \rho_f\big|_{G_{\mathbb{Q}_l}}(\mathrm{Frob}_l^{-1})X\right) = 1 - a_l(f)X + \epsilon(l)l^{k-1}X^2 \quad \text{for all primes } l \nmid Np.$$

Now as Galois representations we have $V_f^* := \mathrm{Hom}(V_f,\mathcal{K}_f) \cong V_{f^*} \otimes_{\mathbb{Z}_p} \mathbb{Z}_p(k-1)$ where $f^*(z) := \sum_{n\geq 1} \overline{a_n(f)}q^n$ denotes the dual cusp form to f. For reasons of normalisation, it will be more convenient to work with V_f^* rather than V_f.

Let M be a positive integer coprime to p. The tensor product $\mathbb{Q}(\mu_M) \otimes_{\mathbb{Q}} \mathbb{Q}_p$ of fields decomposes into a direct product $\prod_{\mathfrak{p}} \mathbb{Q}(\mu_M)_{\mathfrak{p}}$ over the completions at primes $\mathfrak{p} \in \mathrm{Spec}\ \mathbb{Z}[\mu_M]$ lying over p. Each completion is isomorphic to the extension obtained by adjoining $\sqrt[M]{1}$ to the p-adics.

Hypothesis. *Assume that our field K is one of these component $\mathbb{Q}(\mu_M)_{\mathfrak{p}}$'s.*

We can then view the cup product

$$\exp^*_{K(\mu_{p^n}),V_f^*(-j)}\left(x_{p^n} \otimes \zeta_{p^n}^{\otimes -j}\right)^{\sigma} \cup \varphi^{-n}\mathbf{v} \qquad \text{(c.f. Theorem 2.7)}$$

as lying in a copy of the field $\mathcal{K}_f(\mu_{p^n M})$, via the sequence

$$\mathrm{Fil}^0\mathbf{D}_{\mathrm{dR},n}(V_f^*(-j)) \times \mathbf{D}_{\mathrm{cris}}(V_f) \xrightarrow{\cup_{\mathrm{dR}}} \mathbf{D}_{\mathrm{dR},n}(\mathcal{K}_f(-j)) \cong \mathbf{D}_{\mathrm{dR},n}(\mathcal{K}_f) = \mathcal{K}_f(\mu_{p^n M})$$

where the isomorphism shifts the filtration j places.

Remark: If $0 \leq j \leq k-2$ then $\mathrm{Fil}^0\mathbf{D}_{\mathrm{dR},n}(V_f^*(-j)) \cong \mathrm{Fil}^{k-1-j}\mathbf{D}_{\mathrm{dR},n}(V_{f^*})$ is known to be one-dimensional over $\mathcal{K}_f(\mu_{p^n M})$, generated by the dual cusp form f^*.

Let $\mathbf{v}_{k,\epsilon}$ be the element of $\mathbf{D}_{\mathrm{cris},\mathbb{Q}_p}(V_f)^{\varphi=a_p(f)}$ satisfying $f^* \cup \mathbf{v}_{k,\epsilon} = 1$ in $\mathbf{D}_{\mathrm{dR},\mathbb{Q}_p}(\mathcal{K}_f)$. Note that the φ-eigenspace of $\mathbf{D}_{\mathrm{cris},\mathbb{Q}_p}(V_f)$ with eigenvalue $a_p(f)$ is non-empty, by monodromy theorems of Saito [Sa]. The operator $(1-p^{-j-1}\varphi)$ acting on $\mathbf{v}_{k,\epsilon}$ is always invertible, whilst $(1-p^j\varphi^{-1})$ fails to be invertible when $j=0$ and $a_p(f)=1$, i.e. the exceptional zero condition mentioned in §2.1.

Definition 2.8. *To simplify notation, for any $x \in H^1\big(K(\mu_{p^n}), V_f^*(-j)\big)$ we shall write '$\exp_j^*(x)$' for the unique scalar satisfying*

$$\exp_j^*(x) \;=\; \exp^*_{K(\mu_{p^n}),V_f^*(-j)}(x) \;\cup\; \big(1-\delta_{n=0}p^{-j-1}\varphi\big)^{-1}\,\mathbf{v}_{k,\epsilon}$$

where $\delta_{n=0}$ equals one if $n=0$, and $\delta_{n=0}$ equals zero if $n>0$.

The proof of the following result will be given at the end of this section.

Lemma 2.9. *There exists a constant $\theta_{\mathbf{T}} \in \mathbb{Z}$ depending on a choice of lattice $\mathbf{T} \subset V_f^*$ and on the integer M, such that for $0 \leq j \leq k-2$*

$$\exp_j^*\Big(H^1_{\#}\big(K(\mu_{p^n}),\mathbf{T}(-j)\big)\Big) \quad \textit{is contained in} \quad p^{\theta_{\mathbf{T}}-n(j+1)}\mathcal{O}_{\mathcal{K}_f}[\mu_{Mp^n}].$$

Corollary 2.10. *Defining $\mathrm{PR}_{k,\epsilon,K}(\underline{x}) := \mathrm{PR}\left(\underline{x}\otimes\mathbf{v}_{k,\epsilon}\right)$ at all $\underline{x} \in H^1_{\mathrm{Iw}}(K,V_f^*)$, then*

$$\mathrm{PR}_{k,\epsilon,K} : H^1_{\mathrm{Iw}}(K,V_f^*) \;\longrightarrow\; \mathbb{Q}\otimes\mathcal{O}_f[\![G_{\infty,K}]\!][\mu_M]$$

is a homomorphism of finitely generated $\mathbb{Q}\otimes\Lambda_f$-modules.

Proof: It suffices to show that $\mathrm{PR}_{k,\epsilon,K}\Big(\varprojlim_n H^1\big(K(\mu_{p^n}),\mathbf{T}\big)\Big)$ is contained in the module $p^{\theta_{\mathbf{T}}-1}\Lambda_f^{\mathrm{cy}}[\mu_M]$. Let us abbreviate $\mathrm{Gal}\big(K(\mu_{p^n})/K\big)$ with the shorter $\mathcal{G}_{p^n,K}$. Writing out the special value formulae in full, then for all primitive $\psi : \mathcal{G}_{p^n,K} \to \overline{\mathbb{Q}}_p^{\times}$ and $0 \leq j \leq k-2$, we find that $\psi\chi^j_{\mathrm{cy}}\big(\mathrm{PR}_{k,\epsilon,K}(\underline{x})\big)$ equals

$$\frac{j!p^{nj}\left(1-\delta_{\psi=\mathbf{1}}p^j/a_p(f)\right)}{a_p(f)^n}G(\psi^{-1},\zeta_{p^n}) \sum_{\sigma\in\mathcal{G}_{p^n,K}} \psi(\sigma)\exp_j^*\left(x_{p^n}\otimes\zeta_{p^n}^{\otimes -j}\right)^{\sigma}$$

where $\delta_{\psi=\mathbf{1}}$ equals 1 if ψ is trivial, and 0 if ψ is non-trivial. Here we used the fact that $\varphi\mathbf{v}_{k,\epsilon} = a_p(f)\mathbf{v}_{k,\epsilon}$ in tandem with the interpolation Theorem 2.7.

Now $a_p(f)$ is a p-adic unit since f is a p-ordinary eigenform, and it follows that $\left|\frac{j!p^{nj}\left(1-\delta_{\psi=\mathbf{1}}p^j/a_p(f)\right)}{a_p(f)^n}\right|_p$ is bounded above by p^{-nj}. On the other hand,

$$\begin{aligned} G(\psi^{-1},\zeta_{p^n}) \sum_{\sigma\in\mathcal{G}_{p^n,K}} \psi(\sigma)\exp_j^*(\ldots)^{\sigma} &= \psi\left(\sum_{\tau\in\mathcal{G}_{p^n,K}} \tau^{-1}.\zeta_{p^n}^{\tau} \sum_{\sigma\in\mathcal{G}_{p^n,K}} \sigma.\exp_j^*(\ldots)^{\sigma}\right) \\ &= \psi\left(\mathrm{Tr}_{\,\mathcal{G}_{p^n,K}}\left(\zeta_{p^n}\sum_{\sigma\in\mathcal{G}_{p^n,K}} \sigma.\exp_j^*(\ldots)^{\sigma}\right)\right) \end{aligned}$$

and we know from Lemma 2.9 that $p^{n(j+1)-\theta_{\mathbf{T}}}\exp_j^*(...)^\sigma$ is a p-integral element. However $\mathrm{Tr}_{\mathcal{G}_{p^n,K}}\left(\mathcal{O}_{\mathcal{K}_f}[\mu_{Mp^n}]\right)$ is contained in $\#\mathcal{G}_{p^n,K}.\mathcal{O}_{\mathcal{K}_f}[\mu_M] = p^{n-1}\mathcal{O}_{\mathcal{K}_f}[\mu_M]$, from which we may deduce

$$G(\psi^{-1},\zeta_{p^n})\sum_{\sigma\in\mathcal{G}_{p^n,K}}\psi(\sigma)\exp_j^*(...)^\sigma \quad \text{lies in} \quad p^{\theta_{\mathbf{T}}-n(j+1)+(n-1)}\mathcal{O}_{\mathcal{K}_f}[\mu_{Mp^n}]\,.$$

Since $p^{-nj}\times\left|p^{\theta_{\mathbf{T}}-n(j+1)+(n-1)}\right|_p = p^{1-\theta_{\mathbf{T}}}$, the nth-level of $p^{1-\theta_{\mathbf{T}}}\mathrm{PR}_{k,\epsilon,K}(\underline{x})$ in the group ring must be p-integral for every n. The coefficients in the power series expansion of $\mathrm{PR}_{k,\epsilon,K}(\underline{x})$ must therefore be p-bounded, and the result follows. □

In the non-ordinary case the corollary is false, scuppered by the fact that the eigenvalue $a_p(f)$ is no longer a p-adic unit. Nevertheless, there is a nice trick of Kobayashi and Pollack [Ky,Po] which allows one to split $\mathrm{PR}_{2,1,\mathbb{Q}_p}$ into a $\pm$-part. This produces two p-adic L-functions living in the Iwasawa algebra, for modular elliptic curves with good supersingular reduction at $p\geq 5$.

The following argument is rather unenlightening, and could easily be skipped on a first reading.

The Proof of Lemma 2.9

Recall that we are searching for a constant $\theta_{\mathbf{T}}\in\mathbb{Z}$ dependent on the Galois stable lattice $\mathbf{T}\subset V_f^*$ and on the integer M, such that

$$\exp^*_{V_f^*(-j)}\left(H^1_{\#}(K(\mu_{p^n}),\mathbf{T}(-j))\right)\ \cup\ (1-\delta_{n=0}p^{-j-1}\varphi)^{-1}\mathbf{v}_{k,\epsilon}$$

is contained within $p^{\theta_{\mathbf{T}}-n(j+1)}\mathcal{O}_{\mathcal{K}_f}[\mu_{Mp^n}]$ for all integers $n\geq 0$ and $0\leq j\leq k-2$. The notation $H^1_{\#}$ denoted the group of universal traces

$$\bigcap_{m\geq n}\mathrm{cores}_{K(\mu_{p^m})/K(\mu_{p^n})}H^1\big(K(\mu_{p^n}),\,-\,\big)$$

for the cyclotomic p-extension of $K(\mu_{p^n})$. If $n\geq 1$, this is equivalent to showing

$$\exp^*_{V_f^*(-j)}\left(H^1_{\#}(K(\mu_{p^n}),\mathbf{T}(-j))\right)\quad\subset\quad p^{\theta_{\mathbf{T}}-n(j+1)}\mathcal{O}_{\mathcal{K}_f}[\mu_{Mp^n}].t^j f^*$$

as $t^jf^*\cup\mathbf{v}_{k,\epsilon}=1$ viewed inside $\mathbf{D}_{\mathrm{dR},n}\big(\mathcal{K}_f(-j)\big)$. To prove our lemma, we exploit the fact that H^1_{Iw} can be computed in terms of (φ,Γ)-modules introduced by Fontaine.

If $\pi=[\varepsilon]-1$ we write $\mathbf{A}_{\mathbb{Q}_p}$ for the topological closure of $\mathbb{Z}_p[\pi,\frac{1}{\pi}]$ in the ring $W(\widetilde{\mathbf{E}})$, whose topology is then given by neighborhoods $\pi^aW(\widetilde{\mathbf{E}}^+)+p^{b+1}W(\widetilde{\mathbf{E}})$ of 0. Note that $\mathbf{A}_{\mathbb{Q}_p}$ is a complete discrete valuation ring with residue field $\mathbb{F}_p((\varepsilon-1))$. The Frobenius $\varphi:\pi\mapsto(1+\pi)^p-1$ on $\mathbf{A}_{\mathbb{Q}_p}$ extends uniquely to $\mathbf{B}$, the completion of the maximal unramified extension of $\mathbf{A}_{\mathbb{Q}_p}\left[\frac{1}{p}\right]$.

Definition 2.11. *Let V be any de Rham representation, and for simplicity set $L_n = K(\mu_{p^n})$ with $L_\infty = \bigcup_{n\geq 0} L_n$. The '$(\varphi,\Gamma)$-module associated to V' is given by the invariant vectors*

$$D(V) \ := \ \left(V \otimes_{\mathbb{Q}_p} \mathbf{B}\right)^{\mathrm{Gal}\left(\overline{\mathbb{Q}}_p/L_\infty\right)}$$

and upon which the operator φ commutes with the diagonal action of $\mathrm{Gal}(L_\infty/L_0)$.

The advantage of studying $D(V)$ in the context of Iwasawa theory, is justified by the existence of an isomorphism

$$\mathrm{Log}^*_{V^*(1)} : D(V)^{\varphi = p^{-1}\mathrm{Tr}_{\mathbf{B}/\varphi(\mathbf{B})}} \stackrel{\sim}{\longrightarrow} H^1_{\mathrm{Iw}}(L_0, V)$$

which was constructed by Fontaine in never published work – an explicit description for $\mathrm{Log}^*_{V^*(1)}$ will not be needed in the proof.

If $\underline{x} = (x_{p^n})_n \in H^1_{\mathrm{Iw}}(L_0, V)$ we may define

$$\mathrm{Exp}^*_{L_n}(\underline{x}) \ := \sum_{j\in\mathbb{Z}} \exp^*_{L_n, V(-j)}\left(x_{p^n} \otimes \zeta_{p^n}^{\otimes -j}\right),$$

which converges to an element of $L_n(\!(t)\!) \otimes_{L_0} \mathbf{D}_{\mathrm{dR}}(V)$ since $\exp^*_{V(-j)} \equiv 0$ for $j \ll 0$.

Remark: By [CC, Théorème IV.2.1(iii)] there exists a positive integer $m(V)$ such that for all integers $n \geq m(V)$,

$$\mathrm{Exp}^*_{L_n}(\underline{x}) \ = \ p^{-n}\varphi^{-n}\left(\mathrm{Log}^{*\ -1}_{V^*(1)}(\underline{x})\right).$$

The proof of Cherbonnier and Colmez' result is heavily based on overconvergent representation theory, together with an explicit formula of Kato's calculating $\exp^*$.

In our situation $V = V_f^*$, we can even say that $\mathrm{Exp}^*_{L_n}(\underline{x}) \in L_n[\![t]\!] \otimes_{L_0} \mathbf{D}_{\mathrm{dR}}(V)$ because $\mathrm{Fil}^0\left(t^j \mathbf{D}_{\mathrm{dR},n}(V_f^*)\right) = 0$ for $j < 0$. Restricting $\mathrm{Exp}^*_{L_n}$ yields a factorisation

$$\mathrm{Exp}^*_{L_n} : \varprojlim_m H^1(L_m, \mathbf{T}) \stackrel{\mathrm{proj}_n}{\longrightarrow} H^1_{\#}(L_n, \mathbf{T}) \stackrel{\sum_{j\geq 0} \exp^*_{V_f^*(-j)}}{\longrightarrow} L_n[\![t]\!] \otimes_{L_0} \mathbf{D}_{\mathrm{dR}}(V_f^*)$$

and as $H^1_{\#}(L_n, \mathbf{T})$ is a compact, finitely-generated $\mathbb{Z}_p$-module, so is its image under the homomorphism $\mathrm{Exp}^*_{L_n}$. It follows that there exist constants $\theta_n \in \mathbb{Z}$ depending on n, $\mathbf{T}$ and L_0, such that

$$\mathrm{Exp}^*_{L_n}\left(\varprojlim_m H^1(L_m, \mathbf{T})\right) \ \subset \ p^{\theta_n} \sum_{j=0}^{k-2} t^j \mathcal{O}_{\mathcal{K}_f}[\mu_{Mp^n}].f^* \ + \ t^{k-1} L_n[\![t]\!] \otimes_{L_0} \mathbf{D}_{\mathrm{dR}}(V_f^*).$$

To ease congestion we write m_f in place of $m(V_f^*)$. By the above remark, for all integers $n \geq m_f$ one deduces that

$$\begin{aligned}
\mathrm{Exp}^*_{L_n}\left(\varprojlim_m H^1(L_m,\mathbf{T})\right) &= (p\,\varphi)^{-n}\left(\mathrm{Log}^{*\ -1}_{V_f(1)}\left(\varprojlim_m H^1(L_m,\mathbf{T})\right)\right)\\
&= (p\,\varphi)^{-(n-m_f)}\left(\mathrm{Exp}^*_{L_{m_f}}\left(\varprojlim_m H^1(L_m,\mathbf{T})\right)\right)\\
\subset\ (p\,\varphi)^{-(n-m_f)} &\left(p^{\theta_{m_f}}\sum_{j=0}^{k-2} t^j\mathcal{O}_{\mathcal{K}_f}[\mu_{Mp^\infty}].f^* + t^{k-1}L_\infty[\![t]\!]\otimes_{L_0}\mathbf{D}_{\mathrm{dR}}(V_f^*)\right)\\
=\ p^{m_f+\theta_{m_f}}\sum_{j=0}^{k-2} & p^{-n(j+1)+jm_f}\ t^j\mathcal{O}_{\mathcal{K}_f}[\mu_{Mp^\infty}].f^* + t^{k-1}L_\infty[\![t]\!]\otimes_{L_0}\mathbf{D}_{\mathrm{dR}}(V_f^*).
\end{aligned}$$

Focussing only on the powers t^j for $j \in \{0,\ldots,k-2\}$,

$$\exp^*_{V_f^*(-j)}\left(H^1_\#(L_n,\mathbf{T}(-j))\right) \quad\subset\quad p^{(j+1)m_f+\theta_{m_f}-n(j+1)}\mathcal{O}_{\mathcal{K}_f}[\mu_{Mp^n}].t^j f^*$$

at every integer $n \geq m_f$.

Conversely, there are only finitely many (n,j) with $0 \leq j \leq k-2$ and $0 \leq n < m_f$. At these values there must exist a $\widetilde{\theta} \in \mathbb{Z}$ for which

$$\exp^*_{V_f^*(-j)}\left(H^1_\#(L_n,\mathbf{T}(-j))\right) \cup (1-\delta_{n=0}p^{-j-1}\varphi)^{-1}\mathbf{v}_{k,\epsilon}$$

is properly contained within $p^{\widetilde{\theta}}\ \mathcal{O}_{\mathcal{K}_f}[\mu_{Mp^n}]$. Picking the constant $\theta_{\mathbf{T}} = \min\{\theta_{m_f},\widetilde{\theta}\}$ ensures that each of the images $p^{n(j+1)-\theta_{\mathbf{T}}}\exp^*_j\left(H^1_\#(L_n,\mathbf{T}(-j))\right)$ will be p-integral **for all** integer pairs $n \geq 0$ and $0 \leq j \leq k-2$, as desired.

2.4 Norm relations in *K*-theory

In the first section we defined the p-adic L-function, and recalled some well-known conjectures about its behaviour at critical points. The explicit local machinery of Perrin-Riou then allowed the conversion of norm-compatible cocycles into power series, convergent on the open unit disk.

Question. *What is the correct input into this machine, such that the output is the p-adic L-function attached to a Hecke eigenform?*

The answer was provided by Kato [Ka1], who proved the norm-compatibility of zeta-elements lying in K_2 of modular curves. Roughly speaking, we're heading for the following scenario:

$$\begin{array}{ccc}
\Big\{\text{Galois cohomology}\Big\} & \overset{\text{Perrin-Riou's theory}}{\longrightarrow} & \Big\{\text{Iwasawa algebras}\Big\}\\
\text{Kato's zeta-elements} & \mapsto & \text{the } p\text{-adic } L\text{-function.}
\end{array}$$

We start by outlining how Beilinson [Be] originally constructed these elements out of Siegel units, although we will be more concerned with their p-adic properties rather than their archimedean ones.

Let $\mathbb{E}_{\text{univ}}$ denote the universal elliptic curve with $\Gamma(L)$-structure, at level $L \geq 3$. The modular curve $Y(L)_{/\mathbb{Q}}$ without cusps, represents the functor $S \mapsto (E, e_1, e_2)$ classifying isomorphism classes of triples where E is an elliptic curve over S, and (e_1, e_2) is a pair of sections for E over the scheme S generating the torsion $E[L]$. For each $(\alpha, \beta) = \left(\frac{a}{L}, \frac{b}{L}\right) \neq (0,0)$ lying in $\left(\frac{1}{L}\mathbb{Z}/\mathbb{Z}\right)^2$, there is a torsion section

$$\iota_{\alpha,\beta} = a\mathbf{e}_1 + b\mathbf{e}_2 : Y(L) \longrightarrow \mathbb{E}_{\text{univ}} \backslash \text{Ker}[\times c]$$

where c is a prescribed integer prime to 6 and the orders of α, β, and $(\mathbf{e}_1, \mathbf{e}_2)$ forms a $\mathbb{Z}$-basis of $\text{Ker}\left(\times L : \mathbb{E}_{\text{univ}} \to \mathbb{E}_{\text{univ}}\right)$. Henceforth assume that $\gcd(c, 6L) = 1$.

If we choose $\left(\frac{a}{L}, \frac{b}{L}\right) \in \left(\frac{1}{L}\mathbb{Z}/\mathbb{Z}\right)^2 \backslash \{(0,0)\}$, then there exists a unique Siegel unit ${}_c g_{\frac{a}{L},\frac{b}{L}} \in \mathcal{O}\big(Y(L)\big)^\times$ which is the pullback along $\iota_{\frac{a}{L},\frac{b}{L}}$ of the complex function

$${}_c\vartheta(\tau, z) \;=\; q^{\frac{c^2-1}{12}} \, (-t)^{\frac{c-c^2}{2}} \, \frac{\prod_{n\geq 0}(1-q^n t)^{c^2} \prod_{n\geq 1}(1-q^n t^{-1})^{c^2}}{\prod_{n\geq 0}(1-q^n t^c) \prod_{n\geq 1}(1-q^n t^{-c})}$$

on $\mathbb{E}_{\text{univ}}$ – here $q = \exp(2\pi i\tau)$, $\tau \in \mathfrak{H}$ and $t = \exp(2\pi i z)$ with $z \in \mathbb{C}\backslash c^{-1}(\tau\mathbb{Z} + \mathbb{Z})$.

Remark: Alternatively, one can define ${}_c\vartheta$ more algebraically as being the unique function on $\mathbb{E}_{\text{univ}}$ with divisor $c^2.O_{\mathbb{E}_{\text{univ}}} - \text{Ker}(\times c)$, invariant under all norm maps $\text{N}_a : \mathcal{O}\big(\text{Ker}(\times c)\big)^\times \to \mathcal{O}\big(\text{Ker}(\times ac)\big)^\times$. However, we prefer to write down a good old-fashioned q-expansion.

Now fix integers A, B such that $A + B \geq 5$ and $A|L$, $B|L$. The affine modular curve $Y(A, B)$ can be defined as the quotient of $Y(L)$ by the matrix group

$$\left\{ \begin{pmatrix} x & y \\ z & t \end{pmatrix} \in \text{GL}_2(\mathbb{Z}/L\mathbb{Z}) \;\middle|\; x \equiv 1,\; y \equiv 0 (\bmod A) \text{ and } z \equiv 0,\; t \equiv 1 (\bmod B) \right\}.$$

The above description induces a canonical morphism of curves $Y(L) \to Y(A, B)$, so we may regard ${}_c g_{\frac{1}{A},0} \in \mathcal{O}\big(Y(A,1)\big)^\times$ and ${}_d g_{0,\frac{1}{B}} \in \mathcal{O}\big(Y(1,B)\big)^\times$ for $\gcd(cd, 6AB) = 1$.

Definition 2.12. *Taking the symbol of these two functions yields the K_2-element*

$${}_{c,d}z_{A,B} \;=\; \left\{ {}_c g_{1/A,0},\; {}_d g_{0,1/B} \right\} \in K_2\Big(Y(A,B)\Big)$$

whose properties are described at great length in [Be] *and* [Ka1, §2].

If $c, d \neq \pm 1$ then we also have the (possibly) non-integral elements

$$z_{A,B} = \left\{ {}_c g_{1/A,0} \otimes \frac{1}{c^2-1},\; {}_d g_{0,1/B} \otimes \frac{1}{d^2-1} \right\} \in K_2\Big(Y(A,B)\Big) \otimes \mathbb{Q}$$

which may at least be shown to be independent of the initial choice of $c, d \in \mathbb{Z}$. They are related to the integral ones via the formula

$${}_{c,d}z_{A,B} \;=\; \left(c^2 - \begin{pmatrix} c & 0 \\ 0 & 1 \end{pmatrix}^*\right)\left(d^2 - \begin{pmatrix} 1 & 0 \\ 0 & d \end{pmatrix}^*\right) z_{A,B}.$$

Proposition 2.13. [Ka1, Sect.2.4] *The norm homomorphism*

$$K_2\Big(Y(A',B')\Big) \longrightarrow K_2\Big(Y(A,B)\Big) \quad \textit{sends} \quad {}_{c,d}z_{A',B'} \mapsto {}_{c,d}z_{A,B}\,,$$

provided that both $\mathrm{supp}(A') = \mathrm{supp}(A)$ *and* $\mathrm{supp}(B') = \mathrm{supp}(B)$.

Similarly, the element $z_{A',B'}$ is sent to $z_{A,B}$ under the $\mathbb{Q}$-linear extension of this mapping. It follows from the above result, that we obtain a compatible family

$$\Big({}_{c,d}z_{Ap^n,Bp^n}\Big)_n \;\in\; \varprojlim_n K_2\Big(Y(Ap^n,Bp^n)\Big)$$

where the projective limit of K-groups is again taken with respect to these norms. In order to utilise Kato's zeta-elements in arithmetic applications, they first need to be converted into one-cocycles. The major reason for this conversion is that the various Selmer groups, Tate-Shafarevich groups, etc... are best described in terms of the Galois cohomology of the underlying p-adic representations. In view of this we now describe the two-step construction of the realisation map.

Consider the smooth $\mathbb{Z}_p$-sheaf $\underline{H}^1_p$ on $Y(Ap^n,Bp^n)_{\text{ét}}$ defined to be $\mathbb{R}^1\delta^{(n)}_*(\mathbb{Z}_p)$, where the morphism $\delta^{(n)} : \mathbb{E}_{\text{univ}} \to Y(Ap^n,Bp^n)$. Regarding the Tate module of the universal elliptic curve $\mathbb{E}_{\text{univ}}$ as a smooth sheaf on the étale site of $Y(Ap^n,Bp^n)$, Poincaré duality then yields a canonical isomorphism $\mathrm{Ta}_p(\mathbb{E}_{\text{univ}}) \cong \underline{H}^1_p \otimes_{\mathbb{Z}_p} \mathbb{Z}_p(1)$ whence

$$\mathrm{Sym}^{k-2}_{\mathbb{Z}_p}(\underline{H}^1_p) \;\cong\; \mathrm{Sym}^{k-2}_{\mathbb{Z}_p}\Big(\mathrm{Ta}_p(\mathbb{E}_{\text{univ}})\Big)(2-k) \qquad \text{for all } k \geq 2.$$

Remark: For a scheme $\mathcal{X}$ on which p is invertible and for functions $F, G \in \mathcal{O}(\mathcal{X})^\times$, the second Chern class

$$\mathrm{ch}_{2,2} : K_2(\mathcal{X}) \longrightarrow H^2\Big(\mathcal{X}, (\mathbb{Z}_p/p^n)(2)\Big)$$

sends the symbol $\{F,G\}$ to the cup-product $\partial(F) \cup \partial(G)$, where ∂ denotes the boundary map in the 1-cohomology of $0 \longrightarrow (\mathbb{Z}_p/p^n)(1) \longrightarrow \mathcal{O}^\times_{\mathcal{X}} \xrightarrow{p^n} \mathcal{O}^\times_{\mathcal{X}} \longrightarrow 0$.

We now fix integers $j, j' \in \mathbb{Z}$ lying between 1 and $k-1$. For each such choice, one can write down a long sequence of compositions

$$\begin{aligned}
\varprojlim_n K_2\Big(Y(Ap^n,Bp^n)\Big) &\overset{\mathrm{ch}_{2,2}}{\longrightarrow} \varprojlim_n H^2\Big(Y(Ap^n,Bp^n), (\mathbb{Z}_p/p^n)(2)\Big)\\
&\xrightarrow{\cup e_{1,n}^{\otimes j'-1}\otimes e_{2,n}^{\otimes k-j'-1}\otimes \zeta_{p^n}^{\otimes -j}} \varprojlim_n H^2\Big(Y(Ap^n,Bp^n),\, \mathrm{Sym}^{k-2}_{\mathbb{Z}_p}\Big(\mathrm{Ta}_p(\mathbb{E}_{\text{univ}})/p^n\Big)(2-j)\Big)\\
&\overset{\sim}{\longrightarrow} \varprojlim_n H^2\Big(Y(Ap^n,Bp^n),\, \mathrm{Sym}^{k-2}_{\mathbb{Z}_p}\Big(\mathrm{Ta}_p(\underline{H}^1_p)/p^n\Big)(k-j)\Big)\\
&\overset{\mathrm{proj}_*}{\longrightarrow} \varprojlim_n H^2\Big(Y(A,B),\, \Big(\mathrm{Sym}^{k-2}_{\mathbb{Z}_p}(\underline{H}^1_p)/p^n\Big)(k-j)\Big)\\
&\overset{\text{Hochschild-Serre}}{\longrightarrow} \varprojlim_n H^1\Big(\mathbb{Q},\, H^1\Big(Y(A,B)\otimes\overline{\mathbb{Q}},\, \mathrm{Sym}^{k-2}_{\mathbb{Z}_p}(\underline{H}^1_p)/p^n\Big)(k-j)\Big)
\end{aligned}$$

which essentially maps K-theory to the étale cohomology of the modular forms.

The penultimate map is induced by natural projection $Y(Ap^n, Bp^n) \twoheadrightarrow Y(A, B)$. The final arrow arises from the edge homomorphism in the spectral sequence

$$E_2^{a,b} = H^a\Big(\mathbb{Q}, H^b(Y(A,B)\otimes\overline{\mathbb{Q}}, -)\Big) \implies H^{a+b}_{\text{ét}}(Y(A,B), -)$$

together with the general fact that $H^b_{\text{ét}}(Y(A,B)\otimes\overline{\mathbb{Q}}, -) = 0$ for all $b \geq 2$.

2.5 Kato's *p*-adic zeta-elements

To complete the story, it remains to cut out the Galois representations attached to our modular forms. Let f_k be a p-ordinary eigenform of weight $k \geq 2$, level N and nebentypus ϵ. In particular, its dual f_k^* corresponds to a maximal ideal $\mathfrak{m}_{f_k^*}$ of the Hecke algebra $h_k\big(\Gamma_1(N); \mathcal{O}_{f_k}\big)$. We also write $\lambda_{f_k^*} : h_k\big(\Gamma_1(N)\big) \longrightarrow \mathcal{O}_{f_k}$ for the homomorphism that $\mathfrak{m}_{f_k^*}$ induces.

Let $m, a, \mathbf{a} \geq 1$ be chosen so that $m\mathbf{a}|A$. Assume further that the conditions $N|B$, $A|B$, $\text{supp}(A) = \text{supp}(m\mathbf{a})$ and $\text{supp}(B) = \text{supp}(m\mathbf{a}) \cup \text{supp}(N)$ all hold. The mapping $\tau \mapsto \frac{\tau + a}{\mathbf{a}}$ on $\mathfrak{H}$ induces a unique finite, flat morphism of schemes

$$\Psi_{\mathbf{a},a} : Y(A,B) \longrightarrow Y_1(N) \otimes \mathbb{Q}(\mu_m).$$

For example, it is shown in [Ka1, §5.1] that $(\Psi_{\mathbf{a},a})_*\big({}_{c,d}z_{A,B}\big) \in K_2\Big(Y_1(N)\otimes\mathbb{Q}(\mu_m)\Big)$ depends only on the class of a mod $\mathbf{a}$. In Chapter VI we shall lift these constructions to a two-variable setting, and there we choose $(\mathbf{a}, a) = (1, 0)$ throughout. However this choice sometimes causes problems, e.g. at weight two in the p-adic setting.

Note that we can pass to twists of the Galois representation attached to f_k, via

$$\begin{array}{c} H^1\Big(\mathbb{Q},\ H^1\Big(Y(A,B)\otimes\overline{\mathbb{Q}}, \text{Sym}^{k-2}_{\mathbb{Z}_p}(\underline{H}^1_p)\Big)(k-j)\Big) \\ \big\downarrow {\scriptstyle (\Psi_{\mathbf{a},a})_*} \\ H^1\Big(\mathbb{Q},\ H^1\Big(Y_1(N)\otimes\mathbb{Q}(\mu_m)\otimes\overline{\mathbb{Q}},\ \text{Sym}^{k-2}_{\mathbb{Z}_p}(\underline{H}^1_p)\Big)(k-j)\Big) \\ \big\downarrow {\scriptstyle \cong} \\ H^1\Big(\mathbb{Q}(\mu_m),\ H^1\Big(Y_1(N)\otimes\overline{\mathbb{Q}}, \text{Sym}^{k-2}_{\mathbb{Z}_p}(\underline{H}^1_p)\Big)(k-j)\Big) \\ \big\downarrow {\scriptstyle \otimes_{\lambda_{f_k^*}}\mathcal{O}_{f_k}} \\ H^1\Big(\mathbb{Q}(\mu_m),\ H^1\Big(Y_1(N)\otimes\overline{\mathbb{Q}}, \text{Sym}^{k-2}_{\mathbb{Z}_p}(\underline{H}^1_p)\Big)(k-j)\Big) \bmod \mathfrak{m}_{f_k^*}. \end{array}$$

The maximal ideal $\mathfrak{m}_{f_k^*}$ is generated by primitive tensors $T_h\otimes 1 - 1\otimes\overline{a_h(f_k)}$, $h \in \mathbb{N}$. It follows the final cohomology group is none other than $H^1\Big(\mathbb{Q}(\mu_m), \mathbf{T}'(k-j)\Big)$, where $\mathbf{T}'$ is a lattice in the p-adic Galois representation $\rho_{f_k^*} : G_{\mathbb{Q}} \to \text{Aut}(V_{f_k^*})$ associated to the dual cusp form $f_k^* \in \mathcal{S}_k\big(\Gamma_0(N), \epsilon^{-1}\big)$.

Remark: It is worthwhile to take stock, and to summarise where this has reached. Previously, the composition $(\text{H-S}) \circ \text{proj}_* \circ \left(- \cup \mathrm{e}_{1,n}^{\otimes j'-1} \otimes \mathrm{e}_{2,n}^{\otimes k-j'-1} \otimes \zeta_{p^n}^{\otimes -j} \right) \circ \mathrm{ch}_{2,2}$ yielded a map

$$\varprojlim_n K_2\Big(Y(Ap^n, Bp^n)\Big) \longrightarrow \varprojlim_n H^1\Big(\mathbb{Q}, H^1\Big(Y(A,B) \otimes \overline{\mathbb{Q}}, \mathrm{Sym}_{\mathbb{Z}_p}^{k-2}(\underline{H}_p^1)/p^n\Big)(k-j)\Big) \qquad (*)$$

depending on an original choice of integers $(j, j') \in \mathbb{Z} \times \mathbb{Z}$, lying between 1 and $k-1$. Furthermore, we have just seen how applying $(\Psi_{\mathbf{a},a})_*$ then quotienting modulo $\mathfrak{m}_{f_k^*}$, allows us to pass from the latter group to the target $H^1\Big(\mathbb{Q}(\mu_m), \mathbf{T}'(k-j)\Big)$.

Definition 2.14. *For all $j, j' \in \{1, \ldots, k-1\}$, the realisation map*

$$\mathfrak{c}_{k,j,j',a(\mathbf{a}),p} : \varprojlim_n K_2\Big(Y(Ap^n, Bp^n)\Big) \longrightarrow H^1\Big(\mathbb{Q}(\mu_m), \mathbf{T}'(k-j)\Big)$$

is defined as the composition of $(\Psi_{\mathbf{a},a})_*$ mod $\mathfrak{m}_{f_k^*}$ with the homomorphism **(*)**.

To make life simpler, we write $z_m^{(N)}$ for the image of $(z_{Ap^n,Bp^n})_n$ under $\mathfrak{c}_{k,1,j',a(\mathbf{a}),p}$. In particular, each zeta-element $z_m^{(N)}$ lives in the cohomology group

$$\mathbb{Q} \otimes H^1\Big(\mathbb{Q}(\mu_m),\ \mathbf{T}'(k-1)\Big) \ \cong\ H^1\Big(\mathbb{Q}(\mu_m),\ V_{f_k}^*\Big).$$

Let $\mathfrak{F}^{(k)}$ denote the $(k-2)$-nd symmetric power of the pullback of a local system $\underline{H}^1$ on $\mathbb{E}_{\text{univ}} \to Y_1(N)$. For each pair $a, \mathbf{a} \geq 1$ as above, Kato [Ka1, §5.5] considered certain distinguished elements $\delta_{1,N}(k, j', a(\mathbf{a}))$ lying inside $H^1\Big(Y_1(N)(\mathbb{C}),\ \mathfrak{F}^{(k)}\Big)$. To define these elements, let β_1, β_2 be global sections of $\phi_{a(\mathbf{a})}^* \mathrm{Sym}_{\mathbb{Z}}^{k-2}(\underline{H}^1)$ where $\phi_{a(\mathbf{a})} : (0, \infty) \to Y_1(N)$ by sending $y \mapsto \mathrm{pr}\big((yi + a)/\mathbf{a}\big)$ – the stalk of β_1 at y is yi, and the stalk of β_2 at y is 1. Then $\delta_{1,N}(k, j', a(\mathbf{a}))$ is the image of the class of $\Big(\phi_{a(\mathbf{a})},\ \beta_1^{j'-1}\beta_2^{k-j'-1}\Big)$ under the homomorphisms

$$\begin{array}{ccc} H_1\Big(X_1(N)(\mathbb{C}),\ \{\text{cusps}\},\ \mathrm{Sym}_{\mathbb{Z}}^{k-2}(\underline{H}^1)\Big) & \cong & H^1\Big(Y_1(N)(\mathbb{C}),\ \mathrm{Sym}_{\mathbb{Z}}^{k-2}(\underline{H}^1)\Big) \\ & & \Big\downarrow {\scriptstyle \text{trace}} \\ & & H^1\Big(Y_1(N)(\mathbb{C}),\ \mathfrak{F}^{(k)}\Big). \end{array}$$

To state the main result over $\mathbb{Q}$, we now assume:

Hypothesis. *The congruence class $a(\bmod\ \mathbf{a})$ is chosen so $\delta_{1,N}(k, j', a(\mathbf{a}))^{\pm} \neq 0$.*

Let $\Omega_{a(\mathbf{a})}^{\pm}$ denote the real and imaginary periods associated to $\delta_{1,N}(k, j', a(\mathbf{a}))^{\pm}$. Equivalently, the period map sends the dual form f_k^* to the path vector

$$(2\pi i)^{k-2}\, \Omega_{a(\mathbf{a})}^{+} \times \delta_{1,N}(k, j', a(\mathbf{a}))^{+} \quad + \quad (2\pi i)^{k-2}\, \Omega_{a(\mathbf{a})}^{-} \times \delta_{1,N}(k, j', a(\mathbf{a}))^{-}$$

the equality taking place inside the f_k^*-isotypic part of $H^1\Big(Y_1(N)(\mathbb{C}),\ \mathfrak{F}^{(k)}\Big)$.

Theorem 2.15. *(Kato* [Ka1, §12]*)*
(i) The elements $\left(z_{p^m}^{(N)}\right)_m$ *live inside the limit* $\mathbb{Q}\otimes\varprojlim_m H^1\Big(\mathbb{Q}(\mu_{p^m}),\mathbf{T}'(k-1)\Big)$;

(ii) At non-trivial characters $\psi:\mathcal{G}_{p^n,\mathbb{Q}_p}\to\overline{\mathbb{Q}}_p^\times$ *and integers* $j\in\{0,\dots,k-2\}$,

$$\psi\chi_{\rm cy}^j\circ\mathrm{PR}_{k,\epsilon,\mathbb{Q}_p}\left(\mathrm{loc}_p\left(z_{p^m}^{(N)}\right)_m\right) \;=\; \frac{j!\,p^{nj}}{\alpha_p^n}\left(\sum_{m=1}^{p^n}\psi^{-1}(m)\zeta_{p^n}^m\right)\frac{L_{\{N\}}(f_k,\psi,j+1)}{(2\pi i)^j\Omega_{a(\mathbf{a})}^{\pm}};$$

(iii) At trivial ψ *and for each* $j\in\{0,\dots,k-2\}$,

$$\chi_{\rm cy}^j\circ\mathrm{PR}_{k,\epsilon,\mathbb{Q}_p}\left(\mathrm{loc}_p\left(z_{p^m}^{(N)}\right)_m\right) \;=\; \left(1-\frac{p^j}{\alpha_p}\right)\left(1-\epsilon(p)\frac{p^{k-2-j}}{\alpha_p}\right)\frac{L_{\{N\}}(f_k,j+1)}{(2\pi i)^j\Omega_{a(\mathbf{a})}^{\pm}}.$$

As a consequence, the norm-compatible zeta-elements $\left(z_{p^m}^{(N)}\right)_m$ are transformed into the p-adic L-function $\mathbf{L}_{p,\alpha_p,\Omega_{a(\mathbf{a})}^{\pm}}(f_k,s)$ albeit without its Euler factors at primes dividing N. Of course, the choice of periods $\Omega_{a(\mathbf{a})}^{+}$, $\Omega_{a(\mathbf{a})}^{-}$ was intrinsic to the definition of the zeta-elements, and therefore the p-adic L-function they interpolate. One of the aims of Chapter VI is to show how in the two-variable setting, there is a unique choice compatible with the analytic theory of Greenberg and Stevens [GS].

Proof: The papers [Ka1,Ka2] are of about 300 pages in length, so at best we will only sketch the argument here. In fact, provided one is perfectly willing to believe that 'the zeta-element encodes the L-value', all the other work has already been mentioned in this chapter. However, the most difficult and beautiful parts of the proof are encapsulated in this one statement. Let's start by explaining 2.15(i).

Firstly, it follows from the norm-compatibility of the ${}_{c,d}z_{A,B}$'s originating in $K_2\Big(Y(A,B)\Big)$ and the functoriality of the realisation map, that at least the integral elements ${}_{c,d}z_{p^m}^{(N)} \;:=\; \mathfrak{c}_{k,1,j',a(\mathbf{a}),p}\Big({}_{c,d}z_{Ap^n,Bp^n}\Big)_n$ must be vertically compatible. They are related to the non-integral ones, via the formula

$$\left({}_{c,d}z_{p^m}^{(N)}\right)_m \;=\; \left(c^2-c^{k+1-j'}\sigma_c\right)\left(d^2-d^{j'+1}\sigma_d\right).\left(z_{p^m}^{(N)}\right)_m.$$

This means $\left(z_{p^m}^{(N)}\right)_m$ is an element of $\mathbb{Q}\otimes\varprojlim_m H^1\Big(\mathbb{Q}(\mu_{p^m}),\mathbf{T}'(k-1)\Big)$ up to possible simple poles, where the factors $\left(c^2-c^{k+1-j'}\sigma_c\right)$ and $\left(d^2-d^{j'+1}\sigma_d\right)$ vanish. That these poles are purely hypothetical is justified in Appendix A, using technical arguments from Iwasawa theory.

We now shift our attention to the special value formulae predicted in 2.15(ii),(iii). We write $f_k^{(0)}$ for the p-stabilisation of the form f_k, in particular $f_k^{(0)}|U_p=\alpha_p f_k^{(0)}$. Similarly, the φ-operator has eigenvalue $\alpha_p=a_p\left(f_k^{(0)}\right)$ acting on the vector $\mathbf{v}_{k,\epsilon}$.

Applying Corollary 2.10, the image $\mathrm{PR}_{k,\epsilon,\mathbb{Q}_p}\left(\mathrm{loc}_p\left(z_{p^m}^{(N)}\right)_m\right)$ lies in $\mathbb{Q}\otimes\mathcal{O}_{f_k}[\![G_{\infty,\mathbb{Q}_p}]\!]$. In fact Theorem 2.7 implies that $\psi\chi_{\mathrm{cy}}^j \circ \mathrm{PR}_{k,\epsilon,\mathbb{Q}_p}\left(\mathrm{loc}_p\left(z_{p^m}^{(N)}\right)_m\right)$ takes the value

$$\frac{j!p^{nj}}{\alpha_p^n}\left(1-\delta_{\psi=1}\frac{p^j}{\alpha_p}\right)G(\psi^{-1},\zeta_{p^n})\sum_{\sigma\in\mathcal{G}_{p^n,\mathbb{Q}_p}}\psi(\sigma)\exp_j^*\left(z_{p^n}^{(N)}\otimes\zeta_{p^n}^{\otimes -j}\right)^\sigma$$

where $\exp_j^*(-)$ denoted the cup-product $\exp^*_{V^*_{f_k}(-j)}(-)\cup\left(1-\delta_{n=0}p^{-j-1}\varphi\right)^{-1}\mathbf{v}_{k,\epsilon}$. If $n=0$ then $\left(1-\delta_{n=0}p^{-j-1}\varphi\right)^{-1}\mathbf{v}_{k,\epsilon}$ is equal to $\left(1-p^{-j-1}\alpha_p\right)^{-1}\mathbf{v}_{k,\epsilon}$, otherwise it is just the vector $\mathbf{v}_{k,\epsilon}$ when $n>0$.

Key Fact: $$\sum_{\sigma\in\mathcal{G}_{p^n,\mathbb{Q}_p}}\psi(\sigma)\exp^*_{V^*_{f_k}(-j)}\left(z_{p^n}^{(N)}\otimes\zeta_{p^n}^{\otimes -j}\right)^\sigma\cup\mathbf{v}_{k,\epsilon} = \frac{L_{\{Np\}}(f_k,\psi,j+1)}{(2\pi i)^j\Omega^{\pm}_{a(\mathbf{a})}}.$$

We cannot emphasise enough just how difficult this single statement is to prove. At its heart lies an *explicit reciprocity law* which is itself a vast generalisation of the reciprocity laws of Artin, Iwasawa and many others. The details of Kato's argument are well beyond the scope of this book!

However, incorporating this Key Fact into the above equation for the special value at the character $\psi\chi_{\mathrm{cy}}^j$, we arrive at the same formulae described in 2.15(ii),(iii). This completes the proof.

□

The missing Euler factors at the primes dividing N can be restored, however the Euler factor at p is unrecoverable. We should also point out that there are extra sideways compatibility relations, in addition to the vertical ones already mentioned. For those who are familiar with the proof of the classical Iwasawa Main Conjecture, zeta-elements play an analogous rôle in bounding Selmer groups, that the Euler system of cyclotomic units plays in bounding the class group of $\mathbb{Q}(\mu_{p^n})$.

The main work of Kato is contained in the articles [Ka1,Ka2,KKT], although these might be too hard to follow for someone completely new to number theory. For a far more down-to-earth account there is the excellent survey of Scholl [Sch] which treats the case of primitive cusp forms of weight two and trivial character. The exceptional zero conjecture was first stated by Mazur, Tate and Teitelbaum in [MTT], and two markedly different proofs of it can be found inside [GS,KKT]. Finally, for a thorough introduction to the theory of Euler systems there is Rubin's book on the subject [Ru2], and also the papers of Kato and Perrin-Riou [Ka3,PR4].

CHAPTER III

Cyclotomic Deformations of Modular Symbols

The algebraic methods converting norm-compatible families into p-adic L-functions, have profound consequences in arithmetic. However, they also contain a number of unfortunate drawbacks. First of all, the cohomological zeta-elements satisfy some beautiful Euler system relations, yet these relations are lost once the zeta-elements have been converted into Iwasawa functions. Another disadvantage is the choice of periods occurring in the denominator of the interpolation is somewhat arbitrary. Whilst this isn't too much of a problem p-adically, in a two-variable setting it is potentially disastrous. The control theorems of Hida et al are phrased in terms of the parabolic cohomology, and ideally we would like to work with these objects, instead of cruder power series rings. These power series rings don't see the action of the Hecke algebra, and thus cannot satisfy multiplicity one-type results.

The first half of this book proposes a workable solution to each of these problems. Our approach is motivated by the simple observation that the Kato-Beilinson Euler system encodes L-values of a primitive eigenform, over all abelian extensions of the rationals. With a bit of care, one can utilise this huge amount of data to write down explicit symbols, which will then end up as elements inside Betti cohomology. All these notions generalise seamlessly to the setting of two-variable deformation theory. Furthermore, they permit the formulation of a Tamagawa Number Conjecture for the universal nearly-ordinary Galois representation (the second half of this book). For instance at weight two and trivial character, this conjecture is more or less equivalent to the p-adic BSD conjecture mentioned in §2.1.

As a warm-up to the two-variable case, we interpret the machinery of the previous chapter in terms of cyclotomic deformations of modular symbols. One by-product of this point of view, is that we can think of certain Hecke eigenspaces as being generated by the module of zeta-elements. We begin by recalling some standard facts from Shimura's book [Sh].

3.1 $\mathcal{Q}$-continuity

Let $\mathrm{div}\mathbb{P}^1(\mathbb{Q})$ denote the free abelian group generated on the set $\mathbb{Q} \cup \{\infty\}$, and we also write $\mathrm{div}^0\mathbb{P}^1(\mathbb{Q})$ for the subgroup consisting of those divisors of degree zero. It is easy to check the Möbius action of $\mathrm{GL}_2(\mathbb{Q})$ on $\mathrm{div}\mathbb{P}^1(\mathbb{Q})$ leaves $\mathrm{div}^0\mathbb{P}^1(\mathbb{Q})$ stable. Functions on $\mathrm{div}^0\mathbb{P}^1(\mathbb{Q})$ with values in some $\mathrm{GL}_2(\mathbb{Q})$-module will be called symbols, and if they are invariant under a congruence modular subgroup of $\mathrm{SL}_2(\mathbb{Z})$ they will be termed modular symbols.

Fix an integer $k \geq 2$. Let us write $\mathrm{Sym}^{k-2}(A)$ for the space of homogeneous polynomials in two variables X, Y of degree $k-2$, with coefficients in some ring A. We shall adopt the convention that a matrix $\begin{pmatrix} a & b \\ c & d \end{pmatrix} \in \mathrm{GL}_2(A)$ operates on the polynomial $\mathcal{P} \in \mathrm{Sym}^{k-2}(A)$ via the right action

$$\mathcal{P}\Big|\begin{pmatrix} a & b \\ c & d \end{pmatrix}(X,Y) \quad := \quad \mathcal{P}\big(dX - cY, aY - bX\big).$$

In particular, we have carefully normalised so that it's the contravariant action corresponding to the anti-involution $\gamma \mapsto \det(\gamma)\gamma^{-1}$.

The main interest in studying symbols with values in Sym^{k-2} is given by the following example of Eichler and Shimura. Let f be a cusp form in $\mathcal{S}_k\big(\Gamma_1(Np^r)\big)$.

Definition 3.1. *We define* $\xi_f \in \mathrm{Hom}_{\mathbb{Z}}\Big(\mathrm{div}^0\mathbb{P}^1(\mathbb{Q}), \mathrm{Sym}^{k-2}(\mathbb{C})\Big)$ *by its values*

$$\big(\xi_f\big)_{(a)-(b)} \quad := \quad 2\pi i \int_b^a f(z)(zX+Y)^{k-2}dz$$

at all $a, b \in \mathbb{P}^1(\mathbb{Q})$*, integrating along a geodesic joining* b *to* a*.*

A straightforward calculation shows that $\xi_f\big|\gamma = \xi_{f|\gamma}$ for all $\gamma \in \mathrm{GL}_2^+(\mathbb{Q})$, hence ξ_f must be $\Gamma_1(Np^r)$-invariant. If f has Fourier coefficients lying in a number field $\mathbf{F}$, by duality there corresponds a homomorphism

$$\lambda_f : h_k(Np^r; \mathbf{F}) \longrightarrow \mathbf{F}, \quad \lambda_f : T_n \mapsto a_n(f)$$

of the Hecke algebra $h_k(Np^r; \mathbf{F})$ acting on weight $k \geq 2$ cuspidal forms for $\Gamma_1(Np^r)$. It follows that the action of $h_k(Np^r; \mathbf{F})$ on ξ_f must factor through λ_f, whence

$$\xi_f \;\in\; \mathrm{Hom}_{\Gamma_1(Np^r)}\Big(\mathrm{div}^0\mathbb{P}^1(\mathbb{Q}),\; \mathrm{Sym}^{k-2}(\mathbb{C})\Big)[\lambda_f].$$

We shall use the notation $\mathcal{MS}\big(Np^r; \mathrm{Sym}^{k-2}(A)\big)$ as a shorthand for the A-module $\mathrm{Hom}_{\Gamma_1(Np^r)}\Big(\mathrm{div}^0\mathbb{P}^1(\mathbb{Q}), \mathrm{Sym}^{k-2}(A)\Big)$ of modular symbols of weight k, level Np^r. In fact the f-isotypic part of these modular symbols satisfies multiplicity one.

Proposition 3.2. *(Shimura* [Sh]*) For* f *and* λ_f *as above,*

$$\dim_{\mathbf{F}}\mathcal{MS}\Big(Np^r; \mathrm{Sym}^{k-2}(\mathbf{F})\Big)^{\pm}[\lambda_f] \;=\; 1$$

where the $\pm$*-eigenspace is taken with respect to the matrix action of* $\tau = \begin{pmatrix} 1 & 0 \\ 0 & -1 \end{pmatrix}$.

Furthermore $\mathcal{MS}\Big(Np^r; \mathrm{Sym}^{k-2}(\mathcal{O}_{\mathbf{F}})\Big)^{\pm}[\lambda_f]$ *are both free of rank one, over the ring of integers* $\mathcal{O}_{\mathbf{F}}$*.*

One can now fix generators $\eta_f^{\pm}$ of these $\pm$-eigenspaces. Since

$$\mathcal{MS}\Big(Np^r;\mathrm{Sym}^{k-2}(\mathbb{C})\Big)^{\pm}[\lambda_f] \;=\; \mathcal{MS}\Big(Np^r;\mathrm{Sym}^{k-2}(\mathcal{O}_{\mathbf{F}})\Big)^{\pm}[\lambda_f]\otimes_{\mathcal{O}_{\mathbf{F}}}\mathbb{C}$$

there must exist complex periods $\Omega_f^{\pm}\in\mathbb{C}^{\times}$ such that $\xi_f^{\pm}=\Omega_f^{\pm}\eta_f^{\pm}$. It is customary to call the modular symbol $\eta_f=\eta_f^{+}+\eta_f^{-}$ the *algebraic part* of ξ_f.

Remark: Our task then is to see how the maps $\mathrm{PR}_{k,\epsilon}$ behave as $f\in\mathcal{S}_k\big(\Gamma_0(Np^r),\epsilon\big)$ varies in an analytic family. There are two naive approaches. The first is to work in the power series rings $\mathbb{Q}\otimes\Lambda_f$ but these have no action of the Hecke operator U_p. Alternatively we could work instead with algebraic modular symbols, but these have no Λ-action. The compromise is to introduce a cyclotomic deformation of the classical modular symbols, where Hida's control theory can be considered as coexisting side-by-side with Iwasawa theory.

Let $\mathcal{B}$ be any p-adic Banach module, and assume it comes with a right GL_2-action. Ideally we would like to give $\mathrm{Hom}_{\mathbb{Z}}\big(\mathrm{div}^0\mathbb{P}^1(\mathbb{Q}),\mathcal{B}\big)$ the structure of a Λ-module, but in reality the only such action that exists is the trivial one. In the context of this book, it makes far more sense to introduce the notion of $\mathcal{Q}$-continuous symbols (defined directly below) and then write down an explicit Λ-action on these.

First we partition the rationals. Let p be any prime number (even $p=2$ works). For integers M and c satisfying $M\in\mathbb{N}$, $p\nmid M$, $0<c\le M$ and $\gcd(c,M)=1$, set

$$\Sigma_{M,c}:=\left\{\frac{a}{p^nM}\in\mathbb{Q}\ \text{such that}\ n\ge 0,\ \gcd(p^nM,a)=1\ \text{and}\ a\equiv c \mod M\right\}\cup\{0\}.$$

Lemma 3.3. *(i) The rational numbers may be written as the disjoint union*

$$\mathbb{Q} \;=\; \{0\}\cup\bigcup_{p\nmid M}\ \bigcup_{0<c\le M}\ \Sigma_{M,c}\backslash\{0\}.$$

(ii) Each set $\Sigma_{M,c}$ is dense inside $\mathbb{Q}$ under the p-adic topology.

Proof: Part (i) is totally obvious.

To show part (ii) is true, let $x\in\mathbb{Q}$, $x\neq 0$ and write $x=p^{\mathrm{ord}_p x}u$ where $u\in\mathbb{Z}_p^{\times}$. If $x\in\mathbb{Z}_p$, we need an integer a_m satisfying $a_m\equiv c \bmod M$, $\frac{a_m}{M}\equiv p^{\mathrm{ord}_p x}u \bmod p^m$. If $x\notin\mathbb{Z}_p$, we need an integer a_m satisfying $a_m\equiv c \bmod M$, $\frac{a_m}{M}\equiv u \bmod p^{m-\mathrm{ord}_p x}$. By the Chinese remainder theorem, such approximations exist. It follows that

$$\frac{a_m}{p^{\max\{0,-\mathrm{ord}_p x\}}M}\in\Sigma_{M,c}\qquad\text{and}\qquad\left|x-\frac{a_m}{p^{\max\{0,-\mathrm{ord}_p x\}}M}\right|_p\le p^{-m};$$

letting $m\to\infty$, the p-adic density of $\Sigma_{M,c}$ within the rationals is clear. □

Definition 3.4. *A symbol* $\Xi \in \mathrm{Hom}_{\mathbb{Z}}\big(\mathrm{div}^0\mathbb{P}^1(\mathbb{Q}), \mathcal{B}\big)$ *is said to be* $\mathcal{Q}$*-continuous if for each* $M, c \in \mathbb{N}$ *as above, the function* $\mathcal{Q}_{\Xi,M,c} : \Sigma_{M,c} \to \mathcal{B}$ *defined by*

$$\mathcal{Q}_{\Xi,M,c}(r) \;:=\; \Xi_{(r)-(\infty)}\Big| \begin{pmatrix} 1 & r \\ 0 & 1 \end{pmatrix} \qquad \text{for all } r \in \Sigma_{M,c}$$

extends to a continuous function on $\mathbb{Q}_p$.

Each function $\mathcal{Q}_{\Xi,M,c}$ should be thought of as encoding the special values of Ξ corresponding to the tame denominator M and the congruence class $c \bmod M$. For example, if $M = 1$ then $c = 1$ and $\Sigma_{1,1} = \mathbb{Z}[1/p]$. However, for an arbitrary $\mathcal{Q}$-continuous symbol Ξ the continuous functions extending $\mathcal{Q}_{\Xi,M,c}$ and $\mathcal{Q}_{\Xi,M',c'}$ should disagree in general whenever $(M, c) \neq (M', c')$.

Notation: The subspace of $\mathcal{Q}$-continuous symbols with values in $\mathcal{B}$ will be denoted by $\mathrm{Hom}_{\mathcal{Q}\mathrm{cts}}\big(\mathrm{div}^0\mathbb{P}^1(\mathbb{Q}), \mathcal{B}\big)$, and we shall frequently refer to this as *the* $\mathcal{Q}$*-part of the* $\mathcal{B}$*-valued symbols.*

Of course, we may replace $\mathrm{div}^0\mathbb{P}^1(\mathbb{Q})$ in the above definitions with $\mathrm{div}^0\big(S \cup \{\infty\}\big)$ for any set $S \subset \mathbb{Q}$, with appropriate modifications. In a similar vein, we also write $\mathrm{Hom}_{\mathbb{Z}}^{p\text{-adic}}\big(\mathrm{div}^0\mathbb{P}^1(\mathbb{Q}_p), \mathcal{B}\big)$ for the subspace of symbols $\Xi' \in \mathrm{Hom}_{\mathbb{Z}}\big(\mathrm{div}^0\mathbb{P}^1(\mathbb{Q}_p), \mathcal{B}\big)$ with the property that $\Xi'_{(r)-(\infty)}\Big| \begin{pmatrix} 1 & r \\ 0 & 1 \end{pmatrix}$ is continuous in $r \in \mathbb{Q}_p$.

The next result follows immediately from the p-adic density of $S = \Sigma_{M,c}$ in $\mathbb{Q}_p$.

Corollary 3.5. *The inclusion* $\beta_{M,c} : \Sigma_{M,c} \hookrightarrow \mathbb{Q}_p$ *induces an isomorphism*

$$(\beta_{M,c})^* : \mathrm{Hom}_{\mathbb{Z}}^{p\text{-adic}}\big(\mathrm{div}^0\mathbb{P}^1(\mathbb{Q}_p), \mathcal{B}\big) \;\xrightarrow{\sim}\; \mathrm{Hom}_{\mathcal{Q}\mathrm{cts}}\Big(\mathrm{div}^0\big(\Sigma_{M,c} \cup \{\infty\}\big), \mathcal{B}\Big).$$

One major advantage of this isomorphism is that $\mathrm{Hom}_{\mathbb{Z}}^{p\text{-adic}}\big(\mathrm{div}^0\mathbb{P}^1(\mathbb{Q}_p), \mathcal{B}\big)$ comes with a natural action of the completed group ring $\Lambda^{\mathrm{cy}} = \mathbb{Z}_p\big[\big[\mathrm{Gal}\big(\mathbb{Q}(\mu_{p^\infty})/\mathbb{Q}\big)\big]\big]$. This can be pushed forward via $(\beta_{M,c})^*$ to act on $\mathrm{Hom}_{\mathcal{Q}\mathrm{cts}}\Big(\mathrm{div}^0\big(\Sigma_{M,c} \cup \{\infty\}\big), \mathcal{B}\Big)$, and then taking the union over (M, c) endows the $\mathcal{Q}$-part of $\mathrm{Hom}_{\mathbb{Z}}\big(\mathrm{div}^0\mathbb{P}^1(\mathbb{Q}), \mathcal{B}\big)$ with the same Iwasawa action.

To see how this works in detail, first abbreviate $\mathrm{Gal}\big(\mathbb{Q}(\mu_{p^\infty})/\mathbb{Q}\big)$ simply by G_∞. If Ξ is a $\mathcal{Q}$-continuous $\mathcal{B}$-valued symbol, then

$$\Xi_{M,c} \;=\; \Xi\Big|_{\mathrm{div}^0\left(\Sigma_{M,c} \cup \{\infty\}\right)} \;\in\; \mathrm{Hom}_{\mathcal{Q}\mathrm{cts}}\Big(\mathrm{div}^0\big(\Sigma_{M,c} \cup \{\infty\}\big), \mathcal{B}\Big);$$

let's write $\Xi' = \big((\beta_{M,c})^*\big)^{-1}\Big(\Xi_{M,c}\Big)$ for its pre-image in $\mathrm{Hom}_{\mathbb{Z}}^{p\text{-adic}}\big(\mathrm{div}^0\mathbb{P}^1(\mathbb{Q}_p), \mathcal{B}\big)$. For any element $g \in G_\infty$ the map $(\Xi')^g : \mathrm{div}^0\mathbb{P}^1(\mathbb{Q}_p) \to \mathcal{B}$, which is the linear extension to $\mathrm{div}^0\mathbb{P}^1(\mathbb{Q}_p)$ of

$$(\Xi')^g_{(r)-(\infty)} \;:=\; (\Xi')_{\left(r\chi_{\mathrm{cy}}(g)^{-1}\right)-(\infty)}\Bigg| \begin{pmatrix} 1 & r\big(\chi_{\mathrm{cy}}(g)^{-1} - 1\big) \\ 0 & 1 \end{pmatrix} \quad \text{for every } r \in \mathbb{Q}_p$$

also belongs to $\mathrm{Hom}_{\mathbb{Z}}^{p\text{-adic}}\big(\mathrm{div}^0\mathbb{P}^1(\mathbb{Q}_p),\mathcal{B}\big)$. In fact, we have the equality

$$(\Xi')^g_{(r)-(\infty)}\Big|\begin{pmatrix}1 & r\\ 0 & 1\end{pmatrix} \;=\; (\Xi')_{(r\chi_{\rm cy}(g)^{-1})-(\infty)}\Big|\begin{pmatrix}1 & r\chi_{\rm cy}(g)^{-1}\\ 0 & 1\end{pmatrix}$$

therefore the p-adic behaviour of $(\Xi')^g$ is governed entirely by the continuity of Ξ'. Pulling things back to $\mathrm{div}^0\mathbb{P}^1(\mathbb{Q})$, we then write $(g\circ\Xi)\in\mathrm{Hom}_{\mathcal{Q}\mathrm{cts}}\big(\mathrm{div}^0\mathbb{P}^1(\mathbb{Q}),\mathcal{B}\big)$ for the unique symbol satisfying

$$(g\circ\Xi)\Big|_{\mathrm{div}^0\left(\Sigma_{M,c}\cup\{\infty\}\right)} = (\beta_{M,c})^*\Big(\big((\beta_{M,c})^*\big)^{-1}\big(\Xi_{M,c}\big)^g\Big) \quad \text{at all } (M,c) \text{ as before.}$$

Remark The unipotent action $\Xi\mapsto(g\circ\Xi)$ of $g\in G_\infty$ extends by linearity and continuity to an action of the whole of $\Lambda^{\rm cy}$. Note that it makes good sense to speak of $\mathrm{Hom}_{\mathcal{Q}\mathrm{cts}}\Big(\mathrm{div}^0\big(\Sigma_{M,c}\cup\{\infty\}\big),\mathcal{B}\Big)$ as a $\Lambda^{\rm cy}$-module. At the end of the chapter we will exploit this extra structure, in order to calculate the $\Lambda^{\rm cy}$-ranks of the cyclotomic deformation of the space of modular symbols.

Recall that the contravariant action of a matrix $\gamma\in\mathrm{GL}_2(\mathbb{Q})$ on a symbol Ξ produces a symbol $\gamma\circ\Xi$, taking values $\big(\gamma\circ\Xi\big)_D=\big(\Xi_{\gamma D}\big)\big|\gamma$ at every $D\in\mathrm{div}^0\mathbb{P}^1(\mathbb{Q})$. Exploiting the fact that the Hecke operator at p extends in a $\mathcal{Q}$-continuous direction, we now explain how to attach distributions to the $\mathcal{Q}$-part of a symbol.

Proposition 3.6. *Define* $U_p\in\mathrm{End}\Big(\mathrm{Hom}_{\mathbb{Z}}\big(\mathrm{div}^0\mathbb{P}^1(\mathbb{Q}),\mathcal{B}\big)\Big)$ *by*

$$U_p\circ\Xi \;:=\; \sum_{b=0}^{p-1}\begin{pmatrix}1 & b\\ 0 & p\end{pmatrix}\circ\Xi \qquad \textit{for all } \mathcal{B}\textit{-valued symbols } \Xi.$$

(i) If the diagonal part of GL_2 *acts continuously on* $\mathcal{B}$*, then the operator* U_p *leaves the subspace* $\mathrm{Hom}_{\mathcal{Q}\mathrm{cts}}\big(\mathrm{div}^0\mathbb{P}^1(\mathbb{Q}),\mathcal{B}\big)$ *stable.*

(ii) Let $\Xi\in\mathrm{Hom}_{\mathcal{Q}\mathrm{cts}}\big(\mathrm{div}^0\mathbb{P}^1(\mathbb{Q}),\mathcal{B}\big)$ *be any symbol fixed by* $\begin{pmatrix}1 & 1\\ 0 & 1\end{pmatrix}$*, and assume that* U_p *acts invertibly on* Ξ*. Then there is a* $\mathcal{B}$*-valued distribution* μ_Ξ *on the Lie group* $\mathbb{Z}_{p,M}^\times:=\big(\varprojlim_n\mathbb{Z}/p^nM\mathbb{Z}\big)^\times$ *satisfying*

$$\mu_\Xi\big(a+(p^nM)\big) \;=\; \mathcal{Q}_{U_p^{-n}\circ\Xi,M,c}\left(\frac{a}{p^nM}\right)\Bigg|\begin{pmatrix}1 & 0\\ 0 & p\end{pmatrix}^n$$

for all $n\geq 0$, $a\in\mathbb{Z}\cap\mathbb{Z}_{p,M}^\times$ *with* $0<c\leq M$, $c\equiv a \mod M$.

Proof: We start by showing that if Ξ is $\mathcal{Q}$-continuous, then so is the symbol $U_p\circ\Xi$. By the definition of U_p,

$$(U_p\circ\Xi)_{(r)-(\infty)} \;=\; \sum_{b=0}^{p-1}\left(\begin{pmatrix}1 & b\\ 0 & p\end{pmatrix}\circ\Xi\right)_{(r)-(\infty)} \;=\; \sum_{b=0}^{p-1}\Xi_{(\frac{r+b}{p})-(\infty)}\Big|\begin{pmatrix}1 & b\\ 0 & p\end{pmatrix}.$$

If $r = \frac{a}{p^n M}$ lies in a partition $\Sigma_{M,c}$ then $\frac{r+b}{p}$ will lie in the set $\Sigma_{M,c_{r,b}}$ where $c_{r,b} \equiv \frac{a+bM}{\gcd(p,a+bM)} \mod M$ if $n=0$, and $c_{r,b} \equiv a + bp^n M \equiv c \mod M$ if $n > 0$. It follows that

$$\begin{aligned}\mathcal{Q}_{U_p \circ \Xi, M, c}(r) &= \left(\sum_{b=0}^{p-1} \Xi_{\left(\frac{r+b}{p}\right)-(\infty)} \Big| \begin{pmatrix} 1 & b \\ 0 & p \end{pmatrix} \right) \Big| \begin{pmatrix} 1 & r \\ 0 & 1 \end{pmatrix} \\ &= \sum_{b=0}^{p-1} \mathcal{Q}_{\Xi, M, c_{r,b}} \left(\frac{r+b}{p} \right) \Big| \begin{pmatrix} 1 & -\left(\frac{r+b}{p}\right) \\ 0 & 1 \end{pmatrix} \begin{pmatrix} 1 & b \\ 0 & p \end{pmatrix} \begin{pmatrix} 1 & r \\ 0 & 1 \end{pmatrix}\end{aligned}$$

which equals $\sum_{b=0}^{p-1} \mathcal{Q}_{\Xi,M,c_{r,b}} \left(\frac{r+b}{p}\right) \Big| \begin{pmatrix} 1 & 0 \\ 0 & p \end{pmatrix}$. The functions $\mathcal{Q}_{\Xi,M,c_{r,b}} \left(\frac{r+b}{p}\right)$ vary continuously in r, and by assumption $\begin{pmatrix} 1 & 0 \\ 0 & p \end{pmatrix}$ acts continuously on the space $\mathcal{B}$. Thus $\mathcal{Q}_{U_p \circ \Xi, M, c}$ extends to a continuous function on $\mathbb{Q}_p$, which proves (i).

Remark: In particular, observe that $\mathcal{Q}_{U_p \circ \Xi, M, c}(r) = \sum_{b=0}^{p-1} \mathcal{Q}_{\Xi,M,c} \left(\frac{r+b}{p}\right) \Big| \begin{pmatrix} 1 & 0 \\ 0 & p \end{pmatrix}$ whenever $r = \frac{a}{p^n M} \in \Sigma_{M,c}$ with $n > 0$.

To see why part (ii) is true, we must show that μ_Ξ satisfies distribution relations. We begin by evaluating

$$\begin{pmatrix} 1 & 1 \\ 0 & 1 \end{pmatrix} \circ (U_p \circ \Xi) = \sum_{b=0}^{p-1} \begin{pmatrix} 1 & 1 \\ 0 & 1 \end{pmatrix} \begin{pmatrix} 1 & b \\ 0 & p \end{pmatrix} \circ \Xi = \sum_{b=0}^{p-1} \begin{pmatrix} 1 & b \\ 0 & p \end{pmatrix} \begin{pmatrix} 1 & 1 \\ 0 & 1 \end{pmatrix}^p \circ \Xi$$

which equals $U_p \circ \Xi$ because $\begin{pmatrix} 1 & 1 \\ 0 & 1 \end{pmatrix} \circ \Xi = \Xi$. Consequently $U_p^m \circ \Xi$ is also translation invariant, for all $m \in \mathbb{Z}$. It follows directly from Definition 3.4 that

$$\mathcal{Q}_{U_p^{-n} \circ \Xi, M, c}(r) = \mathcal{Q}_{U_p^{-n} \circ \Xi, M, c}(r+1) \qquad \text{at every } r \in \Sigma_{M,c}.$$

Choosing $a' \in \mathbb{Z}$ such that $a' \equiv a \mod p^n M$, we have

$$\begin{aligned}\mu_\Xi\big(a + (p^n M)\big) &= \mathcal{Q}_{U_p^{-n} \circ \Xi, M, c} \left(\frac{a'}{p^n M} + d \right) \Big| \begin{pmatrix} 1 & 0 \\ 0 & p \end{pmatrix}^n \quad \text{where } d = \frac{a - a'}{p^n M} \in \mathbb{Z} \\ &= \mathcal{Q}_{U_p^{-n} \circ \Xi, M, c} \left(\frac{a'}{p^n M} \right) \Big| \begin{pmatrix} 1 & 0 \\ 0 & p \end{pmatrix}^n = \mu_\Xi\big(a' + (p^n M)\big)\end{aligned}$$

so at least μ_Ξ is well-defined. Furthermore

$$\begin{aligned}\sum_{b=0}^{p-1} \mu_\Xi\big(a + bp^n M + (p^{n+1} M)\big) &= \sum_{b=0}^{p-1} \mathcal{Q}_{U_p^{-(n+1)} \circ \Xi, M, c} \left(\frac{a + bp^n M}{p^{n+1} M} \right) \Big| \begin{pmatrix} 1 & 0 \\ 0 & p \end{pmatrix}^{n+1} \\ &= \sum_{b=0}^{p-1} \mathcal{Q}_{U_p^{-1} \circ (U_p^{-n} \circ \Xi), M, c} \left(\frac{\frac{a}{p^n M} + b}{p} \right) \Big| \begin{pmatrix} 1 & 0 \\ 0 & p \end{pmatrix} \begin{pmatrix} 1 & 0 \\ 0 & p \end{pmatrix}^n\end{aligned}$$

and by our remark, we know that $\sum_{b=0}^{p-1} \mathcal{Q}_{U_p^{-1}\circ\ldots,M,c}\left(\frac{r+b}{p}\right)\Big|\begin{pmatrix}1&0\\0&p\end{pmatrix} = \mathcal{Q}_{\ldots,M,c}(r)$ at any $r \in \Sigma_{M,c}$ with $\mathrm{ord}_p(r) < 0$. Therefore, the above equation becomes

$$\sum_{b=0}^{p-1} \mu_\Xi\big(a+bp^nM+(p^{n+1}M)\big) = \mathcal{Q}_{U_p^{-n}\circ\Xi,M,c}\left(\frac{a}{p^nM}\right)\Bigg|\begin{pmatrix}1&0\\0&p\end{pmatrix}^n = \mu_\Xi\big(a+(p^nM)\big)$$

which means that μ_Ξ is finitely additive on the compact open subsets of $\mathbb{Z}_{p,M}^\times$.

The proof of the proposition is complete. □

3.2 Cohomological subspaces of Euler systems

We have two main examples in mind. The first is when our Banach space $\mathcal{B}$ is taken to be the homogenous polynomial ring $\mathrm{Sym}^{k-2}(\mathcal{K}_f)$, for some p-stabilised ordinary newform $f \in \mathcal{S}_k\big(\Gamma_0(Np^r), \epsilon\big)$. The second is when $\mathcal{B}$ is the rigid analytic image of a modular deformation ring. The latter situation is considered in Chapters V–VI, whilst the former occupies the remainder of this chapter.

Before subjecting the reader to the vagaries of the full N-primitive zeta-element Euler system, it makes good sense to review the classical cyclotomic situation first. Throughout we fix primitive m^{th}-roots of unity ζ_m satsifying the compatibilities $(\zeta_{m'})^{m'/m} = \zeta_m$. For example, $\zeta_m = \exp(2\pi i m)$ would be an acceptable choice.

The Euler system of cyclotomic units

Let $m \not\equiv 2 \mod 4$ be a positive integer, and E_m the group of units of $\mathbb{Z}[\mu_m]$. The subgroup of cyclotomic units is defined to be the intersection $C_m = V_m \cap E_m$, where V_m is generated by the set of numbers $\big\{\pm\zeta_m,\ 1-\zeta_m^a \quad \text{such that } 1 \le a \le m-1\big\}$. It was Euler in the mid-eighteenth century, who found the remarkable connection between cyclotomic units, and special values of the classical Riemann zeta-function. For even Dirichlet characters $\psi \neq \mathbf{1}$, he proved the formula

$$L(1,\psi) \quad = \quad -\frac{1}{\mathfrak{f}_\psi} \times \left(\sum_{a=1}^{\mathfrak{f}_\psi} \psi(a)\zeta_{\mathfrak{f}_\psi}^a\right) \times \left(\sum_{b=1}^{\mathfrak{f}_\psi} \overline{\psi}(b)\log\left|1-\zeta_{\mathfrak{f}_\psi}^b\right|\right).$$

At twists by odd characters, $L(1,\psi)$ is related to the $\overline{\psi}$-twisted Bernoulli numbers.

Let l be a prime number not dividing m, and write $\sigma_l \in \mathrm{Gal}\big(\mathbb{Q}(\mu_m)/\mathbb{Q}\big)$ for the Frobenius element sending $\zeta_m \mapsto \zeta_m^l$. From Euclid's algorithm there exist integers u, v satisfying $1 = ul + vm$, and a basic calculation reveals

$$\begin{aligned}
\mathrm{Norm}_{(\mu_{ml})/(\mu_m)}\big(1-\zeta_{ml}\big) &= \mathrm{Norm}_{(\mu_{ml})/(\mu_m)}\big(1-\zeta_m^u\zeta_l^v\big) = \prod_{j=1}^{l-1}\big(1-\zeta_m^u\zeta_l^j\big) \\
&= \frac{\prod_{j=0}^{l-1}\big(1-\zeta_m^u\zeta_l^j\big)}{1-\zeta_m^u} = \frac{1-\zeta_m^{ul}}{1-\zeta_m^u} \\
&= \frac{1-\zeta_m}{1-\zeta_m^{\sigma_l^{-1}}} = \big(1-\sigma_l^{-1}\big).\big(1-\zeta_m\big).
\end{aligned}$$

Conversely, if the prime l does divide m, then

$$\mathrm{Norm}_{(\mu_{ml})/(\mu_m)}(1-\zeta_{ml}) \;=\; \prod_{j=0}^{l-1}\left(1-\zeta_{ml}\zeta_l^j\right) \;=\; 1-\zeta_{ml}^l \;=\; 1-\zeta_m.$$

The formula obtained when we remove a prime from the level ml is commonly called a *sideways relation*, and the formula obtained when the support of ml doesn't change is called a *vertical relation*. The total collection of elements $\bigcup_{m\not\equiv 2(\bmod 4)} C_m$ is known as the Euler system of cyclotomic units.

Coleman proved that for every $\underline{u}=\left(u_{p^n}\right)_{n\geq 1}\in \varprojlim_n \mathbb{Z}_p[\mu_{p^n}]^\times$, there is a unique power series $\mathrm{Col}_{\underline{u}}(X)\in\mathbb{Z}_p[\![X]\!]^\times$ such that $\mathrm{Col}_{\underline{u}}(\zeta_{p^n}-1)=u_{p^n}$ for all $n\in\mathbb{N}$. Moreover, if one makes a choice of $\underline{u}=\left(\frac{1-\zeta_{p^n}^c}{1-\zeta_{p^n}}\right)_{n\geq 1}$ with $p\nmid c$, then at every non-positive critical point

$$\mathcal{U}\circ\log\circ\,\mathrm{Col}_{\underline{u}}\Big((1+p)^j-1\Big) \;=\; (c^j-1)\left(1-p^{j-1}\right)\zeta(1-j)$$

where $\mathcal{U}\in\mathrm{End}\left(\mathbb{Z}_p[\![X]\!]\right)$ is the idempotent $\mathcal{U}g(X)=g(X)-\sum_{j=0}^{p-1}g\left(\zeta_p^j(1+X)-1\right)$. In other words, the image of a norm-compatible system of cyclotomic units under Coleman's homomorphism, yields the special values of the Riemann zeta-function.

Theorem 3.7. [Co] *Defining the mapping* $\widetilde{\mathrm{Log}}$ *as the composition* $\mathcal{U}\circ\log\circ\mathrm{Col}$, *then there is a 4-term exact sequence of* $\mathrm{Gal}\left(\mathbb{Q}_p(\mu_{p^\infty})/\mathbb{Q}_p\right)$*-modules*

$$0\longrightarrow \mu_{p-1}\times\varprojlim_n \mu_{p^n}\longrightarrow \varprojlim_n U_n \xrightarrow{\widetilde{\mathrm{Log}}} \mathbb{Z}_p[\![X]\!]^{\mathcal{U}=0}\longrightarrow \mathbb{Z}_p(1)\longrightarrow 0$$

where U_n *is the unit group* $\mathbb{Z}_p\left[\mu_{p^n}\right]^\times$, *and the right-hand map is* $g\mapsto\left(\zeta_{p^n}^{D\log g(0)}\right)_n$.

This long exact sequence gives a highly satisfactory description of the Iwasawa structure of the local units. In Chapter VI we give a two-variable generalisation of this theorem, to encompass the universal nearly-ordinary Galois representation (c.f. Theorem 6.11). For the present time, we return to our original discussion of p-stabilised Hecke eigenforms, in order to write down a natural analogue $\mathbb{H}^1_{\mathrm{Eul}}$ of the collection of cyclotomic Euler systems $\bigcup_{m\not\equiv 2(\bmod 4)} C_m$ mentioned above.

The Euler system of N-primitive zeta-elements

Let us start by mentioning some extra properties satisfied by the zeta-elements constructed by Kato in [Ka1]. As in the previous chapter, the sheaf $\mathfrak{F}^{(k)}$ denotes the $(k-2)$-nd symmetric power of the pullback of a local system $\underline{H}^1$, on the universal elliptic curve $\mathbb{E}_{\mathrm{univ}}$ over $Y_1(Np^r)$. Similarly, the p-adic sheaf $\mathfrak{F}_p^{(k)}$ was the pullback of $\mathrm{Sym}^{k-2}_{\mathbb{Z}_p}\left(\underline{H}^1\otimes\mathbb{Z}_p\right)$ to $Y_1(Np^r)$. Several results in p-adic Hodge theory due to Faltings, Kato and Hyodo [Fa,Ka4,Hy], allow the identification

$$\mathrm{Fil}^{k-1}\mathbf{D}_{\mathrm{dR}}\left(H^1\left(Y_1(Np^r)\otimes\overline{\mathbb{Q}}_p,\ \mathfrak{F}_p^{(k)}\right)\right) \;=\; M_k\left(X_1(Np^r)\right)\otimes\mathbb{Q}_p$$

where $M_k\left(X_1(Np^r)\right)$ is the space of modular forms $H^0\left(X_1(Np^r),\ \mathrm{coLie}(\overline{\mathbb{E}}_{\mathrm{univ}})^{\otimes k}\right)$.

This is how we considered $\mathrm{Fil}^0\mathbf{D}_{\mathrm{dR}}(V_f^*) \cong \mathrm{Fil}^{k-1}H^1_{\mathrm{dR}}\big(X_1(Np^r)/\mathbb{Q}_p, \mathfrak{F}_p^{(k)}\big) \otimes_{\lambda_{f^*}} \mathcal{K}_f$ previously as being generated by the dual cusp form f^*. The period mapping sends

$$\mathrm{per}_\infty : M_k\big(X_1(Np^r)\big) \otimes \overline{\mathbb{Q}} \longrightarrow H^1\Big(Y_1(Np^r)(\mathbb{C}),\ \mathfrak{F}^{(k)}\Big) \otimes \mathbb{C}$$

and the image of the subspace of cusp forms will be the compactly supported cohomology. Since the Betti realisation has an additional action of the complex conjugation τ, we can fix generators $\gamma^\pm$ of $H^1\Big(Y_1(Np^r)(\mathbb{C}),\ \mathfrak{F}^{(k)}\Big)^\pm[\lambda_{f^*}]$ satisfying $(2\pi i)^{2-k}\mathrm{per}_\infty(f^*) = \Omega_f^+\gamma^+ + \Omega_f^-\gamma^-$.

Remark: In order to evaluate the dual exponential of Kato's zeta-elements, we pass from global to semi-local cohomology via

$$H^1\big(\mathbb{Q}(\mu_m),\ -\big) \xrightarrow{\mathrm{loc}_p} H^1\big(\mathbb{Q}(\mu_m)\otimes\mathbb{Z}_p,\ -\big) \xrightarrow{\sim} \prod_{\mathfrak{P}_m|p} H^1\big(D_{\mathfrak{P}_m},\ -\big)$$

with each $D_{\mathfrak{P}_m}$ – the absolute Galois group of $\mathbb{Q}(\mu_m)_{\mathfrak{P}_m}$ – clearly dependent upon the choice of prime ideal $\mathfrak{P}_m \in \mathrm{Spec}\,\mathbb{Z}[\mu_m]$ lying over p. In this manner

$$\exp^* : H^1\big(\mathbb{Q}(\mu_m)\otimes\mathbb{Z}_p, V_f^*\big) \longrightarrow M_k\big(X_1(Np^r)\big)[\lambda_{f^*}] \otimes \mathbb{Q}(\mu_m) \otimes \mathbb{Z}_p$$

and we shall frequently abuse notation by writing '$\exp^*$' without first mentioning the semi-localisation map at p.

Let $\mathbf{T}$ be the $G_{\mathbb{Q}}$-stable lattice generated by the image of $H^1\Big(Y_1(Np^r)(\mathbb{C}),\ \mathfrak{F}_p^{(k)}\Big)$ inside the vector space V_f^*. The following zeta-elements are slight modifications of those described in [Ka1, Thm 6.6 and Prop 8.12].

Theorem 3.8. *For all integers $n \geq 0$ and every integer $M \geq 1$ coprime to p, there exist modified zeta-elements $\mathbf{z}_{Mp^n} \in H^1_{\text{ét}}\Big(\mathbb{Z}\big[\zeta_{Mp^n}, 1/p\big],\ V_f^*\Big)$ satisfying:*

(i) If $m = Mp^n$ and l is any prime number, then

$$\mathrm{cores}_{(\mu_{ml})/(\mu_m)}\mathbf{z}_{ml} = \begin{cases} \Big(1 - a_l(f)(l\sigma_l)^{-1} + \epsilon(l)l^{k-1}(l\sigma_l)^{-2}\Big).\,\mathbf{z}_m & \text{if } l \nmid pm \\ \mathbf{z}_m & \text{if } l|pm; \end{cases}$$

(ii) For each index $m = Mp^n$, there exists a constant $\nu_M \geq 0$ depending on the support of M, such that $\mathbf{z}_{Mp^n} \in H^1_{\text{ét}}\Big(\mathbb{Z}\big[\zeta_{Mp^n}, 1/p\big],\ p^{-\nu_M}\mathbf{T}\Big)$;

(iii) For any primitive character ψ modulo Mp^n and integer $j \in \{0, \ldots, k-2\}$, choosing $\pm$ to be the sign of $\psi(-1)(-1)^j$ we have

$$\mathrm{per}_\infty\left(\sum_{b\in(\mathbb{Z}/Mp^n\mathbb{Z})^\times} \psi(b)\exp^*\Big(\mathbf{z}_{Mp^n}\otimes\zeta_{p^n}^{\otimes -j}\Big)^{\sigma_b}\right)^\pm = (2\pi i)^{k-2-j}L_{\{p\}}(f,\psi,1+j)\gamma^\pm.$$

The principal difference between the $\mathbf{z}_{Mp^n}$'s and the elements constructed in [Ka1] is that the Euler factors at primes dividing N are present in the above L-values. In this sense, our zeta-elements could be viewed as N-primitive versions of Kato's. Unfortunately, what we gain in primitivity we lose in integrality, because the lattices $p^{-\nu_M}\mathbf{T}$ enlarge with the support of M. Thus the N-primitive zeta-elements won't give finer bounds in the direction of the p-adic Birch, Swinnerton-Dyer Conjecture. The technical details of this modification are written up in Appendix A.

The one-cocycles $\mathbf{z}_m$ themselves should be thought of as the incarnation of the L-function within Galois cohomology. In this chapter, we exploit Theorem 3.8(iii) to reconstruct the algebraic modular symbol η_f associated to each cusp form f. Clearly it is worthwhile to first describe the subspace of étale cohomology where Euler systems naturally reside. Picking primes $\mathfrak{P}_{p^nM} \in \operatorname{Spec} \mathbb{Z}[\mu_{p^nM}]$ which form a compatible sequence of ideals lying over p, we obtain restriction maps

$$\varprojlim_n H^1_{\text{ét}}\Big(\mathbb{Z}\big[\zeta_{Mp^n}, 1/p\big],\, p^{-\nu_M}\mathbf{T}\Big) \quad \stackrel{\text{loc}_p}{\longrightarrow} \quad \prod_{\mathfrak{P}_M|p} H^1_{\text{Iw}}\big(D_{\mathfrak{P}_M}, V_f^*\big)$$

N.B. the $\mathfrak{P}_M$'s never split in the totally ramified extension $\mathbb{Q}(\mu_{Mp^\infty})/\mathbb{Q}(\mu_M)$.

Definition 3.9. *We write $\mathbb{H}^1_{\text{Eul}}(V_f^*)$ for the submodule of elements*

$$\vec{x} = \big(\ldots,\, (x_{Mp^n})_n, \ldots\big) \in \prod_{M\in\mathbb{N}-p\mathbb{N}} \mathbb{Q}_p \otimes_{\mathbb{Z}_p} \left(\varprojlim_n H^1_{\text{ét}}\Big(\mathbb{Z}\big[\zeta_{Mp^n}, 1/p\big],\, \mathbf{T}\Big)\right)$$

satisfying for all $n \geq 0$, $p \nmid M$ and primes l

$$\text{cores}_{(\mu_{Mp^nl})/(\mu_{Mp^n})}\big(x_{Mp^nl}\big) = \begin{cases} \Big(1 - a_l(f)(l\sigma_l)^{-1} + \epsilon(l)l^{k-1}(l\sigma_l)^{-2}\Big).\, x_{Mp^n} & \text{if } l \nmid pM \\ x_{Mp^n} & \text{if } l|pM. \end{cases}$$

Similarly, we write $\mathbb{H}^1_{\text{Eul},p}(V_f^*)$ for the subspace of $\prod_{M\in\mathbb{N}-p\mathbb{N}} H^1_{\text{Iw}}(\mathbb{Q}(\mu_M)\otimes\mathbb{Z}_p, V_f^*)$ which is the exact image of $\mathbb{H}^1_{\text{Eul}}(V_f^*)$ under loc_p.

Thus candidate Euler systems correspond to infinite products of global cocycles, satisfying stringent corestriction conditions. It follows directly from Theorem 3.8(i), we can consider

$$\vec{\mathbf{z}} \;=\; \vec{\mathbf{z}}_{k,\epsilon} \;:=\; \big(\ldots,\, (\mathbf{z}_{Mp^n})_n, \ldots\big) \quad \text{as an element of} \quad \mathbb{H}^1_{\text{Eul}}\big(V_f^*\big)$$

and then applying restriction at the primes over p, means it also lies in $\mathbb{H}^1_{\text{Eul},p}\big(V_f^*\big)$. The cohomology classes in the vector $\vec{\mathbf{z}}$ simultaneously encode the special values $\Big\{L_{\{p\}}\big(f/\mathbb{Q}(\mu_m), 1\big), \ldots, L_{\{p\}}\big(f/\mathbb{Q}(\mu_m), k-1\big)\Big\}$ at all cyclotomic fields $\mathbb{Q}(\mu_m)$. Therefore it is not unreasonable to expect that most of the symbol η_f can be recovered, just by computing the image of $\vec{\mathbf{z}}$ under dual exponential maps.

3.3 The one-variable interpolation

The strategy we adopt is as follows. We define a map $\mathcal{E}_{k,\epsilon}$ converting cohomology classes into $\mathcal{Q}$-continuous symbols. In particular, the N-primitive zeta-element Euler system is converted into the algebraic modular symbol η_f discussed in §3.1. It turns out that the distributions associated to these symbols in Proposition 3.6(ii) are p-bounded measures, and thus correspond to elements of the Iwasawa algebra. We subsequently show in §3.4 that the Euler system dominates the image of $\mathcal{E}_{k,\epsilon}$ as a Λ_f^{cy}-module. This means the behaviour of the interpolating maps as we vary f in an analytic family of eigenforms, should be determined by the variation in η_f.

Previously we fixed primitive m^{th}-roots of unity ζ_m such that $(\zeta_{m'})^{m'/m} = \zeta_m$.

Theorem 3.10. *We write* $\mathcal{E}_{k,\epsilon} : \mathbb{H}^1_{\mathrm{Eul},p}\big(V_f^*\big) \to \mathrm{Hom}_{\mathbb{Z}}\Big(\mathrm{div}^0\mathbb{P}^1(\mathbb{Q}), \mathrm{Sym}^{k-2}(\mathcal{K}_f)\Big)$ *to denote the map with special values*

$$\begin{aligned}\mathcal{E}_{k,\epsilon}(\vec{x})_{\left(\frac{a}{p^nM}\right)-(\infty)} \quad &:= \quad \sum_{j=0}^{k-2} \frac{\Gamma(k-1)}{\Gamma(k-1-j)}(-X)^j \left(Y + \frac{a}{p^nM}X\right)^{k-2-j} \\ &\quad \times \sum_{b\in(\mathbb{Z}/p^nM\mathbb{Z})^\times} \left(\zeta_{p^nM}^a \exp_j^*\left(x_{Mp^n} \otimes \zeta_{p^n}^{\otimes -j}\right)\right)^{\sigma_b}\end{aligned}$$

for all $\vec{x} = \big(\ldots,\ (x_{Mp^n})_n,\ \ldots\big) \in \mathbb{H}^1_{\mathrm{Eul},p}\big(V_f^*\big)$ *and* $\frac{a}{p^nM} \in \mathbb{Q}$ *with* $\gcd(a, M) = 1$. *Then*

(i) The image of $\mathcal{E}_{k,\epsilon}$ *is* $\mathcal{Q}$*-continuous, fixed by* $\begin{pmatrix} 1 & 1 \\ 0 & 1 \end{pmatrix}$ *and free of* Λ_f^{cy}*-torsion;*

(ii) $\mathcal{E}_{k,\epsilon}$ *is a* Λ_f^{cy}*-homomorphism sending* $\vec{\mathbf{z}} = \vec{\mathbf{z}}_{k,\epsilon}$ *to the modular symbol* η_f*;*

(iii) The symbol η_f *is dense in the image, that is*

$$\mathrm{Frac}\big(\Lambda_f^{\mathrm{cy}}\big).\ \eta_f\Bigg|_{\mathrm{div}^0\left(\Sigma_{M,c}\cup\{\infty\}\right)} \quad = \quad \mathrm{Frac}\big(\Lambda_f^{\mathrm{cy}}\big) \otimes_{\Lambda_f^{\mathrm{cy}}} \mathrm{Im}\big(\mathcal{E}_{k,\epsilon}\big)\Bigg|_{\mathrm{div}^0\left(\Sigma_{M,c}\cup\{\infty\}\right)}$$

for every pair (M, c) *such that* M *is coprime to* p *and* $\gcd(c, M) = 1$.

An immediate corollary of 3.10(ii) is that the $\pm$-part of the image of $\mathcal{E}_{k,\epsilon}$ contains $\mathcal{MS}\big(Np^r; \mathrm{Sym}^{k-2}(\mathcal{K}_f)\big)[\lambda_f]^\pm$, as the M-symbol $\eta_f^\pm$ generates these eigenspaces. In fact $\mathrm{Im}(\mathcal{E}_{k,\epsilon})$ should be viewed as the Λ^{cy}-deformation of $\mathcal{MS}\big(Np^r; \mathrm{Sym}^{k-2}\big)$. We shall use an identical approach in Chapter VI, in order to deform the $\mathbf{I}$-adic analogue of the modular symbol η_f along the cyclotomic direction.

It should also be pointed out that in 3.10(iii), it is necessary to restrict to each of the cuspidal partitions $\Sigma_{M,c}$ to make a meaningful statement about the density. The reason why we cannot consider the whole of $\mathrm{div}^0\mathbb{P}^1(\mathbb{Q})$ is that both $\Lambda_f^{\mathrm{cy}}.\eta_f$ and $\mathrm{Im}(\mathcal{E}_{k,\epsilon})$ have infinite rank over Λ_f^{cy}. Fortunately, we know from §3.1 that $\mathrm{Hom}_{\mathcal{Q}\mathrm{cts}}\Big(\mathrm{div}^0\left(\Sigma_{M,c} \cup \{\infty\}\right), -\Big)$ is naturally an Iwasawa module, which allows us to make a precise statement about the rank locally.

Proof of 3.10(i),(ii): Let $\Xi = \mathcal{E}_{k,\epsilon}(\vec{x})$ for some product of cohomology classes $\vec{x} = \big(\ldots, (x_{Mp^n})_n, \ldots\big) \in \mathbb{H}^1_{\mathrm{Eul},p}(V_f^*)$, and without loss of generality assume that $(x_{Mp^n})_n \in \varprojlim_n H^1\Big(\mathbb{Q}(\mu_{Mp^n}) \otimes \mathbb{Z}_p, \mathbf{T}\Big)$ is p-integral, in terms of the lattice $\mathbf{T}$. From the Definition 3.4, if $\frac{a}{p^n M} \in \Sigma_{M,c}$ then

$$\mathcal{Q}_{\Xi,M,c}\left(\frac{a}{p^n M}\right) = \sum_{j=0}^{k-2} \frac{\Gamma(k-1)(-X)^j Y^{k-2-j}}{\Gamma(k-1-j)} \sum_{b \in (\mathbb{Z}/p^n M\mathbb{Z})^\times} \Big(\zeta^a_{p^n M} \exp^*_j\big(x_{Mp^n} \otimes \zeta_{p^n}^{\otimes -j}\big)\Big)^{\sigma_b}$$

and the second summation $\sum_{b \in (\mathbb{Z}/p^n M\mathbb{Z})^\times} (\ldots)^{\sigma_b}$ only depends on the congruence class of $a \mod p^n M$. In particular, $\mathcal{Q}_{\Xi,M,c}\left(\frac{a}{p^n M} + 1\right) = \mathcal{Q}_{\Xi,M,c}\left(\frac{a}{p^n M}\right)$ which is precisely equivalent to the assertion that $\Xi_{(\frac{a}{p^n M}+1)-(\infty)}\Big| \begin{pmatrix} 1 & 1 \\ 0 & 1 \end{pmatrix} = \Xi_{(\frac{a}{p^n M})-(\infty)}$. Because the divisors $\left(\frac{a}{p^n M}\right) - (\infty)$ generate $\mathrm{div}^0\mathbb{P}^1(\mathbb{Q})$ over $\mathbb{Z}$, this is strong enough to imply that our symbol is $\begin{pmatrix} 1 & 1 \\ 0 & 1 \end{pmatrix}$-invariant.

Furthermore, if we choose any $\frac{a'}{p^{n'} M} \in \Sigma_{M,c}$ p-adically close to $\frac{a}{p^n M}$ then clearly $n' = n$ and $a \equiv a' \mod p^C$ for some $C \gg 0$. If $C \geq n$ then $\zeta^a_{p^n M}$ equals $\zeta^{a'}_{p^{n'} M}$ as $a \equiv a' \equiv c \mod M$, and it follows we must have $\mathcal{Q}_{\Xi,M,c}\left(\frac{a}{p^n M}\right) = \mathcal{Q}_{\Xi,M,c}\left(\frac{a'}{p^{n'} M}\right)$. In this way $\mathcal{Q}_{\Xi,M,c}$ extends to a continuous function on $\mathbb{Q}_p$, which implies the $\mathcal{Q}$-continuity of $\Xi = \mathcal{E}_{k,\epsilon}(\vec{x})$.

Remark: We can now invoke Proposition 3.6(i), to verify that the operator U_p exists on the image of $\mathcal{E}_{k,\epsilon}$. By definition its action coincides with that of the usual Hecke operator U_p on $\mathbb{H}^1_{\mathrm{Eul},p}(V_f^*)$, i.e. via the scalar element $\lambda_f(U_p) = a_p(f) \in \mathcal{O}^\times_{\mathcal{K}_f}$. Therefore, we deduce that U_p is actually invertible on $\mathrm{Im}(\mathcal{E}_{k,\epsilon})$.

This is actually the crux of the argument. Applying 3.6(ii), for each $M \in \mathbb{N}$, $p \nmid M$ there is a map

$$\mathrm{Im}(\mathcal{E}_{k,\epsilon}) \longrightarrow \mathrm{Dist}\left(\mathbb{Z}^\times_{p,M}, \mathrm{Sym}^{k-2}(\mathcal{K}_f)\right), \quad \Xi = \mathcal{E}_{k,\epsilon}(\vec{x}) \mapsto \mu_\Xi$$

satisfying for all $n \geq 0$ and $a \in (\mathbb{Z}/p^n M\mathbb{Z})^\times$

$$\begin{aligned} \mu_\Xi\Big(a + (p^n M)\Big) &= \mathcal{Q}_{U_p^{-n}\circ\Xi,M,a}\left(\frac{a}{p^n M}\right)\Bigg| \begin{pmatrix} 1 & 0 \\ 0 & p \end{pmatrix}^n \\ &= \sum_{j=0}^{k-2} \frac{\Gamma(k-1)(-p^n X)^j Y^{k-2-j}}{\Gamma(k-1-j)\, a_p(f)^n} \sum_{b \in (\mathbb{Z}/p^n M\mathbb{Z})^\times} \Big(\zeta^a_{p^n M} \exp^*_j\big(x_{Mp^n} \otimes \zeta_{p^n}^{\otimes -j}\big)\Big)^{\sigma_b} \end{aligned}$$

$$\begin{aligned} \text{whilst} \quad \mu_\Xi\left(\mathbb{Z}^\times_{p,M}\right) &= \Xi_{(0)-(\infty)} - U_p^{-1} \circ \Xi_{(0)-(\infty)}\Bigg| \begin{pmatrix} 1 & 0 \\ 0 & p \end{pmatrix} \\ &= \sum_{j=0}^{k-2} \frac{\Gamma(k-1)(-X)^j Y^{k-2-j}}{\Gamma(k-1-j)} \big(1 - p^j/a_p(f)\big) \sum_{b \in (\mathbb{Z}/M\mathbb{Z})^\times} \exp^*_j(x_M)^{\sigma_b}. \end{aligned}$$

For stupid reasons of normalisation, we are forced to consider the contragredient distribution $\mu_\Xi^\bullet$ defined by $\mu_\Xi^\bullet(z) := \mu_\Xi(z^{-1})$ for every group element $z \in \mathbb{Z}_{p,M}^\times$. In terms of multiplicative characters $\psi \mod p^n M$,

$$\int_{\mathbb{Z}_{p,M}^\times} \psi d\mu_\Xi^\bullet := \sum_{a \mod p^n M} \psi(a)\mu_\Xi^\bullet\Big(a + (p^n M)\Big) = \sum_{a \mod p^n M} \psi^{-1}(a)\mu_\Xi\Big(a + (p^n M)\Big)$$

then plugging in the definition of μ_Ξ, we obtain the special values

$$\int_{\mathbb{Z}_{p,M}^\times} \psi d\mu_\Xi^\bullet = \sum_{j=0}^{k-2} \binom{k-2}{j} X^j Y^{k-2-j} \frac{(-1)^j j!\, p^{nj}}{a_p(f)^n} \left(1 - \delta_{\psi=\mathbf{1}} p^j / a_p(f)\right)$$
$$\times\ G(\psi^{-1}, \zeta_{p^n M}) \sum_{b \in (\mathbb{Z}/p^n M\mathbb{Z})^\times} \psi(b) \exp_j^*\left(x_{Mp^n} \otimes \zeta_{p^n}^{\otimes -j}\right)^{\sigma_b}.$$

Significantly, this is none other than the p-adic interpolation formula encountered in the proof of Corollary 2.10, albeit without the restriction that ψ had p-power conductor. By an identical argument to the demonstration of 2.10, the right-hand side lies in $p^{\theta_{\mathbf{T}}-1}\mathcal{O}_{\mathcal{K}_f}[\mu_{Mp^n}][X, Y]$ and so must be p-bounded.

To summarise so far, the association $\vec{x} \mapsto \mathcal{E}_{k,\epsilon}(\vec{x}) \mapsto \mu_\Xi^\bullet$ defined a sequence of homomorphisms

$$\mathbb{H}^1_{\mathrm{Eul},p}(V_f^*) \xrightarrow{\mathcal{E}_{k,\epsilon}} \mathrm{Hom}_{\mathcal{Q}\mathrm{cts}}\Big(\mathrm{div}^0\mathbb{P}^1(\mathbb{Q}), \mathrm{Sym}^{k-2}(\mathcal{K}_f)\Big)$$
$$\big\downarrow \mu^\bullet$$
$$\mathrm{Dist}\Big(\mathbb{Z}_{p,M}^\times,\ \mathrm{Sym}^{k-2}(\mathcal{K}_f)\Big)$$

at every integer M coprime to p. Perrin-Riou's Theorem 2.7 then tells us that the composition $\mu^\bullet \circ \mathcal{E}_{k,\epsilon}$ is a Λ_f^{cy}-homomorphism. It is an easy exercise to verify that the Λ_f^{cy}-action on $\mathrm{Hom}_{\mathcal{Q}\mathrm{cts}}\Big(\mathrm{div}^0(\Sigma_{M,c} \cup \{\infty\}),\ \mathrm{Sym}^{k-2}(\mathcal{K}_f)\Big)$ is compatible with the maps $\mathcal{E}_{k,\epsilon}$ and $\mu^\bullet$ (note that $g \in G_\infty$ sends a bounded measure $\nu(z)$ to its translate $\nu\big(\chi_{\mathrm{cy}}(g)z\big)$).

Taking the direct sum of the $\mu^\bullet$'s at each M, we obtain a Λ_f^{cy}-homomorphism

$$\mathrm{Im}(\mathcal{E}_{k,\epsilon}) \xrightarrow{\oplus\mu^\bullet} \bigoplus_{M \in \mathbb{N} - p\mathbb{N}} \mathrm{Meas}\Big(\mathbb{Z}_{p,M}^\times,\ \mathrm{Sym}^{k-2}(\mathcal{K}_f)\Big).$$

The right-hand side is Λ^{cy}-free, because each space $\mathrm{Meas}\Big(\mathbb{Z}_{p,M}^\times,\ \mathrm{Sym}^{k-2}(\mathcal{K}_f)\Big)$ is isomorphic to $\phi(pM)$ copies of the power series ring $\mathrm{Sym}^{k-2}(\mathcal{O}_{\mathcal{K}_f})[[T]][p^{-1}]$. Also $\oplus\mu^\bullet$ is injective, as any symbol which maps to zero in $\mathrm{Meas}\Big(\mathbb{Z}_{p,M}^\times, \mathrm{Sym}^{k-2}(\mathcal{K}_f)\Big)$ simultaneously vanishes at all divisors $\left(\frac{a}{p^n M}\right) - (\infty)$, hence must itself be zero. Consequently $\mathcal{E}_{k,\epsilon}$ is a Λ_f^{cy}-homomorphism, and its image is torsion-free.

Remark: To complete the proof of 3.10(i),(ii) it remains to check that $\mathcal{E}_{k,\epsilon}(\vec{\mathbf{z}}) = \eta_f$. Greenberg and Stevens [GS, §4.10] introduced the twist operator on symbols

$$\mathrm{Tw}_\psi \circ \Xi \quad := \quad \sum_{a=0}^{C_\psi - 1} \psi(a) \begin{pmatrix} 1 & a \\ 0 & C_\psi \end{pmatrix} \circ \Xi$$

for multiplicative characters ψ of conductor C_ψ. For example, the special values of the algebraic modular symbol twisted by ψ, are given by

$$\left(\mathrm{Tw}_\psi \circ \eta_f\right)_{(0)-(\infty)} \quad = \quad \sum_{j=0}^{k-2} \binom{k-2}{j} X^j Y^{k-2-j} (-1)^j j! C_\psi^j \, \frac{G(\psi, \zeta_{C_\psi})\, L(f, \psi^{-1}, j+1)}{(2\pi i)^j \, \Omega_f^\pm}$$

according to [GS, 4.12 and Thm 4.14].

To demonstrate why the symbols $\mathcal{E}_{k,\epsilon}(\vec{\mathbf{z}})$ and η_f are the same, we simply show that they possess identical special values under all Dirichlet twists by characters ψ. Writing $C_\psi = p^n M$ where the integer $n = \mathrm{ord}_p C_\psi$, one discovers that

$$\left(\mathrm{Tw}_\psi \circ \mathcal{E}_{k,\epsilon}(\vec{\mathbf{z}})\right)_{(0)-(\infty)} \quad = \quad \sum_{a=0}^{p^n M - 1} \psi(a)\, \mathcal{E}_{k,\epsilon}(\vec{\mathbf{z}})_{\left(\frac{a}{p^n M}\right)-(\infty)} \Bigg| \begin{pmatrix} 1 & a \\ 0 & p^n M \end{pmatrix}$$

$$= \sum_{j=0}^{k-2} \binom{k-2}{j} X^j Y^{k-2-j} (-1)^j j!\, C_\psi^j \sum_{a,b \in (\mathbb{Z}/p^n M\mathbb{Z})^\times} \psi(a) \zeta_{p^n M}^{ab} \exp_j^* \left(\mathbf{z}_{p^n M} \otimes \zeta_{p^n}^{\otimes -j}\right)^{\sigma_b}.$$

The inner summation term $\sum_{a,b}$ above equals

$$G(\psi, \zeta_{C_\psi}) \sum_b \psi^{-1}(b) \exp^*_{V_f^*(-j)} \left(\mathbf{z}_{p^n M} \otimes \zeta_{p^n}^{\otimes -j}\right)^{\sigma_b} \cup \left(1 - \delta_{n=0} p^{-j-1} \varphi\right)^{-1} \mathbf{v}_{k,\epsilon}$$

i.e. $\dfrac{G(\psi, \zeta_{C_\psi})\, L_{\{p\}}(f, \psi^{-1}, j+1)}{(2\pi i)^j \, \Omega_f^\pm \, \left(1 - \delta_{n=0} p^{-j-1} a_p(f)\right)}$ upon applying Theorem 3.8(iii)

(recall that the sign $\pm$ was chosen so that $\psi(-1)(-1)^j = \pm 1$). Moreover, the Euler factor $\left(1 - \delta_{n=0} p^{-j-1} a_p(f)\right)$ is precisely the missing L-factor in $L_{\{p\}}(f, \psi^{-1}, j+1)$. It follows directly from these computations, that

$$\left(\mathrm{Tw}_\psi \circ \mathcal{E}_{k,\epsilon}(\vec{\mathbf{z}})\right)_{(0)-(\infty)} = \sum_{j=0}^{k-2} \binom{k-2}{j} X^j Y^{k-2-j} (-1)^j j! C_\psi^j \, \frac{G(\psi, \zeta_{C_\psi})\, L(f, \psi^{-1}, j+1)}{(2\pi i)^j \, \Omega_f^\pm}$$

which coincides with $\left(\mathrm{Tw}_\psi \circ \eta_f\right)_{(0)-(\infty)}$ by its very definition. In other words

$$\sum_{a=0}^{p^n M - 1} \psi(a) \mathcal{E}_{k,\epsilon}(\vec{\mathbf{z}})_{\left(\frac{a}{p^n M}\right)-(\infty)} \Bigg| \begin{pmatrix} 1 & a \\ 0 & p^n M \end{pmatrix} = \sum_{a=0}^{p^n M - 1} \psi(a) \left(\eta_f\right)_{\left(\frac{a}{p^n M}\right)-(\infty)} \Bigg| \begin{pmatrix} 1 & a \\ 0 & p^n M \end{pmatrix}.$$

Lastly, applying the Fourier inversion formula for the characters ψ of the finite group $(\mathbb{Z}/p^n M\mathbb{Z})^\times$, we deduce that $\mathcal{E}_{k,\epsilon}(\vec{\mathbf{z}})$ coincides with η_f at all divisors $\left(\frac{a}{p^n M}\right) - (\infty)$. Fortunately this is enough to show the general equality $\mathcal{E}_{k,\epsilon}(\vec{\mathbf{z}}) = \eta_f$, and we are done.

□

3.4 Local freeness of the image

In order to complete the proof of Theorem 3.10, we had better study the image of $\mathcal{E}_{k,\epsilon}$ as a module over the Iwasawa algebra. The basic idea is to obtain upper bounds on its Λ^{cy}-ranks by applying ideas from [PR1, §2.3]. We then get lower bounds by using the fact that the Euler system is not torsion, which can be readily deduced from non-vanishing results of Jacquet-Shalika and Rohrlich [JS,Ro]. Luckily the upper and lower bounds coincide, which interestingly enough does not occur when considering non-ordinary cusp forms.

Before embarking on the demonstration, we need to recall some notations from [BK] which have since become standard fare in the literature. If V is an arbitrary p-adic representation of G_K, the group $H^1(K,V)$ classifies isomorphism classes of extensions of $\mathbb{Q}_p$ by V. The cohomological subspace which classifies the de Rham (resp. crystalline) extension classes will be written as $H^1_g(K,V)$ (resp. $H^1_f(K,V)$). Under the local duality

$$H^1(K,V) \times H^1(K,V^*(1)) \xrightarrow{\ \cup\ } H^2(K,\mathbb{Q}_p(1)) \cong \mathbb{Q}_p$$

the orthogonal complement of $H^1_g(K,V)$ is $H^1_e\big(K,V^*(1)\big)$, where $H^1_e(K,V)$ denotes the image of the exponential map $\exp_V$. Similarly, the orthogonal complement of $H^1_f(K,V)$ is the subgroup $H^1_f\big(K,V^*(1)\big)$, so in this sense the H^1_f's are self-dual. For a lattice $\mathbf{T} \subset V$, we shall write $H^1_g(K,\mathbf{T})$ for the inverse image of $H^1_g(K,V)$ under the natural map $H^1(K,\mathbf{T}) \to H^1(K,V)$, with an analogous definition for $H^1_f(K,\mathbf{T})$ and $H^1_e(K,\mathbf{T})$.

Proof of 3.10(iii): We begin with some very general comments about Λ-modules. Let $\mathfrak{Y}$ denote a finitely-generated, compact module over the algebra $\Lambda \cong \mathbb{Z}_p[\![X]\!]$. One way to determine its rank is to estimate the growth in

$$\mathrm{rank}_{\mathbb{Z}_p}\Big(\mathfrak{Y} \otimes_\Lambda \Lambda_{\Gamma_m}\Big) \qquad \text{where `}\,{}_{\Gamma_m}\text{' indicates the } \Big((1+X)^{p^m}-1\Big)\text{-coinvariants.}$$

If $\mathrm{rank}_{\mathbb{Z}_p}\mathfrak{Y}_{\Gamma_m} \approx \lambda_0 \times p^m + O(1)$ for $m \gg 0$, the Λ-rank of $\mathfrak{Y}$ is bounded by λ_0. Several times during the course of the proof, we'll need to calculate ranks of various inverse systems of cohomology groups, and this will prove a handy method.

Let $\Delta_{pM} = (\mathbb{Z}/pM\mathbb{Z})^\times$, and assume $\psi : \Delta_{pM} \to \overline{\mathbb{Q}}^\times$ is a multiplicative character. We write $\mathcal{O}_{K_f,\psi}$ for the extension obtained by adjoining the values of ψ to $\mathcal{O}_{K_f}$. One also employs notation $\varprojlim_n H^1_{/g}$ to denote the quotient of $\varprojlim_n H^1$ by $\varprojlim_n H^1_g$, which is not necessarily the same thing as $\varprojlim_n H^1/H^1_g$!

$$\textbf{Key Claim:} \qquad \mathrm{rank}_{\mathcal{O}_{K_f,\psi}[[\Gamma]]}\left(\varprojlim_n H^1_{/g}\big(\mathbb{Q}(\mu_{Mp^n})\otimes\mathbb{Z}_p,\mathbf{T}\big)^{(\psi)}\right) = 1.$$

The demonstration of this assertion will occupy the rest of this section, but we first explain how Theorem 3.10(iii) follows from it.

Restricting the homomorphism $\mathcal{E}_{k,\epsilon}$ to the component $\varprojlim_n H^1\Big(\mathbb{Q}(\mu_{Mp^n})\otimes\mathbb{Z}_p,\mathbf{T}\Big)$, there is a factorisation

$$\varprojlim_n H^1\Big(\mathbb{Q}(\mu_{Mp^n})\otimes\mathbb{Z}_p,\mathbf{T}\Big) \longrightarrow \varprojlim_n H^1_{/g}\Big(\mathbb{Q}(\mu_{Mp^n})\otimes\mathbb{Z}_p,\mathbf{T}\Big)$$
$$\overset{\mathcal{E}_{k,\epsilon}}{\longrightarrow} \mathrm{Hom}_{\mathcal{Q}\mathrm{cts}}\Big(D_M^0,\mathrm{Sym}^{k-2}(\mathcal{K}_f)\Big)$$

because each $H^1_g \subset \mathrm{Ker}(\exp^*)$, and the map $\mathcal{E}_{k,\epsilon}$ is built out of dual exponentials; here D_M^0 denotes the divisors of degree zero on the set $\bigcup_{\gcd(c,M)=1}\Sigma_{M,c}\cup\{\infty\}$.

Taking ψ-eigenspaces of the above sequence induces

$$\varprojlim_n H^1_{/g}\Big(\mathbb{Q}(\mu_{Mp^n})\otimes\mathbb{Z}_p,\mathbf{T}\Big)^{(\psi)} \overset{\mathcal{E}_{k,\epsilon}}{\longrightarrow} \text{ the } \psi\text{-part of } \mathrm{Hom}_{\mathcal{Q}\mathrm{cts}}\Big(D_M^0,\mathrm{Sym}^{k-2}(\mathcal{K}_f)\Big)$$

where the right-hand term consists of symbols taking values inside $\mathrm{Sym}^{k-2}(\mathcal{K}_{f,\psi})$. It follows immediately from our Key Claim, that

$$\mathrm{rank}_{\mathcal{O}_{\mathcal{K}_{f,\psi}}[[\Gamma]]\otimes\mathbb{Q}}\left(\text{the } \psi\text{-eigenspace of } \mathrm{Im}\left(\mathcal{E}_{k,\epsilon}\right)\Big|_{D_M^0}\right) \le 1.$$

On the other hand, η_f is in the image of $\mathcal{E}_{k,\epsilon}$ and the values of the ψ-twist of the measure $\mu^{\bullet}_{\eta_f}\in\mathrm{Meas}\Big(\mathbb{Z}_{p,M}^{\times},\ \mathrm{Sym}^{k-2}(\mathcal{K}_f)\Big)$ at characters $\Phi:\Gamma/\Gamma^{p^{n-1}}\to\overline{\mathbb{Q}}_p^{\times}$ will be given by the integral $\int_{\mathbb{Z}_{p,M}^{\times}}\psi\Phi d\mu^{\bullet}_{\eta_f}$.

In the proof of 3.10(i),(ii) we saw that when $\psi\Phi\neq\mathbf{1}$, this integral equals

$$\sum_{j=0}^{k-2}\binom{k-2}{j}X^jY^{k-2-j}\frac{(-1)^j j!\,p^{nj}}{a_p(f)^n}G\left((\psi\Phi)^{-1},\zeta_{p^nM}\right)\frac{L(f\otimes\psi,\Phi,j+1)}{(2\pi i)^j\,\Omega_f^{\pm}}.$$

Are these values non-zero?

Theorem 3.11. [JS] *If the weight k is odd or if $j+1\neq\frac{k}{2}$, then the Φ-twisted L-values $L\big(f\otimes\psi,\Phi,j+1\big)$ are all non-zero.*

Theorem 3.12. [Ro] *If the weight k is even and $j+1=\frac{k}{2}$, then all bar finitely many of the Φ-twisted L-values $L\big(f\otimes\psi,\Phi,k/2\big)$ are non-zero at the central point.*

It follows directly that the ψ-part of $\eta_f\Big|_{D_M^0}$ is not $\mathcal{O}_{\mathcal{K}_{f,\psi}}[[\Gamma]]\otimes\mathbb{Q}$-torsion, so the ψ-eigenspace of $\mathrm{Im}\left(\mathcal{E}_{k,\epsilon}\right)\Big|_{D_M^0}$ must be (free) of rank one over this Iwasawa algebra. In summary, we have shown that

$$\mathrm{rank}_{\mathcal{O}_{\mathcal{K}_f}[[\Gamma]][\Delta_{pM}]\otimes\mathbb{Q}}\left(\mathrm{Im}\left(\mathcal{E}_{k,\epsilon}\right)\Big|_{D_M^0}\right) = 1$$

from which part 3.10(iii) follows readily.

Justifying the Key Claim

All that remains is to prove our assertion. Courtesy of the Λ^{cy}-isomorphism $\varprojlim_n H^1\Big(\mathbb{Q}(\mu_{Mp^n})\otimes\mathbb{Z}_p,\mathbf{T}\Big) \stackrel{\sim}{\longrightarrow} \prod_{\mathfrak{P}_M|p}\varprojlim_n H^1\Big(\mathbb{Q}(\mu_{Mp^n})_{\mathfrak{P}_M},\mathbf{T}\Big)$, it would be enough to prove that

$$\mathrm{rank}_{\mathcal{O}_{\mathcal{K}_f,\psi}[[\Gamma]]}\left(\varprojlim_n H^1_{/g}\Big(\mathbb{Q}(\mu_{Mp^n})_{\mathfrak{P}_M},\mathbf{T}\Big)^{(\psi)}\right) = 1$$

at each prime $\mathfrak{P}_M$ lying above p, where ψ now ranges over multiplicative characters of the subgroup $\mathrm{Gal}\Big(\mathbb{Q}(\mu_{Mp^n})_{\mathfrak{P}_M}/\mathbb{Q}_p\Big)\cong\mathrm{Gal}\Big(\mathbb{Q}_p(\mu_{Mp^n})/\mathbb{Q}_p\Big)$.

Let L_∞ be the cyclotomic $\mathbb{Z}_p$-extension of $\mathbb{Q}_p$, and write L_n for the nth-layer subfield of degree $[L_n:\mathbb{Q}_p]=p^n$. We abbreviate the $\mathcal{O}_{\mathcal{K}_f,\psi}$-lattice $\mathbf{T}\otimes\psi^{-1}$ by $\mathbf{T}_\psi$. Writing out the inflation-restriction sequence for the field extension $L_n(\mu_{pM})/L_n$, we obtain the long exact sequence

$$0\to H^1\left(L_n(\mu_{pM})/L_n,\mathbf{T}_\psi^{G_{L_n(\mu_{pM})}}\right)\xrightarrow{\mathrm{infl}} H^1\Big(L_n,\mathbf{T}_\psi\Big)\xrightarrow{\mathrm{rest}}$$
$$H^1\Big(L_n(\mu_{pM}),\mathbf{T}_\psi\Big)^{\mathrm{Gal}(L_n(\mu_{pM})/L_n)}\longrightarrow H^2\left(L_n(\mu_{pM})/L_n,\mathbf{T}_\psi^{G_{L_n(\mu_{pM})}}\right)\longrightarrow\cdots$$

Moreover, passing to the projective limit over n,

$$0\longrightarrow H^1\left(L_\infty(\mu_{pM})/L_\infty,\mathbf{T}_\psi^{G_{L_\infty(\mu_{pM})}}\right)\longrightarrow\varprojlim_n H^1\Big(L_n,\mathbf{T}_\psi\Big)\longrightarrow$$
$$\varprojlim_n H^1\Big(\mathbb{Q}(\mu_{Mp^n})_{\mathfrak{P}_M},\mathbf{T}\Big)^{(\psi)}\longrightarrow H^2\left(L_\infty(\mu_{pM})/L_\infty,\mathbf{T}_\psi^{G_{L_\infty(\mu_{pM})}}\right).$$

The left-hand and right-hand groups are $\mathcal{O}_{\mathcal{K}_f,\psi}[[\Gamma]]$-torsion since they are killed by the number $\#\Delta_{pM}$, hence

$$\mathrm{rank}_{\mathcal{O}_{\mathcal{K}_f,\psi}[[\Gamma]]}\left(\varprojlim_n H^1\Big(\mathbb{Q}(\mu_{Mp^n})_{\mathfrak{P}_M},\mathbf{T}\Big)^{(\psi)}\right) = \mathrm{rank}_{\mathcal{O}_{\mathcal{K}_f,\psi}[[\Gamma]]}\left(\varprojlim_n H^1\Big(L_n,\mathbf{T}_\psi\Big)\right)$$
$$= \dim_{\mathcal{K}_{f,\psi}}\Big(V_f^*\otimes\psi^{-1}\Big) = 2.$$

Remark: It suffices to show that $\varprojlim_n H^1_g(L_n,\mathbf{T}_\psi)$ has rank one over $\mathcal{O}_{\mathcal{K}_f,\psi}[[\Gamma]]$. In [PR1, Prop 2.3.5] Perrin-Riou computed the Iwasawa rank of the crystalline extensions. Because they occur in the kernel of the dual exponential, we will work with de Rham extensions instead, though the method is identical.

Putting $\Gamma_m=\mathrm{Gal}(L_\infty/L_m)$, we have the invariants/coinvariants sequence

$$0\to\varprojlim_n H^1_g(L_n,\mathbf{T}_\psi)^{\Gamma_m}\to\varprojlim_n H^1(L_n,\mathbf{T}_\psi)^{\Gamma_m}\to\varprojlim_n H^1_{/g}(L_n,\mathbf{T}_\psi)^{\Gamma_m}$$
$$\to\left(\varprojlim_n H^1_g(L_n,\mathbf{T}_\psi)\right)_{\Gamma_m}\to\left(\varprojlim_n H^1(L_n,\mathbf{T}_\psi)\right)_{\Gamma_m}\to\left(\varprojlim_n H^1_{/g}(L_n,\mathbf{T}_\psi)\right)_{\Gamma_m}\to 0$$

and as the Γ_m-coinvariants of $\varprojlim_n H^1(L_n, \mathbf{T}_\psi)$ are contained within $H^1(L_m, \mathbf{T}_\psi)$, we obtain a short exact sequence

$$\varprojlim_n H^1_{/g}(L_n, \mathbf{T}_\psi)^{\Gamma_m} \longrightarrow \left(\varprojlim_n H^1_g(L_n, \mathbf{T}_\psi)\right)_{\Gamma_m} \longrightarrow H^1_g(L_m, \mathbf{T}_\psi).$$

The left-most group has bounded $\mathcal{O}_{\mathcal{K}_{f,\psi}}$-rank ρ_1 say, as m varies, so

$$\mathrm{rank}_{\mathcal{O}_{\mathcal{K}_{f,\psi}}}\left(\varprojlim_n H^1_g(L_n, \mathbf{T}_\psi)\right)_{\Gamma_m} \leq \mathrm{rank}_{\mathcal{O}_{\mathcal{K}_{f,\psi}}}\left(H^1_g(L_m, \mathbf{T}_\psi)\right) + \rho_1.$$

On the other hand, if $V_\psi = V_f^* \otimes \psi^{-1}$ then

$$\begin{aligned}\dim_{\mathcal{K}_{f,\psi}} H^1_g(L_m, V_\psi) &= \dim \mathrm{tang}(V_\psi/L_m) + \dim V_\psi^{G_{L_m}} + \dim \mathbf{D}_{\mathrm{cris},L_m}\left(V_\psi^*(1)\right)^{\varphi=1} \\ &\leq [L_m : \mathbb{Q}_p].\dim \mathrm{tang}(V_\psi/\mathbb{Q}_p) + \rho_2 \quad \text{say.}\end{aligned}$$

Thus we may deduce that the $\mathcal{O}_{\mathcal{K}_{f,\psi}}$-rank of $\left(\varprojlim_n H^1_g(L_n, \mathbf{T}_\psi)\right)_{\Gamma_m}$ is bounded above by $p^m . \dim_{\mathcal{K}_{f,\psi}} \mathrm{tang}(V_\psi/\mathbb{Q}_p) + \rho_1 + \rho_2$ (recall $\mathbf{D}_{\mathrm{cris}}$ was Fontaine's crystalline functor, equipped with Frobenius endomorphism φ).

By the compact version of Nakayama's lemma,

$$\mathrm{rank}_{\mathcal{O}_{\mathcal{K}_{f,\psi}}[[\Gamma]]}\left(\varprojlim_n H^1_g(L_n, \mathbf{T}_\psi)\right) \leq \dim_{\mathcal{K}_{f,\psi}} \mathrm{tang}(V_\psi/\mathbb{Q}_p) = 1.$$

To obtain an equality, we shall exploit the fact that V_f^* is a nearly-ordinary Galois representation. There is a one-dimensional $G_{\mathbb{Q}_p}$-stable subspace $\mathrm{F}^+V_f^*$ such that the quotient $V_f^*/\mathrm{F}^+V_f^*$ is unramified, and inertia acts on $\mathrm{F}^+V_f^*$ through $\chi_{\mathrm{cy}}^{k-1}\epsilon$.

Remark: Since any ordinary representation is automatically semistable [Bu, §IV], V_f^* will certainly become semistable over the fields $\mathbb{Q}(\mu_{Mp^n})_{\mathfrak{P}_M}$ for $n \geq \mathrm{ord}_p(C_\epsilon)$. Work of Nekovář in [Ne1, Sect.6] implies the exactness of

$$\begin{aligned}0 \to \mathbf{T}^{G_{\mathbb{Q}(\mu_{Mp^n})\mathfrak{P}_M}} \to (\mathbf{T}/\mathrm{F}^+\mathbf{T})^{G_{\mathbb{Q}(\mu_{Mp^n})\mathfrak{P}_M}} \to H^1_g\left(\mathbb{Q}(\mu_{Mp^n})_{\mathfrak{P}_M}, \mathrm{F}^+\mathbf{T}\right) \to \\ H^1_g\left(\mathbb{Q}(\mu_{Mp^n})_{\mathfrak{P}_M}, \mathbf{T}\right) \to H^1_g\left(\mathbb{Q}(\mu_{Mp^n})_{\mathfrak{P}_M}, \mathbf{T}/\mathrm{F}^+\mathbf{T}\right) \longrightarrow \text{(a finite group)}\end{aligned}$$

where $\mathrm{F}^+\mathbf{T} := \mathrm{F}^+V_f^* \cap \mathbf{T}$.

Taking ψ-eigenspaces and passing to the inverse limit over n, yields (yet another) long exact sequence

$$\begin{aligned}0 = \varprojlim_n (\mathbf{T}/\mathrm{F}^+\mathbf{T})^{G_{\mathbb{Q}(\mu_{Mp^n})\mathfrak{P}_M}} &\longrightarrow \varprojlim_n H^1_g\left(\mathbb{Q}(\mu_{Mp^n})_{\mathfrak{P}_M}, \mathrm{F}^+\mathbf{T}\right)^{(\psi)} \\ \longrightarrow \varprojlim_n H^1_g\left(\mathbb{Q}(\mu_{Mp^n})_{\mathfrak{P}_M}, \mathbf{T}\right)^{(\psi)} &\longrightarrow \varprojlim_n H^1_g\left(\mathbb{Q}(\mu_{Mp^n})_{\mathfrak{P}_M}, \mathbf{T}/\mathrm{F}^+\mathbf{T}\right)^{(\psi)}.\end{aligned}$$

In fact both $\varprojlim_n H^1_g\Big(\mathbb{Q}(\mu_{Mp^n})_{\mathfrak{P}_M}, \mathbf{T}\Big)^{(\psi)}$ and $\varprojlim_n H^1_g\Big(L_n, \mathbf{T}_\psi\Big)$ differ by a group of order dividing $\#\Delta_{pM}$, and the same is true replacing $\mathbf{T}$ with $\mathrm{F}^+\mathbf{T}$ or $\mathbf{T}/\mathrm{F}^+\mathbf{T}$. We may therefore rewrite the above (up to finite groups) as

$$0 \longrightarrow \varprojlim_n H^1_g\left(L_n, \mathrm{F}^+\mathbf{T}_\psi\right) \longrightarrow \varprojlim_n H^1_g\left(L_n, \mathbf{T}_\psi\right) \longrightarrow \varprojlim_n H^1_g\left(L_n, \mathbf{T}_\psi/\mathrm{F}^+\right).$$

The right-hand group is $\mathcal{O}_{\mathcal{K}_{f,\psi}}[[\Gamma]]$-torsion, because its rank is bounded above by $\dim_{\mathcal{K}_{f,\psi}} \mathrm{tang}(V_\psi/\mathrm{F}^+)$ which is zero. Thus our problem of verifying whether

$$\mathrm{rank}_{\mathcal{O}_{\mathcal{K}_{f,\psi}}[[\Gamma]]}\left(\varprojlim_n H^1_g\left(L_n, \mathbf{T}_\psi\right)\right) \overset{??}{=} \dim_{\mathcal{K}_{f,\psi}} \mathrm{tang}(V_\psi/\mathbb{Q}_p) = 1$$

reduces to showing that $\varprojlim_n H^1_g\left(L_n, \mathrm{F}^+\mathbf{T}_\psi\right)$ has rank one.

The natural dualities $H^1\big(L_n, \mathrm{F}^+\mathbf{T}_\psi\big) \times H^1\big(L_n, \mathrm{F}^+\mathbf{T}^*_\psi(1)\big) \longrightarrow \mathcal{O}_{\mathcal{K}_{f,\psi}}$ induced by the cup-product, give rise to

$$\begin{aligned} 0 \longrightarrow \varprojlim_n H^1_g\Big(L_n, \mathrm{F}^+\mathbf{T}_\psi\Big) &\longrightarrow \varprojlim_n H^1\Big(L_n, \mathrm{F}^+\mathbf{T}_\psi\Big) \\ &\longrightarrow \mathrm{Hom}_{\mathbb{Z}_p}\Big(H^1_e\Big(L_\infty, \mathrm{F}^+\mathbf{T}^*_\psi(1)\Big), \mathcal{O}_{\mathcal{K}_{f,\psi}}\Big) \end{aligned}$$

because $H^1_g\big(L_n, \mathrm{F}^+\mathbf{T}_\psi\big)^\perp$ coincides with $H^1_e\big(L_n, \mathrm{F}^+\mathbf{T}^*_\psi(1)\big)$ up to $\mathcal{O}_{\mathcal{K}_{f,\psi}}$-torsion. However, the right-hand term in this sequence must be torsion over the Iwasawa algebra $\mathcal{O}_{\mathcal{K}_{f,\psi}}[[\Gamma]]$, since $H^1_e\Big(L_\infty, \mathrm{F}^+\mathbf{T}^*_\psi(1)\Big) \otimes_{\mathbb{Z}_p} \mathbb{Q}_p$ is a finite-dimensional vector space. Finally, we have shown that

$$\begin{aligned} \mathrm{rank}_{\mathcal{O}_{\mathcal{K}_{f,\psi}}[[\Gamma]]}\left(\varprojlim_n H^1_g\left(L_n, \mathbf{T}_\psi\right)\right) &= \mathrm{rank}_{\mathcal{O}_{\mathcal{K}_{f,\psi}}[[\Gamma]]}\left(\varprojlim_n H^1\left(L_n, \mathrm{F}^+\mathbf{T}_\psi\right)\right) \\ &= \dim_{\mathcal{K}_{f,\psi}}\left(\mathrm{F}^+V^*_f \otimes \psi^{-1}\right) = 1 \end{aligned}$$

and the proof of Theorem 3.10 is thankfully over.

□

Lemma 3.13. *If we write* ∇_{k-2} *for the forgetful map*

$$\mathrm{Hom}_{\mathbb{Z}}\Big(\mathrm{div}^0\mathbb{P}^1(\mathbb{Q}), \mathrm{Sym}^{k-2}(\mathcal{K}_f)\Big) \longrightarrow \mathrm{Hom}_{\mathbb{Z}}\Big(\mathrm{div}^0\mathbb{P}^1(\mathbb{Q}), \mathcal{K}_f\Big)$$

sending a symbol to its evaluation at $(X,Y) = (0,1)$, *then the restriction of* ∇_{k-2} *to the image of* $\mathcal{E}_{k,\epsilon}$ *is a* $\mathcal{K}_f$*-linear isomorphism.*

Proof: Clearly this restriction

$$\Big(\text{the image of } \mathcal{E}_{k,\epsilon}\Big) \overset{\nabla_{k-2}}{\longrightarrow} \Big(\text{the image of } \mathcal{E}_{k,\epsilon}\Big)_{X=0,Y=1}$$

is surjective, the question we must now address is whether it is invertible or not. Equivalently, can one reconstruct a symbol $\Xi = \mathcal{E}_{k,\epsilon}(\vec{x})$ simply by knowing the values Ξ_D at divisors $D \in \mathrm{div}^0\mathbb{P}^1(\mathbb{Q})$, exclusively at the monomial term Y^{k-2}?

Recall that for each $M \geq 1$ coprime to p, we associated a bounded measure $\mu_\Xi(z) \in \text{Meas}\Big(\mathbb{Z}_{p,M}^\times,\ \text{Sym}^{k-2}(\mathcal{K}_f)\Big)$ satisfying

$$\mu_\Xi\big(a + (p^n M)\big) \ = \ a_p(f)^{-n}\ \Xi_{\left(\frac{a}{p^n M}\right)-(\infty)}\Bigg| \begin{pmatrix} 1 & \frac{a}{p^n M} \\ 0 & 1 \end{pmatrix} \begin{pmatrix} 1 & 0 \\ 0 & p \end{pmatrix}^n .$$

Writing this equation backwards means that

$$\Xi_{\left(\frac{a}{p^n M}\right)-(\infty)} = a_p(f)^n \left(\int_{a+(p^n M)} d\mu_\Xi(z) \right) \Bigg| \begin{pmatrix} 1 & -\frac{a}{p^n M} \\ 0 & p^{-n} \end{pmatrix}$$

and it is an elementary fact that $d\mu_\Xi(z) = \big(Y - M^{-1} z_p X\big)^{k-2} . \ d\mu_\Xi(z)_{X=0,Y=1}$ where $z_p : \mathbb{Z}_{p,M}^\times \twoheadrightarrow \mathbb{Z}_p^\times$ is the canonical projection. Applying the binomial theorem,

$$\big(Y - M^{-1} z_p X\big)^{k-2} \quad = \quad \sum_{j=0}^{k-2} \frac{\Gamma(k-1) M^{-j} (-X)^j Y^{k-2-j}}{\Gamma(k-1-j)\Gamma(j+1)} \ \times \ z_p^j$$

whence at the divisor $D = \left(\frac{a}{p^n M}\right) - (\infty)$, the symbol Ξ must equal

$$a_p(f)^n \left(\sum_{j=0}^{k-2} \frac{\Gamma(k-1) M^{-j} (-X)^j Y^{k-2-j}}{\Gamma(k-1-j)\Gamma(j+1)} \int_{a+(p^n M)} z_p^j d\mu_\Xi(z)_{X=0,Y=1} \right) \Bigg| \begin{pmatrix} 1 & -\frac{a}{p^n M} \\ 0 & p^{-n} \end{pmatrix}.$$

The above is an expression involving only $z_p^j \ d\mu_\Xi(z)_{X=0,Y=1} \in \text{Meas}\Big(\mathbb{Z}_{p,M}^\times,\ \mathcal{K}_f\Big)$ within the range $0 \leq j \leq k-2$.

The terms $z_p^j \ d\mu_\Xi(z)_{X=0,Y=1}$ are bounded measures, and themselves are uniquely determined by values of all integrals

$$\int_{a'+(p^{n'} M)} d\mu_\Xi(z)_{X=0,Y=1} \qquad \text{with } \gcd(a', pM) = 1 \text{ and } n' \geq 0.$$

These latter integrals equal $a_p(f)^{-n'} \left(\Xi_{\left(\frac{a'}{p^{n'} M}\right)-(\infty)} \right)_{X=0,Y=1}$ from the definition of $d\mu_\Xi$.

Consequently, the value of our symbol $\Xi = \mathcal{E}_{k,\epsilon}(\vec{x})$ at every $D = \left(\frac{a}{p^n M}\right) - (\infty)$ can be expressed entirely in terms of the special values of $\nabla_{k-2}(\Xi)$ on $\text{div}^0 \mathbb{P}^1(\mathbb{Q})$. $\square$

Remark: This lemma turns out to be a valuable control theorem in disguise, for the cyclotomic deformation of η_f. Be careful, it is **not** true that the restriction

$$\text{Hom}_{\mathcal{Q}\text{cts}}\Big(\text{div}^0 \mathbb{P}^1(\mathbb{Q}), \text{Sym}^{k-2}(\mathcal{K}_f)\Big) \ \stackrel{\nabla_{k-2}}{\longrightarrow} \ \text{Hom}_{\mathcal{Q}\text{cts}}\Big(\text{div}^0 \mathbb{P}^1(\mathbb{Q}), \mathcal{K}_f\Big)$$

is an isomorphism; whilst it is surjective, its kernel is absolutely massive.

CHAPTER IV

A User's Guide to Hida Theory

The philosophy behind deformation theory is straightforward. One starts off with a mathematical problem about an object f_{k_0} say, which may be too hard to solve. The idea is to view this object f_{k_0} as a single member of a family of objects $\{f_k\}$ parametrised by the variable k. With a bit of luck, the variation in k should be analytic (in a suitable sense). One then translates the original problem regarding f_{k_0} into an even bigger question concerning the whole family $\{f_k\}$.

At this point the reader might well protest, surely this has made life harder? The surprising answer – and the real magic of deformation theory – is that by deforming the problem, we have actually made life easier. The underlying reason is that for the analytic family $\{f_k\}$ we have tools and techniques at our disposal, which are unavailable for each individual family member. If one can then solve the bigger problem, the delicate part is to translate back in terms of the original element f_{k_0}. This last step requires what is known as a Control Theory.

The previous chapter saw us develop a cyclotomic control theory for M-symbols. Over the next fifteen pages or so, we give a short (and by no means complete) background on the control theory in the *weight direction*. In the ordinary case, this theory was primarily developed by Hida in a long series of articles [Hi1,Hi2,Hi3]. The non-ordinary case has subsequently been considered by Coleman and others. Whilst some of our theorems here will work in this more general context, the book itself deals exclusively with p-ordinary families.

Our intention is simple – we want to lift the whole of Chapter III to a situation where the principal objects vary in a bounded analytic family. Under this approach classical modular forms will be replaced by Λ-adic modular forms, modular symbols by $\mathbf{I}$-adic modular symbols, and Deligne's representations will be interpolated by the universal Galois representation whose properties we now describe.

4.1 The universal ordinary Galois representation

Fix a positive integer N such that $p \nmid N$. For each $r \in \mathbb{N}$, let $X_1(Np^r)$ be the compactified modular curve associated to the congruence modular group $\Gamma_1(Np^r)$; then $X_1(Np^r)$ has a canonical $\mathbb{Q}$-structure in which the cusp ∞ is rational [Sh]. We set

$$J_r := \text{Jac}\, X_1(Np^r) \qquad \text{and} \qquad \text{Ta}_p(J_\infty) := \varprojlim_{m,r} J_r[p^m]$$

where the inverse limit is induced by the projection maps $X_1(Np^{r+1}) \twoheadrightarrow X_1(Np^r)$, and $[p^m]$ denotes the kernel of multiplication by p^m.

Since each individual J_r is defined over $\mathbb{Q}$, the Tate module $\mathrm{Ta}_p(J_\infty)$ has a natural action of the absolute Galois group $G_\mathbb{Q}$. Also, the Hecke operators

$$T_l := \left\{\Gamma_1(Np^r)\begin{pmatrix}1 & 0\\ 0 & l\end{pmatrix}\Gamma_1(Np^r)\right\} \quad \text{at each prime number } l$$

operate on the completed curves $X_1(Np^r)$ covariantly, and this is compatible with the projection maps. It follows T_n is an endomorphism of $\mathrm{Ta}_p(J_\infty)$ for all $n \in \mathbb{N}$. We shall allow $(\mathbb{Z}/Np^r\mathbb{Z})^\times$ to act on the Tate module $\mathrm{Ta}_p(J_r)$ through the **scaled** diamond operators

$$\langle\, l\,\rangle := l^2\left\{\Gamma_1(Np^r)\, l\gamma_l\, \Gamma_1(Np^r)\right\} \quad \text{with } \gamma_l \equiv \begin{pmatrix}* & *\\ 0 & l\end{pmatrix} \mod Np^r, \;\; l \nmid Np.$$

By linearity and continuity, this extends to an action of the completed group algebra $\mathbb{Z}_p[\![\mathbb{Z}_{p,N}^\times]\!] := \varprojlim_r \mathbb{Z}_p\left[(\mathbb{Z}/Np^r\mathbb{Z})^\times\right]$ on the whole of $\mathrm{Ta}_p(J_\infty)$.

Definition 4.1. *The abstract Hecke ring* $\mathbf{H}_{Np^\infty}$ *is given by*

$$\mathbf{H}_{Np^\infty} := \mathbb{Z}_p\left[\left[\mathbb{Z}_{p,N}^\times\right]\right][T_n;\, n \in \mathbb{N}]$$

hence $\mathbf{H}_{Np^\infty}$ *will be generated over the weight algebra* $\Lambda^{\mathrm{wt}} := \mathbb{Z}_p\left[\left[1 + p\mathbb{Z}_p\right]\right]$ *by* $\left(\mathbb{Z}/Np\mathbb{Z}\right)^\times$ *and the* T_n*'s.*

The Galois action and Hecke correspondences commute with each other on the modular curves $X_1(Np^r)$ for all $r \in \mathbb{N}$, consequently one can consider $\mathrm{Ta}_p(J_\infty)$ as an $\mathbf{H}_{Np^\infty}[G_\mathbb{Q}]$-module.

Unfortunately $\mathrm{Ta}_p(J_\infty)$ is way too big for practical purposes, as it simultaneously encodes information about all cusp forms of non-negative slope at p. Hida [Hi1] first described the piece corresponding to slope zero modular forms, using the ordinary idempotent $\mathbf{e} := \lim_{n\to\infty} U_p^{n!}$ living in the abstract ring $\mathbf{H}_{Np^\infty}$ (by convention, we shall write U_l for the Hecke operator at primes $l|Np$, and T_l for primes $l \nmid Np$). If the superscript ^{ord} denotes the component cut out by $\mathbf{e}$, then there is a natural decomposition

$$\mathrm{Ta}_p(J_\infty) = \mathrm{Ta}_p(J_\infty)^{\mathrm{ord}} \oplus (1 - \mathbf{e}).\mathrm{Ta}_p(J_\infty)$$

where U_p acts on the complement $(1 - \mathbf{e}).\mathrm{Ta}_p(J_\infty)$ topologically nilpotently.

The representation $\mathrm{Ta}_p(J_\infty)^{\mathrm{ord}}$ is still a little too large, so one considers instead

$$\mathbb{T}_\infty := \mathrm{Ta}_p(J_\infty)^{\mathrm{ord}} \cap \mathbf{e}_{\mathrm{prim}}.\left(\mathrm{Ta}_p(J_\infty) \otimes_{\Lambda^{\mathrm{wt}}} \mathrm{Frac}\left(\Lambda^{\mathrm{wt}}\right)\right)$$

with $\mathbf{e}_{\mathrm{prim}}$ denoting the idempotent in $\mathbf{H}_{Np^\infty} \otimes_{\Lambda^{\mathrm{wt}}} \mathrm{Frac}\left(\Lambda^{\mathrm{wt}}\right)$ corresponding to the N-primitive part.

Definition 4.2. *The pro-artinian algebra $\mathbf{H}_{Np^\infty}$ decomposes into a direct sum of local rings, and we denote by $\mathcal{R}$ the local factor through which $\mathbf{H}^{\mathrm{ord}}_{Np^\infty} = \mathbf{e}.\mathbf{H}_{Np^\infty}$ acts on $\mathbb{T}_\infty$.*

The deformation ring $\mathcal{R}$ is a complete local noetherian ring, finite and flat over Λ^{wt} with finite residue field. One can associate a universal character $\Psi_{\mathcal{R}} : G_{\mathbb{Q}_p} \to \mathcal{R}^\times$ via the sequence of compositions

$$\Psi_{\mathcal{R}} : G_{\mathbb{Q}_p} \twoheadrightarrow G^{\mathrm{ab}}_{\mathbb{Q}_p} \overset{\text{C.F.T.}}{\twoheadrightarrow} \mathbb{Z}^\times_{p,N} \hookrightarrow \mathbb{Z}_p\big[\big[\mathbb{Z}^\times_{p,N}\big]\big]^\times \to \mathbf{H}^\times_{Np^\infty} \overset{\text{proj}}{\twoheadrightarrow} \mathcal{R}^\times$$

where the arrow 'C.F.T.' arises from the reciprocity map of local class field theory. The following important result describing the structure of $\mathbb{T}_\infty$, might be thought of as a deformation theory analogue of the statement that the Tate module of an abelian variety has $\mathbb{Z}_p$-rank equal to twice its dimension.

Theorem 4.3. [Hi1,Hi2,MW] *Assume that $p \nmid 6N$. Then*

(i) $\mathbb{T}_\infty$ has rank 2 over $\mathcal{R}$, i.e. $\dim_{\mathcal{F}}(\mathbb{T}_\infty \otimes_{\mathcal{R}} \mathcal{F}) = 2$ where $\mathcal{F} = \mathrm{Frac}(\mathcal{R})$;

(ii) As a $G_{\mathbb{Q}}$-representation $\mathbb{T}_\infty$ is unramified outside of Np, and for $l \nmid Np$ the characteristic polynomial of Frob_l in $\mathcal{R}[X]$ is given by $1 - T_l X + l^{-1}\langle l \rangle X^2$;

(iii) If $\rho_\infty : G_{\mathbb{Q}} \to \mathrm{Aut}_{\mathcal{R}}(\mathbb{T}_\infty)$ gives the Galois action on $\mathbb{T}_\infty$, then

$$\rho_\infty\Big|_{G_{\mathbb{Q}_p}} \sim \begin{pmatrix} \chi_{\mathrm{cy}}\Psi_{\mathcal{R}}\phi_{\mathcal{R}}^{-1} & * \\ 0 & \phi_{\mathcal{R}} \end{pmatrix} \quad \text{where } \Psi_{\mathcal{R}} \text{ is the universal character,}$$

and $\phi_{\mathcal{R}} : G_{\mathbb{Q}_p} \twoheadrightarrow G_{\mathbb{Q}_p}/I_p \to \mathcal{R}^\times$ is the unramified character which sends Frob_p to the image of U_p in $\mathcal{R}$.

In particular, $\mathbb{T}_\infty$ contains a one-dimensional $G_{\mathbb{Q}_p}$-submodule $\mathrm{F}^+\mathbb{T}_\infty$ such that $\mathbb{T}_\infty/\mathrm{F}^+\mathbb{T}_\infty$ is unramified. Whilst $\mathrm{F}^+\mathbb{T}_\infty$ is clearly free of $\mathcal{R}$-torsion, whether or not the quotient $\mathbb{T}_\infty/\mathrm{F}^+\mathbb{T}_\infty$ is also $\mathcal{R}$-free is a very delicate question [Hi1,MW,Ti,TW]. In the next section, we indicate how $\mathbb{T}_\infty$ can be specialised at certain height one prime ideals, to recover the representations associated to classical eigenforms.

4.2 Λ-adic modular forms

The natural dualities between modular forms and Hecke algebras work best in the Λ-adic setting, when we pass to a suitable finite extension of the weight algebra. Let $\mathcal{K}$ denote a finite extension of $\mathbb{Q}_p$. The Banach space of p-adic cusp forms with coefficients in the ring of integers $\mathcal{O}_{\mathcal{K}}$, is the p-adic completion $\overline{\mathcal{S}}(Np^\infty; \mathcal{O}_{\mathcal{K}})$ of

$$\mathcal{S}(Np^\infty; \mathcal{O}_{\mathcal{K}}) := \varinjlim_{j,r} \left(\bigoplus_{k=2}^{j} \mathcal{S}_k(\Gamma_1(Np^r); \mathcal{O}_{\mathcal{K}}) \right)$$

viewed inside the q-expansions $\mathcal{O}_{\mathcal{K}}[[q]]$. The norm that is used here is the sup-norm $\|f\| = \sup_{n\in\mathbb{N}} |a_n(f)|_p$ where $f = \sum_{n=1}^\infty a_n(f) q^n$.

By continuity, the p-adic Hecke algebra $\mathbf{h}(Np^\infty;\mathcal{O}_\mathcal{K}) := \varprojlim_{j,r} h_j(Np^r;\mathcal{O}_\mathcal{K})$ acts faithfully on $\overline{\mathcal{S}}(Np^\infty;\mathcal{O}_\mathcal{K})$, and the association $(f,T) \mapsto a_1(f|T)$ yields a perfect pairing

$$\overline{\mathcal{S}}(Np^\infty;\mathcal{O}_\mathcal{K}) \times \mathbf{h}(Np^\infty;\mathcal{O}_\mathcal{K}) \longrightarrow \mathcal{O}_\mathcal{K}.$$

Cutting out using the ordinary idempotent $\mathbf{e}$ means we can identify $\overline{\mathcal{S}}^{\rm ord}(Np^\infty;\mathcal{O}_\mathcal{K})$ with the $\mathcal{O}_\mathcal{K}$-dual of $\mathbf{h}^{\rm ord}(Np^\infty;\mathcal{O}_\mathcal{K})$, which is known to be free of finite rank over $\mathcal{O}_\mathcal{K}[\![1+p\mathbb{Z}_p]\!]$ by [Hi2,Wi2]. To avoid ambiguity we write $\Lambda_\mathcal{K}^{\rm wt} = \mathcal{O}_\mathcal{K}[\![1+p\mathbb{Z}_p]\!]$ for the weight algebra, not to be confused with the cyclotomic Iwasawa algebra $\Lambda^{\rm cy}$!

Let $\mathcal{L}'$ be some finite algebraic extension of $\mathcal{L}_\mathcal{K} = \mathrm{Frac}(\Lambda_\mathcal{K}^{\rm wt})$, and write $\mathbf{I}$ for the normal closure of $\Lambda_\mathcal{K}^{\rm wt}$ in $\mathcal{L}'$. By enlarging $\mathcal{K}$ if necessary, we may assume that $\mathbf{I}\cap\overline{\mathbb{Q}}_p = \mathcal{O}_\mathcal{K}$. We further suppose that $\mathbf{I}$ satisfies:

Hypothesis(Diag). *The space of ordinary* $\mathbf{I}$*-adic cusp forms*

$$\mathbb{S}_\mathbf{I}^{\rm ord} \;:=\; \overline{\mathcal{S}}^{\rm ord}(Np^\infty;\mathcal{O}_\mathcal{K})\otimes_{\Lambda_\mathcal{K}^{\rm wt}}\mathbf{I}$$

has a normalised basis consisting of common eigenforms for all Hecke operators.

As the components of $\mathbf{h}^{\rm ord}(Np^\infty;\mathcal{O}_\mathcal{K})\otimes_{\Lambda_\mathcal{K}^{\rm wt}}\mathcal{L}_\mathcal{K}$ are finite algebraic extensions of $\mathcal{L}_\mathcal{K}$ we can always achieve this condition, by ensuring $\mathcal{L}'$ contains the union of all the isomorphic images of these components.

Question. *How does one pass from* $\mathbf{I}$*-adic to classical modular forms?*

Let $\epsilon: 1+p\mathbb{Z}_p \to \overline{\mathbb{Q}}_p^\times$ be a character of finite order, so it yields a corresponding homomorphism $\Lambda_\mathcal{K}^{\rm wt} \stackrel{\epsilon}{\to} \overline{\mathbb{Q}}_p$ of algebras. Pick an integer $k\ge 2$ which is the weight. We call an element $P = P_{k,\epsilon}$ of Spec $\mathbf{I}(\overline{\mathbb{Q}}_p)$ an *algebraic point of type* (k,ϵ), if the restriction of $P_{k,\epsilon}$ to $\Lambda_\mathcal{K}^{\rm wt}$ is induced by the map

$$u_0 \;\mapsto\; u_0^k\epsilon(u_0) \qquad \text{where } u_0 \text{ generates } 1+p\mathbb{Z}_p \text{ topologically.}$$

Henceforth we shall write Spec $\mathbf{I}(\overline{\mathbb{Q}}_p)^{\rm alg}$ for the set of all algebraic specialisations. Alternatively, we can identify these specialisations with height one prime ideals of the integral domain $\mathbf{I}$, and frequently interchange these two viewpoints without further comment.

Once more enlarging $\mathcal{K}$ (if necessary), we make:

Hypothesis(Arith). Spec $\mathbf{I}(\mathcal{O}_\mathcal{K})^{\rm alg}$ *is Zariski dense in* Spec $\mathbf{I}(\overline{\mathbb{Q}}_p)$.

Theorem 4.4. [Hi2] *For any point* $P\in$ Spec $\mathbf{I}(\mathcal{O}_\mathcal{K})^{\rm alg}$ *algebraic of type* (k,ϵ), *there is a canonical isomorphism*

$$\mathbf{h}^{\rm ord}(Np^\infty;\mathcal{O}_\mathcal{K})\otimes_{\Lambda_\mathcal{K}^{\rm wt}}\mathbf{I}/P \;\cong\; \mathbf{e}.h_k(\Phi_r,\epsilon;\mathcal{O}_\mathcal{K})$$

where $\Phi_r = \Gamma_0(Np^r)\cap\Gamma_1(Np)$, *with* $r\ge 1$ *chosen so that* $\mathrm{Ker}(\epsilon) = (1+p\mathbb{Z}_p)^{p^{r-1}}$.

Results of this sort are called 'control theorems' as they relate specialisations of objects over deformation rings, with the objects these deformations interpolate. Throughout the remainder of this book we will encounter numerous theorems of this type, be they for eigenforms, modular symbols, Euler systems or Selmer groups. We begin by restating the above result in terms of cusp forms.

Suppose we are given an $\mathbf{I}$-adic cusp form $\mathbf{f} \in \mathbb{S}_{\mathbf{I}}^{\mathrm{ord}}$, and assume that (Diag) holds. In particular, its q-expansion $\sum_{n\geq 1} a_n(\mathbf{f})q^n$ can be viewed as lying inside of $\mathbf{I}[[q]]$. Let $P \in \mathrm{Spec}\,\mathbf{I}(\mathcal{O}_{\mathcal{K}})^{\mathrm{alg}}$ be algebraic of type (k, ϵ). If we denote by

$$\lambda : \mathbf{h}^{\mathrm{ord}}(Np^{\infty}; \mathcal{O}_{\mathcal{K}}) \otimes_{\Lambda_{\mathcal{K}}^{\mathrm{wt}}} \mathbf{I} \longrightarrow \mathbf{I}$$

the corresponding $\mathbf{I}$-algebra homomorphism satisfying $\lambda(T_n) = a_n(\mathbf{f})$ for all $n \in \mathbb{N}$, then the composition

$$\lambda_P := P \circ \lambda : \mathbf{h}^{\mathrm{ord}}(Np^{\infty}; \mathcal{O}_{\mathcal{K}}) \otimes_{\Lambda_{\mathcal{K}}^{\mathrm{wt}}} \mathbf{I} \longrightarrow \mathcal{O}_{\mathcal{K}}$$

factors through $h_k(\Phi_r, \epsilon; \mathcal{O}_{\mathcal{K}})$ by Hida's control theorem.

Remark: The dual $h_k(\Phi_r, \epsilon; \mathcal{O}_{\mathcal{K}})^*$ is identified with $\mathcal{S}_k(\Phi_r, \epsilon; \mathcal{O}_{\mathcal{K}})$, and it follows that there must exist a cusp form $\mathbf{f}_P \in \mathcal{S}_k(\Phi_r, \epsilon; \mathcal{O}_{\mathcal{K}})$ whose q-coefficients satisfy $a_n(\mathbf{f}_P) = \lambda_P(T_n)$ at every positive integer n. The $a_n(\mathbf{f}_P)$'s are themselves algebraic numbers, which implies $\mathbf{f}_P$ is indeed a classical modular form.

Recalling that $\mathbf{h}^{\mathrm{ord}}(Np^{\infty}; \mathcal{O}_{\mathcal{K}})$ is an $\mathcal{O}_{\mathcal{K}}[[\mathbb{Z}_{p,N}^{\times}]]$-algebra, we have a sequence

$$(\mathbb{Z}/Np\mathbb{Z})^{\times} \longrightarrow \mathbb{Z}_{p,N}^{\times} \hookrightarrow \mathcal{O}_{\mathcal{K}}[[\mathbb{Z}_{p,N}^{\times}]]^{\times} \longrightarrow \mathbf{h}^{\mathrm{ord}}(Np^{\infty}; \mathcal{O}_{\mathcal{K}})^{\times} \xrightarrow{P\circ\lambda} \mathcal{O}_{\mathcal{K}}^{\times}$$

and $\xi : (\mathbb{Z}/Np\mathbb{Z})^{\times} \longrightarrow \mathcal{O}_{\mathcal{K}}^{\times}$ will denote the multiplicative character obtained by taking the composition of all these maps.

Definition 4.5. *The homomorphism* $\lambda : \mathbf{h}^{\mathrm{ord}}(Np^{\infty}; \mathcal{O}_{\mathcal{K}}) \otimes_{\Lambda_{\mathcal{K}}^{\mathrm{wt}}} \mathbf{I} \to \mathbf{I}$ *is called primitive, if either of the following two equivalent conditions are met:*

(i) There is a point $P \in \mathrm{Spec}\,\mathbf{I}(\mathcal{O}_{\mathcal{K}})^{\mathrm{alg}}$ *of type* (k, ϵ) *such that* $\mathbf{f}_P$ *is a primitive form of conductor* Np^r *with* $\mathrm{Ker}(\epsilon) = (1 + p\mathbb{Z}_p)^{p^{r-1}}$;

(ii) At every point $P \in \mathrm{Spec}\,\mathbf{I}(\mathcal{O}_{\mathcal{K}})^{\mathrm{alg}}$ *of type* (k, ϵ) *such that* $\mathrm{ord}_p\left(C_{\epsilon\xi\omega^{-k}}\right) \neq 0$, *we have that* $\mathbf{f}_P$ *is a primitive form of conductor* Np^r.

What implications does this algebraic control theory have for $\mathbb{T}_{\infty}$, the universal ordinary Galois representation? To find a suitable answer, we are naturally led to consider correspondences:

Big Galois representations	$\xleftrightarrow{\text{specialising}}$	Deligne's Galois representations
$\rho_{\infty} : G_{\mathbb{Q}} \to \mathrm{Aut}(\mathbb{T}_{\infty})$		$\rho_f : G_{\mathbb{Q}} \to \mathrm{Aut}(V_f)$

where the cuspidal eigenforms f are varying analytically in the ordinary family $\mathbf{f}$.

Let's fix an algebraic point $P = P_{k_0,\epsilon} \in \mathrm{Spec}\, \mathbf{I}(\mathcal{O}_{\mathcal{K}})^{\mathrm{alg}}$ of type (k_0, ϵ) with $k_0 \geq 2$. For any $t \in \mathbb{Z}_p^\times$, the power series representing the map $s \mapsto \epsilon(t)\omega^{k_0}(t)t^s$ is clearly convergent on the unit disk. Its associated Mellin transform

$$\mathrm{Mell}_{k_0,\epsilon} : \mathcal{O}_{\mathcal{K}}[\![1 + p\mathbb{Z}_p]\!] \quad \longrightarrow \quad \mathcal{O}_{\mathcal{K}}[[s - k_0]]$$

is an analytic homomorphism, between $\Lambda_{\mathcal{K}}^{\mathrm{wt}}$ and the affinoid $\mathcal{K}$-algebra of the disk.

Since the localisation $\mathbf{I}_{P_{k_0,\epsilon}}$ is a discrete valuation ring which is unramified over the weight algebra [Hi1, Corr 1.4], Hensel's lemma implies the transform extends uniquely to $\widetilde{\mathrm{Mell}}_{k_0,\epsilon} : \mathbf{I}_{P_{k_0,\epsilon}} \to \mathcal{K}[\![s - k_0]\!]$. This gives rise to a $\Lambda_{\mathcal{K}}^{\mathrm{wt}}$-homomorphism

$$\widetilde{\mathrm{Mell}}_{k_0,\epsilon} : \mathcal{L}' \quad \longrightarrow \quad \mathcal{K}(\!(s - k_0)\!)$$

upon passing to the field of fractions. Let $\mathbb{U}_{k_0}$ be the intersection of the domains of convergence of $\widetilde{\mathrm{Mell}}_{k_0,\epsilon}(x)$ as x ranges over all elements of $\mathbf{I}$.

Remark: Because $\mathbf{I}$ is of finite type over the weight algebra $\Lambda_{\mathcal{K}}^{\mathrm{wt}}$, it follows that $\mathbb{U}_{k_0}$ must be an open p-adic disk centred on k_0. We conclude that the Mellin transform maps the ring $\mathbf{I}$ to the space of rigid analytic functions, convergent on the disk $\mathbb{U}_{k_0}$. Somewhat abusively, the notation $\mathcal{K}_{\mathbb{U}_{k_0}}\langle\!\langle s \rangle\!\rangle$ will denote the latter rigid space.

For a primitive $\lambda : \mathbf{h}^{\mathrm{ord}}(Np^\infty; \mathcal{O}_{\mathcal{K}}) \otimes_{\Lambda_{\mathcal{K}}^{\mathrm{wt}}} \mathbf{I} \to \mathbf{I}$ corresponding to the family $\mathbf{f} \in \mathbb{S}_{\mathbf{I}}^{\mathrm{ord}}$, let us define the rank two $\mathbf{I}$-module

$$\mathbb{T}_\infty[\lambda] \quad := \quad \Big\{ \underline{w} \in \mathbb{T}_\infty \otimes_{\Lambda^{\mathrm{wt}}} \mathbf{I} \ \text{ such that } T_h^{\mathcal{D}} \underline{w} = \lambda(T_h)\underline{w} \text{ for all } h \geq 1 \Big\}.$$

Here $T_h^{\mathcal{D}}$ denotes the dual Hecke operator to T_h, e.g. on $\mathrm{Ta}_p(J_r)$ we have

$$T_l^{\mathcal{D}} \quad = \quad T_l \circ \begin{pmatrix} l & 0 \\ 0 & l^{-1} \end{pmatrix}^* \qquad \text{for all primes } l \nmid Np^r.$$

Clearly $\mathbb{T}_\infty[\lambda]$ is Galois stable, as both the Hecke algebra and $G_{\mathbb{Q}}$-actions commute. Furthermore, evaluating the transform $\widetilde{\mathrm{Mell}}_{k_0,\epsilon}$ at an integral weight $k \geq 2$ in $\mathbb{U}_{k_0}$ induces homomorphisms from $\mathbf{I}$ to $\mathcal{K}$, whence

$$G_{\mathbb{Q}} \overset{\rho_\infty[\lambda]}{\longrightarrow} \mathrm{Aut}_{\mathbf{I}}(\mathbb{T}_\infty[\lambda]) \;=\; \mathrm{GL}_2(\mathbf{I}) \overset{\left(\widetilde{\mathrm{Mell}}_{k_0,\epsilon}\right)_{s=k}}{\longrightarrow} \mathrm{GL}_2(\mathcal{K}).$$

In this $G_{\mathbb{Q}}$-representation, each Frobenius element at $l \nmid Np$ has trace $a_l(\mathbf{f}_{P_{k,\epsilon}})$ and determinant $\epsilon\xi\omega^{k_0-k}(l)l^{k-1}$ by Theorem 4.3(ii). It follows that Frob_l has the same characteristic polynomial as given by its action on (the contragredient of) Deligne's representation, associated to the eigenform $\mathbf{f}_{P_{k,\epsilon}} \in \mathcal{S}_k\Big(\Gamma_0\big(Np^r\big), \epsilon\xi\omega^{k_0-k}\Big)$.

By the Chebotarev density theorem these two Galois representations must be equivalent, so in the notation of §2.3

$$\left(\widetilde{\mathrm{Mell}}_{k_0,\epsilon}\right)_{s=k} \circ \rho_\infty[\lambda] \;\sim\; \left(\rho_{\mathbf{f}_{P_{k,\epsilon}}}\right)^{\vee} \qquad \text{for all } k \in \mathbb{U}_{k_0}.$$

In other words, we may consider each specialisation '$\mathbb{T}_\infty[\lambda] \mod P_{k,\epsilon}$' as a Galois stable lattice in the dual space $V^*_{\mathbf{f}_{P_{k,\epsilon}}}$.

Remark: The action of $T_h^{\mathcal{D}}$ on $V^*_{\mathbf{f}_{P_{k,\epsilon}}}$ is through the scalar $a_h(\mathbf{f}_{P_{k,\epsilon}}) = P_{k,\epsilon} \circ \lambda(T_h)$. It follows that the Hecke action on $\mathbb{T}_\infty[\lambda]$ is compatible with specialisation at $P_{k,\epsilon}$, which explains our slightly counter-intuitive definition of $\mathbb{T}_\infty[\lambda]$ above.

This is just what was needed – we now have at our disposal, a universal object which the Euler systems should lift to, namely the first Galois cohomology of $\mathbb{T}_\infty$. Moreover, for triples (λ, k_0, ϵ) and positive integers $k \in \mathbb{U}_{k_0}$, there is a plentiful supply of specialisation maps to the cohomology of Deligne's p-adic representations, where zeta-elements habitually reside.

4.3 Multiplicity one for I-adic modular symbols

We have already seen in Theorem 3.10, that Kato's Euler systems can be mapped to the algebraic modular symbol associated to a cusp form. It is natural to ask if a Λ-adic Euler system (whatever that is) can be converted into something analogous. Basically at all arithmetic points $P_{k,\epsilon}$, we are aiming to complete the diagram

$$\begin{array}{ccc} \Lambda\text{-adic zeta-elements} & \dashrightarrow & ??? \\ \big\downarrow \text{mod } P_{k,\epsilon} & & \big\downarrow \text{mod } P_{k,\epsilon} \\ p\text{-adic zeta-elements} & \xrightarrow{\mathcal{E}_{k,\epsilon}} & \text{algebraic modular symbol.} \end{array}$$

In unpublished work, Mazur showed (under suitable hypotheses) that there is a universal **I**-adic modular symbol, governing the behaviour of the L-values

$$\frac{L(\mathbf{f}_{P_{k,\epsilon}}, s)}{\text{period}} \qquad \text{for every } (s,k) \in \mathbb{Z} \times \mathbb{Z} \text{ with } k \in \mathbb{U}_{k_0},\ 1 \le s \le k-1.$$

Such an object would seem to be a good candidate for the rôle of the **???** above. Moreover, these L-values are themselves interpolated by the two-variable functions of Kitagawa and Greenberg-Stevens, which will be introduced in the next section. Most of the proofs may be found in Kitagawa's PhD thesis, and we won't bother reproducing them here.

We begin by mimicking the constructions of $\mathcal{S}(Np^\infty; \mathcal{O}_\mathcal{K})$ and $\overline{\mathcal{S}}(Np^\infty; \mathcal{O}_\mathcal{K})$.

Definition 4.6. *For any integer $k \ge 2$, the space $\overline{\mathcal{MS}}\left(Np^\infty; \mathrm{Sym}^{k-2}(\mathcal{O}_\mathcal{K})\right)$ of p-adic modular symbols denotes the completion of*

$$\mathcal{MS}\left(Np^\infty; \mathrm{Sym}^{k-2}(\mathcal{O}_\mathcal{K})\right) \;:=\; \varinjlim_r \mathcal{MS}\left(Np^r; \mathrm{Sym}^{k-2}(\mathcal{O}_\mathcal{K})\right)$$

with respect to the p-adic metric.

The abstract Hecke algebra $\mathbf{H}_{Np^\infty}$ acts on both these spaces, and one may cut out their ordinary parts via the idempotent $\mathbf{e}$. After doing this the dependence on the weight k becomes superfluous, because from [Ki, Thm 5.3] there is an isomorphism

$$\nabla_{k-2} : \overline{\mathcal{MS}}^{\mathrm{ord}}\left(Np^\infty; \mathrm{Sym}^{k-2}(\mathcal{O}_\mathcal{K})\right) \xrightarrow{\sim} \overline{\mathcal{MS}}^{\mathrm{ord}}\left(Np^\infty; \mathcal{O}_\mathcal{K}\right) \otimes \mathcal{X}_{k-2}$$

where the right-hand space has had its matrix action twisted by the character

$$\mathcal{X}_{k-2} : \begin{pmatrix} a & b \\ c & d \end{pmatrix} \mapsto d^{k-2}.$$

Definition 4.7.
(i) The module $\mathcal{UM}(\mathcal{O}_{\mathcal{K}})$ is defined to be the $\mathcal{O}_{\mathcal{K}}$-Banach dual of $\overline{\mathcal{MS}}\,(Np^{\infty};\mathcal{O}_{\mathcal{K}})$ viewed with the **dual** *Hecke action to the latter space.*

(ii) If we let $\mathcal{UM}^{\mathrm{ord}}(\mathcal{O}_{\mathcal{K}}) = \mathbf{e}.\mathcal{UM}(\mathcal{O}_{\mathcal{K}})$, then

$$\mathcal{MS}^{\mathrm{ord}}(\mathbf{I}) \;\; := \;\; \mathrm{Hom}_{\Lambda_{\mathcal{K}}^{\mathrm{wt}}}\Big(\mathcal{UM}^{\mathrm{ord}}(\mathcal{O}_{\mathcal{K}}),\, \mathbf{I}\Big)$$

denotes the space of **I***-adic modular symbols.*

It is perhaps more common to introduce the universal p-adic modular symbols $\mathcal{UM}(\mathcal{O}_{\mathcal{K}})$ as an inverse limit of relative homology groups

$$\varprojlim_{r} H_1\Big(X_1(Np^r),\, \{\mathrm{cusps}\},\, \mathcal{O}_{\mathcal{K}}\Big).$$

Our definition is equivalent, however we should really use the superscript $^{\mathcal{D}}$ to indicate that the Hecke action on $\mathcal{UM}(\mathcal{O}_{\mathcal{K}})$ has been inverted.

Remark: The natural pairing

$$\langle\!\langle\, ,\, \rangle\!\rangle_{\mathbf{I}} : \mathcal{MS}^{\mathrm{ord}}(\mathbf{I}) \times \mathcal{UM}^{\mathrm{ord}}(\mathcal{O}_{\mathcal{K}}) \otimes_{\Lambda_{\mathcal{K}}^{\mathrm{wt}}} \mathbf{I} \longrightarrow \mathbf{I}$$

is perfect, because the ordinary part $\mathcal{UM}^{\mathrm{ord}}(\mathcal{O}_{\mathcal{K}})$ is free of finite rank over $\Lambda_{\mathcal{K}}^{\mathrm{wt}}$ [Ki, Prop 5.7]. We should also bear in mind Proposition 3.2, which stated that the f-isotypic part of $\mathcal{MS}\big(Np^r; \mathrm{Sym}^{k-2}(\mathcal{K}_f)\big)$ is always two-dimensional over $\mathcal{K}_f$.

It is then pertinent to ask:

Question. *What rank has the* **f***-isotypic part of $\mathcal{MS}^{\mathrm{ord}}(\mathbf{I})$ as a module over* **I***?*

Recall that $\lambda : \mathbf{h}^{\mathrm{ord}}(Np^{\infty};\mathcal{O}_{\mathcal{K}}) \otimes_{\Lambda_{\mathcal{K}}^{\mathrm{wt}}} \mathbf{I} \longrightarrow \mathbf{I}$ was the **I**-algebra homomorphism corresponding to $\mathbf{f} \in \mathbb{S}_{\mathbf{I}}^{\mathrm{ord}}$. Let's make more precise the word *isotypic* in this context. One defines $\mathcal{UM}^{\mathrm{ord}}(\mathcal{O}_{\mathcal{K}})^{\pm}[\lambda]$ as the projection of $\mathcal{UM}^{\mathrm{ord}}(\mathcal{O}_{\mathcal{K}})^{\pm} \otimes_{\Lambda_{\mathcal{K}}^{\mathrm{wt}}} \mathbf{I}$ to the first factor in the decomposition

$$\begin{aligned}\mathcal{UM}^{\mathrm{ord}}&(\mathcal{O}_{\mathcal{K}})^{\pm} \otimes_{\Lambda_{\mathcal{K}}^{\mathrm{wt}}} \mathcal{L}' \\ &= \;\; \mathcal{UM}^{\mathrm{ord}}(\mathcal{O}_{\mathcal{K}})^{\pm} \otimes_{\Lambda_{\mathcal{K}}^{\mathrm{wt}}} \mathcal{L}'[\lambda] \;\oplus\; \mathrm{Ker}(\lambda'_{\mathcal{L}'})\mathcal{UM}^{\mathrm{ord}}(\mathcal{O}_{\mathcal{K}})^{\pm} \otimes_{\Lambda_{\mathcal{K}}^{\mathrm{wt}}} \mathcal{L}'\end{aligned}$$

where $\lambda'_{\mathcal{L}'} : \mathbf{H}^{\mathrm{ord}}(Np^{\infty};\mathcal{O}_{\mathcal{K}}) \otimes_{\Lambda_{\mathcal{K}}^{\mathrm{wt}}} \mathcal{L}' \to \mathcal{L}'$ is the homomorphism induced by λ.

Hypothesis(Rk1). *$\mathcal{UM}^{\mathrm{ord}}(\mathcal{O}_{\mathcal{K}})^{\pm}[\lambda]$ are free of rank one over* $\mathbf{I}$.

Hypothesis(UF). *The domain* $\mathbf{I}$ *has unique factorisation.*

The advantage of assuming either one of (Rk1) or (UF) holds is that

$$\mathcal{MS}^{\mathrm{ord}}(\mathbf{I})^{\pm}[\lambda] \ := \ \Big\{ \Xi \in \mathcal{MS}^{\mathrm{ord}}(\mathbf{I})^{\pm} \quad \text{such that } \Xi|T = \lambda'_{\mathcal{L}'}(T)\Xi \\ \text{for every } T \in \mathbf{H}^{\mathrm{ord}}(Np^{\infty};\mathcal{O}_{\mathcal{K}}) \otimes_{\Lambda^{\mathrm{wt}}_{\mathcal{K}}} \mathbf{I} \Big\}$$

has no option but to be free of rank one over $\mathbf{I}$. In other words, there is a generating $\mathbf{I}$-adic modular symbol in each $\pm$-eigenspace of the $\mathbf{f}$-isotypic component.

The proof of this fact in [Ki, Lemma 5.11] uses Hida's control theory to establish that $\mathcal{MS}^{\mathrm{ord}}(\mathcal{L}')^{\pm}[\lambda]$ is one-dimensional. Thus we may regard $\mathcal{MS}^{\mathrm{ord}}(\mathbf{I})^{\pm}[\lambda]$ as a lattice in an $\mathcal{L}'$-line. Since it is the dual of a finitely-generated $\mathbf{I}$-module it is reflexive, and either hypothesis will then imply it is free of the desired rank.

Corollary 4.8. *Under either (Rk1) or (UF),* $\mathcal{MS}^{\mathrm{ord}}(\mathbf{I})^{\pm}[\lambda]$ *is free of rank one.*

Are these reasonable hypotheses to make, i.e. do they occur frequently in nature?

If there is a unique p-stabilised ordinary newform of tame conductor N then $\mathbf{I} = \Lambda^{\mathrm{wt}}$, and condition (UF) will certainly be satisfied as Λ^{wt} is a factorial ring. Alternatively, if λ is primitive and there exists no congruence between $\mathbf{f}$ and the space of Eisenstein series, then condition (Rk1) has been verified in numerous cases. Recall that $\xi : (\mathbb{Z}/Np\mathbb{Z})^{\times} \longrightarrow \mathcal{O}_{\mathcal{K}}^{\times}$ denoted the character of the homomorphism λ. Then condition (Rk1) is known to be true when:

(i) $\xi\Big|_{(\mathbb{Z}/p\mathbb{Z})^{\times}} = \omega^n$ provided $n \neq 1, 2$ [MW,Ti];

(ii) $\xi\Big|_{(\mathbb{Z}/p\mathbb{Z})^{\times}} = \omega$ [Hi1];

(iii) $\xi\Big|_{(\mathbb{Z}/p\mathbb{Z})^{\times}} = \omega^2$ and $N = 1$ [Mz1].

N.B. All three results above require that the local component through which λ factorises is Gorenstein.

Henceforth we shall assume that at least one of (Rk1) or (UF) is true.

Notation: Let $\Xi_\lambda = \Xi_\lambda^+ + \Xi_\lambda^-$ where $\Xi_\lambda^{\pm}$ generate the $\pm$-eigenspace $\mathcal{MS}^{\mathrm{ord}}(\mathbf{I})^{\pm}[\lambda]$; in particular, both Ξ_λ^+ and Ξ_λ^- are well-defined up to units in $\mathbf{I}^{\times}$.

The element Ξ_λ is usually termed the *universal* $\mathbf{I}$*-adic modular symbol*, yet this is a slight misnomer since it clearly depends on the initial choice of primitive λ. Nevertheless, it is the elusive object **???** alluded to at the very start of this section. In what way do these universal elements influence the arithmetic of each $\mathbf{f} \in \mathbb{S}_{\mathbf{I}}^{\mathrm{ord}}$?

Definition 4.9. *For any algebraic point* $P = P_{k,\epsilon} \in \text{Spec } \mathbf{I}(\mathcal{O}_{\mathcal{K}})^{\text{alg}}$ *of type* (k,ϵ)*, the periods* $\text{Per}^{\pm}_{\mathbf{I},\lambda_P} \in \mathcal{O}_{\mathcal{K}}$ *are the p-adic numbers satisfying the equation*

$$\nabla^{-1}_{k-2}\Big((P \circ \Xi^{\pm}_{\lambda}) \otimes \mathcal{X}_{k-2} \Big) \quad = \quad \text{Per}^{\pm}_{\mathbf{I},\lambda_P} \times \eta^{\pm}_{\mathbf{f}_P}.$$

The equality itself is viewed inside $\mathcal{MS}(Np^r; \text{Sym}^{k-2})^{\pm}[\lambda_{\mathbf{f}_P}]$*, with* $\eta_{\mathbf{f}_P}$ *being the algebraic modular symbol associated to the eigenform* $\mathbf{f}_P$ *back in §3.1.*

We shall shortly see how these **I**-adic periods turn up in both the L-functions of Kitagawa and Greenberg-Stevens, measuring the failure of Ξ_{λ} to specialise exactly to a generator of the eigenspace $\mathcal{MS}(Np^r; \text{Sym}^{k-2})[\lambda_{\mathbf{f}_P}]$. The following result from [Ki, 5.12] confirms that under either hypothesis, these error terms are well behaved.

Proposition 4.10. *(i) Under condition (Rk1), the* $\text{Per}^{\pm}_{\mathbf{I},\lambda_P}$*'s are p-adic units;*

(ii) If (UF) holds, the function $P \mapsto \left|\text{Per}^{\pm}_{\mathbf{I},\lambda_P}\right|_p$ *is locally constant of finite range.*

Proof: For a fixed choice of sign $\pm$, let $E^{\pm}_{\lambda}$ denote the ideals of **I** defined by

$$E^{\pm}_{\lambda} \ := \ \langle\!\langle \mathcal{MS}^{\text{ord}}(\mathbf{I})^{\pm}[\lambda],\ \mathcal{UM}^{\text{ord}}(\mathcal{O}_{\mathcal{K}}) \otimes_{\Lambda^{\text{wt}}_{\mathcal{K}}} \mathbf{I} \rangle\!\rangle_{\mathbf{I}}.$$

If $P \in \text{Spec } \mathbf{I}(\mathcal{O}_{\mathcal{K}})^{\text{alg}}$ is algebraic of type (k,ϵ), it follows that

$$P \circ E^{\pm}_{\lambda} \ = \ \Big\langle P \circ \Xi^{\pm}_{\lambda},\ \mathcal{UM}^{\text{ord}}(\mathcal{O}_{\mathcal{K}})^{\pm} \Big\rangle \quad \text{are non-zero ideals of } \mathcal{O}_{\mathcal{K}}$$

where $P \circ \Xi^{\pm}_{\lambda}$ lie in $\overline{\mathcal{MS}}^{\text{ord}}(Np^{\infty}; \mathcal{O}_{\mathcal{K}})^{\pm}[\lambda_P]$, and $\langle\,,\,\rangle$ is the pairing between $\overline{\mathcal{MS}}^{\text{ord}}$ and its $\mathcal{O}_{\mathcal{K}}$-dual.

We now observe that the $\pm$-part of $\eta_{\mathbf{f}_P}$ is an $\mathcal{O}_{\mathcal{K}}$-base for

$$\mathcal{MS}\left(Np^r; \text{Sym}^{k-2}(\mathcal{O}_{\mathcal{K}})\right)^{\pm}[\lambda_{\mathbf{f}_P}] \ = \ \overline{\mathcal{MS}}^{\text{ord}}\left(Np^{\infty}; \text{Sym}^{k-2}(\mathcal{O}_{\mathcal{K}})\right)^{\pm}[\lambda_P]$$

whence the modular symbol $\nabla_{k-2}\big(\eta_{\mathbf{f}_P}\big)^{\pm}$ must generate $\overline{\mathcal{MS}}^{\text{ord}}(Np^{\infty}; \mathcal{O}_{\mathcal{K}})^{\pm}[\lambda_P]$. As a direct consequence,

$$\Big\langle \overline{\mathcal{MS}}^{\text{ord}}(Np^{\infty}; \mathcal{O}_{\mathcal{K}})^{\pm}[\lambda_P],\ \mathcal{UM}^{\text{ord}}(\mathcal{O}_{\mathcal{K}})^{\pm} \Big\rangle \ = \ \mathcal{O}_{\mathcal{K}}.$$

It follows that in our equality $P \circ \Xi^{\pm}_{\lambda} \ = \ \text{Per}^{\pm}_{\mathbf{I},\lambda_P} \times \nabla_{k-2}\big(\eta_{\mathbf{f}_P}\big)^{\pm}$ above, the numbers $\text{Per}^{\pm}_{\mathbf{I},\lambda_P}$ are principal generators of the ideal $P \circ E^{\pm}_{\lambda}$.

If condition (Rk1) is true then $E^{\pm}_{\lambda} = \mathbf{I}$, and therefore the periods $\text{Per}^{\pm}_{\mathbf{I},\lambda_P} \in \mathcal{O}^{\times}_{\mathcal{K}}$. If condition (UF) holds, then $\mathbf{I}/E^{\pm}_{\lambda}$ is a pseudo-null **I**-module and hence is finite; the order of the map $P \mapsto \#\left(\mathcal{O}_{\mathcal{K}}/\text{Per}^{\pm}_{\mathbf{I},\lambda_P}\mathcal{O}_{\mathcal{K}}\right)$ must be bounded by $\#(\mathbf{I}/E^{\pm}_{\lambda})$.

□

4.4 Two-variable p-adic L-functions

The plan of this section is as follows. We begin by introducing the two-variable L-function in terms of p-adic measures. We next explain its relationship with an 'improved p-adic L-function' constructed by Stevens. This connection lies at the heart of the proof of the trivial zero formula, and we briefly sketch the argument. Finally, we reinterpret these concepts in terms of the arithmetic of an abelian field and then make some conjectures in the non-abelian case.

Note the hypotheses and notations are kept the same as the previous section.

Theorem 4.11. [GS,Ki] *For all primitive triples (λ, k_0, ϵ), there exists a unique measure $\mu_{\mathbf{f}}$ on the two-dimensional Lie group $\mathbb{Z}_{p,M}^{\times} \times \mathbb{Z}_p$ satisfying*

$$\int_{\mathbb{Z}_{p,M}^{\times} \times \mathbb{Z}_p^{\times}} \psi^{-1}(x)\, x^j\, \epsilon\left(\frac{y}{x}\right) \cdot \left(\frac{y}{x}\right)^{k-k_0} d\mu_{\mathbf{f}}(x,y) \;=\; \frac{j! M^j p^{nj}\left(1 - \psi^{-1}(p) p^j / a_p\left(\mathbf{f}_{P_{k,\epsilon}}\right)\right)}{a_p\left(\mathbf{f}_{P_{k,\epsilon}}\right)^n}$$
$$\times\ \mathrm{Per}_{\mathbf{I},\lambda_{P_{k,\epsilon}}}^{\psi(-1)(-1)^j} \times \frac{G(\psi^{-1}, \zeta_{p^n M}) L\left(\mathbf{f}_{P_{k,\epsilon}}, \psi, j+1\right)}{(2\pi i)^j\, \Omega_{\mathbf{f}_{P_{k,\epsilon}}}^{\psi(-1)(-1)^j}}$$

at arithmetic points $P_{k,\epsilon} \in \mathrm{Spec}\, \mathbf{I}(\mathcal{O}_{\mathcal{K}})^{\mathrm{alg}}$, where the positive integers k lie in $\mathbb{U}_{k_0}$. Here $j \in \{0, \ldots, k-2\}$, and ψ denotes a character modulo $p^n M$ with $n = \mathrm{ord}_p C_\psi$.

Switching between measures and power series is permissible, so the distribution $\mu_{\mathbf{f}}$ transforms naturally into an element of the two-variable algebra $\mathcal{K}_{\mathbb{U}_{k_0}}\langle\langle w \rangle\rangle [[\mathbb{Z}_{p,M}^{\times}]]$. Note that the affinoid $\mathcal{K}$-algebra over the closed disk $\mathbb{U}_{k_0}$, was defined as

$$\mathcal{K}_{\mathbb{U}_{k_0}}\langle\langle w \rangle\rangle \;:=\; \left\{ h(w) \in \mathcal{K}[[w]] \quad \text{such that } h \text{ converges on the whole of } \mathbb{U}_{k_0} \right\}.$$

In fact quite a lot more is true about these analytic distributions.

In Chapter VI we shall outline a purely algebraic construction of these elements. Under various hypotheses, we'll show how there exist power series inside $\mathbf{I}[[\mathbb{Z}_{p,M}^{\times}]]$ whose images under the transforms $\widetilde{\mathrm{Mell}}_{k_0,\epsilon}$ yield the above measures. One should also point out that they are uniquely determined by these special value formulae. The reason is that characters of the form ψx^j are dense in $\mathrm{Hom}\left(\mathbb{Z}_{p,M}^{\times}, \mathbb{C}_p^{\times}\right)$, and under assumption (Arith) the algebraic points $P_{k,\epsilon}$ densely populate $\mathrm{Spec}\, \mathbf{I}(\overline{\mathbb{Q}}_p)$.

Definition 4.12. [GS,Ki] *For primitive triples (λ, k_0, ϵ), the integral*

$$\mathbf{L}_{p,\,\mathbb{Q}(\mu_M)}^{\mathrm{GS}}(\mathbf{f}, \psi, w, s) \;:=\; \int_{\mathbb{Z}_{p,M}^{\times} \times \mathbb{Z}_p^{\times}} \psi^{-1}(x) < x >^{s-1} \epsilon\left(\frac{y}{x}\right) \frac{< y >^{w-k_0}}{< x >^{w-k_0}}\, d\mu_{\mathbf{f}}(x,y)$$

defines a two-variable p-adic L-function attached to the p-ordinary family.

Clearly this two-variable object also depends on the choice of character ϵ, as well as the choice of base weight $k_0 \geq 2$. We nevertheless omit them from the notation.

The power series $\mathbf{L}^{\mathrm{GS}}_{p,\,\mathbb{Q}(\mu_M)}(\mathbf{f},\psi,w,s)$ encodes quite a large amount of arithmetic. Its first nice property is that it interpolates the p-adic L-functions attached to each specialisation of the family $\mathbf{f} \in \mathbb{S}^{\mathrm{ord}}_{\mathbf{I}}$. More precisely,

$$\mathbf{L}^{\mathrm{GS}}_{p,\,\mathbb{Q}(\mu_M)}(\mathbf{f},\psi,k,s) \quad = \quad \mathrm{Per}^{\pm}_{\mathbf{I},\lambda_{P_{k,\epsilon}}} \times \mathbf{L}_{p,\alpha_p,\Omega^{\pm}_{\mathbf{f}_{P_{k,\epsilon}}}}(\mathbf{f}_{P_{k,\epsilon}},\psi,s)$$

at all integer weights $k \in \mathbb{U}_{k_0}$, $k \geq 2$ where $\pm$ denotes the sign of the character ψ. This formula follows immediately from the behaviour of $\mu_{\mathbf{f}}$ in Theorem 4.11.

Secondly, $\mathbf{L}^{\mathrm{GS}}_{p,\,\mathbb{Q}(\mu_M)}$ satisfies a functional equation linking values at s with $w-s$. To make life easier, one now assumes that the tame level N divides the integer M. If $k \in \mathbb{U}_{k_0}$ is a positive integral weight, then by [GS, Corr 5.17]

$$\begin{aligned}\mathbf{L}^{\mathrm{GS}}_{p,\,\mathbb{Q}(\mu_M)}\Big(\mathbf{f},\psi,k,s\Big) \quad &= \quad -\sqrt{\epsilon\omega^{k_0-2}(-N)}\ \psi^{-1}(-N) < -N >^{k/2-s} \\ &\qquad \times\ \mathbf{L}^{\mathrm{GS}}_{p,\,\mathbb{Q}(\mu_M)}\Big(\mathbf{f},\epsilon\omega^{k_0-2}\psi^{-1},k,k-s\Big)\end{aligned}$$

where the square root is chosen so that $\sqrt{\epsilon\omega^{k_0-2}(-N)}\ < -N >^{k/2-1}$ represents the action of the W_N-operator locally about k_0.

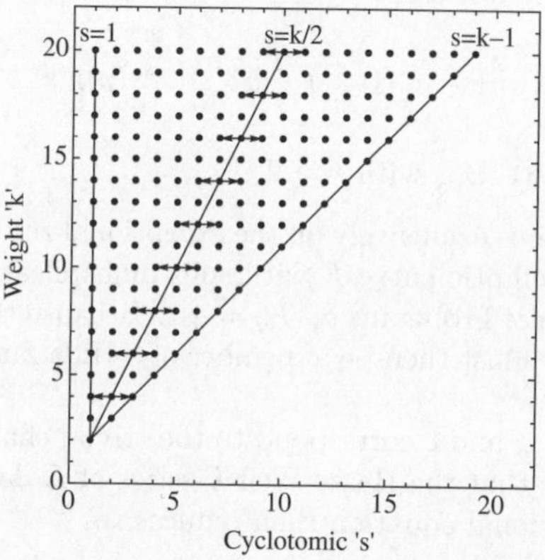

Figure 4.1 The critical triangle interpolating the L-values of $\mathbf{f}$.

There are three critical lines in the (s,k)-plane of principal arithmetic importance: the vertical line $s=1$, the central line $s=k/2$ and the boundary line $s=k-1$. Actually that is a slight exaggeration – there are really only two, since the boundary line can always be reflected (via the functional equation) into the vertical line $s=1$. In the arithmetic applications that are considered later on in Chapters VII, VIII, IX, X, our main objectives are to study behaviour of Selmer groups along the line $s=1$, and of course, along the line of symmetry in the functional equation.

Sketch of the exceptional zero formula.
In §2.1 we introduced the p-adic L-function attached to a modular elliptic curve E. When the curve had split multiplicative reduction over $\mathbb{Q}_p$, then $\mathbf{L}_p(E,s)$ possessed a trivial zero at $s=1$, irrespective of whether the complex L-series vanished too. Mazur, Tate and Teitelbaum [MTT] discovered a formula for $\mathbf{L}_p'(E,1)$ numerically, but did not manage to prove it back in 1986.

The eventual proof came six years later, and was due to Greenberg and Stevens. Since their work was published there have been two other independent proofs, each utilising markedly different techniques. However we give an outline of the original, as it introduces the following fundamental object into the mix.

Definition 4.13. [GS] *For a fixed integer $s_0 < k_0$, the improved p-adic L-series is*

$$\mathbf{L}^{\mathrm{imp}}_{p,\,\mathbb{Q}(\mu_M),s_0}(\mathbf{f},\psi,w) \quad := \quad \int_{\mathbb{Z}^\times_{p,M}\times\mathbb{Z}_p} \psi^{-1}(x) < x >^{s_0-1} \beta_{w,k_0,\epsilon}\left(\frac{y}{x}\right) d\mu_{\mathbf{f}}(x,y)$$

$$\textit{where the function } \beta_{w,k_0,\epsilon}(z) \;=\; \begin{cases} \epsilon(z) < z >^{w-k_0} & \textit{if } z\in\mathbb{Z}_p^\times \\ < z >^{w-k_0} & \textit{if } \epsilon \textit{ is trivial and } z\notin\mathbb{Z}_p^\times \\ 0 & \textit{otherwise.}\end{cases}$$

It can be shown that the two-variable L-function admits a factorisation, into the product of rigid analytic functions

$$\mathbf{L}^{\mathrm{GS}}_{p,\,\mathbb{Q}(\mu_M)}\Big(\mathbf{f},\psi,k,s_0\Big) \;=\; \Big(1-a_p\big(\mathbf{f}_{P_{k,\epsilon}}\big)^{-1}\psi^{-1}\omega^{1-s_0}(p)p^{s_0-1}\Big)\times \mathbf{L}^{\mathrm{imp}}_{p,\,\mathbb{Q}(\mu_M),s_0}(\mathbf{f},\psi,k)$$

at all integral weights $k\in\mathbb{U}_{k_0}$ with $k\geq 2$.

Remark: Let's now focus exclusively on the exceptional zero situation. This means we are considering an elliptic curve E with split multiplicative reduction at $p\geq 5$. In particular, the trace of Frobenius $a_p(E)=+1$. Because this is a p-adic unit, the associated newform f_E must then be a member of a Hida family $\mathbf{f}$ at weight $k_0=2$.

It follows that both ψ and ϵ correspond to the trivial character. Without loss of generality, assume also that the Hasse-Weil L-series of E does not vanish at $s=1$. The two-variable functional equation then reduces to

$$\mathbf{L}^{\mathrm{GS}}_p\Big(\mathbf{f},k,s\Big) \quad = \quad -< -N >^{k/2-s} \;\times\; \mathbf{L}^{\mathrm{GS}}_p\Big(\mathbf{f},k,k-s\Big)$$

as the sign in the complex functional equation equals one whenever $L(E,1)\neq 0$. As an immediate corollary, the function $\mathbf{L}^{\mathrm{GS}}_p$ vanishes identically along the line of symmetry $s=k/2$ in the functional equation.

Therefore, its Taylor series expansion about $(s,k)=(1,2)$ looks like

$$\mathbf{L}^{\mathrm{GS}}_p\Big(\mathbf{f},k,s\Big) \quad\approx\quad \mathfrak{c}\times\left((s-1)-\frac{k-2}{2}\right) + \;\cdots \qquad (*)$$

where the error terms ... are of quadratic or higher order, and $\mathfrak{c}$ is some constant.

Conversely $\mathbf{L}_p^{\mathrm{GS}}$ splits into an Euler factor and improved L-function, namely

$$\mathbf{L}_p^{\mathrm{GS}}\Big(\mathbf{f}, k, 1\Big) \quad = \quad \Big(1 - a_p\big(\mathbf{f}_{P_k}\big)^{-1}\Big) \times \mathbf{L}_{p,1}^{\mathrm{imp}}(\mathbf{f}, k). \qquad (**)$$

Differentiating **(*)** with respect to s along the line $k = 2$, one readily deduces that $\mathbf{L}_p'(E,1) = \frac{\mathrm{d}}{\mathrm{d}s}\mathbf{L}_p^{\mathrm{GS}}\Big(\mathbf{f}, 2, s\Big)$ is equal to the constant $\mathfrak{c}$. On the other hand, differentiating **(**)** with respect to k

$$-\frac{1}{2}\,\mathfrak{c} \quad = \quad 0 + \frac{\mathrm{d}\, a_p\big(\mathbf{f}_{P_k}\big)^{-1}}{\mathrm{d}k}\Bigg|_{k=2} \times \mathbf{L}_{p,1}^{\mathrm{imp}}(\mathbf{f}, 2) \quad = \quad -2a_p'\big(\mathbf{f}_{P_2}\big) \times \frac{L(E,1)}{\Omega_E^+}.$$

Finally, the derivative of the Hecke eigenvalue $a_p\big(\mathbf{f}_{P_k}\big)$ at $k = 2$ is well known to coincide with $-\frac{1}{2} \times$ the value of the local $\mathcal{L}$-invariant of the elliptic curve at p.

Greenberg-Stevens Formula. *If E has split multiplicative reduction at $p \geq 5$,*

$$\mathbf{L}_p'(E,1) \quad = \quad \frac{\log_p(\mathbf{q}_E)}{\mathrm{ord}_p(\mathbf{q}_E)} \times \frac{L(E,1)}{\Omega_E^+} \qquad \textit{where } \mathbf{q}_E \textit{ denotes the Tate period of } E_{/\mathbb{Q}_p}.$$

This result is really quite astounding – the derivative (in a purely p-adic sense) of a rigid analytic function, turns into the ratio of two transcendental numbers in $\mathbb{C}$. It is germane to mention the above equation does not represent $0 = 0$ in disguise! In fact using delicate bounds from analytic number theory, it was shown by various people in [StÉ] that the quantity $\log_p(\mathbf{q}_E)$ is always non-trivial, thus verifying an old conjecture of Manin.

Example 4.14 Consider the elliptic curve whose Weierstrass equation is

$$E \;:\; y^2 + y \;=\; x^3 - x^2 - 10x - 20$$

which is more commonly known as the modular curve $X_0(11)$ in the standard texts. At $p = 11$ it has split multiplicative reduction, so the q-coefficient $a_{11}(E) = +1$. The space $\mathcal{S}_2\big(\Gamma_0(11)\big)$ is one-dimensional over $\mathbb{C}$, generated by a single newform

$$f_{X_0(11)} \quad = \quad q\prod_{n\geq 1}\Big(\big(1-q^n\big)\big(1-q^{11n}\big)\Big)^2 \qquad \text{with rational coefficients.}$$

Setting the tame level $N = 1$, it follows that $\mathbb{T}_\infty$ must be rank two over $\mathbb{Z}_p[\![1+p\mathbb{Z}_p]\!]$. Here $f_{X_0(11)}$ lifts to a family $\mathbf{f} \in \Lambda^{\mathrm{wt}}[\![q]\!]$ – each specialisation is the 11-stabilisation of a newform $\mathbf{f}_k$ at weight $k \equiv 2 \bmod 10$, unless $k = 2$ in which case $\mathbf{f}_2 = f_{X_0(11)}$. Finally, the Greenberg-Stevens formula confirms that

$$\begin{aligned}\mathbf{L}_{11}'(E,1) \quad &= \quad \frac{\log_{11}\big(\mathbf{q}_{X_0(11)}\big)}{\mathrm{ord}_{11}\big(\mathbf{q}_{X_0(11)}\big)} \times \frac{L(X_0(11),1)}{\Omega_{X_0(11)}^+}\\ &= \quad 10\times 11 + 9\times 11^2 + 5\times 11^3 + 11^4 + 8\times 11^5 + 3\times 11^6 + \cdots\end{aligned}$$

and consequently $\mathbf{L}_{11}(E,s)$ has a simple zero at $s = 1$.

Speculations over number fields.
Let $\mathbf{F}$ be a number field so $[\mathbf{F}:\mathbb{Q}]<\infty$, and assume $p\geq 5$ does not ramify in $\mathcal{O}_{\mathbf{F}}$. The discriminant of $\mathbf{F}$ is written as $\mathfrak{d}_{\mathbf{F}}$, and will thus be an integer coprime to p. For an elliptic modular form f, its archimedean period over $\mathbf{F}$ is taken to be

$$\Omega_f^{\pm}(\mathbf{F}) \;=\; \left(\Omega_f^{\pm}\right)^{r_1} \times \left(\Omega_f^{+}\,\Omega_f^{-}\right)^{r_2}$$

where r_1 is the number of real embeddings $\mathbf{F}\hookrightarrow\mathbb{R}$, and r_2 denotes the number of conjugate pairs of complex embeddings.

We are interested in extending these notions inside a p-ordinary deformation. Let $\lambda:\mathbf{h}^{\mathrm{ord}}(Np^{\infty};\mathcal{O}_{\mathcal{K}})\otimes_{\Lambda_{\mathcal{K}}^{\mathrm{wt}}}\mathbf{I}\longrightarrow\mathbf{I}$ be the $\mathbf{I}$-algebra homomorphism corresponding to a family $\mathbf{f}\in\mathbb{S}_{\mathbf{I}}^{\mathrm{ord}}$. Assume as well that $P_{k,\epsilon}$ is an algebraic point of type (k,ϵ). It seems logical and natural to define

$$\mathrm{Per}^{\pm}_{\mathbf{I},\lambda_{P_{k,\epsilon}}}(\mathbf{F}) \;:=\; \left(\mathrm{Per}^{\pm}_{\mathbf{I},\lambda_{P_{k,\epsilon}}}\right)^{r_1} \times \left(\mathrm{Per}^{+}_{\mathbf{I},\lambda_{P_{k,\epsilon}}}\,\mathrm{Per}^{-}_{\mathbf{I},\lambda_{P_{k,\epsilon}}}\right)^{r_2}$$

bearing in mind the definition of the archimedean version above.

Remark: For simplicity, we further suppose that $\mathbf{F}$ is an abelian extension of $\mathbb{Q}$ i.e. that the Galois group $\mathrm{Gal}(\mathbf{F}/\mathbb{Q})$ is commutative. Its characters

$$X_{\mathbf{F}} \;=\; \mathrm{Hom}\Big(\mathrm{Gal}(\mathbf{F}/\mathbb{Q}),\,\mathbb{G}_{\mathrm{mult}}\Big)$$

are non-canonically isomorphic to the original group, and there are $[\mathbf{F}:\mathbb{Q}]$ of them.

Definition 4.15. *We define the improved p-adic L-function of $\mathbf{f}$ over $\mathbf{F}$*

$$\mathbf{L}_p^{\mathrm{imp}}(\mathbf{f}/\mathbf{F},w,s_0) \;:=\; \prod_{\psi\in X_{\mathbf{F}}} \mathbf{L}^{\mathrm{imp}}_{p,\,\mathbb{Q}(\mu_{C_\psi}),s_0}(\mathbf{f},\psi,w)$$

at the integer point $s_0\leq k_0-1$, which is an analytic function of $w\in\mathbb{U}_{k_0}$.

What arithmetical data is this improved L-function interpolating?

Lemma 4.16. *Pick any integral weight $k\in\mathbb{U}_{k_0}$ greater than 1. Then*

$$\mathbf{L}_p^{\mathrm{imp}}(\mathbf{f}/\mathbf{F},k,s_0) \;=\; \frac{\Gamma(s_0)^{[\mathbf{F}:\mathbb{Q}]}\,\sqrt{|\mathfrak{d}_{\mathbf{F}}|}}{i^{r_2(1-2s_0)}\,\mathfrak{d}_{\mathbf{F}}^{1-s_0}} \times \mathrm{Per}^{\pm}_{\mathbf{I},\lambda_{P_{k,\epsilon}}}(\mathbf{F}) \times \frac{L\Big(\mathbf{f}_{P_{k,\epsilon}}/\mathbf{F},\,s_0\Big)}{(2\pi i)^{[\mathbf{F}:\mathbb{Q}](s_0-1)}\,\Omega^{\pm}_{\mathbf{f}_{P_{k,\epsilon}}}(\mathbf{F})}$$

where the sign $\pm$ has the parity of $(-1)^{s_0-1}$.

In the abelian case we are considering, this lemma does no more than encode the properties of each $\mathbf{L}^{\mathrm{imp}}_{p,\,\mathbb{Q}(\mu_{C_\psi}),s_0}(\mathbf{f},\psi,w)$ at an arithmetic integral weight $w=k$. However it is worthwhile to pursue this point of view, because it generalises well to non-abelian extensions of $\mathbb{Q}$.

Proof: Each conductor C_ψ is prime to p. Taking the product of the ψ-twisted functions $\mathbf{L}^{\mathrm{imp}}_{p,\,\mathbb{Q}(\mu_{C_\psi}),s_0}(\mathbf{f},\psi,w)$'s over $X_{\mathbf{F}}$, the left-hand side becomes equal to

$$\prod_{\psi\in X_{\mathbf{F}}} (s_0-1)!\, C_\psi^{s_0-1} \times \mathrm{Per}^{\psi(-1)(-1)^{s_0-1}}_{\mathbf{I},\lambda_{P_{k,\epsilon}}} \times \frac{G(\psi^{-1},\zeta_{C_\psi})L\left(\mathbf{f}_{P_{k,\epsilon}},\psi,s_0\right)}{(2\pi i)^{s_0-1}\,\Omega^{\psi(-1)(-1)^{s_0-1}}_{\mathbf{f}_{P_{k,\epsilon}}}}$$

at an arithmetic weight $w=k$.

By the conductor-discriminant formula, one knows that $\prod_{\psi\in X_{\mathbf{F}}} C_\psi = (-1)^{r_2}\mathfrak{d}_{\mathbf{F}}$. Moreover, it is a standard fact from algebraic number theory

$$\prod_{\psi\in X_{\mathbf{F}}} G(\psi^{-1},\zeta_{C_\psi}) \quad = \quad \begin{cases} \sqrt{|\mathfrak{d}_{\mathbf{F}}|} & \text{if } \mathbf{F} \text{ is totally real,} \\ i^{[\mathbf{F}:\mathbb{Q}]/2}\sqrt{|\mathfrak{d}_{\mathbf{F}}|} & \text{otherwise.} \end{cases}$$

Lastly, if $\mathbf{F}$ is totally real then $r_2=0$, and the period with parity $(-1)^{s_0-1}$ turns up all the time in the product. Conversely, if $\mathbf{F}$ is complex then $r_1=0$, and both the $+$ and $-$ periods will occur with equal multiplicity in the product over ψ.

The lemma follows easily. □

Proceeding to the harder case, we now assume the extension $\mathbf{F}/\mathbb{Q}$ is *non-abelian.* In particular, the group $\mathrm{Gal}(\mathbf{F}/\mathbb{Q})$ is no longer commutative. Whilst the right-hand side of the formula in Definition 4.15 is certainly valid, it no longer gives the p-adic L-function over $\mathbf{F}$; rather it is the corresponding L-function over $\mathbf{F}\cap\mathbb{Q}^{\mathrm{ab}}$.

This leaves us in something of a quandary.

Conjecture 4.17. *Let* $\mathbf{F}$ *denote a non-abelian extension with finite degree over* $\mathbb{Q}$*. There should exist an improved p-adic L-function* $\mathbf{L}^{\mathrm{imp}}_p(\mathbf{f}/\mathbf{F},w,s_0)$ *analytic in* w*, which interpolates the critical values*

$$\frac{\Gamma(s_0)^{[\mathbf{F}:\mathbb{Q}]}\ \sqrt{|\mathfrak{d}_{\mathbf{F}}|}}{i^{r_2(1-2s_0)}\ \mathfrak{d}_{\mathbf{F}}^{1-s_0}} \times \mathrm{Per}^{\pm}_{\mathbf{I},\lambda_{P_{k,\epsilon}}}(\mathbf{F}) \times \frac{L\left(\mathbf{f}_{P_{k,\epsilon}}/\mathbf{F},\ s_0\right)}{(2\pi i)^{[\mathbf{F}:\mathbb{Q}](s_0-1)}\,\Omega^{\pm}_{\mathbf{f}_{P_{k,\epsilon}}}(\mathbf{F})}$$

at every positive integral weight $k\in\mathbb{U}_{k_0}$.

There is a liberal dose of optimism underlying this prediction.

Our main piece of evidence is that the conjecture is true in the abelian case. However, there are a couple of situations where one can say something interesting. If the field $\mathbf{F}$ is totally real, the existence should follow from properties of p-adic families of Hilbert modular forms. Likewise if $\mathbf{F}$ is an imaginary quadratic extension of a totally real field, we have the CM-theory to fall back on.

Finally, we mention that the conjecture has been proven by the author for field extensions of the form $\mathbf{F}_n = \mathbb{Q}\left(\mu_{l^n},\sqrt[l^n]{\Delta}\right)$, where $l\neq 2$ is a prime number and Δ is an l-power free integer coprime to Np. The method combines the fact that these L-values are obtained via twisting by two-dimensional Artin representations over $\mathbb{Q}$, together with analyticity results from the automorphic theory.

CHAPTER V

Crystalline Weight Deformations

Our ultimate aim is to develop a two-variable interpolation theory for M-symbols. Because this is a rather demanding objective, it turns out that we need to subdivide the task into two distinct components. As is commonly the case in number theory the first of these is a purely local problem, whilst the second is essentially global (all arguments which are global in nature have been confined to the next chapter). Here we work exclusively over finite extensions of the p-adic numbers.

Let us begin by outlining in more detail, the local issues that need to be resolved. If one reviews the constructions in Chapter III carefully, we discover the cornerstone of our one-variable interpolation map $\mathcal{E}_{k,\epsilon}$ to be the dual Bloch-Kato exponential. Roughly speaking, the dual exponentials send

$$\begin{array}{rcl} \text{étale cohomologies} & \xrightarrow{\exp^*} & \text{cotangent spaces} \\ p\text{-adic zeta-elements} & \mapsto & L\text{-values of modular forms}, \end{array}$$

without being overly pedantic about the objects appearing in the above picture. When transferring to the two-variable setting, there are three notions to deform: étale cohomology, cotangent modules, and lastly the dual exponential map.

The layout of the fifth chapter is as follows. We commence by discussing how to make proper sense of Galois cohomology, admitting coefficients inside a complete Noetherian local ring. The obvious example is the universal deformation ring $\mathcal{R}$. For discrete modules there is no problem, however if the modules are compact then we need to use continuous étale cohomology in the sense described by Jannsen. Indeed the exposition closely mirrors the article [Ja].

Once the reader is happy with continuous cochains, we spend a couple of sections deforming the above diagram along the weight-axis. This is not as easy as it sounds. Unfortunately, the cotangent spaces behave particularly badly as vector bundles over the weight space. This in turn causes significant problems when evaluating the deformed exponential at arithmetic primes in $\mathcal{R}$. The final solution is based heavily upon Iovita and Stevens' definition of an affinoid analogue of the functor $\mathbf{D}_{\mathrm{dR}}(-)$. We also exploit some extra facts that hold for nearly-ordinary representations.

Finally, we shall conclude by explaining how the kernel of each deformed dual exponential, can be identified with a family of local points on a pro-jacobian variety. This generalises certain duality theorems of Bloch and Kato [BK, Sect 3] to the setting of an ordinary deformation – the proofs themselves may be found in [DS].

5.1 Cohomologies over deformation rings

Let $\mathcal{R}$ denote the image of the Hecke algebra inside the endomorphisms of $\mathbb{T}_\infty$. The only property of the deformation ring that we actually exploit, is that $\mathcal{R}$ is a complete local Noetherian ring, with finite residue field $\mathbb{F}_q$ and maximal ideal $\mathfrak{m}_{\mathcal{R}}$. We are interested in studying the cohomologies of modules of finite-type over $\mathcal{R}$.

Suppose that G is a profinite group. If M is a free $\mathcal{R}$-module of rank n with a continuous G-action, then there is a homomorphism

$$\rho_M : G \longrightarrow \mathrm{GL}_n(\mathcal{R}) \qquad \text{depending upon fixing an } \mathcal{R}\text{-basis for } M.$$

By convention, one usually writes $\underline{\mathrm{Rep}}_{\mathcal{R}}(G)$ for the category of continuous $\mathcal{R}$-adic representations of the group G.

In particular if both modules M, N belong to $\underline{\mathrm{Rep}}_{\mathcal{R}}(G)$, so does $\mathrm{Hom}_{\mathcal{R}}(M, N)$. The G-action on $\mathrm{Hom}_{\mathcal{R}}(M, N)$ is prescribed by the formula

$$\vartheta^g(m) \;=\; g\big(\vartheta(g^{-1}m)\big) \qquad \text{where } \vartheta \in \mathrm{Hom}_{\mathcal{R}}(M, N) \text{ and } m \in M.$$

One writes $H^i(G, M) = H^i_{\mathrm{cont}}(G, M)$ for the cohomology groups defined using continuous cochains in [Ja].

Remark: Given a short exact sequence $0 \to A \to B \to C \to 0$ of $\mathcal{R}[G]$-modules, we obtain a long exact sequence in G-cohomology

$$\dots \longrightarrow H^i(G, A) \longrightarrow H^i(G, B) \longrightarrow H^i(G, C) \stackrel{\partial}{\longrightarrow} H^{i+1}(G, A) \longrightarrow \cdots.$$

It is therefore important to establish sufficient conditions under which the above groups are finitely-generated over the ring $\mathcal{R}$.

There are two examples we have in mind. The first occurs if $G = \mathrm{Gal}(\mathbf{F}_\Sigma/\mathbf{F})$ where $\mathbf{F}$ is a number field, Σ is a finite set of places, and $\mathbf{F}_\Sigma$ denotes the maximal algebraic extension of $\mathbf{F}$ unramified outside Σ. The second example is when the group $G = \mathrm{Gal}(\overline{\mathbf{F}}_\nu/\mathbf{F}_\nu)$, where $\mathbf{F}_\nu$ is the completion of $\mathbf{F}$ at some place $\nu \in \Sigma$.

Both of these possibilities satisfy the following criterion.

Definition 5.1. *The profinite group G is said to obey the 'p-finiteness condition' if for every open subgroup H inside G, there are only finitely many continuous homomorphisms from H to $\mathbb{F}_p$.*

In fact this finiteness condition can be stated in a number of equivalent ways, but the above is perhaps the most convenient. Assume from now on that G satisfies the p-finiteness assumption.

Lemma 5.2. *(Tate) If Y is a finitely-generated $\mathcal{R}$-submodule of $H^i(G, M)$, then the quotient $H^i(G, M)/Y$ contains no non-trivial $\mathfrak{m}_{\mathcal{R}}$-divisible submodules.*

Proposition 5.3. *Each of the $H^i(G,M)$'s is a finitely-generated $\mathcal{R}$-module.*

Proof: The argument reduces easily to the p-adic case, but we include it anyway. Assume that $\mathfrak{m}_{\mathcal{R}}$ is generated over $\mathcal{R}$ by the minimal set of elements $\{e_1, \ldots, e_t\}$. The maximal ideal of $\mathcal{R}/(e_t)$ is clearly equal to $(e_1, \ldots, e_{t-1})$, and by induction we can assume the result holds for G-cohomology over the quotient $\mathcal{R}/(e_t)$.

As M has finite rank over $\mathcal{R}$, there is a tautological exact sequence

$$0 \longrightarrow M[e_t] \longrightarrow M \longrightarrow M/M[e_t] \longrightarrow 0$$

where $M[e_t]$ is the kernel of multiplication by e_t. Taking G-cohomology yields the long exact sequence

$$H^i(G, M[e_t]) \longrightarrow H^i(G,M) \longrightarrow H^i(G, M/M[e_t]) \stackrel{\partial}{\longrightarrow} H^{i+1}(G, M[e_t]).$$

Clearly $M[e_t]$ is a finitely-generated $\mathcal{R}/(e_t)$-module. It follows from our inductive assumption, both $H^i\big(G, M[e_t]\big)$ and $H^{i+1}\big(G, M[e_t]\big)$ are of finite-type over $\mathcal{R}/(e_t)$. Thus it is enough to prove that $H^i\big(G, M/M[e_t]\big)$ is finitely-generated.

Put $M_k = M/M[e_t^k]$ so there is an direct system $M_1 \twoheadrightarrow M_2 \twoheadrightarrow M_3 \twoheadrightarrow M_4 \twoheadrightarrow \cdots$ To show $H^i\big(G, M_1\big)$ is finitely-generated over the quotient ring $\mathcal{R}/(e_t)$, by the above argument we must show $H^i\big(G, M_2\big)$ is finitely-generated over $\mathcal{R}/(e_t^2)$, etc. Because M is a Noetherian $\mathcal{R}$-module, the chain $M[e_t]$, $M[e_t^2]$, $M[e_t^3]$, ... eventually stabilises, whence $M_k = M_{k+1} = M_{k+2} = \cdots$ for a large enough index $k \gg 0$. Replacing M by M_k, we may assume that $M[e_t]$ is trivial.

It follows from the G-cohomology of $0 \longrightarrow M \stackrel{\times e_t}{\longrightarrow} M \longrightarrow M/e_t.M \longrightarrow 0$ that there are injections

$$\frac{H^i(G,M)}{e_t.H^i(G,M)} \hookrightarrow H^i\Big(G, M/e_t.M\Big) \qquad \text{for all integers } i \geq 0.$$

However $H^i\Big(G, M/e_t.M\Big)$ is of finite-type over $\mathcal{R}/(e_t)$ by our inductive assumption, whence the quotient module $H^i\big(G,M\big)/e_t.H^i\big(G,M\big)$ is too.

If the set $\{\widetilde{v_1}, \ldots, \widetilde{v_r}\}$ generates this quotient of the ith-cohomology over $\mathcal{R}/(e_t)$, then we can lift to a set of representatives $\{v_1, \ldots, v_r\}$ for which $\widetilde{v_j} = v_j \mod e_t$. Let Y be the $\mathcal{R}$-submodule generated by $\{v_1, \ldots, v_r\}$. Then

$$H^i\big(G,M\big) \;=\; Y \,+\, e_t.\, H^i\big(G,M\big)$$

and consequently $H^i\big(G,M\big)/Y$ must be $\mathfrak{m}_{\mathcal{R}}$-divisible.

Finally applying Lemma 5.2, we deduce that there is an equality $H^i\big(G,M\big) = Y$. The latter object is generated over $\mathcal{R}$ by $\{v_1, \ldots, v_r\}$ which is a finite set, and the result follows immediately.

□

It is sometimes convenient to describe big cohomology groups, as projective limits of standard cohomology. If the module $M \in \underline{\mathrm{Rep}}_{\mathcal{R}}(G)$, let us write $M_n = M/\mathfrak{m}_{\mathcal{R}}^n M$. By [Ja, 2.1] there is an exact sequence

$$0 \longrightarrow \varprojlim{}^1 H^{i-1}(G, M_n) \longrightarrow H^i(G, M) \longrightarrow \varprojlim H^i(G, M_n) \longrightarrow 0.$$

Because G satisfies the p-finiteness assumption, the $H^{i-1}(G, M_n)$'s are finite groups. Consequently the projective system $\left\{H^i(G, M_n)\right\}_n$ will fulfill the Mittag-Leffler condition, where of course M_n is viewed with the discrete topology.

Corollary 5.4. *There are isomorphisms $H^i(G, M) \cong \varprojlim_n H^i(G, M_n)$ for all $i \geq 0$.*

It is worthwhile mentioning a useful tool from commutative algebra, that we shall continually employ throughout the text.

Let $\mathcal{P}$ be a non-zero prime ideal in $\mathrm{Spec}(\mathcal{R})$. The quotient ring $\mathcal{R}/\mathcal{P}$ will also be a complete Noetherian local ring (CNL for short), with residue characteristic p. If $M \in \underline{\mathrm{Rep}}_{\mathcal{R}}(G)$ is an $\mathcal{R}$-module of finite rank, then taking the specialisation at $\mathcal{P}$ supplies us with a functor

$$\underline{\mathrm{Rep}}_{\mathcal{R}}(G) \longrightarrow \underline{\mathrm{Rep}}_{\mathcal{R}/\mathcal{P}}(G), \qquad M \mapsto M/\mathcal{P}M.$$

Moreover, it induces homomorphisms from $H^i(G, M)$ to the group $H^i(G, M/\mathcal{P}M)$.

In this way, we are able to compare the behaviour of objects defined over the deformation ring $\mathcal{R}$, with their p-adic specialisations over discrete valuation rings. At arithmetic points $P \in \mathrm{Spec}(\mathcal{R})$ the latter cohomologies should be related to familiar household objects, such as the Selmer groups and Tate-Shafarevich groups attached to abelian varieties, modular forms, etc ...

Dualities: compact and discrete.

In the sequel, there are basically three sorts of dual that we shall need to consider. The first of these is the standard dual. Let M be an $\mathcal{R}$-module of finite rank d. Then we define

$$M^* \ := \ \mathrm{Hom}_{\mathcal{R}}(M, \mathcal{R})$$

which is also an $\mathcal{R}$-module of rank d. If M is compact, then so too is M^*.

The Pontrjagin dual of M is the group of continuous homomorphisms

$$M^\vee \ := \ \mathrm{Hom}_{\mathrm{cont}}(M, \mathbb{Q}_p/\mathbb{Z}_p)$$

where $\mathbb{Q}_p/\mathbb{Z}_p$ is endowed with the discrete topology. In particular, the module $M^\vee$ is discrete rather than compact. Furthermore, there is a canonical isomorphism between M and its double dual $M^{\vee\vee}$.

Finally, the Kummer dual of M is taken to be χ_{cy}-twisted group of continuous homomorphisms

$$A_M \ := \ \mathrm{Hom}_{\mathrm{cont}}(M, \mu_{p^\infty})$$

where the roots of unity μ_{p^∞} are viewed as a G-module with the discrete topology.

Question. *How are these three notions of duality related, and what pairings do they induce on the G-cohomology of the module M?*

Let us first recall that $\mathcal{R}$ is Gorenstein if it is a CNL with finite injective dimension, or equivalently its Ext-groups trivialise via

$$\mathrm{Ext}^i_{\mathcal{R}}(\mathbb{F}_q,\mathcal{R}) \quad\cong\quad \begin{cases} 0 & \text{if } i < \text{inj.dim}(\mathcal{R}) \\ \mathbb{F}_q & \text{if } i = \text{inj.dim}(\mathcal{R}). \end{cases}$$

Lemma 5.5. *If $\mathcal{R}$ is a Gorenstein ring, then* $A_M \cong M^*(1)\otimes_{\mathbb{Z}_p}\mathbb{Q}_p/\mathbb{Z}_p$.

Proof: By its very definition,

$$A_M \;=\; \mathrm{Hom}_{\mathrm{cont}}\big(M,\mu_{p^\infty}\big) \;=\; \mathrm{Hom}_{\mathbb{Z}_p}\big(M,\mathbb{Z}_p(1)\big)\otimes_{\mathbb{Z}_p}\mathbb{Q}_p/\mathbb{Z}_p.$$

However, there is an isomorphism

$$\mathrm{Hom}_{\mathcal{R}}\big(M,\mathcal{R}\big)\otimes_{\mathcal{R}}\mathrm{Hom}_{\mathbb{Z}_p}\big(\mathcal{R},\mathbb{Z}_p\big) \;\cong\; \mathrm{Hom}_{\mathbb{Z}_p}\big(M,\mathbb{Z}_p\big), \quad x\otimes y\mapsto y\circ x.$$

Therefore the result is true provided the dual $\mathrm{Hom}_{\mathbb{Z}_p}\big(\mathcal{R},\mathbb{Z}_p\big)$ is free of rank one. This is certainly the case whenever $\mathcal{R}$ is Gorenstein. □

As we have already discussed in Chapter IV, the deformation rings that arise from the local components of p-adic Hecke algebras tend to be complete intersections, and the utility of the above result becomes obvious. Nevertheless, it's not always necessary to automatically assume $\mathcal{R}$ possesses the Gorenstein property.

Suppose now that $N\in\underline{\mathrm{Rep}}_{\mathbb{Z}_p}(G)$, whence the product $N\otimes_{\mathbb{Z}_p}\mathcal{R}\in\underline{\mathrm{Rep}}_{\mathcal{R}}(G)$. There is a natural map

$$H^i\big(G,N\big)\otimes\mathcal{R} \;\stackrel{-\otimes 1}{\longrightarrow}\; H^i\big(G,N\otimes\mathcal{R}\big)$$

which is an isomorphism when $\mathcal{R}$ is flat over $\mathbb{Z}_p$, and the group G satisfies the p-finiteness hypothesis.

If $N=\mathbb{Z}_p(1)$ and $G=G_K$ denotes the Galois group of a local field K, then class field theory produces a canonical isomorphism $\mathrm{inv}_K : H^2\big(G_K,\mathcal{R}(1)\big)\stackrel{\sim}{\longrightarrow}\mathcal{R}$. For any $M\in\underline{\mathrm{Rep}}_{\mathcal{R}}(G)$, the duality $M\times M^*(1)\longrightarrow\mathcal{R}(1)$ induces a cup-product pairing on local cohomology

$$H^1\big(G_K,M\big)\;\times\;H^1\big(G_K,M^*(1)\big)\;\longrightarrow\;\mathcal{R}$$

which in general, is **not** perfect (the left-hand and right-hand kernels comprise the $\mathcal{R}$-torsion submodules).

Unfortunately, there is no satisfactory analogue of this pairing when G is the absolute Galois group of a global field. The best available substitute is due to Poitou and Tate, which is something we shall touch on in forthcoming chapters.

Proposition 5.6. *The groups $H^i(G, A_M)$ are cofinitely generated over the ring $\mathcal{R}$ if either $G = \mathrm{Gal}(\mathbf{F}_\Sigma/\mathbf{F})$, or alternatively $G = \mathrm{Gal}(\overline{\mathbf{F}}_\nu/\mathbf{F}_\nu)$ for some place $\nu \in \Sigma$.*

Proof: If $G = \mathrm{Gal}(\overline{\mathbf{F}}_\nu/\mathbf{F}_\nu)$, the result follows immediately from the perfect pairing

$$H^i(G, M) \times H^{2-i}(G, A_M) \longrightarrow \mathbb{Q}_p/\mathbb{Z}_p \quad \text{with } i = 0, 1, 2$$

in tandem with Proposition 5.3.

On the other hand, if $G = \mathrm{Gal}(\mathbf{F}_\Sigma/\mathbf{F})$ then define

$$\mathcal{H}^i_\Sigma(X) \;:=\; \mathrm{Ker}\left(H^i(\mathbf{F}_\Sigma/\mathbf{F}, X) \overset{\oplus \mathrm{res}_\nu}{\longrightarrow} \bigoplus_{\nu\in\Sigma} H^i(\overline{\mathbf{F}}_\nu/\mathbf{F}_\nu, X) \right)$$

where the coefficients $X = M$ or A_M. When $i \geq 3$ these objects are all zero anyway. However if $i = 1$ or 2, there is a perfect pairing of $\mathcal{R}$-modules

$$\mathcal{H}^{3-i}_\Sigma(M) \times \mathcal{H}^i_\Sigma(A_M) \longrightarrow \mathbb{Q}_p/\mathbb{Z}_p$$

between the compact and discrete versions.

By definition $H^i(\mathbf{F}_\Sigma/\mathbf{F}, A_M)$ fits into an exact sequence, sandwiched between the groups $\mathcal{H}^i_\Sigma(A_M)$ and $\bigoplus_{\nu\in\Sigma} H^i(\overline{\mathbf{F}}_\nu/\mathbf{F}_\nu, A_M)$. The former is cofinitely generated by the above pairing, and the latter is cofinitely generated via our previous argument. □

Remarks: (a) Let $\lambda : \mathbf{h}^{\mathrm{ord}}(Np^\infty; \mathcal{O}_\mathcal{K}) \otimes_{\Lambda^{\mathrm{wt}}_\mathcal{K}} \mathbf{I} \to \mathbf{I}$ be a primitive homomorphism and consider the eigenspace $\mathbb{T}_\infty[\lambda]$, which is a rank two $G_\mathbb{Q}$-representation over $\mathbf{I}$. One can ask identical questions about the finite generation of its G-cohomology. Fortunately all the results we have stated over the deformation ring $\mathcal{R}$ work equally well over the algebra $\mathbf{I}$, with minimal adjustments.

Hint: Exploit the fact both $\mathcal{R}$ and $\mathbf{I}$ are of finite-type over the weight algebra.

(b) There is no reason to restrict solely to considering CNL's such as those above. For example, the one-variable Tate algebra $\mathcal{K}\langle\langle s\rangle\rangle$ whilst not actually a local ring, is a Banach algebra complete with respect to the sup-norm. Analogous results affirming that the G-cohomology of objects in $\underline{\mathrm{Rep}}_{\mathcal{K}\langle\langle s\rangle\rangle}(G)$ are finitely-generated, can then be found in the PhD thesis [Sm, Sect 2.3] of P. Smith.

(c) More generally, recall that the analytic transform $\widetilde{\mathrm{Mell}}_{k_0,\epsilon} : \mathbf{I} \longrightarrow \mathcal{K}_{\mathbb{U}_{k_0}}\langle\langle s\rangle\rangle$ takes values in the affinoid $\mathcal{K}$-algebra of power series convergent on the disk $\mathbb{U}_{k_0}$. To deduce similar sorts of finite generation results for objects inside $\underline{\mathrm{Rep}}_{\mathcal{K}_{\mathbb{U}_{k_0}}\langle\langle s\rangle\rangle}(G)$, it is simply a matter of reducing to part (b). This is achieved by performing a transformation of affine space which sends k_0 to 0, and scales the radii suitably.

5.2 p-Ordinary deformations of B_{cris} and $\mathbf{D}_{\mathrm{cris}}$

After this exposition on coefficient rings, let's focus instead on crystalline matters. Most of Chapter III was devoted to a parametrisation of cohomology via symbols. For a p-stabilised ordinary newform $f \in \mathcal{S}_k(\Gamma_0(Np^r), \epsilon)$, we defined interpolations

$$\mathcal{E}_{k,\epsilon} : \mathbb{H}^1_{\mathrm{Eul},p}(V_f^*) \longrightarrow \mathrm{Hom}_{\mathcal{Q}\mathrm{cts}}\Big(\mathrm{div}^0\mathbb{P}^1(\mathbb{Q}),\ \mathrm{Sym}^{k-2}(\mathcal{K}_f)\Big)$$

which depended fundamentally on a choice of base vector $\mathbf{v}_{k,\epsilon} \in \mathbf{D}_{\mathrm{cris},\mathbb{Q}_p}(V_f)^{\varphi=a_p(f)}$. For example, changing $\mathbf{v}_{k,\epsilon}$ by an element of $\mathcal{K}_f$ scales $\mathcal{E}_{k,\epsilon}$ by the same factor.

We now further assume that our eigenforms lie in a Hida family $\mathbf{f} \in \mathbb{S}^{\mathrm{ord}}_{\mathbf{I}}$, which must correspond to some primitive homomorphism $\lambda : \mathbf{h}^{\mathrm{ord}}(Np^\infty; \mathcal{O}_\mathcal{K}) \otimes_{\Lambda^{\mathrm{wt}}_\mathcal{K}} \mathbf{I} \to \mathbf{I}$. In this section, we will address the following problem.

Question. *Is there a universal element '$\mathbf{v}_{\mathbf{I},\lambda}$' which specialises at algebraic points $P_{k,\epsilon}$ of type (k,ϵ), to the element $\mathbf{v}_{k,\epsilon}$ associated to each form $\mathbf{f}_{P_{k,\epsilon}}$?*

To find such an element, we are led to consider the family of $\mathrm{Frac}\,(\mathbf{I}/P)$-vector spaces $\mathbf{D}_{\mathrm{cris},\mathbb{Q}_p}(V_{\mathbf{f}_P})^{\varphi=a_p(f)}$, and to study how these vary with $P \in \mathrm{Spec}\,\mathbf{I}$.

Remember the weight algebra is homomorphic to the affinoid algebra of the disk, via the transform $\Lambda^{\mathrm{wt}} = \mathbb{Z}_p[\![1+p\mathbb{Z}_p]\!] \longrightarrow \mathbb{Z}_p\langle\langle s\rangle\rangle$ associating to any $u \in 1+p\mathbb{Z}_p$ the power series representing $s \mapsto u^s$.

Definition 5.7.
(i) We define $\mathbb{A}^{\mathrm{wt}}_{\mathrm{cris}}$ by the completed tensor product

$$\mathbb{A}^{\mathrm{wt}}_{\mathrm{cris}} \ := \ \varprojlim_m \Big(A_{\mathrm{cris}}/p^m A_{\mathrm{cris}} \otimes_{\mathbb{Z}/p^m\mathbb{Z}} \mathbb{Z}/p^m\mathbb{Z}[s]\Big),$$

analogously $\mathbb{B}^{+,\mathrm{wt}}_{\mathrm{cris}} := \mathbb{A}^{\mathrm{wt}}_{\mathrm{cris}}[1/p]$ *and* $\mathbb{B}^{\mathrm{wt}}_{\mathrm{cris}} := \mathbb{B}^{+,\mathrm{wt}}_{\mathrm{cris}}[t^{-1}]$.
(ii) If Ω is a finite étale extension of Λ^{wt}, then set

$$\mathbb{A}^{\Omega}_{\mathrm{cris}} := \mathbb{A}^{\mathrm{wt}}_{\mathrm{cris}} \otimes_{\mathbb{Z}_p[\![1+p\mathbb{Z}_p]\!]} \Omega, \quad \mathbb{B}^{+,\Omega}_{\mathrm{cris}} := \mathbb{A}^{\Omega}_{\mathrm{cris}}[1/p] \quad \text{and} \quad \mathbb{B}^{\Omega}_{\mathrm{cris}} := \mathbb{B}^{+,\Omega}_{\mathrm{cris}}[t^{-1}]$$

which all have the structure of topological $\big(\varphi, G_{\mathbb{Q}_p}, A_{\mathrm{cris}}, \Omega\big)$-modules.

In fact, we may interpret $\mathbb{B}^{\mathrm{wt}}_{\mathrm{cris}}$ as the space of power series of the form $\sum_{n=0}^\infty a_n s^n$ with $a_n \in B_{\mathrm{cris}}$ tending to 0 as $n \to \infty$. The Frobenius action is then described by

$$\varphi\left(\sum_{n=0}^\infty a_n s^n\right) \ = \ \sum_{n=0}^\infty \varphi(a_n) s^n \qquad \text{for all } \sum_{n=0}^\infty a_n s^n \in \mathbb{B}^{\mathrm{wt}}_{\mathrm{cris}}.$$

Proposition 5.8. *Under Hypothesis(Diag) of §4.2, if we define*

$$\mathbb{D}^{\mathbf{I}}_{\mathrm{cris}}(\mathbb{T}^*_\infty) \;:=\; H^0_{\mathrm{cont}}\Big(G_{\mathbb{Q}_p},\, \mathbb{T}^*_\infty \otimes_{\Lambda^{\mathrm{wt}}} \mathbb{B}^{\mathbf{I}}_{\mathrm{cris}}\Big)^{\varphi=U_p} \quad \text{with } \mathbb{T}^*_\infty = \mathrm{Hom}_{\mathcal{R}}(\mathbb{T}_\infty, \mathcal{R})$$

then $\mathbb{D}^{\mathbf{I}}_{\mathrm{cris}}(\mathbb{T}^*_\infty)[\lambda]$ *is a rank one* $\mathbf{I}[1/p]$*-module, on which* φ *acts through* $\lambda(U_p) \in \mathbf{I}^\times$*. Furthermore, if* $\mathcal{R}$ *is Gorenstein then* $\mathbb{D}^{\mathbf{I}}_{\mathrm{cris}}(\mathbb{T}^*_\infty)[\lambda]$ *will be free of* $\mathbf{I}[1/p]$*-torsion.*

The answer to the question posed before, must therefore be in the affirmative. Given a fixed base weight $k_0 \geq 2$ and character ϵ, evaluating the Mellin transform at an integer $k \in \mathbb{U}_{k_0}$ yields a specialisation

$$\left(\widetilde{\mathrm{Mell}_{k_0,\epsilon}}\right)_{s=k} : \mathbb{D}^{\mathbf{I}}_{\mathrm{cris}}(\mathbb{T}^*_\infty)[\lambda] \longrightarrow \mathbf{D}_{\mathrm{cris},\mathbb{Q}_p}\big(V_{\mathbf{f}_{P_{k,\epsilon}}}\big)^{\varphi=U_p}.$$

Because $\mathbf{v}_{\mathbf{I},\lambda}$ generates an $\mathbf{I}[1/p]$-line inside the eigenspace $\mathbb{D}^{\mathbf{I}}_{\mathrm{cris}}(\mathbb{T}^*_\infty)[\lambda]$, it follows directly that $\mathbf{v}_{\mathbf{I},\lambda}$ mod P must be equal to $\mathbf{v}_{k,\epsilon}$, up to a scalar.

Notice that the dimension of $\mathbf{D}_{\mathrm{cris},\mathbb{Q}_p}(V_{\mathbf{f}_P})$ is either 2 or 1 depending on whether $V_{\mathbf{f}_P}$ is crystalline or not, whilst the rank of $\mathbb{D}^{\mathbf{I}}_{\mathrm{cris}}(\mathbb{T}^*_\infty)[\lambda]$ is always 1 no matter what. Equivalently, the crystalline part of the representation associated to the Λ^{wt}-adic family is *generically* one-dimensional. This appears to be a common phenomenon i.e. the rank of a family of objects on the whole tends to be smaller than the dimensions of the individual objects themselves.

Proof: Fix a primitive triple (λ, k_0, ϵ). We shall commence by treating the case where the local component through which $\mathbf{H}^{\mathrm{ord}}_{Np^\infty}$ operates on $\mathbb{T}_\infty[\lambda]$ is Gorenstein. In §4.2 we saw that as a $G_{\mathbb{Q}_p}$-representation, $\left(\widetilde{\mathrm{Mell}_{k_0,\epsilon}}\right)_{s=k} \circ \rho_\infty[\lambda]$ has the form

$$\begin{pmatrix} \widetilde{\chi} <\chi_{\mathrm{cy}}>^k \phi_k^{-1} & * \\ 0 & \phi_k \end{pmatrix} \quad \text{at every integer } k \in \mathbb{U}_{k_0},\ k \geq 2$$

where $\widetilde{\chi} = \chi_{\mathrm{cy}}^{-1}\xi\epsilon$, and ϕ_k is the unramified character sending Frob_p to $a_p(\mathbf{f}_{P_{k,\epsilon}})$.

Tensoring the contragredient representation $\rho^\vee_\infty[\lambda]$ by the topological ring $\mathbb{B}^{\mathrm{wt}}_{\mathrm{cris}}$ and taking its $G_{\mathbb{Q}_p}$-invariants, we obtain the exact sequence

$$0 \to \Big(\mathbb{B}^{\mathrm{wt}}_{\mathrm{cris}} \otimes_{\Lambda^{\mathrm{wt}}} \mathbf{I}(\Phi)\Big)^{G_{\mathbb{Q}_p}} \xrightarrow{\mathcal{J}_-} \Big(\mathbb{B}^{\mathrm{wt}}_{\mathrm{cris}} \otimes_{\Lambda^{\mathrm{wt}}} \mathbb{T}^*_\infty[\lambda]\Big)^{G_{\mathbb{Q}_p}} \xrightarrow{\mathcal{J}_+} \Big(\mathbb{B}^{\mathrm{wt}}_{\mathrm{cris}} \otimes_{\Lambda^{\mathrm{wt}}} \mathbf{I}(\Theta\Phi^{-1})\Big)^{G_{\mathbb{Q}_p}}.$$

Here $\Phi : G_{\mathbb{Q}_p}/I_p \to \mathbf{I}^\times$ denotes the unramified character sending Frob_p^{-1} to $\lambda(U_p)$. Also $\Theta := \widetilde{\chi}^{-1}\Psi^{-1}$ where the universal cyclotomic character Ψ has the expansion

$$\Psi \;=\; \exp\big(s \log(<\chi_{\mathrm{cy}}>)\big) \;=\; \sum_{n=0}^{\infty} \frac{\log^n(<\chi_{\mathrm{cy}}>)}{\Gamma(n+1)}\, s^n \quad \text{at all } s \in \mathbb{Z}_p.$$

Lastly we have written $\mathcal{J}_-$ above to denote the dual of quotienting modulo $\mathrm{F}^+\mathbb{T}_\infty$, and similarly $\mathcal{J}_+$ for the dual map to the natural inclusion $\mathrm{F}^+\mathbb{T}_\infty \hookrightarrow \mathbb{T}_\infty$.

Remark: We once more observe how the Mellin transform maps the ring $\mathbf{I}$ to the space of power series with coefficients in $\mathcal{K}$, convergent on the closed disk $\mathbb{U}_{k_0}$. Performing the affine transformation $s \mapsto k_0 + s \times \text{radius}(\mathbb{U}_{k_0})$, without loss of generality we'll assume that $\widetilde{\text{Mell}}_{k_0,\epsilon}$ takes values inside the affinoid algebra $\mathcal{K}\langle\langle s\rangle\rangle$ of the unit disk over the valuation field $\mathcal{K}$.

We make the following three assertions:

(a) $\left(\mathbb{B}^{\text{wt}}_{\text{cris}} \otimes_{\Lambda^{\text{wt}}} \mathbf{I}(\Phi)\right)^{G_{\mathbb{Q}_p}}$ is isomorphic to $\left(\widehat{\mathbb{Q}}_p^{\text{nr}}\langle\langle s\rangle\rangle \otimes_{\mathbb{Z}_p\langle\langle s\rangle\rangle} \mathbf{I}(\Phi)\right)^{\text{Frob}_p=1}$;

(b) $\left(\widehat{\mathbb{Q}}_p^{\text{nr}}\langle\langle s\rangle\rangle \otimes_{\mathbb{Z}_p\langle\langle s\rangle\rangle} \mathbf{I}(\Phi)\right)^{\text{Frob}_p=1}$ is a free $\mathbf{I}[1/p]$-module of rank one;

(c) $\left(\mathbb{B}^{\text{wt}}_{\text{cris}} \otimes_{\Lambda^{\text{wt}}} \mathbf{I}(\Theta\Phi^{-1})\right)^{G_{\mathbb{Q}_p}}$ is zero.

Taking (a),(b),(c) on trust for the moment, it follows from our exact sequence above that $\mathbb{D}^{\mathbf{I}}_{\text{cris}}(\mathbb{T}^*_\infty)[\lambda] = \left(\left(\mathbb{B}^{\text{wt}}_{\text{cris}} \otimes_{\Lambda^{\text{wt}}} \mathbb{T}^*_\infty[\lambda]\right)^{G_{\mathbb{Q}_p}}\right)^{\varphi=U_p}$ must be free of rank one in the Gorenstein situation. In the case where the local component through which $\mathbf{H}^{\text{ord}}_{Np^\infty}$ acts on $\mathbb{T}_\infty[\lambda]$ is not Gorenstein, the whole argument works fine except $\left(\mathbb{T}_\infty/\text{F}^+\mathbb{T}_\infty\right)[\lambda]$ will no longer be $\mathbf{I}$-free, which means that $\mathbb{D}^{\mathbf{I}}_{\text{cris}}(\mathbb{T}^*_\infty)[\lambda]$ may inherit some non-trivial $\mathbf{I}[1/p]$-torsion.

The filtration on $\mathbb{B}^{\text{wt}}_{\text{cris}}$ is a little unforgiving, instead we prefer to use the Λ^{wt}-adic period rings introduced by Iovita and Stevens [IS].

Definition 5.9. *If $\mathfrak{D}_n$ is the integral structure $W(\widetilde{\mathbf{E}}^+)/\text{Ker}(\theta)^n$ on $B^+_{\text{dR}}/t^n B^+_{\text{dR}}$,*

$$\mathbb{B}^{+,\text{wt}}_{\text{dR}} := \varprojlim_n \left(\varprojlim_m \left(\mathfrak{D}_n/p^m\mathfrak{D}_n \otimes_{\mathbb{Z}/p^m\mathbb{Z}} \mathbb{Z}/p^m\mathbb{Z}[s]\right) \otimes_{\mathbb{Z}_p} \mathbb{Q}_p\right)$$

and similarly $\mathbb{B}^{\text{wt}}_{\text{dR}} := \mathbb{B}^{+,\text{wt}}_{\text{dR}}[t^{-1}]$ is the affinoid analogue of B_{dR}.

Let us start by establishing the truth of assertion (a). Since the character Φ is unramified, $\left(\mathbb{B}^{\text{wt}}_{\text{cris}} \otimes_{\Lambda^{\text{wt}}} \mathbf{I}(\Phi)\right)^{G_{\mathbb{Q}_p}}$ is exactly the same thing as $\left(\mathbb{B}^{+,\text{wt}}_{\text{cris}} \otimes_{\Lambda^{\text{wt}}} \mathbf{I}(\Phi)\right)^{G_{\mathbb{Q}_p}}$. When viewed as homomorphisms at the level of power series coefficients, the natural injections $\widehat{\mathbb{Q}}_p^{\text{nr}} \hookrightarrow B^+_{\text{cris}} \hookrightarrow B^+_{\text{dR}}$ induce containments

$$\left(\widehat{\mathbb{Q}}_p^{\text{nr}}\langle\langle s\rangle\rangle \otimes_{\Lambda^{\text{wt}}} \mathbf{I}(\Phi)\right)^{G_{\mathbb{Q}_p}} \hookrightarrow \left(\mathbb{B}^{+,\text{wt}}_{\text{cris}} \otimes_{\Lambda^{\text{wt}}} \mathbf{I}(\Phi)\right)^{G_{\mathbb{Q}_p}} \hookrightarrow \left(\mathbb{B}^{+,\text{wt}}_{\text{dR}} \otimes_{\Lambda^{\text{wt}}} \mathbf{I}(\Phi)\right)^{G_{\mathbb{Q}_p}}.$$

Tensoring the sequence $0 \to \text{Fil}^1\mathbb{B}^{+,\text{wt}}_{\text{dR}} \to \mathbb{B}^{+,\text{wt}}_{\text{dR}} \to \mathbb{C}_p\langle\langle s\rangle\rangle \to 0$ by $\mathbf{I}(\Phi)$ and taking $G_{\mathbb{Q}_p}$-invariants, we have an inclusion

$$\left(\mathbb{B}^{+,\text{wt}}_{\text{dR}} \otimes_{\Lambda^{\text{wt}}} \mathbf{I}(\Phi)\right)^{G_{\mathbb{Q}_p}} \hookrightarrow \left(\mathbb{C}_p\langle\langle s\rangle\rangle \otimes_{\mathbb{Z}_p\langle\langle s\rangle\rangle} \mathbf{I}(\Phi)\right)^{G_{\mathbb{Q}_p}} = \left(\widehat{\mathbb{Q}}_p^{\text{nr}}\langle\langle s\rangle\rangle \otimes_{\mathbb{Z}_p\langle\langle s\rangle\rangle} \mathbf{I}(\Phi)\right)^{G_{\mathbb{Q}_p}/I_p}$$

because $\left(\text{Fil}^1\mathbb{B}^{+,\text{wt}}_{\text{dR}} \otimes_{\Lambda^{\text{wt}}} \mathbf{I}(\Phi)\right)^{G_{\mathbb{Q}_p}}$ is zero. Thus the module $\left(\mathbb{B}^{+,\text{wt}}_{\text{cris}} \otimes_{\Lambda^{\text{wt}}} \mathbf{I}(\Phi)\right)^{G_{\mathbb{Q}_p}}$ coincides with $\left(\widehat{\mathbb{Q}}_p^{\text{nr}}\langle\langle s\rangle\rangle \otimes_{\mathbb{Z}_p\langle\langle s\rangle\rangle} \mathbf{I}(\Phi)\right)^{G_{\mathbb{Q}_p}/I_p}$, which in turn proves (a).

Statement (b) is shown in far greater generality in the preprint [IS]. For the sake of completeness we include a short proof, which was pointed out to us by Iovita. Assume that $\Phi(\mathrm{Frob}_p) \in \mathcal{O}_{\mathcal{K}}\langle\langle s\rangle\rangle^\times$ has the power series expansion $\sum_{n=0}^\infty a_n s^n$. We need to find a series $\mathcal{G}_\Phi \in \widehat{\mathbb{Q}}_p^{\mathrm{nr}}\langle\langle s\rangle\rangle$ with Taylor expansion $\mathcal{G}_\Phi = \sum_{n=0}^\infty c_n s^n$ satisfying $\sum_{n=0}^\infty c_n s^n = \Phi(\mathrm{Frob}_p)\sum_{n=0}^\infty c_n^{\mathrm{Frob}_p} s^n$.

This amounts to solving the system of equations

$$\left\{ c_m = \sum_{i+j=m} c_i^{\mathrm{Frob}_p} a_j \quad \text{for all } m \geq 0 \right\} \text{ where } c_m \in \widehat{\mathbb{Q}}_p^{\mathrm{nr}} \text{ and } \lim_{m\to\infty} c_m = 0.$$

We can use induction. If $m = 0$ then we need to find a $c_0 \in \left(\widehat{\mathbb{Q}}_p^{\mathrm{nr}}\right)^\times$ such that $c_0 = c_0^{\mathrm{Frob}_p} a_0$, and its existence follows immediately from the surjectivity of the map $\left(\widehat{\mathbb{Q}}_p^{\mathrm{nr}}\right)^\times \twoheadrightarrow \left(\widehat{\mathbb{Q}}_p^{\mathrm{nr}}\right)^\times$, $x \mapsto \frac{x}{x^{\mathrm{Frob}_p}}$.

Let's now hypothesise we have determined scalars $c_0, c_1, \ldots, c_m \in \widehat{\mathbb{Q}}_p^{\mathrm{nr}}$ satisfying the first m equations above, and in addition $\mathrm{val}(c_i) \geq \min_{b+d=i,d>0}\{\mathrm{val}(c_b a_d)\}$. We must search for a c_{m+1} which solves the $(m+1)^{\mathrm{st}}$ equation

$$c_{m+1} = c_{m+1}^{\mathrm{Frob}_p} a_0 + \sum_{i+j=m+1, i\neq m+1} c_i^{\mathrm{Frob}_p} a_j.$$

Since $a_0 = c_0/c_0^{\mathrm{Frob}_p}$ one can rewrite this as

$$\frac{c_{m+1}}{c_0} - \left(\frac{c_{m+1}}{c_0}\right)^{\mathrm{Frob}_p} = \frac{1}{c_0}\sum_{i+j=m+1, i\neq m+1} c_i^{\mathrm{Frob}_p} a_j.$$

Because the homomorphism $\widehat{\mathbb{Q}}_p^{\mathrm{nr}} \twoheadrightarrow \widehat{\mathbb{Q}}_p^{\mathrm{nr}}$, $x \mapsto x - x^{\mathrm{Frob}_p}$ is surjective, we can certainly find such a solution c_{m+1}. Moreover, adjusting c_{m+1}/c_0 by an element of $\mathbb{Q}_p$ if necessary, one may then assume that $\mathrm{val}(c_{m+1}) = \mathrm{val}(c_{m+1}/c_0)$ equals $\mathrm{val}\left(\frac{c_{m+1}}{c_0} - \left(\frac{c_{m+1}}{c_0}\right)^{\mathrm{Frob}_p}\right)$. As an immediate corollary,

$$\mathrm{val}(c_{m+1}) = \mathrm{val}\left(\sum_{i+j=m+1, j>0} c_i^{\mathrm{Frob}_p} a_j\right) \geq \min_{i+j=m+1, j>0}\{\mathrm{val}(c_i a_j)\}.$$

The fact that $\mathrm{val}(a_n) > 0$ for all $n \geq 1$ and $\lim_{n\to\infty} a_n = 0$, together with our inductive hypothesis, implies that $\lim_{n\to\infty} c_n = 0$ too.

This procedure supplies us a unit $\mathcal{G}_\Phi = \sum_{n=0}^\infty c_n s^n \in \widehat{\mathbb{Q}}_p^{\mathrm{nr}}\langle\langle s\rangle\rangle^\times$ living inside of $\left(\widehat{\mathbb{Q}}_p^{\mathrm{nr}}\langle\langle s\rangle\rangle \otimes_{\mathbb{Z}_p\langle\langle s\rangle\rangle} \mathbf{I}(\Phi)\right)^{\mathrm{Frob}_p=1}$. For an arbitrary $\mathcal{H}_\Phi \in \left(\widehat{\mathbb{Q}}_p^{\mathrm{nr}}\langle\langle s\rangle\rangle \otimes_{\mathbb{Z}_p\langle\langle s\rangle\rangle} \mathbf{I}(\Phi)\right)^{\mathrm{Frob}_p=1}$ the quotient series $\mathcal{H}_\Phi/\mathcal{G}_\Phi$ belongs to $\left(\widehat{\mathbb{Q}}_p^{\mathrm{nr}}\langle\langle s\rangle\rangle \otimes_{\mathbb{Z}_p\langle\langle s\rangle\rangle} \mathbf{I}\right)^{\mathrm{Frob}_p=1} = \mathbb{Q}_p\langle\langle s\rangle\rangle \otimes_{\mathbb{Z}_p\langle\langle s\rangle\rangle} \mathbf{I}$, which is of rank one over $\mathbf{I}[1/p]$. The rank of $\left(\widehat{\mathbb{Q}}_p^{\mathrm{nr}}\langle\langle s\rangle\rangle \otimes_{\mathbb{Z}_p\langle\langle s\rangle\rangle} \mathbf{I}(\Phi)\right)^{\mathrm{Frob}_p=1}$ must therefore also be one, and the proof of (b) is complete.

It remains to justify claim (c). We remark that multiplication by the element $t.\mathcal{G}_\Phi \in \mathbb{B}_{\rm cris}^{+,\rm wt}$ yields an isomorphism

$$\left(\mathbb{B}_{\rm cris}^{\rm wt} \otimes_{\Lambda^{\rm wt}} \mathbf{I}(\Theta\Phi^{-1})\right)^{G_{\mathbb{Q}_p}} \cong \left(\mathbb{B}_{\rm cris}^{\rm wt} \otimes_{\Lambda^{\rm wt}} \mathbf{I}\big((\xi\epsilon)^{-1}\Psi^{-1}\big)\right)^{G_{\mathbb{Q}_p}}.$$

If $L' = \overline{\mathbb{Q}}_p^{\,\mathrm{Ker}(\xi\epsilon)}$ is the field cut out by the finite character $\xi\epsilon$, clearly the right-hand space lies inside of $H^0_{\rm cont}\big(G_{L'}, \mathbb{B}^{\rm wt}_{\rm cris}(\Psi^{-1})\big) \otimes_{\Lambda^{\rm wt}} \mathbf{I}$. We claim that the larger space $H^0_{\rm cont}\big(G_{L'}, \mathbb{B}^{\rm wt}_{\rm dR}(\Psi^{-1})\big) \otimes_{\Lambda^{\rm wt}} \mathbf{I}$ is trivial.

Assume α is an element of $H^0_{\rm cont}\big(G_{L'}, \mathbb{B}^{\rm wt}_{\rm dR}(\Psi^{-1})\big)$ whence $\alpha \in \big(\mathrm{Fil}^j\mathbb{B}^{\rm wt}_{\rm dR}(\Psi^{-1})\big)^{G_{L'}}$ for some sufficiently small $j \in \mathbb{Z}$. As $\mathrm{Fil}^j\mathbb{B}^{\rm wt}_{\rm dR}/\mathrm{Fil}^{j+1}\mathbb{B}^{\rm wt}_{\rm dR} \cong \mathbb{C}_p\langle\langle s\rangle\rangle(j)$, the sequence

$$0 \to \left(\mathrm{Fil}^{j+1}\mathbb{B}^{\rm wt}_{\rm dR}(\Psi^{-1})\right)^{G_{L'}} \to \left(\mathrm{Fil}^{j}\mathbb{B}^{\rm wt}_{\rm dR}(\Psi^{-1})\right)^{G_{L'}} \to \left(\mathbb{C}_p\langle\langle s\rangle\rangle \otimes \big(\chi^j_{\rm cy}\Psi^{-1}\big)\right)^{G_{L'}}$$

is exact. The right-hand space is zero by a calculation of P. Smith [Sm, Prop 2.22]. Consequently α actually lay inside of $\big(\mathrm{Fil}^{j+1}\mathbb{B}^{\rm wt}_{\rm dR}(\Psi^{-1})\big)^{G_{L'}}$. Repeating the process ad infinitum and shrewdly observing that $\bigcap_{i\geq j}\mathrm{Fil}^i\mathbb{B}^{\rm wt}_{\rm dR} = \{0\}$, we conclude that our original element α was 0. It follows $H^0_{\rm cont}\big(G_{L'}, \mathbb{B}^{\rm wt}_{\rm dR}(\Psi^{-1})\big)$ must itself be trivial, and we are done.

□

5.3 Constructing big dual exponentials

The algebraic interpolations $\mathcal{E}_{k,\epsilon}$ were built out of Bloch-Kato exponentials maps, or more precisely their dual versions. To attempt the same approach within an analytically varying family, it is therefore natural to seek out an $\mathbf{I}$-adic analogue of these homomorphisms. We are very grateful to Iovita and Stevens for revealing the following result to us, which is the key to defining the big dual exponential.

Proposition 5.10. [IS, §3.1] *There exists a fundamental exact sequence*

$$0 \longrightarrow \mathbb{Q}_p\langle\langle s\rangle\rangle \longrightarrow \left(\mathbb{B}^{\rm wt}_{\rm cris}\right)^{\varphi=1} \longrightarrow \mathbb{B}^{\rm wt}_{\rm dR}/\mathbb{B}^{+,\rm wt}_{\rm dR} \longrightarrow 0$$

generalising the usual p-adic sequence of Fontaine-Messing [FM].

Actually their sequence involves the rigid deformation of the period ring $B_{\rm max}$ instead of $B_{\rm cris}$, but the two proofs are absolutely identical because $B_{\rm max}^{\varphi=1} = B_{\rm cris}^{\varphi=1}$.

Unfortunately we cannot apply their algebraic Fourier transforms directly to $\mathbb{T}_\infty[\lambda]$, because the Galois representations they consider are the *contragredients* of those occuring in this book. For technical reasons, the resulting dual exponential

$$\mathrm{EXP}^* : H^1\big(G, \mathbb{T}_\infty[\lambda]\big) \longrightarrow \left(\mathbb{T}_\infty[\lambda] \otimes \mathbb{B}^{+,\rm wt}_{\rm dR}\right)^G$$

might very well fail to specialise to its p-adic counterparts $\exp^*_{V^*_{\mathbf{f}_P}}$ at a dense set of algebraic points $P \in \mathrm{Spec}\,\mathbf{I}(\mathcal{O}_\mathcal{K})^{\rm alg}$.

Remark: By a stroke of good fortune, the dual representation $\mathbb{T}^*_\infty[\lambda]$ is endowed with an unramified submodule of rank one over $\mathbf{I}$. As we shall shortly discover, cup-products of EXP* with vectors in the crystalline deformation space $\mathbb{D}^{\mathbf{I}}_{\mathrm{cris}}(\mathbb{T}^*_\infty)$ behave well at their arithmetic specialisations.

Let L be any finite extension of $\mathbb{Q}_p$, and for the present moment choose $i \in \mathbb{Z}$. Tensoring the fundamental sequence above by $\mathbf{I}(\Phi\chi^i_{\mathrm{cy}})$ then taking G_L-cohomology, we obtain an $\mathbf{I}$-adic exponential (boundary) map

$$\mathrm{EXP}_{\Phi\chi^i_{\mathrm{cy}}} : \Big(\mathbf{I}(\Phi\chi^i_{\mathrm{cy}}) \otimes_{\Lambda^{\mathrm{wt}}} \mathbb{B}^{\mathrm{wt}}_{\mathrm{dR}}/\mathbb{B}^{+,\mathrm{wt}}_{\mathrm{dR}}\Big)^{G_L} \longrightarrow H^1\Big(L,\, \mathbf{I}(\Phi) \otimes_{\mathbb{Z}_p} \mathbb{Q}_p(i)\Big).$$

This homomorphism is characterised by the commutativity of the diagram

$$\begin{array}{ccc}
\Big(\mathbf{I}(\Phi\chi^i_{\mathrm{cy}}) \otimes_{\Lambda^{\mathrm{wt}}} \mathbb{B}^{\mathrm{wt}}_{\mathrm{dR}}/\mathbb{B}^{+,\mathrm{wt}}_{\mathrm{dR}}\Big)^{G_L} & \xrightarrow{\mathrm{EXP}_{\Phi\chi^i_{\mathrm{cy}}}} & H^1\Big(L,\, \mathbf{I}(\Phi) \otimes_{\mathbb{Z}_p} \mathbb{Q}_p(i)\Big) \\
\otimes_{\mathbf{I}} \mathbf{I}/P \big\downarrow & & \big\downarrow \otimes_{\mathbf{I}} \mathbf{I}/P \\
\Big(\mathcal{K}_{\mathbf{f}_P}(\phi_k^{-1}\chi^i_{\mathrm{cy}}) \otimes_{\mathcal{K}_{\mathbf{f}_P}} B_{\mathrm{dR}}/B^+_{\mathrm{dR}}\Big)^{G_L} & \xrightarrow{\exp_{\mathcal{K}_{\mathbf{f}_P}(\phi_k^{-1})(i)}} & H^1\Big(L,\, \mathcal{K}_{\mathbf{f}_P}(\phi_k^{-1}\chi^i_{\mathrm{cy}})\Big) \\
\big(\mathcal{J}_- \bmod P\big)_* \big\downarrow & & \big\downarrow \big(\mathcal{J}_- \bmod P\big)_* \\
\Big(V_{\mathbf{f}_P}(i) \otimes_{\mathcal{K}_{\mathbf{f}_P}} B_{\mathrm{dR}}/B^+_{\mathrm{dR}}\Big)^{G_L} & \xrightarrow{\exp_{V_{\mathbf{f}_P}(i)}} & H^1\Big(L,\, V_{\mathbf{f}_P}(i)\Big)
\end{array}$$

at algebraic points $P \in \mathrm{Spec}\,\mathbf{I}$ of type (k,ϵ).

Here we have identified $\mathcal{K}_{\mathbf{f}_P}(\phi_k^{-1})$ with the unramified subspace of $V_{\mathbf{f}_P}$, via the short exact sequence of $G_{\mathbb{Q}_p}$-modules

$$0 \longrightarrow \mathcal{K}_{\mathbf{f}_P}\Big(\phi_k^{-1}\Big) \overset{\mathcal{J}_- \bmod P}{\longrightarrow} V_{\mathbf{f}_P} \overset{\mathcal{J}_+ \bmod P}{\longrightarrow} \mathcal{K}_{\mathbf{f}_P}\Big((\xi\epsilon)^{-1}\chi^{1-k}_{\mathrm{cy}}\omega^k\phi_k\Big) \longrightarrow 0$$

which are also finite-dimensional $\mathbb{Q}_p$-vector spaces.

Convention: For each primitive homomorphism $\lambda : \mathbf{h}^{\mathrm{ord}}(Np^\infty;\mathcal{O}_\mathcal{K}) \otimes_{\Lambda^{\mathrm{wt}}_\mathcal{K}} \mathbf{I} \to \mathbf{I}$, we now make a choice of generator $\mathbf{v}_{\mathbf{I},\lambda}$ of the Hecke eigenspace

$$\mathbb{D}^{\mathbf{I}}_{\mathrm{cris}}(\mathbb{T}^*_\infty)[\lambda] \;=\; \bigg(\Big(\mathbb{T}^*_\infty[\lambda] \otimes_{\Lambda^{\mathrm{wt}}} \mathbb{B}^{\mathrm{wt}}_{\mathrm{cris}}\Big)^{G_{\mathbb{Q}_p}}\bigg)^{\varphi=U_p} \quad \text{over } \mathbf{I}[1/p].$$

An elementary calculation shows $\Big(\mathbf{I}(\Phi\chi^i_{\mathrm{cy}}) \otimes_{\Lambda^{\mathrm{wt}}} \mathbb{B}^{\mathrm{wt}}_{\mathrm{dR}}/\mathbb{B}^{+,\mathrm{wt}}_{\mathrm{dR}}\Big)^{G_L}$ is zero when $i \leq 0$, so it makes good sense to take the integer $i \geq 1$. In particular $(t^{-i}\mathbf{v}_{\mathbf{I},\lambda}) \bmod \mathbb{B}^{+,\mathrm{wt}}_{\mathrm{dR}}$ will generate a one-dimensional $\mathbf{I}$-submodule in $\Big(\mathbf{I}(\Phi\chi^i_{\mathrm{cy}}) \otimes_{\Lambda^{\mathrm{wt}}} \mathbb{B}^{\mathrm{wt}}_{\mathrm{dR}}/\mathbb{B}^{+,\mathrm{wt}}_{\mathrm{dR}}\Big)^{G_L}$.

Once the vector $\mathbf{v}_{\mathbf{I},\lambda}$ is picked, the rest of the constructions are fairly canonical. We denote by '$\mathrm{EXP}_{L,i}$' the compositions

$$\Big(\mathbf{I}(\Phi) \otimes_{\Lambda^{\mathrm{wt}}} \mathbb{B}^{\mathrm{wt}}_{\mathrm{cris}}\Big)^{G_L} \overset{.t^{-i}}{\hookrightarrow} \Big(\mathbf{I}(\Phi\chi^i_{\mathrm{cy}}) \otimes_{\Lambda^{\mathrm{wt}}} \mathbb{B}^{\mathrm{wt}}_{\mathrm{dR}}\Big)^{G_L} \overset{\bmod \mathbb{B}^{+,\mathrm{wt}}_{\mathrm{dR}}}{\longrightarrow} \Big(\mathbf{I}(\Phi\chi^i_{\mathrm{cy}}) \otimes_{\Lambda^{\mathrm{wt}}} \mathbb{B}^{\mathrm{wt}}_{\mathrm{dR}}/\mathbb{B}^{+,\mathrm{wt}}_{\mathrm{dR}}\Big)^{G_L}$$

$$\overset{\mathrm{EXP}_{\Phi\chi^i_{\mathrm{cy}}}}{\longrightarrow} H^1\Big(L,\, \mathbf{I}(\Phi) \otimes_{\mathbb{Z}_p} \mathbb{Q}_p(i)\Big) \overset{(\mathcal{J}_-)_*}{\longrightarrow} H^1\Big(L,\, \mathbb{T}^*_\infty[\lambda] \otimes_{\mathbb{Z}_p} \mathbb{Q}_p(i)\Big).$$

It follows that $\mathrm{EXP}_{L,i}(\mathbf{I}.\mathbf{v}_{\mathbf{I},\lambda})$ will be an $\mathbf{I}$-line inside $H^1\Big(L, \mathbb{T}^*_\infty[\lambda] \otimes_{\mathbb{Z}_p} \mathbb{Q}_p(i)\Big)$ which is of finite type over $\mathbf{I}[1/p]$. Consequently, there exist constants $\theta_{\mathbb{T}_\infty,\lambda,i} \in \mathbb{Z}$ depending only on $\mathbb{T}_\infty$, λ and $i \geq 1$, such that

$$\mathrm{EXP}_{L,i}(\mathbf{I}.\mathbf{v}_{\mathbf{I},\lambda}) \quad \subset \quad p^{\theta_{\mathbb{T}_\infty,\lambda,i}}.\Big(\text{image of } H^1\big(L, \mathbb{T}^*_\infty(i)[\lambda]\big)\Big).$$

The reason why $\theta_{\mathbb{T}_\infty,\lambda,i}$ is independent of the ground field L is because the square

$$\begin{array}{ccc}
\mathbf{I}.\mathbf{v}_{\mathbf{I},\lambda} & \overset{\mathrm{EXP}_{\mathbb{Q}_p,i}}{\longrightarrow} & H^1\Big(\mathbb{Q}_p, \mathbb{T}^*_\infty[\lambda] \otimes_{\mathbb{Z}_p} \mathbb{Q}_p(i)\Big) \\
{\scriptstyle .t^{-i} \bmod \mathbb{B}^{+,\mathrm{wt}}_{\mathrm{dR}}} \big\downarrow & & \big\downarrow {\scriptstyle \mathrm{res}_{L/\mathbb{Q}_p}} \\
\Big(\mathbf{I}(\Phi\chi^i_{\mathrm{cy}}) \otimes_{\Lambda^{\mathrm{wt}}} \mathbb{B}^{\mathrm{wt}}_{\mathrm{dR}}/\mathbb{B}^{+,\mathrm{wt}}_{\mathrm{dR}}\Big)^{G_L} & \overset{(\mathcal{J}_-)_* \circ \mathrm{EXP}_{\Phi\chi^i_{\mathrm{cy}}}}{\longrightarrow} & H^1\Big(L,\ \mathbb{T}^*_\infty[\lambda] \otimes_{\mathbb{Z}_p} \mathbb{Q}_p(i)\Big)
\end{array}$$

commutes, which means $\mathrm{EXP}_{L,i}(\mathbf{I}.\mathbf{v}_{\mathbf{I},\lambda})$ lies in the restriction of $H^1\big(\mathbb{Q}_p, \mathbb{T}^*_\infty(i)[\lambda]\big)$ up to a power of p.

From now on we'll just stick $i = 1$, and sometimes drop it from the notation.

Remark: Consider the following set-up. Suppose we are given an $\mathbf{I}$-module $\mathbb{D}$ which is of finite type, and a free $\mathbf{I}$-module $\mathbb{M}$ with a continuous $\mathrm{Gal}(\overline{L}/L)$-action. We shall also assume the existence of a homomorphism $\pi : \mathbb{D} \to H^1\big(L, \mathbb{M}^*(1)\big)$. Upon fixing an element $\underline{v} \in \mathbb{D}$, then diagrammatically

$$\begin{array}{ccccc}
H^1\big(L, \mathbb{M}^*(1)\big) & \overset{\pi}{\longleftarrow} & \mathbb{D} & & \\
\times & & & & \\
H^1\big(L, \mathbb{M}\big) & \overset{\pi^*}{\dashrightarrow} & \mathrm{Hom}\big(\mathbb{D}, \mathbf{I}\big) & \overset{\text{evaluation at } \underline{v}}{\longrightarrow} & \mathbf{I} \\
\big\downarrow {\scriptstyle \mathrm{inv}_L \circ \cup_{\mathbf{I}}} & & & & \\
\mathbf{I} & & & &
\end{array}$$

and we obtain a map $\pi^*_{\underline{v}} : H^1\big(L, \mathbb{M}\big) \longrightarrow \mathbf{I}$ defined by $\pi^*_{\underline{v}}(x) = \big(\pi^*(x)\big)(\underline{v})$.

For example, if we put $\mathbb{D} = \Big(\mathbf{I}(\Phi) \otimes_{\Lambda^{\mathrm{wt}}} \mathbb{B}^{\mathrm{wt}}_{\mathrm{cris}}\Big)^{G_L}$, $\mathbb{M} = \mathbb{T}_\infty[\lambda]$ and $\pi = \mathrm{EXP}_{L,1}$ then $\pi^*_{\underline{v}}(x) = \mathrm{inv}_L\big(x \cup_{\mathbf{I}} \mathrm{EXP}_{L,1}(\underline{v})\big)$. Moreover at each point $P \in \mathrm{Spec}\, \mathbf{I}(\mathcal{O}_{\mathcal{K}})^{\mathrm{alg}}$,

$$\begin{aligned}
\pi^*_{\underline{v}}(x) \bmod P \quad &= \quad \mathrm{inv}_L\Big(x \cup_{\mathbf{I}} \mathrm{EXP}_{L,1}(\underline{v})\Big) \bmod P \\
&= \quad \mathrm{inv}_L\Big(x \bmod P \cup \mathrm{EXP}_{L,1}(\underline{v}) \bmod P\Big) \\
&= \quad \mathrm{inv}_L\Big(x \bmod P \cup \exp_{V_{\mathbf{f}_P}(1)}\big(t^{-1}\underline{v} \bmod P\big)\Big) \\
&= \quad \mathrm{Tr}_{L/\mathbb{Q}_p}\Big(\exp^*_{V^*_{\mathbf{f}_P}}\ (x \bmod P) \cup_{\mathrm{dR}} t^{-1}\underline{v} \bmod P\Big).
\end{aligned}$$

This formula hopefully provides ample motivation for the following.

Definition 5.11. *(i) Let us fix a base weight* k_0*, and a character* ϵ *of finite order. For all algebraic points* $P \in \mathrm{Spec}\,\mathbf{I}(\mathcal{O}_{\mathcal{K}})^{\mathrm{alg}}$ *of type* (k,ϵ) *with* $k \in \mathbb{U}_{k_0}$*, one defines*

$$\mathrm{EXP}^*_{\mathbb{T}_\infty,L} \mod P : H^1\Big(L, \mathbb{T}_\infty[\lambda] \otimes_{\mathbf{I}} \mathbf{I}/P\Big) \longrightarrow L.\mathrm{Frac}\big(\mathbf{I}/P\big)$$

by forming the cup-product

$$\big(\mathrm{EXP}^*_{\mathbb{T}_\infty,L} \bmod P\big)(x) \quad := \quad \exp^*_{V^*_{\mathbf{f}_P}}(x) \cup_{\mathrm{dR}} \big(1-\delta_{\mathrm{nr}}\varphi\big)^{-1}.\big(t^{-1}.\mathbf{v}_{\mathbf{I},\lambda} \mod P\big)$$

with δ_{nr} *equal to one if* $L/\mathbb{Q}_p$ *is unramified, and equal to zero otherwise.*

Rather than dualise our $\mathbf{I}$-adic exponential directly, it is more agreeable to glue along prime ideals $P \in \mathrm{Spec}\,\mathbf{I}$. We are then free to modify the definition by the Euler factor $(1-\varphi)^{-1}$ for unramified extensions $L/\mathbb{Q}_p$. In this way $\mathrm{EXP}^*_{\mathbb{T}_\infty,L} \bmod P$ ends up lying inside $\mathbf{D}_{\mathrm{dR},L}(\mathcal{K}_{\mathbf{f}_P})$ rather than $\mathrm{cotang}(V^*_{\mathbf{f}_P}/L)$, which has extremely unpleasant behaviour as a line bundle over the weight-space $\mathbb{U}_{k_0}$.

Lemma 5.12. *Assume* $\mathcal{O}_L$ *has a power integral* $\mathbb{Z}_p$*-basis* $\big\{1,\beta,\beta^2,\ldots,\beta^{[L:\mathbb{Q}_p]-1}\big\}$. *Then at arithmetic points* P *above,* $\mathrm{EXP}^*_{\mathbb{T}_\infty,L} \bmod P$ *sends* $H^1\Big(L,\mathbb{T}_\infty[\lambda]\otimes_{\mathbf{I}}\mathbf{I}/P\Big)$ *to the* $\mathbb{Z}_p$*-lattice* $p^{\theta_{\mathbb{T}_\infty,\lambda}-\mathrm{ord}_p[L:\mathbb{Q}_p]}.\mathcal{O}_{\mathcal{K}_{\mathbf{f}_P}}[\beta]$.

Proof: Extending $\mathrm{EXP}^*_{\mathbb{T}_\infty,L} \bmod P$ by scalars, we find its image is contained in

$$\frac{1}{\#G_\beta}\sum_{\sigma\in G_\beta}\left(\exp^*_{V^*_{\mathbf{f}_P}}\Big(H^1\big(L,\mathbb{T}_\infty[\lambda]\otimes_{\mathbf{I}}\mathbf{I}/P\big)[\beta]\Big)\cup_{\mathrm{dR}}(1-\delta_{\mathrm{nr}}\varphi)^{-1}.(t^{-1}.\mathbf{v}_{\mathbf{I},\lambda} \mod P)\right)^{\sigma}$$

where $G_\beta = \mathrm{Gal}\big(\mathcal{K}_{\mathbf{f}_P}(\beta)/\mathcal{K}_{\mathbf{f}_P}\big)$. This expression further simplifies to become

$$\frac{\big(1-\delta_{\mathrm{nr}}p^{-1}\lambda(U_p)\big)^{-1}}{\#G_\beta}\mathrm{inv}_L\Big(H^1\big(L,\mathbb{T}_\infty[\lambda]\otimes_{\mathbf{I}}\mathbf{I}/P\big)[\beta]\cup\exp_{V_{\mathbf{f}_P}(1)}\big(t^{-1}.\mathbf{v}_{\mathbf{I},\lambda} \mod P\big)\Big).$$

However, we know $\exp_{V_{\mathbf{f}_P}(1)}\big(t^{-1}.\mathbf{v}_{\mathbf{I},\lambda} \mod P\big)$ is equal to $\mathrm{EXP}_{L,1}(\mathbf{v}_{\mathbf{I},\lambda}) \mod P$, and the latter quantity lies in $p^{\theta_{\mathbb{T}_\infty,\lambda}}.\Big(\text{the image of } H^1\big(L,\mathbb{T}^*_\infty(1)[\lambda]\big)\Big)$ modulo P. We deduce that $\big(\mathrm{EXP}^*_{\mathbb{T}_\infty,L} \bmod P\big)\Big(H^1\big(L,\mathbb{T}_\infty[\lambda]\otimes_{\mathbf{I}}\mathbf{I}/P\big)\Big)$ is a subset of

$$\begin{aligned}&\frac{p^{\delta_{\mathrm{nr}}}}{\#G_\beta}\ \mathrm{inv}_L\Big(H^1\big(L,\mathbb{T}_\infty[\lambda]\otimes_{\mathbf{I}}\mathbf{I}/P\big)[\beta]\ \cup\ p^{\theta_{\mathbb{T}_\infty,\lambda}}.H^1\big(L,\mathbb{T}^*_\infty(1)[\lambda]\otimes_{\mathbf{I}}\mathbf{I}/P\big)\Big)\\ &= p^{\delta_{\mathrm{nr}}+\theta_{\mathbb{T}_\infty,\lambda}-\mathrm{ord}_p(\#G_\beta)}.\ \mathrm{inv}_L\Big(H^1\big(L,\mathbb{T}_\infty[\lambda]\big)[\beta]\ \cup_{\mathbf{I}}\ H^1\big(L,\mathbb{T}^*_\infty(1)[\lambda]\big)\Big)\otimes_{\mathbf{I}}\mathbf{I}/P\end{aligned}$$

which is none other than $p^{\delta_{\mathrm{nr}}+\theta_{\mathbb{T}_\infty,\lambda}-\mathrm{ord}_p(\#G_\beta)}\mathbf{I}/P[\beta]$.

Since $\#G_\beta$ divides $[L:\mathbb{Q}_p]$, it follows that $\big(\mathrm{EXP}^*_{\mathbb{T}_\infty,L} \bmod P\big)(\ldots)$ belongs to $p^{\theta_{\mathbb{T}_\infty,\lambda}-\mathrm{ord}_p[L:\mathbb{Q}_p]}\mathcal{O}_{\mathcal{K}_{\mathbf{f}_P}}[\beta]$. Consequently our map modulo P is 'almost p-integral', the value $\theta_{\mathbb{T}_\infty,\lambda}-\mathrm{ord}_p[L:\mathbb{Q}_p]$ being independent of P.

□

Under Hypothesis(Arith) of §4.2, the points $P \in \mathrm{Spec}\ \mathbf{I}(\mathcal{O}_{\mathcal{K}})^{\mathrm{alg}}$ above of type (k,ϵ) densely populate Spec $\mathbf{I}(\overline{\mathbb{Q}}_p)$, thus there is a diagonal embedding $\mathbf{I} \hookrightarrow \bigoplus_P \mathbf{I}/P$. Suppose that M is an $\mathbf{I}$-module of finite type. Clearly the $\mathbf{I}$-torsion submodule of $M - M[P^\infty]$, modulo P injects into the p^∞-torsion of the specialisation $M \otimes_{\mathbf{I}} \mathbf{I}/P$. In particular, there is a natural map $\big(M - M[P^\infty]\big)_{/\mathbf{I}\text{-tors}} \longrightarrow \Big(M \otimes_{\mathbf{I}} \mathbf{I}/P\Big)_{/\mathbb{Z}_p\text{-tors}}$ induced on the quotients.

The relevant situation here is when M is taken to be the $\mathbf{I}$-module $H^1\big(L, \mathbb{T}_\infty[\lambda]\big)$. By the previous discussion, one obtains a diagram

$$\begin{array}{ccc} H^1\big(L,\mathbb{T}_\infty[\lambda]\big)_{/\mathbf{I}\text{-tors}} & \dashrightarrow & \bigoplus_P H^1\big(L,\mathbb{T}_\infty[\lambda]\otimes_{\mathbf{I}}\mathbf{I}/P\big)_{/\mathbb{Z}_p\text{-tors}} \\ & & \Big\downarrow \oplus \mathrm{EXP}^*_{\mathbb{T}_\infty,L} \bmod P \\ p^{\theta_{\mathbb{T}_\infty,\lambda}-\mathrm{ord}_p[L:\mathbb{Q}_p]}\mathbf{I}[\beta] & \longrightarrow & \bigoplus_P p^{\theta_{\mathbb{T}_\infty,\lambda}-\mathrm{ord}_p[L:\mathbb{Q}_p]}\mathbf{I}/P\ [\beta]. \end{array}$$

Definition 5.11. *(ii) For a normal field extension $L = \mathbb{Q}_p(\beta)$, we define*

$$\mathrm{EXP}^*_{\mathbb{T}_\infty,L} : H^1\left(L,\mathbb{T}_\infty[\lambda]\right) \longrightarrow p^{\theta_{\mathbb{T}_\infty,\lambda}-\mathrm{ord}_p[L:\mathbb{Q}_p]}\mathbf{I}[\beta] \ \subset\ L.\mathbf{I}[1/p]$$

by restricting $\oplus\mathrm{EXP}^*_{\mathbb{T}_\infty,L}$ mod P *in the above diagram, to the left-sided column. Alternatively,* $\mathrm{EXP}^*_{\mathbb{T}_\infty,L}$ *is the unique* $\mathbf{I}$*-homomorphism making*

$$\begin{array}{ccccc} H^1\big(L,\mathbb{T}_\infty[\lambda]\big) & & \xrightarrow{\mathrm{EXP}^*_{\mathbb{T}_\infty,L}} & & L.\mathbf{I}[1/p] \\ \Big\downarrow \bmod P & & & & \Big\downarrow \otimes_{\mathbf{I}}\mathbf{I}/P \\ H^1\left(L,V^*_{\mathbf{f}_P}\right) & \xrightarrow{\exp^*_{V^*_{\mathbf{f}_P}}} & \mathrm{cotang}(V^*_{\mathbf{f}_P}/L) & \xrightarrow{\cup(1-\delta_{\mathrm{nr}}\varphi)^{-1}(t^{-1}.\mathbf{v}_{\mathbf{I},\lambda} \bmod P)} & \mathbf{D}_{\mathrm{dR},L}(\mathcal{K}_{\mathbf{f}_P}) \end{array}$$

commute, for each $P \in \mathrm{Spec}\ \mathbf{I}(\mathcal{O}_{\mathcal{K}})^{\mathrm{alg}}$ as before.

(iii) Similarly, for all integers $m \geq 1$ we write

$$\mathrm{EXP}^*_{\mathbb{T}_\infty} : H^1\Big(\mathbb{Q}(\mu_m)\otimes\mathbb{Z}_p,\mathbb{T}_\infty[\lambda]\Big) \xrightarrow{\sim} \prod_{\mathfrak{P}_m|p} H^1\Big(D_{\mathfrak{P}_m},\mathbb{T}_\infty[\lambda]\Big) \xrightarrow{\mathrm{EXP}^*_{\mathbb{T}_\infty,\mathfrak{P}_m}} \mathbf{I}\otimes\mathbb{Q}(\mu_m)$$

for the induced maps on the semi-local cohomology.

In the p-adic context, the dual exponential map played the rôle of courier for the special value of the L-function of the modular form. Morally, one might then expect the big dual exponential map to transport arithmetic for the whole Hida family. Indeed this turns out to be what happens, and it is worthwhile to analyse the local properties of $\mathrm{EXP}^*_{\mathbb{T}_\infty}$ in some detail.

5.4 Local dualities

In order to properly calculate the kernel of $\mathrm{EXP}^*_{\mathbb{T}_\infty}$, we shall take some inspiration and guidance from what happens in the classical case. Let p be a prime number. Throughout this section, K will denote a finite extension of the p-adic numbers $\mathbb{Q}_p$. For an abelian variety $A_{/K}$, the classical Kummer map

$$\partial_{\mathrm{kum}} : \varprojlim_m A(K)/p^m A(K) \hookrightarrow H^1\Big(\mathrm{Gal}\left(\overline{K}/K\right),\ \varprojlim_m A[p^m]\Big)$$

explicitly sends a point $P \pmod{p^m}$ to the class of the one-cochain $\sigma \mapsto Q^\sigma - Q$ where $Q \in A(\overline{K})$ is any point satisfying $p^m Q = P$.

As we already discussed in Example 2.5, the image of this map ∂_{kum} is the geometric part $H^1_g\Big(\mathrm{Gal}\left(\overline{K}/K\right),\ \mathrm{Ta}_p(A)\Big)$ consisting of those cocycles which vanish after tensoring by the ring of p-adic periods B_{dR}. Furthermore, the kernel of the dual Lie group exponential

$$\exp^* : H^1\Big(\mathrm{Gal}\left(\overline{K}/K\right),\ \mathrm{Ta}_p(A)\Big) \longrightarrow \mathrm{cotang}_K(A)$$

is precisely the subgroup $H^1_g\Big(\mathrm{Gal}\left(\overline{K}/K\right),\ \mathrm{Ta}_p(A)\Big)$. As Bloch and Kato showed, these notions generalise seamlessly to de Rham representations.

Remark: Hopefully this digression gives us some idea what we should be aiming for in a Λ^{wt}-adic context. Presumably the kernel of $\mathrm{EXP}^*_{\mathbb{T}_\infty}$ must be related to a group of local points, but it is not really clear where one might find such points. We first try to find solutions to the following local problems:

Question 1. *How do we associate K-rational points to big Galois representations?*

Question 2. *Can we describe the image of these points inside étale cohomology?*

Question 3. *Is the kernel of* $\mathrm{EXP}^*_{\mathbb{T}_\infty}$ *the image of these points under Kummer?*

Question 4. *What is the rank of this image over the deformation ring $\mathcal{R}$?*

In the case of ordinary families of modular forms, we are in a good position to resolve all four issues positively. Intriguingly, the answers themselves strongly suggest similar theorems hold for arrays of points on admissible subspaces.

Our strategy is to consider compatible sequences of K-rational points lying on $\mathrm{Jac}X_1(Np^r)_{r\geq 1}$, and then cut out their ordinary parts utilising the same Hecke idempotents that define $\mathbb{T}_\infty$. This produces a family of points with Λ^{wt}-action, which can then be mapped into the cohomology group $H^1_{\mathrm{cont}}\Big(\mathrm{Gal}(\overline{K}/K),\ \mathbb{T}_\infty\Big)$. This subgroup constitutes the image of the Kummer map mentioned above.

The technical tool underpinning our arguments, is a version of Tate local duality for cohomologies taking affinoid coefficients. We shall work with representations over power series rings, possessing numerous specialisations coming from geometry. For example, the duality works equally well for non-ordinary families of cusp forms. This puts us in the strange position of being able to answer the third and fourth questions, even when we do not have an obvious set of K-rational points at hand. In the nearly-ordinary case, we compute that the rank of the image of the Kummer map is always half the rank of the cohomology $H^1(K, \mathbb{T}_\infty)$.

Let's begin by working rigidly. For a closed disk $\mathbb{U}_{k_0} \subset \mathbb{Z}_p$ centred on an integral weight k_0, we previously wrote $\mathcal{K}_{\mathbb{U}_{k_0}}\langle\langle s\rangle\rangle$ for the $\mathcal{K}$-points of the affinoid over $\mathbb{U}_{k_0}$. Throughout this section, we shall routinely drop the subscript $_{\mathbb{U}_{k_0}}$ from the notation. We are interested in studying a certain category of local Galois representation, which will be explained directly below.

Recall first that the universal cyclotomic character $\Psi : G_K \to \mathbb{Z}_p\langle\langle s\rangle\rangle^\times$ has the Taylor series expansion

$$\Psi \;=\; \exp\left(s \log(<\chi_{\rm cy}>)\right) \quad = \quad \sum_{n=0}^{\infty} \frac{\log^n(<\chi_{\rm cy}>)}{\Gamma(n+1)} s^n \qquad \text{at all } s \in \mathbb{Z}_p$$

where $< >$ was the projection to the principal units. Let $K_\infty = \bigcup_{n\geq 0} K_n$ denote the cyclotomic $\mathbb{Z}_p$-extension of K where the $n^{\rm th}$-layer has degree $[K_n : K] = p^n$.

Definition 5.13. *Assume* $\mathbb{W}$ *is a free* $\mathbb{Q}_p\langle\langle s\rangle\rangle$*-module equipped with a* G_K*-action. Then* $\mathbb{W}$ *is said to be* **pseudo-geometric** *if it satisfies the conditions:*

(i) there exists $n(\mathbb{W}) \geq 0$ *and a decomposition of* $\mathrm{Gal}\left(\overline{K}/K_{n(\mathbb{W})}\right)$*-modules*

$$\mathbb{W} \otimes_{\mathbb{Q}_p\langle\langle s\rangle\rangle} \mathbb{C}_p\langle\langle s\rangle\rangle[s^{-1}] \;\cong\; \bigoplus_{i,j\in\mathbb{Z}} \mathbb{C}_p\langle\langle s\rangle\rangle[s^{-1}](\Psi^i \chi_{\rm cy}^j)^{\oplus e_{i,j}}$$

with all but finitely many of the non-negative integers $e_{i,j}$ *equal to zero;*

(ii) $\mathbb{W}_k = \mathbb{W}/(s-(k-k_0))\mathbb{W}$ *is a de Rham representation at infinitely many distinct weights* $k \in \mathbb{U}_{k_0} \cap \mathbb{N}$.

N.B. If the representation was global rather than local, in addition we would demand that it be unramified outside a finite set of places, and that the characteristic polynomials of the Frob_l's were suitably compatible.

The examples we have in mind are the two-dimensional Galois representations arising from p-ordinary families of cusp forms. Let $\lambda : \mathbf{h}^{\rm ord}(Np^\infty; \mathcal{O}_\mathcal{K}) \otimes_{\Lambda_\mathcal{K}^{\rm wt}} \mathbf{I} \to \mathbf{I}$ be the primitive $\mathbf{I}$-algebra homomorphism corresponding to a Hida family $\mathbf{f} \in \mathbb{S}_\mathbf{I}^{\rm ord}$. By the structure Theorem 4.3(iii),

$$\left(\widetilde{\mathrm{Mell}_{k_0,\epsilon}}\right)_{s=k} \circ \rho_\infty[\lambda]\Bigg|_{G_{\mathbb{Q}_p}} \qquad \text{has the form} \qquad \begin{pmatrix} \omega^{k_0-k}\xi\epsilon\chi_{\rm cy}^{k-1}\phi_k^{-1} & * \\ 0 & \phi_k \end{pmatrix}$$

at every integer $k \in \mathbb{U}_{k_0}$ with $k \geq 2$.

If we write $\mathbb{V}$ for the Mellin transform of the $G_{\mathbb{Q}_p}$-representation $\mathbb{T}_\infty[\lambda]$, there is a decomposition of $\mathrm{Gal}\left(\overline{K}/K_{n(\mathbb{V})}\right)$-modules

$$\mathbb{V}\otimes_{\mathcal{K}\langle\langle s\rangle\rangle}\mathbb{C}_p\langle\langle s\rangle\rangle[s^{-1}] \quad\cong\quad \mathbb{C}_p\langle\langle s\rangle\rangle[s^{-1}] \,\oplus\, \mathbb{C}_p\langle\langle s\rangle\rangle[s^{-1}]\left(\Psi\chi_{\mathrm{cy}}^{k_0-1}\right);$$

here the integer $n(\mathbb{V})$ is chosen so that $K_{n(\mathbb{V})}$ contains the field $K_\infty\cap\overline{\mathbb{Q}}_p^{\mathrm{Ker}(\xi\epsilon)}$.

Remark: Iovita and Stevens have shown that the local Galois representations arising from Coleman families of p-adic modular forms of positive slope $< k_0-1$, are also pseudo-geometric in the above sense. Their proof is in turn based on Sen's ramification theory, together with some clever localisation arguments [IS, Thm 2.3].

Returning once more to the general situation, assume $\mathbb{W}$ is any big local Galois representation of the form occurring in Definition 5.13. Our strategy is to mimic the definition of the cohomology groups H^1_e and H^1_g which was given in [BK, §3.1]. Let us begin by associating a candidate set of K-rational points to $\mathbb{W}$, and to its Kummer dual $\mathbb{W}^*(1)=\mathrm{Hom}\Big(\mathbb{W},\ \mathbb{Q}_p\langle\langle s\rangle\rangle(1)\Big)$, as follows.

Definition 5.14. *(a) We define $H^1_{\mathcal{G}}(K,\mathbb{W})$ to be the $\mathbb{Q}_p\langle\langle s\rangle\rangle$-saturation of the group*

$$H^1_{\mathcal{G}}(K,\mathbb{W})_0 \;:=\; \mathrm{Ker}\Big(H^1(K,\mathbb{W}) \;\overset{-\otimes 1}{\longrightarrow}\; H^1\big(K,\mathbb{W}\otimes_{\mathbb{Q}_p\langle\langle s\rangle\rangle}\mathbb{B}_{\mathrm{dR}}^{\mathrm{wt}}\big)\Big).$$

(b) We also define $H^1_{\mathcal{E}}(K,\mathbb{W}^*(1))$ to be the $\mathbb{Q}_p\langle\langle s\rangle\rangle$-saturation of the group

$$H^1_{\mathcal{E}}(K,\mathbb{W}^*(1))_0 \;:=\; \mathrm{Ker}\Big(H^1(K,\mathbb{W}^*(1)) \;\overset{-\otimes 1}{\longrightarrow}\; H^1\big(K,\mathbb{W}^*(1)\otimes_{\mathbb{Q}_p\langle\langle s\rangle\rangle}(\mathbb{B}_{\mathrm{cris}}^{\mathrm{wt}})^{\varphi=1}\big)\Big).$$

As we'll shortly see, when $\mathbb{W}$ is the nearly ordinary representation associated to $\mathbf{f}$ the group $H^1_{\mathcal{G}}(K,\mathbb{W})$ above may be identified with a coherent family of K-rational points, living on the pro-jacobian variety $\varprojlim_r \mathrm{Jac}\big(X_1(Np^r)\big)$.

Conversely, the group $H^1_{\mathcal{E}}(K,\mathbb{W}^*(1))$ is closely connected to the big exponential. Our starting point is the Fundamental Exact Sequence

$$0\to\mathbb{Q}_p\langle\langle s\rangle\rangle\to(\mathbb{B}_{\mathrm{cris}}^{\mathrm{wt}})^{\varphi=1}\overset{\mathrm{mod}\ \mathbb{B}_{\mathrm{dR}}^{+,\mathrm{wt}}}{\longrightarrow}\mathbb{B}_{\mathrm{dR}}^{\mathrm{wt}}/\mathbb{B}_{\mathrm{dR}}^{+,\mathrm{wt}}\to 0\quad\text{(c.f. Proposition 5.10).}$$

Tensoring this by $\mathbb{W}^*(1)$ and taking G_K-invariants, one obtains the long exact sequence in cohomology

$$0\to\mathbb{W}^*(1)^{G_K}\longrightarrow\Big(\mathbb{W}^*\otimes_{\mathbb{Q}_p\langle\langle s\rangle\rangle}(\mathbb{B}_{\mathrm{cris}}^{\mathrm{wt}})^{\varphi=p}\Big)^{G_K}\longrightarrow\Big(\mathbb{W}^*\otimes_{\mathbb{Q}_p\langle\langle s\rangle\rangle}\mathbb{B}_{\mathrm{dR}}^{\mathrm{wt}}/\mathrm{Fil}^1\mathbb{B}_{\mathrm{dR}}^{\mathrm{wt}}\Big)^{G_K}$$
$$\overset{\partial}{\longrightarrow}\;H^1(K,\mathbb{W}^*(1))\;\overset{-\otimes 1}{\longrightarrow}\;H^1\Big(K,\mathbb{W}^*(1)\otimes_{\mathbb{Q}_p\langle\langle s\rangle\rangle}(\mathbb{B}_{\mathrm{cris}}^{\mathrm{wt}})^{\varphi=1}\Big).$$

In this manner $H^1_{\mathcal{E}}(K,\mathbb{W}^*(1))$ ends up being identified with the $\mathbb{Q}_p\langle\langle s\rangle\rangle$-saturation of the image of the boundary exponential ∂.

It is often useful to know the ranks of $H^1_{\mathcal{E}}(K,\mathbb{W}^*(1))$ and $H^1_{\mathcal{G}}(K,\mathbb{W})$ over $\mathbb{Q}_p\langle\langle s\rangle\rangle$, for various pseudo-geometric representations $\mathbb{W}$. This task is readily accomplished provided we know the $\mathbb{Q}_p$-ranks of enough p-adic specialisations of these objects. Firstly, we employ the invariant map $\mathrm{inv}_K : H^2(K,\mathbb{Z}_p(\chi_{\mathrm{cy}})) \to \mathbb{Z}_p$ of local class field theory to furnish us with a pairing

$$H^1(K,\mathbb{W}) \times H^1(K,\mathbb{W}^*(1)) \xrightarrow{\cup} H^2\Big(K,\ \mathbb{Q}_p\langle\langle s\rangle\rangle(\chi_{\mathrm{cy}})\Big) \overset{\mathrm{inv}_K}{\longrightarrow} \mathbb{Q}_p\langle\langle s\rangle\rangle.$$

The following affinoid analogue of the statement that H^1_g is perpendicular to H^1_e, was proved by the author and P. Smith.

Proposition 5.15. [DS, Prop 1] *Under the local pairing above, the cohomology subgroups $H^1_{\mathcal{G}}(K,\mathbb{W})$ and $H^1_{\mathcal{E}}(K,\mathbb{W}^*(1))$ are the exact annihilators of each other.*

The proof is rather lengthy, and is written up elsewhere. We shall now explain how to apply this result in order to compute the rank of both $H^1_{\mathcal{G}}$ and $H^1_{\mathcal{E}}$.

Let us start by making the observation that $H^1(K,\mathbb{W})_{s=k-k_0} \cong H^1(K,\mathbb{W}_k)$ and $H^1(K,\mathbb{W}^*(1))_{s=k-k_0} \cong H^1(K,\mathbb{W}^*_k(1))$ at all but finitely many bad weights $k \in \mathbb{U}_{k_0}$. Restricting only to those good weights $k \in \mathbb{U}_{k_0} \cap \mathbb{N}_{\geq 2}$ for which $\mathbb{W}_k$ is de Rham, one easily checks the injectivity of $H^1_{\mathcal{G}}(K,\mathbb{W})_{s=k-k_0} \to H^1_g(K,\mathbb{W}_k)$. Consequently,

$$\begin{aligned}\dim_{\mathbb{Q}_p} H^1_g(K,\mathbb{W}_k) &\geq \mathrm{rank}_{\mathbb{Q}_p\langle\langle s\rangle\rangle} H^1_{\mathcal{G}}(K,\mathbb{W}) \\ &= \mathrm{rank}_{\mathbb{Q}_p\langle\langle s\rangle\rangle} H^1(K,\mathbb{W}) - \mathrm{rank}_{\mathbb{Q}_p\langle\langle s\rangle\rangle} H^1_{\mathcal{E}}(K,\mathbb{W}^*(1))\end{aligned}$$

upon exploiting the fact that $H^1_{\mathcal{G}}$ and $H^1_{\mathcal{E}}$ are mutual annihilators.

The induced map $H^1_{\mathcal{E}}(K,\mathbb{W}^*(1))_{s=k-k_0} \to H^1_e(K,\mathbb{W}^*_k(1))$ is also injective, hence

$$\begin{aligned}\mathrm{rank}_{\mathbb{Q}_p\langle\langle s\rangle\rangle} H^1_{\mathcal{G}}(K,\mathbb{W}) &= \dim_{\mathbb{Q}_p} H^1(K,\mathbb{W}_k) - \mathrm{rank}_{\mathbb{Q}_p\langle\langle s\rangle\rangle} H^1_{\mathcal{E}}(K,\mathbb{W}^*(1)) \\ &\geq \dim_{\mathbb{Q}_p} H^1(K,\mathbb{W}_k) - \dim_{\mathbb{Q}_p} H^1_e(K,\mathbb{W}^*_k(1)) \\ &= \dim_{\mathbb{Q}_p} H^1_g(K,\mathbb{W}_k) \qquad \text{by [BK, Prop 3.8].}\end{aligned}$$

Thus every inequality sign above is really an equality, and we have shown

Corollary 5.16. *At almost all weights $k \in \mathbb{U}_{k_0} \cap \mathbb{N}_{\geq 2}$ for which $\mathbb{W}_k$ is de Rham,*

(*i*) $\mathrm{rank}_{\mathbb{Q}_p\langle\langle s\rangle\rangle} H^1_{\mathcal{G}}(K,\mathbb{W}) = \dim_{\mathbb{Q}_p} H^1_g(K,\mathbb{W}_k)$ *and*

(*ii*) $\mathrm{rank}_{\mathbb{Q}_p\langle\langle s\rangle\rangle} H^1_{\mathcal{E}}(K,\mathbb{W}^*(1)) = \dim_{\mathbb{Q}_p} H^1_e(K,\mathbb{W}^*_k(1))$.

This dimension result enables us to link $H^1_{\mathcal{G}}$ to the ordinary part of $\mathrm{Jac}X_1(Np^r)_{r\geq 1}$. We should point out even though we have consistently assumed the prime $p \geq 5$ the calculations in this section are equally valid at $p=3$, with the proviso that the reader still believes Theorem 4.3 holds true. We must confess that we are slightly unsure as to the status of this result at the prime 3, but suspect it has been proved when the tame level $N=1$.

Associating points to big Galois representations.
Let us review some classical Kummer theory on abelian varieties. For pairs of integers $m, r \in \mathbb{N}$, the multiplication by p^m endomorphism on the p-divisible group Jac $X_1(Np^r)$ induces a tautological exact sequence $0 \to J_r[p^m] \to J_r \overset{\times p^m}{\to} J_r \to 0$. Upon taking its Galois invariants, we obtain the long exact sequence

$$0 \to J_r(K)[p^m] \to J_r(K) \overset{\times p^m}{\to} J_r(K) \overset{\partial_{r,m}}{\to} H^1\big(K, J_r[p^m]\big) \to H^1\big(K, J_r\big)[p^m] \to 0$$

in cohomology, the boundary map $\partial_{r,m}$ injecting $J_r(K)/p^m$ into $H^1\big(K, J_r[p^m]\big)$. Applying the functors $\varprojlim_m$ and $\varprojlim_r$ yields a level-compatible Kummer map

$$\varprojlim_{r,m} \partial_{r,m} : J_\infty(K) \,\widehat{\otimes}\, \mathbb{Z}_p \ \hookrightarrow \ H^1\big(K, \mathrm{Ta}_p(J_\infty)\big) \quad \text{as a map of } \mathbf{H}_{Np^\infty}\text{-modules.}$$

Definition 5.17. *We define $\mathbf{X}_{\mathbb{T}_\infty}(K)$ to be the pre-image of the local points*

$$\mathbf{e}_{\mathrm{prim}} \cdot \left(\left(\mathbf{e}_{\mathrm{ord}} \cdot \varprojlim_{r,m} \partial_{r,m} \Big(J_\infty(K) \,\widehat{\otimes}\, \mathbb{Z}_p \Big) \right) \otimes_{\Lambda^{\mathrm{wt}}} \mathrm{Frac}(\Lambda^{\mathrm{wt}}) \right) \subset H^1\big(K, \mathbb{T}_\infty\big) \otimes_{\mathcal{R}} \mathcal{F}$$

under the canonical homomorphism $H^1\big(K, \mathbb{T}_\infty\big) \overset{-\otimes 1}{\longrightarrow} H^1\big(K, \mathbb{T}_\infty\big) \otimes_{\mathcal{R}} \mathcal{F}$.
In particular, the condition $\mathbf{X}_{\mathbb{T}_\infty}(K)$ is clearly $\mathcal{R}$-saturated inside of $H^1\big(K, \mathbb{T}_\infty\big)$.

Bearing in mind Bloch-Kato's result for abelian varieties in the p-adic situation, it makes good sense to speculate how the image of the level-compatible Kummer map is related to

$$H^1_{\mathcal{G}}\big(K, \mathbb{T}_\infty\big) \ = \ \mathrm{Ker}\Big(H^1\big(K, \mathbb{T}_\infty\big) \overset{(-\otimes 1)\otimes 1}{\longrightarrow} H^1\big(K, \mathbb{T}_\infty \otimes_{\Lambda^{\mathrm{wt}}} \mathbb{B}^{\mathrm{wt}}_{\mathrm{dR}}\big) \otimes_{\mathcal{R}} \mathcal{F} \Big).$$

The following revelation should come as no surprise.

Theorem 5.18. [DS, Thm 3] *We have the exact equality $\mathbf{X}_{\mathbb{T}_\infty}(K) = H^1_{\mathcal{G}}\big(K, \mathbb{T}_\infty\big)$, and the rank of this $\mathcal{R}$-module equals the degree $[K : \mathbb{Q}_p]$.*

Proof: Fix a topological generator u_0 of $\Gamma = 1 + p\mathbb{Z}_p$, and recall again the weight algebra $\Lambda^{\mathrm{wt}} = \mathbb{Z}_p[\![\Gamma]\!]$. For all integers $r \geq 1$, we shall use the notation $\Lambda^{\mathrm{wt}}_{\Gamma_r}$ for the module of coinvariants $\mathbb{Z}_p[\![\Gamma]\!] \big/ \big(u_0^{p^{r-1}} - 1\big)$. In particular $\mathbb{T}_\infty \otimes_{\Lambda^{\mathrm{wt}}} \Lambda^{\mathrm{wt}}_{\Gamma_r}$ lies inside $\mathrm{Ta}_p(J_r)$, and corresponds to the Tate module of an ordinary p-divisible subgroup of Jac $X_1(Np^r)$ as outlined in [Hi1, §9].

It is immediate from the commutative diagram

$$\begin{array}{ccccccc}
0 \longrightarrow & H^1_{\mathcal{G}}\big(K, \mathbb{T}_\infty\big)_0 & \longrightarrow & H^1\big(K, \mathbb{T}_\infty\big) & \overset{-\otimes 1}{\longrightarrow} & H^1\big(K, \mathbb{T}_\infty \otimes_{\Lambda^{\mathrm{wt}}} \mathbb{B}^{\mathrm{wt}}_{\mathrm{dR}}\big) \\
& {\scriptstyle \otimes_{\Lambda^{\mathrm{wt}}} \Lambda^{\mathrm{wt}}_{\Gamma_r}} \downarrow & & {\scriptstyle \otimes_{\Lambda^{\mathrm{wt}}} \Lambda^{\mathrm{wt}}_{\Gamma_r}} \downarrow & & {\scriptstyle \otimes_{\Lambda^{\mathrm{wt}}} \Lambda^{\mathrm{wt}}_{\Gamma_r}} \downarrow \\
0 \longrightarrow & H^1_g\big(K, \mathrm{Ta}_p(J_r)\big) & \longrightarrow & H^1\big(K, \mathrm{Ta}_p(J_r)\big) & \overset{-\otimes 1}{\longrightarrow} & H^1\big(K, \mathrm{Ta}_p(J_r) \otimes_{\mathbb{Z}_p} B_{\mathrm{dR}}\big)
\end{array}$$

with exact rows, that $H^1_{\mathcal{G}}\big(K, \mathbb{T}_\infty\big)_0 \otimes_{\Lambda^{\mathrm{wt}}} \Lambda^{\mathrm{wt}}_{\Gamma_r}$ is contained inside of $H^1_g\big(K, \mathrm{Ta}_p(J_r)\big)$.

Passing to the limit over r, there is an injection

$$H^1_{\mathcal{G}}(K,\mathbb{T}_\infty)_0 \quad \hookrightarrow \quad H^1(K,\mathbb{T}_\infty) \cap \left(\varprojlim_r H^1_g(K,\mathrm{Ta}_p(J_r))\right).$$

On the other hand, for our jacobian varieties we may identify the local points $\varprojlim_m \partial_{r,m}(J_r(K)\widehat{\otimes}\mathbb{Z}_p)$ with the cohomology group $H^1_g(K,\mathrm{Ta}_p(J_r))$ via [BK, §3.11]. It then follows from the basic definition of $\mathbf{X}_{\mathbb{T}_\infty}$ that

$$\mathbf{e}_{\mathrm{prim}}\cdot\left(\left(\mathbf{e}_{\mathrm{ord}}\cdot\varprojlim_r H^1_g(K,\mathrm{Ta}_p(J_r))\right)\otimes_{\Lambda^{\mathrm{wt}}}\mathrm{Frac}(\Lambda^{\mathrm{wt}})\right) = \mathbf{X}_{\mathbb{T}_\infty}(K)\otimes_{\Lambda^{\mathrm{wt}}}\mathrm{Frac}(\Lambda^{\mathrm{wt}}).$$

Since both $H^1_{\mathcal{G}}$ and $\mathbf{X}_{\mathbb{T}_\infty}$ are Λ^{wt}-saturated, by the previous two statements we must have the containment

$$H^1_{\mathcal{G}}(K,\mathbb{T}_\infty) \quad \hookrightarrow \quad \mathbf{X}_{\mathbb{T}_\infty}(K) \qquad \text{as an embedding of } \mathcal{R}\text{-modules.}$$

We proceed by showing that these two modules have the same rank.

Remark: Let's now pick a primitive triple (λ,k_0,ϵ). As we have already discussed, if $\mathbb{V}=\widetilde{\mathrm{Mell}}_{k_0,\epsilon}\left(\mathbb{T}_\infty[\lambda]\right)$ then $\mathbb{V}$ is two-dimensional over the affinoid algebra $\mathcal{K}\langle\langle s\rangle\rangle$. As a local $G_{\mathbb{Q}_p}$-module it is pseudo-geometric. Indeed for all $k\in\mathbb{U}_{k_0}$ we have that

$$\mathbb{V}_k \;:=\; \left(\mathbb{V}/(s-(k-k_0))\mathbb{V}\right) \qquad \text{is a de Rham } \mathcal{K}[G_K]\text{-module,}$$

and $\mathbb{V}_k\otimes_{\mathcal{K}}\mathbb{C}_p \cong \mathbb{C}_p\oplus\mathbb{C}_p(k-1)$ will be its associated Hodge-Tate decomposition. The module $\mathbb{V}$ over the disk $\mathbb{U}_{k_0}\subset\mathbb{Z}_p$ interpolates the Galois representations of the Hida family $\left\{\mathbf{f}_{P_{k,\epsilon}}\right\}_{k\in\mathbb{U}_{k_0}}$. More precisely, $\mathbb{V}_k$ is equivalent to the (contragredient) representation $V^*_{\mathbf{f}_{P_{k,\epsilon}}}$ associated to each eigenform $\mathbf{f}_{P_{k,\epsilon}}\in\mathcal{S}_k\left(Np^r,\xi\epsilon\omega^{k_0-k}\right)$.

By Corollary 5.16 we know that for all positive integers $k\in\mathbb{U}_{k_0}$ as above,

$$\mathrm{rank}_{\mathcal{K}\langle\langle s\rangle\rangle}H^1_{\mathcal{G}}(K,\mathbb{V}) \;=\; \dim_{\mathcal{K}} H^1_g\left(K,V^*_{\mathbf{f}_{P_{k,\epsilon}}}\right).$$

To compute the right-hand side, observe that the dual exponential map induces an isomorphism

$$\exp^*: H^1_{/g}\left(K,V^*_{\mathbf{f}_{P_{k,\epsilon}}}\right) \;\stackrel{\sim}{\longrightarrow}\; \mathrm{Fil}^0\mathbf{D}_{\mathrm{dR}}\left(V^*_{\mathbf{f}_{P_{k,\epsilon}}}\right).$$

The cotangent space is generated by the dual form $\mathbf{f}^*_{P_{k,\epsilon}}\in\mathcal{S}_k\left(Np^r,(\xi\epsilon)^{-1}\omega^{k-k_0}\right)$. In particular, the dimension of $H^1_g\left(K,V^*_{\mathbf{f}_{P_{k,\epsilon}}}\right)$ equals $[K:\mathbb{Q}_p]$ since

$$\begin{aligned}\dim_{\mathcal{K}}H^1_g\left(K,V^*_{\mathbf{f}_{P_{k,\epsilon}}}\right) &= \dim_{\mathcal{K}}H^1\left(K,V^*_{\mathbf{f}_{P_{k,\epsilon}}}\right) - \dim_{\mathcal{K}}\mathrm{Fil}^0\mathbf{D}_{\mathrm{dR}}\left(V^*_{\mathbf{f}_{P_{k,\epsilon}}}\right)\\ &= 2[K:\mathbb{Q}_p] - [K:\mathbb{Q}_p] \;=\; [K:\mathbb{Q}_p].\end{aligned}$$

The $\mathbf{I}$-module $\mathbb{T}_\infty[\lambda]$ is completely covered by these $\mathbb{V}$'s, for varying base weights k_0 and characters ϵ. We therefore deduce the rank of $H^1_{\mathcal{G}}(K, \mathbb{T}_\infty)$ is exactly $[K : \mathbb{Q}_p]$.

On the other hand, the $\mathcal{R}$-rank of $\mathbf{X}_{\mathbb{T}_\infty}(K)$ equals

$$\dim_{\mathcal{F}}\left(\mathbf{e}_{\mathrm{prim}}\cdot\left(\left(\mathbf{e}_{\mathrm{ord}}\cdot\varprojlim_{r} H^1_g\big(K, \mathrm{Ta}_p(J_r)\big)\right)\otimes_{\mathcal{R}}\mathcal{F}\right)\right) \quad \text{with } \mathcal{F} = \mathrm{Frac}(\mathcal{R}).$$

The latter is bounded above by the $\mathcal{K}$-dimension of its specialisations $H^1_g\left(K, V^*_{\mathbf{f}_{P_{k,\epsilon}}}\right)$ at arithmetic points $P_{k,\epsilon} \in \mathrm{Spec}\ \mathbf{I}(\mathcal{O}_{\mathcal{K}})^{\mathrm{alg}}$ of type (k, ϵ). However, we now know these latter quantities to be none other than $[K : \mathbb{Q}_p]$. We have proven that

$$[K : \mathbb{Q}_p] \;=\; \mathrm{rank}_{\mathcal{R}}\, H^1_{\mathcal{G}}(K, \mathbb{T}_\infty) \;\le\; \mathrm{rank}_{\mathcal{R}}\, \mathbf{X}_{\mathbb{T}_\infty}(K) \;\le\; [K : \mathbb{Q}_p]$$

which implies the statement of the theorem.

□

Remarks: (i) In later chapters, we will use these families of p-adic points $\mathbf{X}_{\mathbb{T}_\infty}(-)$ as a handy method of defining Selmer groups, within a big Galois representation. In general, these objects tend to be ever so slightly smaller than their counterparts constructed (by Greenberg) using the filtration $\mathrm{F}^+\mathbb{T}_\infty$ and quotient $\mathbb{T}_\infty/\mathrm{F}^+\mathbb{T}_\infty$.

(ii) Given that $H^1_{\mathcal{G}}(K, \mathbb{T}_\infty)$ is identified with the local points $\mathbf{X}_{\mathbb{T}_\infty}(K)$ under 5.18, it is sensible to inquire whether there exists a similar result for $H^1_{\mathcal{E}}(K, \mathbb{T}^*_\infty(1))$. Unfortunately, we are at a loss where to look for an analogue " $\mathbf{X}_{\mathbb{T}^*_\infty(1)}(K)$ " (?) because $\mathbb{T}^*_\infty(1)$ is not the projective limit of Tate modules of abelian varieties.

(iii) Unsurprisingly, there are semi-local versions of all these concepts/definitions. For example, if $\mathbf{F}$ is a number field then we define

$$\mathbf{X}_{\mathbb{T}_\infty}(\mathbf{F}\otimes\mathbb{Z}_p) \;\subset\; H^1_{\mathrm{cont}}(\mathbf{F}\otimes\mathbb{Z}_p,\ \mathbb{T}_\infty)$$

to be the largest $\mathcal{R}$-submodule, which restricts inside the local condition $\mathbf{X}_{\mathbb{T}_\infty}(\mathbf{F}_{\mathfrak{P}})$ at every prime $\mathfrak{P} \in \mathrm{Spec}\ \mathcal{O}_{\mathbf{F}}$ lying above p.

CHAPTER VI

Super Zeta-Elements

The groundwork has been laid with the local theory, and it is time to think global. At the core of the argument is a desire to make maximum use of the arithmetic encoded in the full zeta-element Euler system. In the existing theories, one starts off with a norm-compatible family, then churns out a p-adic L-function as output. However this means we are actually throwing away a large amount of information, completely ignoring the sideways relations, not to mention the level-compatibility.

We are aiming for an existence and uniqueness result relating zeta-elements with modular symbols. Surprisingly the existence is quite easy to establish, and does not require any of the notions of $\mathcal{Q}$-continuity, nor the control theory of Chapter IV. Perversely, establishing uniqueness is a major headache! In this respect we need to make precise exactly what is meant by 'uniqueness'. For example, the two-variable p-adic L-function was itself only well-defined up to an $\mathbf{I}$-adic unit (because $\Xi_{\lambda}^{\pm}$ was). It is therefore unreasonable to expect any better for our algebraic interpolations. Nevertheless, the periods present in our work are unique, up to this initial choice.

The plan of this chapter is as follows. We begin by lifting the zeta-element construction to the universal weight-deformed representation $\mathbb{T}_{\infty}$. This is a fairly natural thing to do, but strangely the details have never been written up anywhere. Section 6.2 contains one of the major theorems in this book, and establishes a two-variable interpolation from étale cohomology, into the space of $\mathbf{I}$-adic symbols. Under this identification, the weight-deformed zeta-elements are essentially mapped to the universal $\mathbf{I}$-adic modular symbol of §4.3.

As a nice consequence, these results help develop a two-variable Iwasawa theory of modular forms (we shall also mention some related work of Ochiai and Fukaya). A key corollary is the existence of a Λ-adic analogue of Coleman's exact sequence in the cyclotomic theory, this time over deformation rings of Krull dimension three. The specialisation of this four-term exact sequence at weight one, is conjecturally connected with the existence of certain Stark S-units in abelian number fields. Using Pontrjagin duality and the complete intersection properties of Hecke algebras, we compute the torsion submodule contained within the kernel of these maps.

Our recommendation to the reader brought up on a diet of MTV, is to read only the statements of Theorems 6.2 and 6.4, and then skip straight to the start of §6.3. The proofs of these two results are rather unenlightening and technical, they can be omitted without compromising any of the arithmetic applications.

6.1 The $\mathcal{R}$-adic version of Kato's theorem

Central to our cyclotomic deformation of the space $\mathcal{MS}\big(Np^r;\mathrm{Sym}^{k-2}(\mathcal{K}_f)\big)$ was the transformation of the Euler systems $\vec{\mathbf{z}}$ into the algebraic modular symbol η_f. The method itself relied on the Λ^{cy}-homomorphism

$$\mathcal{E}_{k,\epsilon}:\mathbb{H}^1_{\mathrm{Eul},p}\big(V_f^*\big)\longrightarrow \mathrm{Hom}_{\mathbb{Q}\mathrm{cts}}\Big(\mathrm{div}^0\mathbb{P}^1(\mathbb{Q}),\mathrm{Sym}^{k-2}\big(\mathcal{K}_f\big)\Big)\quad\text{of }\S3.3.$$

To play the same game within a p-ordinary family, we should first try to deform $\vec{\mathbf{z}}$ over weight-space. Assuming that this plan all works out fine, where should one expect such 'super Euler systems' to live?

Definition 6.1. *(i) We write $\mathbb{H}^1_{\mathrm{Eul}}(\mathbb{T}_\infty)$ for the $\mathcal{R}$-submodule of elements*

$$\vec{\mathfrak{X}}=\big(\ldots,(\mathfrak{X}_{Mp^n})_n,\ldots\big)\in\prod_{M\in\mathbb{N}-p\mathbb{N}}\varprojlim_n H^1_{\text{ét}}\Big(\mathbb{Z}\big[\zeta_{Mp^n},1/p\big],\mathbb{T}_\infty\Big)$$

satisfying for all $n\geq 0$, $p\nmid M$ and primes l

$$\mathrm{cores}_{(\mu_{Mp^nl})/(\mu_{Mp^n})}\big(\mathfrak{X}_{Mp^nl}\big)$$
$$=\begin{cases}\left(1-T_l^{\mathcal{P}}\,(l\sigma_l)^{-1}+l\begin{pmatrix}l&0\\0&l^{-1}\end{pmatrix}^*(l\sigma_l)^{-2}\right).\,\mathfrak{X}_{Mp^n} & \text{if } l\nmid pMN\\ \big(1-T_l^{\mathcal{P}}\,(l\sigma_l)^{-1}\big).\,\mathfrak{X}_{Mp^n} & \text{if } l\nmid pM \text{ but } l|N\\ \mathfrak{X}_{Mp^n} & \text{if } l\,|pM\end{cases}$$

where $\begin{pmatrix}l&0\\0&l^{-1}\end{pmatrix}^*$ denotes the action of the matrix $\begin{pmatrix}l&0\\0&l^{-1}\end{pmatrix}\in\mathrm{GL}_2(\mathbb{Z}_{p,N})$ on $\mathbb{T}_\infty$.

(ii) Similarly, one denotes by

$$\mathbb{H}^1_{\mathrm{Eul},p}(\mathbb{T}_\infty)\subset\prod_{M\in\mathbb{N}-p\mathbb{N}}\varprojlim_n H^1\Big(\mathbb{Q}(\mu_{Mp^n})\otimes\mathbb{Z}_p,\mathbb{T}_\infty\Big)$$

the image of $\mathbb{H}^1_{\mathrm{Eul}}(\mathbb{T}_\infty)$ under the semi-localisation map loc_p.

As we shall shortly discover, for every primitive $\lambda:\mathbf{h}^{\mathrm{ord}}(Np^\infty;\mathcal{O}_\mathcal{K})\otimes_{\Lambda^{\mathrm{wt}}_{\mathcal{K}}}\mathbf{I}\to\mathbf{I}$, fixed base weight $k_0\geq 2$ and character ϵ, there is an abundance of specialisations

$$\mathbb{H}^1_{\mathrm{Eul}}\big(\mathbb{T}_\infty[\lambda]\big)\overset{\mathrm{mod}\,P}{\longrightarrow}\mathbb{H}^1_{\mathrm{Eul}}\big(V^*_{\mathbf{f}_P}\big)\quad\text{and}\quad\mathbb{H}^1_{\mathrm{Eul},p}\big(\mathbb{T}_\infty[\lambda]\big)\overset{\mathrm{mod}\,P}{\longrightarrow}\mathbb{H}^1_{\mathrm{Eul},p}\big(V^*_{\mathbf{f}_P}\big)$$

at almost all arithmetic points $P\in\mathrm{Spec}\,\mathbf{I}(\mathcal{O}_\mathcal{K})^{\mathrm{alg}}$ of type (k,ϵ), $k\in\mathbb{U}_{k_0}$.

Let $\widetilde{\mathcal{R}}=\mathcal{R}\otimes_{\Lambda^{\mathrm{wt}}}\mathrm{Frac}(\Lambda^{\mathrm{wt}})$, where $\mathcal{R}$ was the image of the Hecke algebra inside of the endomorphisms $\mathrm{End}_{\Lambda^{\mathrm{wt}}}\big(\mathbb{T}_\infty\big)$.

Theorem 6.2. *There exist canonical products of global cohomology classes*

$$\vec{\mathfrak{Z}} = \left(\ldots,(\mathfrak{Z}_{Mp^n})_n,\ldots\right) \in \overline{\mathcal{MS}}^{\mathrm{ord}}(Np^\infty) \otimes_{\mathbf{h}^{\mathrm{ord}}(Np^\infty)} \mathbb{H}^1_{\mathrm{Eul}}(\mathbb{T}_\infty) \otimes_{\mathcal{R}} \widetilde{\mathcal{R}}$$

with each $\mathbf{f}$*-isotypic component* $\vec{\mathfrak{Z}}_{[\lambda]}$ *of* $\vec{\mathfrak{Z}}$ *satisfying:*

(a) $\vec{\mathfrak{Z}}_{[\lambda]} \bmod P_{k,\epsilon}$ *belongs to* $\mathcal{MS}^{\mathrm{ord}}(Np^r;\mathcal{O}_{\mathcal{K}}) \otimes_{\mathcal{O}_{\mathcal{K}}} \mathbb{H}^1_{\mathrm{Eul}}\left(V^*_{\mathbf{f}_{P_{k,\epsilon}}}\right)$, *and*

(b) $\vec{\mathfrak{Z}}_{[\lambda]} \bmod P_{k,\epsilon} \;=\; \nabla_{k-2}(\gamma^*) \otimes \vec{\mathbf{z}}_{k,\epsilon}$

at all algebraic points $P_{k,\epsilon} \in \mathrm{Spec}\,\mathbf{I}(\mathcal{O}_{\mathcal{K}})^{\mathrm{alg}}$ *of type* (k,ϵ).

N.B. here $\vec{\mathbf{z}}_{k,\epsilon} \in \mathbb{H}^1_{\mathrm{Eul}}\left(V^*_{\mathbf{f}_{P_{k,\epsilon}}}\right)$ denoted the N-primitive Euler systems associated to the p-stabilised newform $\mathbf{f}_{P_{k,\epsilon}} \in \mathcal{S}_k\left(\Gamma_0(Np^r),\epsilon\xi\omega^{-k}\right)$, introduced in Theorem 3.8. Moreover $\gamma = \gamma_{\mathbf{f}^*_{P_{k,\epsilon}}}$ was the path for which

$$(2\pi i)^{2-k}\mathrm{per}_\infty\left(\mathbf{f}^*_{P_{k,\epsilon}}\right) \;=\; \Omega^+_{\mathbf{f}_{P_{k,\epsilon}}}\gamma^+ + \Omega^-_{\mathbf{f}_{P_{k,\epsilon}}}\gamma^-;$$

its dual vector γ^* fulfills the condition $\langle\gamma,\gamma^*\rangle = 1$, under the Poincaré pairing

$$\langle,\rangle : H^1\left(Y_1(Np^r)(\mathbb{C}),\,\mathfrak{F}^{(k)}_p\right) \times H^1_c\left(Y_1(Np^r)(\mathbb{C}),\,\mathfrak{F}^{(k)}_p\right)\left[\lambda_{\mathbf{f}^*_{P_{k,\epsilon}}}\right] \to \mathcal{O}_{\mathcal{K}}.$$

In order to obtain uniqueness for the super zeta-elements $\vec{\mathfrak{Z}}$, it is necessary to first tensor $\mathbb{H}^1_{\mathrm{Eul}}$ by the space $\overline{\mathcal{MS}}^{\mathrm{ord}}$. Otherwise one obtains a Λ^{wt}-adic Euler system with ill-defined periods.

Proof: In contrast to Kato's original ideas which lie very deep indeed, the Λ^{wt}-adic construction is relatively easy, being a formal consequence of the p-adic situation. Recall from Chapter III that we first considered Beilinson's elements in the K-theory of modular curves, and then showed how they satisfy various compatibility relations under the norm. Using dirty tricks these elements can be mapped into Galois cohomology, whereupon the norm relations translate into the Euler system relations. Finally, using hard techniques from p-adic Hodge theory (explicit reciprocity laws) these elements were shown by Kato [Ka2] to be related to values of the L-function on the critical strip.

Our task is to show that there is a Λ^{wt}-adic route whereby these cohomology classes may be obtained. For the reader's benefit, recall that we chose integers $c,d,A,B \in \mathbb{Z}$ with $\gcd(cd,6AB)=1$, and zeta-elements satisfying

$${}_{c,d}z_{A,B} \;=\; \left\{\,{}_cg_{1/A,0},\ {}_dg_{0,1/B}\right\} \in K_2\left(Y(A,B)\right)$$

– here the Siegel units ${}_cg_{-,0}$ and ${}_dg_{0,-}$ were pullbacks of ϑ-functions on $\mathbb{E}_{\mathrm{univ}}$. Moreover, if $c,d \neq \pm 1$ then we also have non-integral elements $z_{A,B}$, related via

$${}_{c,d}z_{A,B} \;=\; \left(c^2 - \begin{pmatrix} c & 0 \\ 0 & 1\end{pmatrix}^*\right)\left(d^2 - \begin{pmatrix} 1 & 0 \\ 0 & d\end{pmatrix}^*\right) z_{A,B}.$$

Question. *How can these elements be turned into 1-cocycles whose coefficients take values in the big Galois representation $\mathbb{T}_\infty \otimes_{\mathcal{R}} \widetilde{\mathcal{R}}$?*

The argument has four steps in total. We commence by defining a map $\mathfrak{c}^{\mathrm{univ}}_{\infty,p}$ from the double projective limit $\varprojlim_{r,m} K_2\big(Y(Mp^{n+m}, MNp^{n+r+m})\big)$, targeting the Galois cohomology group $H^1\big(\mathrm{Gal}(\overline{\mathbb{Q}}/\mathbb{Q}(\mu_{Mp^n})), \mathbb{T}_\infty\big) \otimes_{\mathcal{R}} \widetilde{\mathcal{R}}$. In the second step, we normalise our constructions via a canonically defined path vector inside $\mathcal{UM}^{\mathrm{ord}}$. Next we show that modulo height one primes, the images of the elements ${}_{c,d}z_{-,-}$ under $\mathfrak{c}^{\mathrm{univ}}_{\infty,p}$ are very close to being the p-adic Euler systems attached to eigenforms. Finally, we remove the dependence on the integers c, d to obtain what we surmise to be the true super zeta-elements.

Remark: T. Fukaya [Fu] only requires a single projective limit in her theory, but we will need the double limit here as we are cutting out a big Galois representation. In her situation she avoids using Galois cohomology completely, when constructing a universal zeta-modular form.

Step 1: Passing from K_2 to étale cohomology.
We begin by outlining how to construct a zeta-element in the cohomology group $H^1(\mathbb{Q}(\mu_{Mp^n}), \mathbb{T}_\infty) \otimes_{\mathcal{R}} \widetilde{\mathcal{R}}$ where as usual M is coprime to p, and the integer $n \geq 0$. Using first Proposition 2.13, if $\gcd(cd, 6MNp) = 1$ we can consider the elements

$$ {}_{c,d}z_{Mp^{n+m}, MNp^{n+r+m}} \quad \text{as belonging to} \quad \varprojlim_r \varprojlim_m K_2\Big(Y\big(Mp^{n+m}, MNp^{n+r+m}\big)\Big) $$

where both limits are taken with respect to the norm homomorphism in K-theory. The second Chern class induces

$$ \varprojlim_{r,m} K_2(Y(Mp^{n+m}, MNp^{n+r+m})) \xrightarrow{\mathrm{ch}_{2,2}} \varprojlim_{r,m} H^2_{\text{ét}}(Y(Mp^{n+m}, MNp^{n+r+m}), \mathbb{Z}/p^m\mathbb{Z}(2)) $$

and then twisting via $\otimes\zeta_{p^m}^{\otimes -1}$ means we actually end up lying inside of the limit $\varprojlim_{r,m} H^2_{\text{ét}}\big(Y(Mp^{n+m}, MNp^{n+r+m}), \mathbb{Z}/p^m\mathbb{Z}(1)\big)$.

The natural projection $\varprojlim_m Y(Mp^{n+m}, MNp^{n+r+m}) \xrightarrow{\mathrm{proj}} Y(Mp^n, MNp^{n+r})$ gives rise to an application

$$ \varprojlim_{r,m} H^2_{\text{ét}}\big(Y(Mp^{n+m}, MNp^{n+r+m}), \mathbb{Z}/p^m\mathbb{Z}(1)\big) \xrightarrow{\mathrm{proj}_*} $$
$$ \varprojlim_{r,m} H^2_{\text{ét}}\big(Y(Mp^n, MNp^{n+r}), \mathbb{Z}/p^m\mathbb{Z}(1)\big) $$

with the right-hand group clearly equalling $\varprojlim_r H^2_{\text{ét}}\big(Y(Mp^n, MNp^{n+r}), \mathbb{Z}_p(1)\big)$. Under this barrage of arrows, the reader should bear in mind that all the maps being applied are, in fact, functorial and canonical (except the twist which depends upon the choice of compatible system $(\zeta_{p^m})_m$).

In order to enter Galois cohomology, as in §2.5 we appeal to the Hochschild-Serre spectral sequence

$$E_2^{i,j} = H^i\Big(\mathbb{Q},\, H^j_{\text{ét}}\big(Y(-,-)\otimes\overline{\mathbb{Q}},\,\ldots\big)\Big) \implies H^{i+j}_{\text{ét}}\big(Y(-,-),\,\ldots\big).$$

Because $H^j_{\text{ét}}\big(Y(-,-)\otimes\overline{\mathbb{Q}},...\big)=0$ for all $j\geq 2$, the edge homomorphism yields

$$H^2_{\text{ét}}\big(Y(Mp^n, MNp^{n+r}), \mathbb{Z}_p(1)\big) \xrightarrow{\partial} H^1\big(\mathbb{Q}, H^1_{\text{ét}}\big(Y(Mp^n, MNp^{n+r})\otimes\overline{\mathbb{Q}}, \mathbb{Z}_p(1)\big)\big)$$

which is identified with $H^1\Big(\mathbb{Q}(\mu_{Mp^n}), H^1_{\text{ét}}\big(Y(Mp^n, MNp^{n+r})\otimes_{\mathbb{Q}(\mu_{Mp^n})}\overline{\mathbb{Q}}, \mathbb{Z}_p(1)\big)\Big)$.

Remark: To paraphrase so far, the composition $\partial\circ\text{proj}_*\circ(\otimes\zeta_{p^m}^{\otimes -1})\circ\text{ch}_{2,2}$ sent

$$\varprojlim_{r,m} K_2\big(Y(Mp^{n+m}, MNp^{n+r+m})\big) \longrightarrow$$
$$H^1\Big(\mathbb{Q}(\mu_{Mp^n}),\, \varprojlim_r H^1_{\text{ét}}\big(Y(Mp^n, MNp^{n+r})\otimes_{\mathbb{Q}(\mu_{Mp^n})}\overline{\mathbb{Q}}, \mathbb{Z}_p(1)\big)\Big)$$

as a homomorphism of abstract Hecke modules.

Let us now see how to change the space that our coefficients lie in, from the inverse system $\varprojlim_r H^1_{\text{ét}}\big(Y(Mp^n, MNp^{n+r})\otimes_{\mathbb{Q}(\mu_{Mp^n})}\overline{\mathbb{Q}},\ \mathbb{Z}_p(1)\big)$ into the universal Galois representation $\mathbb{T}_\infty\otimes_{\mathcal{R}}\widetilde{\mathcal{R}}$.

For each $a, \mathbf{a}\geq 1$ such that $Mp^n\mathbf{a}\geq 2$, the mapping $\tau\mapsto\frac{\tau+a}{\mathbf{a}}$ on $\mathfrak{H}$ induced a unique finite, flat morphism

$$\Psi_{\mathbf{a},a} : Y(Mp^n\mathbf{a}, N') \longrightarrow Y_1(Np^r)\otimes\mathbb{Q}(\mu_{Mp^n})$$

with $N'\geq 1$ any integer such that $Np^r|N'$, $Mp^n|N'$ and $\text{supp}(N')=\text{supp}(MNp\mathbf{a})$. As we previously mentioned,

$$(\Psi_{\mathbf{a},a})_*\big(\,{}_{c,d}z_{Mp^n\mathbf{a},N'}\big)\ \in\ K_2\Big(Y_1(Np^r)\otimes\mathbb{Q}(\mu_{Mp^n})\Big)$$

depends only on the class of a mod $\mathbf{a}$. To simplify the exposition enormously we take $(\mathbf{a},a)=(1,0)$ throughout, although the reader can make the general case work by replacing M by $M\mathbf{a}$ if they so wish. In terms of étale realisations,

$$H^1\Big(\mathbb{Q}(\mu_{Mp^n}),\, \varprojlim_r H^1_{\text{ét}}\big(Y(Mp^n, MNp^{n+r})\otimes_{\mathbb{Q}(\mu_{Mp^n})}\overline{\mathbb{Q}},\ \mathbb{Z}_p(1)\big)\Big)$$
$$\xrightarrow{(\Psi_{1,0})_*}\quad H^1\Big(\mathbb{Q}(\mu_{Mp^n}),\, \varprojlim_r H^1_{\text{ét}}\big(Y_1(Np^r)\otimes\overline{\mathbb{Q}},\ \mathbb{Z}_p(1)\big)\Big)$$

as $\Psi_{1,0}$ is compatible with varying $r\geq 1$.

We will now switch lattices. The Manin-Drinfeld principle gives us a unique splitting of the inclusion

$$\mathbb{Q}\otimes H^1_{\text{ét}}\big(X_1(Np^r)\otimes\overline{\mathbb{Q}},\ \mathbb{Z}_p(1)\big)\ \hookrightarrow\ \mathbb{Q}\otimes H^1_{\text{ét}}\big(Y_1(Np^r)\otimes\overline{\mathbb{Q}},\ \mathbb{Z}_p(1)\big),$$

the constant of integrality depending crucially on the Hecke algebra at level Np^r.

However the number of cusps of $X_1(Np^r)$ equals $\frac{1}{2}\sum_{0<d|Np^r}\phi(d)\phi(Np^r/d)$, which behaves like $O(p^r)$ as we increase r. Therefore upon passage to the limit over $r \geq 1$, we can only construct a section after first tensoring over Λ^{wt} by its field of fractions (c.f. the construction of $\mathbf{e}_{\mathrm{prim}}$ in [Hi1,Hi2]).

Since $H^1_{\text{ét}}\Big(X_1(Np^r)\otimes\overline{\mathbb{Q}},\ \mathbb{Z}_p(1)\Big)$ is none other than $\mathrm{Ta}_p\Big(\mathrm{Jac}\,X_1(Np^r)\Big)$, this is equivalent to splitting the injection

$$\mathrm{Frac}(\Lambda^{\mathrm{wt}})\otimes_{\Lambda^{\mathrm{wt}}}\varprojlim_r \mathrm{Ta}_p(J_r) \quad\hookrightarrow\quad \mathrm{Frac}(\Lambda^{\mathrm{wt}})\otimes_{\Lambda^{\mathrm{wt}}}\varprojlim_r H^1_{\text{ét}}\Big(Y_1(Np^r)\otimes\overline{\mathbb{Q}},\ \mathbb{Z}_p(1)\Big).$$

Writing $\mathbf{e}_{\text{M-D}}$ for the corresponding projector, we get an induced map

$$H^1\Big(\mathbb{Q}(\mu_{Mp^n}),\ \varprojlim_r H^1_{\text{ét}}\left(Y_1(Np^r)\otimes\overline{\mathbb{Q}},\ \mathbb{Z}_p(1)\right)\Big) \overset{(\mathbf{e}_{\text{M-D}})_*}{\longrightarrow}$$
$$H^1\Big(\mathbb{Q}(\mu_{Mp^n}),\ \mathrm{Ta}_p(J_\infty)\otimes_{\Lambda^{\mathrm{wt}}}\mathrm{Frac}(\Lambda^{\mathrm{wt}})\Big).$$

Lastly, using the idempotent $\mathbf{e}_{\mathrm{prim}}.\mathbf{e}\in\mathbf{H}_{Np^\infty}\otimes_{\Lambda^{\mathrm{wt}}}\mathrm{Frac}(\Lambda^{\mathrm{wt}})$ allows us to cut out the direct summand $H^1\Big(\mathbb{Q}(\mu_{Mp^n}),\mathbb{T}_\infty\otimes_{\Lambda^{\mathrm{wt}}}\mathrm{Frac}(\Lambda^{\mathrm{wt}})\Big)$ from the above.

Definition 6.3.
(i) We construct the homomorphism

$$\mathfrak{c}^{\mathrm{univ}}_{\infty,p} : \varprojlim_{r,m} K_2\Big(Y(Mp^{n+m}, MNp^{n+r+m})\Big) \longrightarrow H^1\Big(\mathbb{Q}(\mu_{Mp^n}),\mathbb{T}_\infty\otimes_{\mathcal{R}}\widetilde{\mathcal{R}}\Big)$$

by setting $\mathfrak{c}^{\mathrm{univ}}_{\infty,p} := (\mathbf{e}_{\mathrm{prim}}.\ \mathbf{e}.\ \mathbf{e}_{\text{M-D}})_*\circ(\Psi_{1,0})_*\circ\partial\circ(\pi_{m=0})_*\circ(\otimes\zeta_{p^m}^{\otimes -1})\circ ch_{2,2}$.

(ii) One defines the $\mathcal{R}$-adic one-cocycle ${}_{c,d}\mathfrak{Z}^{\dagger}_{Mp^n}$ *to be the image of the elements* $\left({}_{c,d}z_{Mp^{n+m},MNp^{n+r+m}}\right)_{r,m}$ *under* $\mathfrak{c}^{\mathrm{univ}}_{\infty,p}$, *with* $c,d\neq\pm1$, $\gcd(cd,6MNp)=1$.

Step 2: The universal path vector.
The good news is that we now possess a Λ^{wt}-adic zeta-element at our disposal. The bad news is that it's not the correct element for the statement of Theorem 6.2. Specialisations of the cocycle ${}_{c,d}\mathfrak{Z}^{\dagger}_{Mp^n}$ modulo height one primes differ from the N-primitive elements in two distinct ways. The first is that they have this annoying dependence on the integers c,d (see Step 4). The second is that their values under the dual exponential give the wrong periods, namely $\Omega^{\pm}_{0(1)}$ rather than $\Omega^{\pm}_f$.

Let's address the latter problem. Our remedy is to find an object which measures the difference between these two periods, and the natural place to look should be the space of universal p-adic modular symbols.

Lemma B.1. *There exists a vector* $\underline{\delta}^{\mathrm{univ}\ *}_{1,Np^\infty}$ *in the space* $\overline{\mathcal{MS}}^{\mathrm{ord}}(Np^\infty)$, *such that*

$$\left(\underline{\delta}^{\mathrm{univ}\ *}_{1,Np^\infty}\otimes{}_{c,d}\mathfrak{Z}^{\dagger}_{Mp^n}\right)[\lambda]\ \bmod P_{k,\epsilon} \quad=\quad \nabla_{k-2}\left(\gamma^*_{\mathbf{f}^*_{P_{k,\epsilon}}}\right)\otimes{}_{c,d}\mathbf{z}_{Mp^n}$$

for all primitive $\lambda:\mathbf{h}^{\mathrm{ord}}(Np^\infty;\mathcal{O}_{\mathcal{K}})\otimes_{\Lambda^{\mathrm{wt}}_{\mathcal{K}}}\mathbf{I}\rightarrow\mathbf{I}$, *and points* $P_{k,\epsilon}\in\mathrm{Spec}\,\mathbf{I}(\mathcal{O}_{\mathcal{K}})^{\mathrm{alg}}$.

Note that ${}_{c,d}\mathbf{z}_{Mp^n} \in H^1\Big(\mathbb{Q}(\mu_{Mp^n}), V^*_{\mathbf{f}_{P_{k,\epsilon}}}\Big)$ are the p-integral but (c,d)-dependent zeta-elements, associated to the eigenform $\mathbf{f}_{P_{k,\epsilon}}$. We defer the proof of this lemma to Appendix B as it is rather tedious.

Notation: Let's write ${}_{c,d}\mathfrak{Z}_{Mp^n}$ for the path-normalised element $\underline{\delta}^{\text{univ}\ *}_{1,Np^\infty} \otimes {}_{c,d}\mathfrak{Z}^\dagger_{Mp^n}$ belonging to $\mathcal{MS}^{\text{ord}} \otimes_{\mathbf{h}^{\text{ord}}(Np^\infty)} H^1\Big(\mathbb{Q}(\mu_{Mp^n}), \mathbb{T}_\infty\Big) \otimes_{\mathcal{R}} \widetilde{\mathcal{R}}$.

The behaviour of ${}_{c,d}\mathfrak{Z}_{Mp^n,[\lambda]}$ modulo height one prime ideals, implies these $\mathbf{I}$-adic elements satisfy the properties required in Theorem 6.2(b).

Step 3: Establishing the vertical and sideways relations.
There remain two tasks to complete. The first is to show that the sequence $\Big(\dots, \big({}_{c,d}\mathfrak{Z}_{Mp^n}\big)_n, \dots\Big)$ belongs to the space of $\mathcal{R}$-adic Euler systems (c.f. 6.2(a)). The second and final task, is to remove this silly dependence on the integers c, d.

We assume λ and P are chosen so that $\mathbf{f}_P \in \mathcal{S}_k\big(\Gamma_0(Np^{r'}), \epsilon\xi\omega^{-k}\big)$ for some $r' \in \mathbb{N}$. Returning once more to Definition 2.14, for each pair of integers $j, j' \in \{1, \dots, k-1\}$ there exists a p-adic realisation map

$$\mathfrak{c}_{k,j,j',0(1),p} : \varprojlim_m K_2\Big(Y\big(Mp^{n+m}, Mp^{n+m}Np^{r'}\big)\Big) \longrightarrow H^1\Big(\mathbb{Q}(\mu_{Mp^n}), \mathbf{T}'(k-j)\Big)$$

where the lattice $\mathbf{T}'$ was generated by the image of $H^1_{\text{ét}}\Big(Y_1(Np^{r'}) \otimes \mathbb{Q},\ \mathfrak{F}_p^{(k)}\Big)$ inside the contragredient realisation. In particular, if the couple $(j, j') = (1, k-1)$ then the target of $\mathfrak{c}_{k,1,k-1,0(1),p}$ ends up inside $V_{\mathbf{f}^*_P}(k-1) \cong V^*_{\mathbf{f}_P}$.

Remark: The following diagram commutes

$$\begin{array}{ccc}
\varprojlim_r \varprojlim_m K_2\Big(Y\big(Mp^{n+m}, MNp^{n+r+m}\big)\Big) & \overset{\mathfrak{c}^{\text{univ}}_{\infty,p}}{\longrightarrow} & H^1\Big(\mathbb{Q}(\mu_{Mp^n}), \mathbb{T}_\infty \otimes_{\mathcal{R}} \widetilde{\mathcal{R}}\Big) \\
\Big\downarrow \text{projection to } r'\text{-th level} & & \Big\downarrow \otimes_{\mathbf{h}^{\text{ord}},\lambda} \mathbf{I}/P \\
\varprojlim_m K_2\Big(Y\big(Mp^{n+m}, Mp^{n+m}Np^{r'}\big)\Big) & \overset{\mathfrak{c}_{k,1,k-1,0(1),p}}{\longrightarrow} & H^1\Big(\mathbb{Q}(\mu_{Mp^n}), V^*_{\mathbf{f}_P}\Big)
\end{array}$$

where the right-most arrow is induced by the composition $P \circ \lambda$.

The commutativity of the above square is clear from the following reasoning. Firstly, there is a natural correspondence between $\mathbf{f}^*_P$ and the maximal ideal $\mathfrak{m}_{\mathbf{f}^*_P}$ of the Hecke algebra, generated by the primitive tensors

$$\Big\{T_h^{\mathcal{D}} \otimes 1 - 1 \otimes a_h(\mathbf{f}_P) \in h_k\big(\Phi_{r'}, \epsilon; \mathcal{O}_{\mathcal{K}}\big) \ \text{ for every } \ h \in \mathbb{N}\Big\}.$$

However, the above set constitutes the image under $P \in \text{Spec } \mathbf{I}(\mathcal{O}_{\mathcal{K}})^{\text{alg}}$ of the precise relations cutting out the $\mathbf{f}$-isotypic component of $\mathbb{T}_\infty$, namely

$$\Big\{\underline{w} \in \mathbb{T}_\infty \otimes_{\Lambda^{\text{wt}}} \mathbf{I} \ \text{ such that } \ \Big(T_h^{\mathcal{D}} \otimes 1 - 1 \otimes a_h(\mathbf{f})\Big).\underline{w} \ = \ \underline{0} \ \text{ for every } h \in \mathbb{N}\Big\}.$$

The functoriality of $\mathfrak{c}^{\text{univ}}_{\infty,p}$ and $\mathfrak{c}_{k,1,k-1,0(1),p}$ implies the diagram is commutative.

The vertical Euler system relations are trivial to check. Proposition 2.13 already yielded the norm-compatibility of $\left({}_{c,d}z_{-,-}\right)_{m,r}$, then taking the limit

$$\mathfrak{c}^{\text{univ}}_{\infty,p}\Big(\left({}_{c,d}z_{Mp^{n+m},MNp^{n+r+m}}\right)_{m,r}\Big)_n \quad \text{belongs to} \quad \varprojlim_n H^1\Big(\mathbb{Q}(\mu_{Mp^n}),\mathbb{T}_\infty\Big)\otimes_{\mathcal{R}}\widetilde{\mathcal{R}}.$$

The sideways relations are themselves proved in [Ka1, Prop 8.10], and we refer the reader to this article for further details.

To complete Step 3, it remains to show the Euler factors involved when removing primes in the $\mathcal{R}$-adic situation, coincide modulo P with their p-adic counterparts. Recall that the sideways relations were

$$\begin{cases} \left(1 - T_l^{\mathcal{D}}\,(l\sigma_l)^{-1} + l\begin{pmatrix} l & 0 \\ 0 & l^{-1}\end{pmatrix}^{*}(l\sigma_l)^{-2}\right) & \text{if } l \nmid pMN \\ \left(1 - T_l^{\mathcal{D}}\,(l\sigma_l)^{-1}\right) & \text{if } l \nmid pM \text{ but } l|N. \end{cases}$$

From the discussion in Chapter IV, at each prime l the dual Hecke operator $T_l^{\mathcal{D}}$ acts on $\mathbb{T}_\infty[\lambda]$ via the scalar $\lambda(T_l)$, which is equal modulo P to the eigenvalue $a_l(\mathbf{f}_P)$. Similarly, if the prime number $l \nmid Np$ then

$$\begin{pmatrix} l & 0 \\ 0 & l^{-1}\end{pmatrix}^{*} = \frac{(T_l^{\mathcal{D}})^2 - T_{l^2}^{\mathcal{D}}}{l}$$

operates on $\mathbb{T}_\infty[\lambda]$ through the unscaled diamond action, which coincides mod P with the p-unit $\epsilon\xi\omega^{-k}(l)l^{k-2}$.

It follows that under each $P = P_{k,\epsilon} \in \text{Spec}\,\mathbf{I}(\mathcal{O}_{\mathcal{K}})^{\text{alg}}$, these relations become

$$\begin{cases} \left(1 - a_l(\mathbf{f}_P)(l\sigma_l)^{-1} + \epsilon\xi\omega^{-k}(l)l^{k-1}(l\sigma_l)^{-2}\right) & \text{if } l \nmid pMN \\ \left(1 - a_l(\mathbf{f}_P)(l\sigma_l)^{-1}\right) & \text{if } l \nmid pM \text{ but } l|N. \end{cases}$$

As a corollary, $\mathbb{H}^1_{\text{Eul}}\big(\mathbb{T}_\infty[\lambda]\big)$ mod P injects into $\mathbb{H}^1_{\text{Eul}}\big(V^*_{\mathbf{f}_P}\big)$ at arithmetic points P. Therefore at all primes l, the sideways relations are compatible with specialisation.

Step 4: Jettisoning the integers c,d.
To obtain canonical elements, removing the (c,d)-dependence may potentially cause unwanted singularities – our task is to show these poles are purely hypothetical. We observe from [Ka1, 13.11(1)] that the ${}_{c,d}\mathbf{z}_{Mp^n}$'s are related to the N-primitive zeta-elements, via the formula

$$\Big({}_{c,d}\mathbf{z}_{Mp^n}\Big)_n = \Big(c^2 - c^2.\sigma_c\Big)\Big(d^2 - d^k\widetilde{\epsilon}(d).\sigma_d\Big)\prod_{l|N}\Big(1 - a_l\big(\mathbf{f}_{P_{k,\epsilon}}\big).(l\sigma_l)^{-1}\Big).\Big(\mathbf{z}_{Mp^n}\Big)_n$$

where the finite character $\widetilde{\epsilon} = \epsilon\xi\omega^{-k}$. As usual σ_c (resp. σ_d, σ_l) denotes the element of $\text{Gal}\big(\mathbb{Q}(\mu_{p^\infty})/\mathbb{Q}\big)$ that maps to c (resp. d, l) under the pth-cyclotomic character.

Therefore to obtain true $\mathcal{R}$-adic Euler systems, we need to establish that

$$\left(\mathfrak{Z}^{\dagger}_{p^nM}\right)_n := \left(c^2 - c^2.\sigma_c\right)^{-1}\left(d^2 - \langle d\rangle.\sigma_d\right)^{-1}\prod_{l|N}\left(1 - T_l^{\mathcal{D}}.(l\sigma_l)^{-1}\right)^{-1}\cdot\left({}_{c,d}\mathfrak{Z}^{\dagger}_{p^nM}\right)_n$$

exists as a bona fide member of the projective limit $\varprojlim_n H^1\Big(\mathbb{Q}(\mu_{Mp^n}), \mathbb{T}_\infty\Big)\otimes_{\mathcal{R}}\widetilde{\mathcal{R}}$. In other words $\left(\mathfrak{Z}^{\dagger}_{p^nM}\right)_n$ should have no poles in the cyclotomic variable.

Question. *Along which contours in the (s,k)-plane, could these poles possibly lie?*

There are essentially three cases we'll examine, corresponding to the factor at c, the factor at d, and factors at the various primes l dividing N. To visualise their zeros in the plane, one considers general characters of the form $\psi_{\beta,s} = \omega^{\beta} < - >^{s}$ where the class $\beta \in \mathbb{Z}/(p-1)\mathbb{Z}$, and the continuous variable $s \in \mathbb{Z}_p$.

Let $\lambda : \mathbf{h}^{\mathrm{ord}}(Np^{\infty};\mathcal{O}_{\mathcal{K}})\otimes_{\Lambda^{\mathrm{wt}}_{\mathcal{K}}}\mathbf{I}\rightarrow\mathbf{I}$ be primitive corresponding to some $\mathbf{f}\in\mathbb{S}^{\mathrm{ord}}_{\mathbf{I}}$, with associated character $\xi : \left(\mathbb{Z}/Np\mathbb{Z}\right)^{\times} \longrightarrow \mathcal{O}^{\times}_{\mathcal{K}}$. Assume that $k\in\mathbb{Z}_p$ and the specialisation $P_{k,\epsilon}:\mathbf{I}\rightarrow\mathcal{O}_{\mathcal{K}}$ sends $u_0 \mapsto u_0^k\epsilon(u_0)$, which need not be algebraic.

Case I – The factor at c:

$$\psi_{\beta,s}\times P_{k,\epsilon}\left(c^2 - c^2.\sigma_c\right) = \left(c^2 - c^2.\omega^{\beta}(c) < c >^{s}\right)$$

which vanishes when $\beta\equiv 0 \bmod p-1$, along the line $s=0$, $k\in\mathbb{Z}_p$;

Case II – The factor at d:

$$\begin{aligned}\psi_{\beta,s}\times P_{k,\epsilon}\left(d^2 - \langle d\rangle.\sigma_d\right) &= \left(d^2 - d^k\widetilde{\epsilon}(d)\omega^{\beta}(d) < d >^{s}\right)\\ &= d^2\left(1 - \epsilon\xi\omega^{\beta-2}(d) < d >^{s+k-2}\right)\end{aligned}$$

which vanishes when $\xi = \omega^{2-\beta}$ and ϵ is trivial, along the line $s = 2-k$;

Case III – The factors at primes $l|N$:

Pick a base weight k_0, and consider the transform $\mathrm{Mell}_{k_0,\epsilon}$ mapping the ring $\mathbf{I}$ to the rigid analytic functions in the variable k, convergent on the closed disk $\mathbb{U}_{k_0}$. Then

$$\psi_{\beta,s}\times\mathrm{Mell}_{k_0,\epsilon}\left(1 - T_l^{\mathcal{D}}.(l\sigma_l)^{-1}\right) = \left(1 - \mathrm{Mell}_{k_0,\epsilon}(\lambda\circ U_l)\omega^{-1-\beta}(l) < l >^{-1-s}\right).$$

If $k'\in\mathbb{U}_{k_0}$ is a positive integer, $\mathrm{Mell}_{k_0,\epsilon}(\lambda\circ U_l)\Big|_{k=k'}$ is the lth-eigenvalue of a classical cusp form of weight k', and moreover it is new at l. The only possible slopes at l are therefore $k'-2$, $(k'-1)/2$ or else the coefficient is zero.

Thus $\mathrm{Mell}_{k_0,\epsilon}(\lambda\circ U_l)$ must equal $c_1 < l >^{k-2}$, $c_2 < l >^{\frac{k-1}{2}}$ or is identically zero. It follows that the above Euler factor vanishes either along the line $s=k-3$, or along the line $2s = k-3$, or not at all.

Remark: By Hypothesis(Diag), we have an inclusion

$$\varprojlim_n H^1\Big(\mathbb{Q}(\mu_{Mp^n}), \mathbb{T}_\infty\Big) \otimes_{\mathcal{R}} \widetilde{\mathcal{R}} \hookrightarrow \bigoplus_{\text{prim've } \lambda} \varprojlim_n H^1\Big(\mathbb{Q}(\mu_{Mp^n}), \mathbb{T}_\infty[\lambda]\Big) \otimes_{\mathbf{I}} \mathcal{L}',$$

and the right-hand side injects into

$$\bigoplus_{\text{prim've } \lambda} \bigoplus_{\text{alg'c } P_{k,\epsilon}} \varprojlim_n H^1\Big(\mathbb{Q}(\mu_{Mp^n}), \mathbb{T}_\infty[\lambda]\Big) \otimes_{\mathbf{I}} \mathbf{I}/P_{k,\epsilon} \otimes_{\mathbb{Z}_p} \mathbb{Q}_p$$

under our (Arith) condition.

The poles in $\left(\mathfrak{Z}^{\dagger}_{p^nM,[\lambda]}\right)_n$ will trace out a rigid analytic variety in the (s,k)-plane. Let $\mathcal{V}$ be an irreducible component. From our analysis in cases (I), (II) and (III), we see that $\mathcal{V}$ must either be

(a) one of the rational lines $s = 0$, $s = 2 - k$, $s = k - 3$ or $2s = k - 3$; or
(b) a single point $\mathcal{V} = (s_1, k_1)$.

If (a) occurs then at integral weights $k \geq 2$, we would expect each family $\left(\mathbf{z}_{Mp^n}\right)_n$ attached to $\mathbf{f}_{P_{k,\epsilon}}$ to have poles at $s = 0$, $s = 2 - k$, $s = k - 3$ or $2s = k - 3$ respectively. However this is impossible, because by Theorem 3.8(ii)

$$\left(\mathbf{z}_{Mp^n}\right)_n \in p^{-\nu_M} . \varprojlim_n H^1_{\text{ét}}\Big(\mathbb{Z}\big[\zeta_{Mp^n}, 1/p\big],\ \mathbb{T}_\infty[\lambda] \otimes_{\mathbf{I}} \mathbf{I}/P_{k,\epsilon}\Big) \quad \text{has no poles.}$$

If (b) occurs, the pole $\mathcal{V} = (s_1, k_1)$ must then lie on one of the rational lines above. If k_1 were a positive integer then s_1 would belong to $\mathbb{Z}[1/2]$, and we would have already detected the pole in the corresponding $\left(\mathbf{z}_{Mp^n}\right)_n$. Thus k_1 is not arithmetic. Instead choose any sequence $\left(s_1^{(1)}, k_1^{(1)}\right), \left(s_1^{(2)}, k_1^{(2)}\right), \left(s_1^{(3)}, k_1^{(3)}\right), \ldots$ satisfying

(b1) $k_1^{(m)} \in \mathbb{Z}$, $k_1^{(m)} \geq 2$ and $k_1^{(m)} \in \mathbb{U}_{k_0}$;
(b2) $s_1^{(m)} \in \mathbb{Z}$ and $1 \leq s_1^{(m)} \leq k_1^{(m)} - 1$;
(b3) $\lim_{m\to\infty} \left(s_1^{(m)}, k_1^{(m)}\right) = (s_1, k_1)$ in the p-adic topology.

Write $\left(\mathbf{z}_{Mp^n}\left(\mathbf{f}_{P_{k_1^{(m)},\epsilon}}\right) \otimes \zeta_{p^n}^{\otimes - s_1^{(m)}}\right)_n$ for the Tate twist of the zeta-family cut out by the specialisation $\mathrm{Mell}_{k_0,\epsilon} \circ \lambda\Big|_{k = k_1^{(m)}}$. Again by Theorem 3.8(ii), we know that $p^{\nu_M} \times$ (this zeta-family) is p-integral, where the error term $\nu_M = \nu_M\left(\mathbf{f}_{P_{k_1^{(m)},\epsilon}}, \mathbf{T}\right)$ is defined in Appendix A.

Lemma B.2. *The $\nu_M\left(\mathbf{f}_{P_{k,\epsilon}}, \mathbf{T}\right)$'s are locally constant over weight-space.*

Therefore there exists a neighborhood $\mathbb{D}_{k_1}$ and an upper bound ν_{M,k_0,k_1}, such that $\nu_M\left(\mathbf{f}_{P_{k_1^{(m)},\epsilon}}, \mathbf{T}\right) \leq \nu_{M,k_0,k_1}$ for all $k_1^{(m)} \in \mathbb{D}_{k_1} \cap \mathbb{Z}_{\geq 2}$. As a consequence, all the families $p^{\nu_{M,k_0,k_1}} . \left(\mathbf{z}_{Mp^n}\left(\mathbf{f}_{P_{k_1^{(m)},\epsilon}}\right) \otimes \zeta_{p^n}^{\otimes - s_1^{(m)}}\right)_n$ are p-integral for every $m \gg 0$. But these families should also explode as $\left(s_1^{(m)}, k_1^{(m)}\right)$ approaches the hypothetical pole at $\mathcal{V} = (s_1, k_1)$, which supplies our contradiction!

□

6.2 A two-variable interpolation

In Chapter III we established a fundamental link between the space of p-adic Euler systems, and the space of $\mathcal{Q}$-continuous symbols taking coefficients in Sym^{k-2}. Under this identification, the Kato-Beilinson Euler system transformed into the algebraic modular associated to an eigenform. This is a one-variable phenomenon.

Our aim now is to extend these ideas to a two-variable setting, the first variable being cyclotomic, and the second corresponding to a p-ordinary weight-deformation. Before immersing ourselves in the gory details, let us list all conditions/assumptions we have made so far:

- Recall first that we fixed a rational prime $p \nmid 6N$, and also a system of primitive m^{th}-roots of unity ζ_m satisfying $(\zeta_{m'})^{m'/m} = \zeta_m$ for all $m|m'$.
- We then chose the field $\mathcal{K}$ and the integral domain $\mathbf{I}$ large enough so that both Hypotheses(Diag) and (Arith) were true, which meant we could decompose $\mathbb{T}_\infty$ into classical Hecke eigenspaces.
- Further, for a primitive $\mathbf{I}$-algebra homomorphism $\lambda : \mathbf{h}^{\mathrm{ord}}(Np^\infty; \mathcal{O}_\mathcal{K}) \otimes_{\Lambda^{\mathrm{wt}}_\mathcal{K}} \mathbf{I} \to \mathbf{I}$ corresponding to $\mathbf{f} \in \mathbb{S}^{\mathrm{ord}}_{\mathbf{I}}$, we saw that either the conditions (Rk1) or (UF) ensured that $\mathcal{MS}^{\mathrm{ord}}(\mathbf{I})^\pm[\lambda]$ were free of rank one.
- Finally, the construction of $\mathrm{EXP}^*_{\mathbb{T}_\infty}$ required us to pick generators $\mathbf{v}_{\mathbf{I},\lambda}$ for an $\mathbf{I}$-line in the rank one $\mathbf{I}[1/p]$-module $\mathbb{D}^{\mathbf{I}}_{\mathrm{cris}}(\mathbb{T}^*_\infty)[\lambda]$ – there did not seem to be a canonical choice for these, unless the homomorphism λ is of CM-type.

The main technical result of this book is the following.

Theorem 6.4. *There exists a unique $\Lambda^{\mathrm{wt}}[\![G_\infty]\!]$-homomorphism*

$$\mathcal{E}_\infty : \overline{\mathcal{MS}}^{\mathrm{ord}} \otimes_{\mathbf{h}^{\mathrm{ord}}} \mathbb{H}^1_{\mathrm{Eul},p}(\mathbb{T}_\infty) \otimes_{\mathcal{R}} \widetilde{\mathcal{R}} \longrightarrow \mathrm{Hom}_{\mathcal{Q}\mathrm{cts}}\Big(\mathrm{div}^0\mathbb{P}^1(\mathbb{Q}), \mathcal{L}'\Big)$$

such that for triples (λ, k_0, ϵ), the following diagram

$$\begin{array}{ccc}
\overline{\mathcal{MS}}^{\mathrm{ord}} \otimes_{\mathbf{h}^{\mathrm{ord}}} \mathbb{H}^1_{\mathrm{Eul},p}(\mathbb{T}_\infty)[\lambda] & \xrightarrow{\mathcal{E}^\pm_\infty} & \mathrm{Hom}_{\mathcal{Q}\mathrm{cts}}\Big(\mathrm{div}^0\mathbb{P}^1(\mathbb{Q}), \mathcal{L}'\Big) \\
\mathrm{mod}\ P \Big\downarrow & & \Big\downarrow \nabla^{-1}_{k-2}\big(-\otimes_{\mathbf{I}}\mathbf{I}/P\otimes\mathcal{X}_{k-2}\big) \\
\overline{\mathcal{MS}}^{\mathrm{ord}} \otimes_{P\circ\lambda} \mathbb{H}^1_{\mathrm{Eul},p}\left(V^*_{\mathbf{f}_P}\right) & \xrightarrow{\mathrm{Per}^\pm_{\mathbf{I},\lambda_P} \times \mathcal{E}^\pm_{k,\epsilon,\gamma}} & \mathrm{Hom}_{\mathcal{Q}\mathrm{cts}}\Big(\mathrm{div}^0\mathbb{P}^1(\mathbb{Q}), \mathrm{Sym}^{k-2}(\mathcal{K}_{\mathbf{f}_P})\Big)
\end{array}$$

commutes at almost all algebraic points $P \in \mathrm{Spec}\,\mathbf{I}(\mathcal{O}_\mathcal{K})^{\mathrm{alg}}$ of type (k, ϵ), $k \in \mathbb{U}_{k_0}$.

Furthermore, the map $\mathcal{E}_\infty$ sends the super Euler system $\vec{\mathfrak{Z}}$ to the space $\mathcal{MS}^{\mathrm{ord}}(\mathcal{L}')$, and in particular

$$\mathcal{E}_\infty\Big(\vec{\mathfrak{Z}}_{[\lambda]}\Big)^\pm \ = \ \Xi^\pm_\lambda \quad \textit{for each primitive } \lambda : \mathbf{h}^{\mathrm{ord}}(Np^\infty; \mathcal{O}_\mathcal{K}) \otimes_{\Lambda^{\mathrm{wt}}_\mathcal{K}} \mathbf{I} \to \mathbf{I}.$$

Remark: We point out in the statement of the theorem above, the compositions

$$\overline{\mathcal{MS}}^{\mathrm{ord}}(Np^{\infty}) \otimes_{P\circ\lambda} \mathbb{H}^1_{\mathrm{Eul},p}\left(V^*_{\mathbf{f}_P}\right)$$

$$\Big\downarrow \ \mathrm{id}\otimes\mathcal{E}_{k,\epsilon}$$

$$\mathcal{MS}^{\mathrm{ord}}(Np^r;\mathcal{O}_{\mathcal{K}})[P\circ\lambda]\otimes_{\mathcal{O}_{\mathcal{K}}}\mathrm{Hom}_{\mathcal{Q}\mathrm{cts}}\Big(\mathrm{div}^0\mathbb{P}^1(\mathbb{Q}), \mathrm{Sym}^{k-2}(\mathcal{K}_{\mathbf{f}_P})\Big)$$

$$\Big\downarrow \ \left\langle \gamma_{\mathbf{f}^*_{P_{k,\epsilon}}}, \nabla^{-1}_{k-2}(-)\right\rangle\otimes \mathrm{id}$$

$$\mathrm{Hom}_{\mathcal{Q}\mathrm{cts}}\Big(\mathrm{div}^0\mathbb{P}^1(\mathbb{Q}),\ \mathrm{Sym}^{k-2}(\mathcal{K}_{\mathbf{f}_P})\Big)$$

have been abbreviated by the much shorter version '$\mathcal{E}_{k,\epsilon,\gamma}$'. Hopefully the meaning is self-evident, and won't cause subsequent confusion.

The homomorphism $\mathcal{E}_\infty$ must be uniquely determined under our (Arith) criterion. The image of $\mathcal{E}_\infty$ should be considered as the cyclotomic deformation of $\mathcal{MS}^{\mathrm{ord}}(\mathbf{I})$, since each of the symbols $\Xi^{\pm}_{\lambda}$ generate the $\pm$-part of its $\mathbf{f}$-isotypic component. Identifying the cohomology classes $\vec{\mathfrak{Z}}_{[\lambda]}$ with the $\mathbf{I}$-adic modular symbol Ξ_λ, plays exactly the same rôle as the identification of $\vec{\mathbf{z}}$ with η_f in Theorem 3.10.

In general, there appears to be no way of defining the elements $\mathfrak{Z}_{p^nM}$ without first tensoring by $\widetilde{\mathcal{R}}$. Unfortunately the idempotent $\mathbf{e}_{\text{M-D}}.\mathbf{e}_{\text{prim}}$ only exists inside $\mathbf{H}_{Np^\infty}\otimes_{\Lambda^{\mathrm{wt}}}\mathrm{Frac}(\Lambda^{\mathrm{wt}})$, corresponding to removal of a finite number of bad weights. Nevertheless, if we assume that the residual representation $\overline{\rho_\infty} = \rho_\infty \bmod \mathbf{m}_{\mathcal{R}}$ is absolutely irreducible, one can bypass the fraction field of Λ^{wt} altogether and manufacture $\mathcal{R}$-integral zeta-elements.

Proof: The argument consists of five separate stages. We begin by writing down a map which is 'very nearly $\mathcal{E}_\infty$' using our deformed dual exponential $\mathrm{EXP}^*_{\mathbb{T}_\infty}$, then show it is a $\Lambda^{\mathrm{wt}}[\![G_\infty]\!]$-homomorphism. Next we examine how this map behaves modulo height one prime ideals of $\mathbf{I}$.

In the third step, close attention is payed to specialisations of the $\mathcal{R}$-adic Euler system $\vec{\mathfrak{Z}}$, and we establish that its images always yield classical modular symbols. To lift information modulo primes back up to $\mathcal{MS}^{\mathrm{ord}}$, we require an integrality statement which is proven in the fourth stage. Finally, as $\mathrm{EXP}^*_{\mathbb{T}_\infty}$ depended on a choice of crystalline generators $\mathbf{v}_{\mathbf{I},\lambda}$, we remove these dependencies to obtain $\mathcal{E}_\infty$.

Step 1: The prototype link.
Our initial attempt at an interpolation closely follows the constructions in §3.3. Let $\mathcal{E}^{\dagger}_\infty : \mathbb{H}^1_{\mathrm{Eul},p}(\mathbb{T}_\infty) \to \mathrm{Hom}_{\mathbb{Z}}\left(\mathrm{div}^0\mathbb{P}^1(\mathbb{Q}), \mathbf{I}\left[\frac{1}{p}\right]\right)$ denote the map, whose special values are given by

$$\mathcal{E}^{\dagger}_\infty(\vec{\mathfrak{X}})_{\left(\frac{a}{p^nM}\right)-(\infty)} \ := \sum_{b\in(\mathbb{Z}/p^nM\mathbb{Z})^\times} \Big(\zeta^a_{p^nM}\mathrm{EXP}^*_{\mathbb{T}_\infty}(\mathfrak{X}_{p^nM})\Big)^{\sigma_b}$$

for sequences $\vec{\mathfrak{X}} = \big(\ldots,(\mathfrak{X}_{p^nM})_n,\ldots\big) \in \mathbb{H}^1_{\mathrm{Eul},p}(\mathbb{T}_\infty)$ with $\frac{a}{p^nM}\in\mathbb{Q}$, $\gcd(a,M)=1$.

Since GL_2 acts trivially on $\mathcal{B} = \mathbf{I}[1/p]$, we deduce that for all $\frac{a}{p^n M} \in \Sigma_{M,c}$

$$\mathcal{Q}_{\mathcal{E}^{\dagger}_{\infty}(\vec{\mathfrak{X}}),M,c}\left(\frac{a}{p^n M}\right) \quad = \quad \mathcal{E}^{\dagger}_{\infty}(\vec{\mathfrak{X}})_{(\frac{a}{p^n M})-(\infty)}.$$

The $\mathcal{Q}$-continuity and $\begin{pmatrix}1 & 1\\ 0 & 1\end{pmatrix}$-invariance of $\mathcal{E}^{\dagger}_{\infty}(\vec{\mathfrak{X}})$ can then be easily checked from the defining formula for $\mathcal{E}^{\dagger}_{\infty}(\vec{\mathfrak{X}})_{(\frac{a}{p^n M})-(\infty)}$.

Remark: By Proposition 3.6(i), we know that the operator U_p must exist on $\mathrm{Im}(\mathcal{E}^{\dagger}_{\infty})$, and its action coincides with that of the usual Hecke operator on $\mathbb{T}_\infty$. Moreover $\mathbb{T}_\infty$ is properly contained in the ordinary locus $\mathrm{Ta}_p(J_\infty)^{\mathrm{ord}} = \mathbf{e}.\mathrm{Ta}_p(J_\infty)$, thus U_p will be invertible on the image.

For all positive integers M coprime to p, it follows from 3.6(ii) there is a map

$$\mathrm{Im}(\mathcal{E}^{\dagger}_{\infty}) \longrightarrow \mathrm{Dist}\left(\mathbb{Z}^{\times}_{p,M}, \mathbf{I}[1/p]\right), \qquad \mathcal{E}^{\dagger}_{\infty}(\vec{\mathfrak{X}}) \mapsto \mu_{\mathcal{E}^{\dagger}_{\infty}(\vec{\mathfrak{X}})}$$

where for each $n \geq 0$ and $a \in (\mathbb{Z}/p^n M\mathbb{Z})^{\times}$,

$$\begin{aligned}\mu_{\mathcal{E}^{\dagger}_{\infty}(\vec{\mathfrak{X}})}\left(a + (p^n M)\right) &= \mathcal{Q}_{U_p^{-n}\circ\mathcal{E}^{\dagger}_{\infty}(\vec{\mathfrak{X}}),M,a}\left(\frac{a}{p^n M}\right)\Bigg|\begin{pmatrix}1 & 0\\ 0 & p\end{pmatrix}^n \\ &= U_p^{-n}\left(\sum_{b\in(\mathbb{Z}/p^n M\mathbb{Z})^{\times}}\left(\zeta^a_{p^n M}\mathrm{EXP}^*_{\mathbb{T}_\infty}(\mathfrak{X}_{p^n M})\right)^{\sigma_b}\right).\end{aligned}$$

Let's now restrict to an eigenspace for a primitive $\lambda : \mathbf{h}^{\mathrm{ord}}(Np^\infty; \mathcal{O}_\mathcal{K}) \otimes_{\Lambda^{\mathrm{wt}}_\mathcal{K}} \mathbf{I} \to \mathbf{I}$. One calculates that at non-trivial characters ψ modulo $p^n M$, the contragredient distribution takes the special values

$$\int_{\mathbb{Z}^{\times}_{p,M}} \psi d\mu^{\bullet}_{\mathcal{E}^{\dagger}_{\infty}(\vec{\mathfrak{X}}_{[\lambda]})} = \lambda(U_p)^{-n} G(\psi^{-1}, \zeta_{p^n M}) \sum_{b\in(\mathbb{Z}/p^n M\mathbb{Z})^{\times}} \psi(b)\,\mathrm{EXP}^*_{\mathbb{T}_\infty}(\mathfrak{X}_{[\lambda],p^n M})^{\sigma_b}.$$

The natural question to consider is whether these particular values are p-bounded, i.e. is $\mu^{\bullet}_{\mathcal{E}^{\dagger}_{\infty}(\vec{\mathfrak{X}}_{[\lambda]})}$ a bounded measure?

From Definition 5.11(ii),(iii) we know that $\mathrm{EXP}^*_{\mathbb{T}_\infty}\left(H^1(\mathbb{Q}(\mu_{Mp^n}) \otimes \mathbb{Z}_p, \mathbb{T}_\infty[\lambda])\right)$ lives inside $p^{\theta_{\mathbb{T}_\infty,\lambda} - \mathrm{ord}_p[\mathbb{Q}(\mu_{Mp^n}):\mathbb{Q}]}\mathbf{I} \otimes \mathbb{Z}[\mu_{p^n M}]$, and the constant $\theta_{\mathbb{T}_\infty,\lambda}$ is dependent only on the primitive homorphism λ. Observe too that $\int_{\mathbb{Z}^{\times}_{p,M}} \psi d\mu^{\bullet}_{\mathcal{E}^{\dagger}_{\infty}(\vec{\mathfrak{X}}_{[\lambda]})}$ equals

$$\lambda(U_p)^{-n}\ \psi\left(\mathrm{Tr}_{\mathbb{Q}(\mu_{Mp^n})/\mathbb{Q}}\left(\zeta_{p^n M}\sum_{b\in(\mathbb{Z}/p^n M\mathbb{Z})^{\times}} \sigma_b.\ \mathrm{EXP}^*_{\mathbb{T}_\infty}(\mathfrak{X}_{[\lambda],p^n M})^{\sigma_b}\right)\right).$$

The latter term must then be contained inside of the module $\lambda(U_p)^{-n}p^{\theta_{\mathbb{T}_\infty,\lambda}}\mathbf{I}_\psi$ since the trace map satisfies $\mathrm{Tr}_{\mathbb{Q}(\mu_{Mp^n})/\mathbb{Q}} : \mathbf{I} \otimes \mathbb{Z}[\mu_{p^n M}] \twoheadrightarrow p^{\mathrm{ord}_p[\mathbb{Q}(\mu_{Mp^n}):\mathbb{Q}]}\,\mathbf{I}$.

However the scalar $\lambda(U_p)^{-n}$ is a unit of $\mathbf{I}$, implying that $p^{-\theta_{\mathbb{T}_\infty,\lambda}}\int_{\mathbb{Z}_{p,M}^\times}\psi d\mu^\bullet_{\mathcal{E}^\dagger_\infty(\vec{\mathfrak{X}}_{[\lambda]})}$ is p-integral for all characters ψ, i.e. that $\mu^\bullet_{\mathcal{E}^\dagger_\infty(\vec{\mathfrak{X}}_{[\lambda]})}$ is indeed a bounded measure. Consequently, the associations $\vec{\mathfrak{X}}\mapsto\mathcal{E}^\dagger_\infty(\vec{\mathfrak{X}})\mapsto\mu^\bullet_{\mathcal{E}^\dagger_\infty(\vec{\mathfrak{X}})}$ define maps

$$\begin{array}{ccc}\mathbb{H}^1_{\mathrm{Eul},p}(\mathbb{T}_\infty) & \xrightarrow{\mathcal{E}^\dagger_\infty} & \mathrm{Hom}_{\mathcal{Q}\mathrm{cts}}\Big(\mathrm{div}^0\mathbb{P}^1(\mathbb{Q}),\mathbf{I}[1/p]\Big)\\ & & \big\downarrow \mu^\bullet\\ & & \mathrm{Dist}\Big(\mathbb{Z}_{p,M}^\times,\ \mathbf{I}[1/p]\Big)\end{array}$$

and these are Λ^{cy}-homomorphisms, because the G_∞-actions on the three spaces are all compatible (which is why we chose the contragredient distribution instead of the standard one).

Taking the sum over those M coprime to p, we obtain an injection

$$\mathrm{Im}(\mathcal{E}^\dagger_\infty)\ \overset{\oplus\mu^\bullet}{\hookrightarrow}\ \bigoplus_{M\in\mathbb{N}-p\mathbb{N}}\mathrm{Meas}\Big(\mathbb{Z}_{p,M}^\times,\ \mathbf{I}[1/p]\Big).$$

The right-hand side is free because each space $\mathrm{Meas}\Big(\mathbb{Z}_{p,M}^\times,\ \mathbf{I}[1/p]\Big)$ is isomorphic to $\phi(pM)$ copies of the ring $\mathbf{I}[[T]][\frac{1}{p}]$. Therefore $\mathcal{E}^\dagger_\infty$ is a $\Lambda^{\mathrm{wt}}[\![G_\infty]\!]$-homomorphism and its image is torsion-free.

Step 2: Specialisations at arithmetic points.
In order to reconcile our prototype interpolation with its p-adic cousins, we examine the behaviour of $\mathcal{E}^\dagger_\infty$ modulo a prime ideal of $\mathbf{I}$, corresponding to an integral weight. Let P denote an algebraic point of type (k,ϵ), and suppose the cusp $\frac{a}{p^nM}\in\Sigma_{M,c}$. One computes using Definition 5.11(i) that

$$\begin{aligned}&\mathcal{E}^\dagger_\infty(\vec{\mathfrak{X}}_{[\lambda]})_{(\frac{a}{p^nM})-(\infty)} \bmod P\\ &= \sum_{b\in(\mathbb{Z}/p^nM\mathbb{Z})^\times}\Big(\zeta^a_{p^nM}\ \mathrm{EXP}^*_{\mathbb{T}_\infty}(\mathfrak{X}_{[\lambda],p^nM}) \bmod P\Big)^{\sigma_b}\\ &= \sum_{b\in(\mathbb{Z}/p^nM\mathbb{Z})^\times}\Big(\zeta^a_{p^nM}\ \exp^*_{V^*_{\mathbf{f}_P}}(\mathfrak{X}_{[\lambda],p^nM} \bmod P)\cup(1-\delta_{\mathrm{nr}}\varphi)^{-1}.(t^{-1}.\mathbf{v}_{\mathbf{I},\lambda} \bmod P)\Big)^{\sigma_b}.\end{aligned}$$

Recall from the previous chapter, that $\mathbf{v}_{\mathbf{I},\lambda} \bmod P$ lies in $\mathbf{D}_{\mathrm{cris},\mathbb{Q}_p}(V_{\mathbf{f}_P})^{\varphi=a_p(\mathbf{f}_P)}$. The latter space was generated by the element $\mathbf{v}_{k,\epsilon}$ where $f^*\cup\mathbf{v}_{k,\epsilon}=1$.

Notation: Let's write '$\mathcal{C}_{\mathbf{f}_P}$' for the scalars satisfying $\mathbf{v}_{\mathbf{I},\lambda} \bmod P = \mathcal{C}_{\mathbf{f}_P}\times\mathbf{v}_{k,\epsilon}$.

Bearing in mind 2.8, we then have equalities

$$\begin{aligned}\mathcal{E}^\dagger_\infty(\vec{\mathfrak{X}}_{[\lambda]})_{(\frac{a}{p^nM})-(\infty)} \bmod P &= \mathcal{C}_{\mathbf{f}_P}\sum_{b\in(\mathbb{Z}/p^nM\mathbb{Z})^\times}\Big(\zeta^a_{p^nM}\exp^*_{V^*_{\mathbf{f}_P}}(\mathfrak{X}_{[\lambda],p^nM} \bmod P)\\ &\qquad\cup t^{-1}.\big(1-\delta_{n=0}p^{-1}\varphi\big)^{-1}.\mathbf{v}_{k,\epsilon}\Big)^{\sigma_b}\\ &= \mathcal{C}_{\mathbf{f}_P}\sum_{b\in(\mathbb{Z}/p^nM\mathbb{Z})^\times}\Big(\zeta^a_{p^nM}\ \exp^*_0(\mathfrak{X}_{[\lambda],p^nM} \bmod P)\Big)^{\sigma_b}.\end{aligned}$$

The right-hand side is precisely

$$\mathcal{C}_{\mathbf{f}_P} \;\times\; \left(\text{our formula for } \mathcal{E}_{k,\epsilon}\left(\vec{\mathfrak{X}}_{[\lambda]} \bmod P\right)_{\left(\frac{a}{p^n M}\right)-(\infty)}\right)$$

evaluated at $(X,Y) = (0,1)$ – see Theorem 3.10 for the formula defining $\mathcal{E}_{k,\epsilon}$. However degree zero divisors of the form $\left(\frac{a}{p^n M}\right) - (\infty)$ generate $\mathrm{div}^0\mathbb{P}^1(\mathbb{Q})$ over $\mathbb{Z}$; we deduce that

$$\mathcal{E}_\infty^\dagger\left(\vec{\mathfrak{X}}_{[\lambda]}\right) \bmod P \quad = \quad \mathcal{C}_{\mathbf{f}_P} \times \mathcal{E}_{k,\epsilon}\left(\vec{\mathfrak{X}}_{[\lambda]} \bmod P\right)_{X=0,Y=1}$$

as an equality of symbols.

After twisting the matrix action by $\mathcal{X}_{k-2}$, we can use the inverse isomorphism to Lemma 3.13 to recover a Sym^{k-2}-valued symbol. In other words

$$\nabla_{k-2}^{-1}\left(\left(\mathcal{E}_\infty^\dagger\left(\vec{\mathfrak{X}}_{[\lambda]}\right) \bmod P\right) \otimes \mathcal{X}_{k-2}\right) \;=\; \mathcal{C}_{\mathbf{f}_P} \times \mathcal{E}_{k,\epsilon}\left(\vec{\mathfrak{X}}_{[\lambda]} \bmod P\right). \qquad (*)$$

Question. *How can we relate these scalars $\mathcal{C}_{\mathbf{f}_P}$ to the structure of the universal* $\mathbf{I}$*-adic modular symbol Ξ_λ at a prime ideal P?*

To find the answer, we'll look for a more geometric interpretation of our map $\mathcal{E}_\infty^\dagger$. Fortunately, most of the calculation can be undertaken at weight two.

Step 3: The Euler system under $\mathcal{E}_\infty^\dagger$.
For every integer $r \geq 1$, we will set $\Gamma_r = (1 + p\mathbb{Z}_p)^{p^{r-1}}$ and also $\omega_r = u_0^{p^{r-1}} - 1$. The Γ_r-coinvariants $(\mathbb{T}_\infty)_{\Gamma_r} := \mathbb{T}_\infty/\omega_r\mathbb{T}_\infty$ may be considered as lying in $\mathrm{Ta}_p(J_r)$ where $J_r = \mathrm{Jac}\ X_1(Np^r)$. The jacobian variety $\mathrm{Jac}\ X_1(Np^r)$ can be decomposed into quotient abelian varieties A_f, corresponding to normalised cuspidal eigenforms f of weight 2 and level dividing Np^r.

If f is not p-ordinary then the image of $(\mathbb{T}_\infty)_{\Gamma_r}$ under the quotient map to $V_f^* = \mathrm{Ta}_p(A_f) \otimes \mathbb{Q}$ must be zero, because the Hecke eigenvalue at p is not a unit. If f is not N-primitive, again the image of $(\mathbb{T}_\infty)_{\Gamma_r}$ in V_f^* must be zero as $\mathbb{T}_\infty$ is cut out using the idempotent $\mathbf{e}_{\mathrm{prim}}$.

That leaves only eigenforms f which are both p-ordinary and N-primitive, in particular $f \in \mathcal{S}_2^{\mathrm{ord}}(\Gamma_1(Np^{r'}))$ for some $r' \leq r$. Such an eigenform is automatically p-stabilised, so must lie in a Hida family $\mathbf{f} \in \mathbb{S}_{\mathbf{I}}^{\mathrm{ord}}$ corresponding to a primitive homomorphism $\lambda : \mathbf{h}^{\mathrm{ord}}(Np^\infty; \mathcal{O}_\mathcal{K}) \otimes_{\Lambda_\mathcal{K}^{\mathrm{wt}}} \mathbf{I} \to \mathbf{I}$ by duality. If ϵ is the character of f restricted to the group $1+p\mathbb{Z}_p$, there exists an algebraic point $P_{2,\epsilon} \in \mathrm{Spec}\ \mathbf{I}(\mathcal{O}_{\mathcal{K}_f})^{\mathrm{alg}}$ of type $(2, \epsilon)$ such that $\mathbf{f}_{P_{2,\epsilon}} = f$.

Its specialisation to the $\mathbf{f}$-isotypic component of $\mathbb{T}_\infty$, factorises through

$$\mathbb{T}_\infty[\lambda] \overset{\bmod \omega_r}{\twoheadrightarrow} \left(\mathbb{T}_\infty[\lambda]\right)_{\Gamma_r} \hookrightarrow \mathrm{Ta}_p(J_r)[\lambda \bmod \omega_r] \twoheadrightarrow \mathrm{Ta}_p(A_f) \hookrightarrow V^*_{\mathbf{f}_{P_{2,\epsilon}}}.$$

Let $\mathcal{O}_{r,\epsilon}$ denote the finite integral extension of $\mathbb{Z}_p$, containing $\mathcal{O}_{\mathcal{K}_f}$ for the p-ordinary eigenform f above, and also the values of the finite order character ϵ modulo $\Gamma_{r'}$. Taking Γ_r-coinvariants of our map $\mathcal{E}_\infty^\dagger$ gives rise to the commutative diagram

$$\begin{array}{ccccc}
\mathbb{H}^1_{\mathrm{Eul},p}(\mathbb{T}_\infty)[\lambda] & \overset{\mathcal{E}_\infty^\dagger}{\longrightarrow} & \mathrm{Hom}\left(\mathrm{div}^0, \mathbf{I}\left[\frac{1}{p}\right]\right) & & \\
\Big\downarrow{\scriptstyle \mathrm{mod}\ \omega_r} & & \Big\downarrow{\scriptstyle \otimes_{\Lambda^{\mathrm{wt}}}\Lambda^{\mathrm{wt}}_{\Gamma_r}} & & \\
\mathbb{H}^1_{\mathrm{Eul},p}(\mathbb{T}_\infty)_{\Gamma_r}[\lambda \,\mathrm{mod}\ \omega_r] & \overset{\mathcal{E}_\infty^\dagger\ \mathrm{mod}\ \omega_r}{\longrightarrow} & \mathrm{Hom}\left(\mathrm{div}^0, \mathbf{I}_{\Gamma_r}\left[\frac{1}{p}\right]\right) & \hookrightarrow & \bigoplus_{\epsilon:\Gamma_1/\Gamma_r \to \overline{\mathbb{Q}}_p^\times} \mathrm{Hom}\left(\mathrm{div}^0, \mathbf{I}_{\Gamma_r}\left[\frac{1}{p}\right]\right)^{(\epsilon)} \\
\Big\downarrow & & & & \Big\downarrow{\scriptstyle \left(\otimes_{\mathbb{Z}_p\langle\langle w\rangle\rangle, w\mapsto 2}\mathbb{Z}_p\right)\circ \mathrm{proj}_{[\epsilon]}} \\
\mathbb{H}^1_{\mathrm{Eul},p}\left(\mathrm{Ta}_p(A_{\mathbf{f}_{P_{2,\epsilon}}})\otimes\mathbb{Q}\right) & \overset{\mathcal{C}_{\mathbf{f}_{P_{2,\epsilon}}}\times\mathcal{E}_{2,\epsilon}}{\longrightarrow} & \mathrm{Hom}\left(\mathrm{div}^0, \mathcal{K}_{\mathbf{f}_{P_{2,\epsilon}}}\right) & \hookrightarrow & \mathrm{Hom}\left(\mathrm{div}^0, \mathcal{O}_{r,\epsilon}[1/p]\right)
\end{array}$$

the transition from top to bottom row being induced by the prime ideal $P_{2,\epsilon}$.

Remark: Let us not forget that the $\mathcal{R}$-adic Euler system only becomes canonically defined when viewed as an element of $\overline{\mathcal{MS}}^{\mathrm{ord}}(Np^\infty)\otimes_{\mathbf{h}^{\mathrm{ord}}(Np^\infty)}\mathbb{H}^1_{\mathrm{Eul}}(\mathbb{T}_\infty)\otimes_{\mathcal{R}}\widetilde{\mathcal{R}}$. In the obvious way, our map $\mathcal{E}_\infty^\dagger$ induces

$$\overline{\mathcal{MS}}^{\mathrm{ord}}\otimes_{\mathbf{h}^{\mathrm{ord}}}\mathbb{H}^1_{\mathrm{Eul},p}(\mathbb{T}_\infty)[\lambda] \overset{\mathrm{id}\otimes\mathcal{E}_\infty^\dagger}{\longrightarrow} \overline{\mathcal{MS}}^{\mathrm{ord}}\otimes_{\mathbf{h}^{\mathrm{ord}},\lambda}\mathrm{Hom}_{\mathcal{Q}\mathrm{cts}}\left(\mathrm{div}^0\mathbb{P}^1(\mathbb{Q}), \mathcal{L}'\right)$$

and we shall henceforth write $\mathcal{E}_\infty^{\dagger\dagger}$ for the above $\mathbf{I}[\![G_\infty]\!]$-homomorphism.

Even though $\vec{\mathfrak{Z}}$ may not be $\mathcal{R}$-integral, at least for some non-zero element $\wp\in\Lambda^{\mathrm{wt}}$ satisfying $\wp.\mathbf{e}_{\text{M-D}}.\mathbf{e}_{\mathrm{prim}}\in\mathbf{H}_{Np^\infty}$, we can say that $\wp\vec{\mathfrak{Z}}\in\overline{\mathcal{MS}}^{\mathrm{ord}}\otimes_{\mathbf{h}^{\mathrm{ord}}}\mathbb{H}^1_{\mathrm{Eul},p}(\mathbb{T}_\infty)$. After a brisk chase around our diagram, one determines that

$$\begin{aligned}
\mathrm{proj}_{[\epsilon]}\left(\mathcal{E}_\infty^{\dagger\dagger}(\wp\vec{\mathfrak{Z}}_{[\lambda]})\ \mathrm{mod}\ \omega_r\right)_{s\mapsto 2} &= \mathcal{C}_{\mathbf{f}_{P_{2,\epsilon}}}\times\mathrm{id}\otimes\mathcal{E}_{2,\epsilon}\left(\wp\vec{\mathfrak{Z}}_{[\lambda]}\ \mathrm{mod}\ P_{2,\epsilon}\right)\\
&= \mathcal{C}_{\mathbf{f}_{P_{2,\epsilon}}}\times\left(\wp\ \mathrm{mod}\ P_{2,\epsilon}\right)\times\mathrm{id}\otimes\mathcal{E}_{2,\epsilon}\left(\gamma^*\otimes\vec{\mathbf{z}}_{2,\epsilon}\right)
\end{aligned}$$

the last equality following from 6.2(b). Furthermore, our one-variable interpolation in Theorem 3.10(ii) tells us that $\mathcal{E}_{2,\epsilon,\gamma}\left(\gamma^*\otimes\vec{\mathbf{z}}_{2,\epsilon}\right)=\eta_{\mathbf{f}_{P_{2,\epsilon}}}$.

It follows that $\left\langle\gamma,\ \mathrm{proj}_{[\epsilon]}(\ldots)\right\rangle$ above must be contained inside the $\mathbf{f}_{P_{2,\epsilon}}$-isotype $\mathbf{e}.\mathrm{Hom}_{\Gamma_1(Np^r)}\left(\mathrm{div}^0\mathbb{P}^1(\mathbb{Q}),\mathcal{K}_{\mathbf{f}_{P_{2,\epsilon}}}\right)[\lambda_{\mathbf{f}_{P_{2,\epsilon}}}]$ because $\eta^\pm_{\mathbf{f}_{P_{2,\epsilon}}}$ generates its $\pm$-eigenspaces. Hida's control theory allows the identification

$$\mathbf{e}.\mathrm{Hom}_{\Gamma_1(Np^r)}\left(\mathrm{div}^0\mathbb{P}^1(\mathbb{Q}),\mathcal{O}_{\mathcal{K}}\right)\ \cong\ \overline{\mathcal{MS}}^{\mathrm{ord}}\left(Np^\infty;\mathcal{O}_{\mathcal{K}}\right)^{\Gamma_r}.$$

As a corollary, $\mathrm{proj}_{[\epsilon]}\left(\mathcal{E}_\infty^{\dagger\dagger}\left(\wp\vec{\mathfrak{Z}}_{[\lambda]}\right)\ \mathrm{mod}\ \omega_r\right)_{s\mapsto 2}$ lies in $\overline{\mathcal{MS}}^{\mathrm{ord}}\left(Np^\infty;\mathcal{O}_{r,\epsilon}\right)^{\Gamma_r}\otimes\mathbb{Q}$, after pairing it first with γ.

Step 4: Duality and Integrality.
Let us do better and make a precise statement about the integrality of $\mathrm{Im}(\mathcal{E}^{\dagger\dagger}_{\infty})$. For any non-torsion elements $\underline{\delta}^{\pm}$ of the $\mathcal{O}_{\mathcal{K}}$-module $\mathcal{UM}^{\mathrm{ord}}(\mathcal{O}_{\mathcal{K}})^{\pm}[\lambda]$, we shall write $\mathcal{E}^{\dagger\dagger}_{\infty,\underline{\delta}}$ for the composition

$$\overline{\mathcal{MS}}^{\mathrm{ord}} \otimes_{\mathbf{h}^{\mathrm{ord}}} \mathbb{H}^1_{\mathrm{Eul},p}(\mathbb{T}_\infty)[\lambda] \overset{\langle \underline{\delta}^+ + \underline{\delta}^-,\, \mathcal{E}^{\dagger\dagger}_{\infty}\rangle}{\longrightarrow} \mathrm{Hom}_{\mathbb{Q}\mathrm{cts}}\Big(\mathrm{div}^0\mathbb{P}^1(\mathbb{Q}),\, \mathbf{I}\big[1/p\big]\Big).$$

By Definition 5.11(ii), the big dual exponential $\mathrm{EXP}^*_{\mathbb{T}_\infty,L}$ becomes p-integral after multiplication by the number $p^{\mathrm{ord}_p[L:\mathbb{Q}_p]-\theta_{\mathbb{T}_\infty,\lambda}}$. It follows directly from the various formulae defining $\mathcal{E}^{\dagger}_{\infty}$, $\mathcal{E}^{\dagger\dagger}_{\infty}$, and $\mathcal{E}^{\dagger\dagger}_{\infty,\underline{\delta}}$ that the image of $\mathcal{E}^{\dagger\dagger}_{\infty,\underline{\delta}}$ (for the λ-eigenspace) is contained inside $p^{\theta_{\mathbb{T}_\infty,\lambda}}.\mathrm{Hom}_{\mathbb{Q}\mathrm{cts}}\big(\mathrm{div}^0\mathbb{P}^1(\mathbb{Q}), \mathbf{I}\big)$.

Taking the sum over characters ϵ mod Γ_r, we deduce

$$\mathcal{E}^{\dagger\dagger}_{\infty,\underline{\delta}}\left(\wp\vec{\mathfrak{Z}}_{[\lambda]}\right) \bmod \omega_r \quad \text{lies in} \quad p^{\theta_{\mathbb{T}_\infty,\lambda}}. \bigoplus_{\epsilon:\Gamma_1/\Gamma_r\to\overline{\mathbb{Q}}_p^\times} \overline{\mathcal{MS}}^{\mathrm{ord}}\Big(Np^\infty; \mathcal{O}_{r,\epsilon}\Big)^{\Gamma_r}\big[P_{2,\epsilon}\circ\lambda\big]. \tag{**}$$

Key Claim: The $\mathbf{I}$-valued symbol $p^{-\theta_{\mathbb{T}_\infty,\lambda}}\mathcal{E}^{\dagger}_{\infty}\left(\wp\vec{\mathfrak{Z}}_{[\lambda]}\right)$ belongs to the λ-eigenspace of the module $\mathcal{MS}^{\mathrm{ord}}(\mathbf{I})$.

To justify this assertion, we need to piece information mod ω_r together as r varies. We will then show that $\mathcal{E}^{\dagger}_{\infty}\Big(\wp\vec{\mathfrak{Z}}_{[\lambda]}\Big)^{\pm}$ lie along the $\mathcal{L}'$-line spanned by the universal $\mathbf{I}$-adic modular symbol, which was introduced in §4.3.

The $\mathcal{O}_{r,\epsilon}$-module $\overline{\mathcal{MS}}^{\mathrm{ord}}(Np^\infty;\mathcal{O}_{r,\epsilon})^{\Gamma_r}$ is reflexive (in fact it's free of finite rank), so dualising twice

$$\begin{aligned}\overline{\mathcal{MS}}^{\mathrm{ord}}\Big(Np^\infty;\mathcal{O}_{r,\epsilon}\Big)^{\Gamma_r} &= \mathrm{Hom}_{\mathcal{O}_{r,\epsilon}}\Bigg(\mathrm{Hom}_{\mathcal{O}_{r,\epsilon}}\Big(\overline{\mathcal{MS}}^{\mathrm{ord}}(Np^\infty;\mathcal{O}_{r,\epsilon}),\mathcal{O}_{r,\epsilon}\Big)_{\Gamma_r},\mathcal{O}_{r,\epsilon}\Bigg)\\ &\cong \mathrm{Hom}_{\mathcal{O}_{r,\epsilon}}\Big(\big(\mathcal{UM}^{\mathrm{ord}}(\mathcal{O}_{r,\epsilon})^{\mathcal{D}}\big)_{\Gamma_r},\mathcal{O}_{r,\epsilon}\Big)\end{aligned}$$

N.B. the Banach space $\mathcal{UM}(\mathcal{O}_{r,\epsilon})$ is always viewed with the **dual** Hecke action. Let us now observe the inclusion $\mathcal{O}_{\mathcal{K}}\subset\mathcal{O}_{r,\epsilon}$ gives rise to a natural restriction

$$\mathrm{Hom}_{\mathcal{O}_{r,\epsilon}}\Big(\big(\mathcal{UM}^{\mathrm{ord}}(\mathcal{O}_{r,\epsilon})^{\mathcal{D}}\big)_{\Gamma_r},\mathcal{O}_{r,\epsilon}\Big) \overset{\mathrm{Res}^{\mathcal{O}_{r,\epsilon}}_{\mathcal{O}_{\mathcal{K}}}}{\longrightarrow} \mathrm{Hom}_{\mathcal{O}_{\mathcal{K}}}\Big(\big(\mathcal{UM}^{\mathrm{ord}}(\mathcal{O}_{\mathcal{K}})^{\mathcal{D}}\big)_{\Gamma_r},\mathcal{O}_{r,\epsilon}\Big).$$

Exploiting the fact that $\mathcal{UM}^{\mathrm{ord}}(\mathcal{O}_{\mathcal{K}})$ is of finite type over $\Lambda^{\mathrm{wt}}_{\mathcal{K}}=\mathcal{O}_{\mathcal{K}}[[\Gamma_1]]$,

$$\begin{aligned}&\mathrm{Hom}_{\mathcal{O}_{\mathcal{K}}}\Big(\big(\mathcal{UM}^{\mathrm{ord}}(\mathcal{O}_{\mathcal{K}})^{\mathcal{D}}\big)_{\Gamma_r},\,\mathcal{O}_{r,\epsilon}\Big)\\ &\qquad\cong \mathrm{Hom}_{\mathcal{O}_{\mathcal{K}}[\Gamma_1/\Gamma_r]}\Big(\big(\mathcal{UM}^{\mathrm{ord}}(\mathcal{O}_{\mathcal{K}})^{\mathcal{D}}\big)_{\Gamma_r},\,\mathcal{O}_{r,\epsilon}[\Gamma_1/\Gamma_r]\Big)^{\square}\\ &\qquad= \mathrm{Hom}_{\Lambda^{\mathrm{wt}}_{\mathcal{K}}}\Big(\mathcal{UM}^{\mathrm{ord}}(\mathcal{O}_{\mathcal{K}})^{\mathcal{D}},\,\mathcal{O}_{r,\epsilon}[\Gamma_1/\Gamma_r]\Big)^{\square}\end{aligned}$$

where the superscript $\square$ indicates the Γ_1-action has been inverted.

Via **(**)** one views each ϵ-component of $\mathcal{E}^{\dagger\dagger}_{\infty,\underline{\delta}}\left(\wp\vec{\mathfrak{Z}}_{[\lambda]}\right)$ mod ω_r, as an element of

$$p^{\theta_{\mathbb{T}_\infty,\lambda}}.\mathrm{Hom}_{\Lambda^{\mathrm{wt}}_{\mathcal{K}}}\Big(\mathcal{UM}^{\mathrm{ord}}(\mathcal{O}_{\mathcal{K}})^{\mathcal{D}},\ \mathcal{O}_{r,\epsilon}[\Gamma_1/\Gamma_r]\Big)^{\Box}[\lambda \bmod \omega_r]$$

whose integrality is bounded independent of r.

By its very definition, we know $\mathcal{E}^{\dagger\dagger}_{\infty,\underline{\delta}}\left(\wp\vec{\mathfrak{Z}}_{[\lambda]}\right)$ mod ω_r is a $\mathbf{I}_{\Gamma_r}\left[\frac{1}{p}\right]$-valued symbol, from which we conclude

$$\mathcal{E}^{\dagger\dagger}_{\infty,\underline{\delta}}\left(\wp\vec{\mathfrak{Z}}_{[\lambda]}\right) \bmod \omega_r \ \in\ p^{\theta_{\mathbb{T}_\infty,\lambda}}.\mathrm{Hom}_{\Lambda^{\mathrm{wt}}_{\mathcal{K}}}\Big(\mathcal{UM}^{\mathrm{ord}}(\mathcal{O}_{\mathcal{K}})^{\mathcal{D}},\mathbf{I}_{\Gamma_r}\Big)^{\Box}[\lambda \bmod \omega_r].$$

Finally, $\mathcal{UM}^{\mathrm{ord}}(\mathcal{O}_{\mathcal{K}})^{\mathcal{D}}$ must clearly coincide with $\mathcal{UM}^{\mathrm{ord}}(\mathcal{O}_{\mathcal{K}})^{\Box}$ as a $\Lambda^{\mathrm{wt}}_{\mathcal{K}}$-module. Passing to the projective limit over $r \geq 1$, we have verified that $\mathcal{E}^{\dagger\dagger}_{\infty,\underline{\delta}}$ sends the scalar multiple $\wp\vec{\mathfrak{Z}}_{[\lambda]}$ to the rank two $\mathbf{I}$-module

$$p^{\theta_{\mathbb{T}_\infty,\lambda}}.\,\mathrm{Hom}_{\Lambda^{\mathrm{wt}}_{\mathcal{K}}}\Big(\mathcal{UM}^{\mathrm{ord}}(\mathcal{O}_{\mathcal{K}})^{\Box},\ \mathbf{I}\Big)^{\Box}[\lambda] \quad\cong\quad p^{\theta_{\mathbb{T}_\infty,\lambda}}.\,\mathcal{MS}^{\mathrm{ord}}(\mathbf{I})[\lambda]$$

as our Key Claim asserted.

Step 5: Renormalisation.
Let us review what we have just learnt, to calculate the $\mathcal{C}_{\mathbf{f}_P}$'s at algebraic points. In contrast to the previous step, we now insist that $\underline{\delta}^+$, $\underline{\delta}^-$ are 'chosen maximally' in the following sense. Under Hypothesis (Rk1) the components $\mathcal{UM}^{\mathrm{ord}}(\mathcal{O}_{\mathcal{K}})^{\pm}[\lambda]$ are free of rank one over the ring $\mathbf{I}$, and we simply choose $\underline{\delta}^{\pm}$ to generate them. However if (UF) holds there might be some torsion, in which case we pick $\underline{\delta}^{\pm}$ so the $\mathbf{I}$-torsion modules $\mathcal{UM}^{\mathrm{ord}}(\mathcal{O}_{\mathcal{K}})^{\pm}[\lambda]\big/\mathbf{I}.\underline{\delta}^{\pm}$ are as small as possible (which amounts to discarding yet more bad weights in the final theorem).

We observe under our (Rk1) condition, at all $P \in \mathrm{Spec}\ \mathbf{I}(\mathcal{O}_{\mathcal{K}})^{\mathrm{alg}}$ of type (k,ϵ) we have equalities

$$\begin{aligned}\Big\langle \overline{\mathcal{MS}}^{\mathrm{ord}}(Np^{\infty};\mathcal{O}_{\mathcal{K}})^{\pm}[\lambda_P],\ \underline{\delta}^{\pm} \bmod P\Big\rangle \\ = \quad \mathcal{O}_{\mathcal{K}} \quad = \quad \Big\langle \overline{\mathcal{MS}}^{\mathrm{ord}}(Np^{\infty};\mathrm{Sym}^{k-2}(\mathcal{O}_{\mathcal{K}}))^{\pm}[\lambda_P],\ \gamma^{\pm}_{\mathbf{f}^*_{P_{k,\epsilon}}}\Big\rangle.\end{aligned}$$

If (UF) holds true instead, then we replace *all* with the words *all but finitely many.* In any case, there exist units $u^{\pm}_P \in \mathcal{O}^{\times}_{\mathcal{K}}$ such that pairing an element of $\overline{\mathcal{MS}}^{\mathrm{ord}}$ with $\gamma^{\pm}$ differs (in ratio) from pairing it with $\underline{\delta}^{\pm}$ mod P, by the fixed p-adic number $u^{\pm}_P$. This follows because both paths generate the same principal $\mathcal{O}_{\mathcal{K}}$-ideal.

Remark: We have some freedom in choosing $\underline{\delta}^{\pm}$, and may adjust them by an $\mathbf{I}$-unit. Without loss of generality, assume these elements are picked so the equality

$$\langle \gamma^{\pm},\ \nabla^{-1}_{k-2}(\mathbf{x})\rangle \;=\; \langle \underline{\delta}^{\pm} \bmod P,\ \mathbf{x}\rangle \quad \text{for every } \mathbf{x} \in \overline{\mathcal{MS}}^{\mathrm{ord}}(Np^{\infty};\mathcal{O}_{\mathcal{K}})^{\pm}[\lambda_P]$$

holds at all (resp. all but finitely many) arithmetic points P.

We now piece these disparate strands together, to deduce the interpolation theorem. Firstly, the above remark is equivalent to the commutativity of the square

$$\begin{array}{ccc} \overline{\mathcal{MS}}^{\mathrm{ord}}\otimes_{\mathbf{h}^{\mathrm{ord}},\lambda}\mathrm{Hom}_{\mathcal{Q}\mathrm{cts}}\Big(\mathrm{div}^0\mathbb{P}^1(\mathbb{Q}),\mathcal{L}'\Big) & \overset{\langle\underline{\delta}^{\pm},-\rangle}{\longrightarrow} & \mathrm{Hom}_{\mathcal{Q}\mathrm{cts}}\Big(\mathrm{div}^0\mathbb{P}^1(\mathbb{Q}),\mathcal{L}'\Big) \\ \Big\downarrow {\scriptstyle \mathrm{id}\otimes\nabla^{-1}_{k-2}\left(-\otimes_{\mathbf{I}}\mathbf{I}/P\otimes\mathcal{X}_{k-2}\right)} & & \Big\downarrow {\scriptstyle \nabla^{-1}_{k-2}\left(-\otimes_{\mathbf{I}}\mathbf{I}/P\otimes\mathcal{X}_{k-2}\right)} \\ \overline{\mathcal{MS}}^{\mathrm{ord}}[\lambda_P]\otimes_{\mathcal{O}_{\mathcal{K}}}\mathrm{Hom}_{\mathcal{Q}\mathrm{cts}}\Big(\mathrm{div}^0,\mathrm{Sym}^{k-2}\Big) & \overset{\langle\gamma^{\pm},-\rangle}{\longrightarrow} & \mathrm{Hom}_{\mathcal{Q}\mathrm{cts}}\Big(\mathrm{div}^0,\mathrm{Sym}^{k-2}\Big) \end{array}$$

at all (resp. all but finitely many) arithmetic points $P=P_{k,\epsilon}\in\mathrm{Spec}\ \mathbf{I}(\mathcal{O}_{\mathcal{K}})^{\mathrm{alg}}$. Secondly, our period relation **(*)** may be reinterpreted as saying the diagram

$$\begin{array}{ccc} \overline{\mathcal{MS}}^{\mathrm{ord}}\otimes_{\mathbf{h}^{\mathrm{ord}}}\mathbb{H}^1_{\mathrm{Eul},p}(\mathbb{T}_\infty)[\lambda] & \overset{\mathcal{E}^{\dagger\dagger}_{\infty}}{\longrightarrow} & \overline{\mathcal{MS}}^{\mathrm{ord}}\otimes_{\mathbf{h}^{\mathrm{ord}},\lambda}\mathrm{Hom}_{\mathcal{Q}\mathrm{cts}}\Big(\mathrm{div}^0\mathbb{P}^1(\mathbb{Q}),\mathcal{L}'\Big) \\ {\scriptstyle \mathrm{mod}\ P}\Big\downarrow & & \Big\downarrow {\scriptstyle \mathrm{id}\otimes\nabla^{-1}_{k-2}\left(-\otimes_{\mathbf{I}}\mathbf{I}/P\otimes\mathcal{X}_{k-2}\right)} \\ \overline{\mathcal{MS}}^{\mathrm{ord}}\otimes_{P\circ\lambda}\mathbb{H}^1_{\mathrm{Eul},p}\left(V^*_{\mathbf{f}_P}\right) & \overset{\mathcal{C}_{\mathbf{f}_P}\times(\mathrm{id}\otimes\mathcal{E}_{k,\epsilon})}{\longrightarrow} & \overline{\mathcal{MS}}^{\mathrm{ord}}(Np^\infty;\mathcal{O}_{\mathcal{K}})[\lambda_P] \\ & & \otimes_{\mathcal{O}_{\mathcal{K}}}\mathrm{Hom}_{\mathcal{Q}\mathrm{cts}}\Big(\mathrm{div}^0\mathbb{P}^1(\mathbb{Q}),\mathrm{Sym}^{k-2}\Big) \end{array}$$

always commutes. Splicing these two commutative squares together, we conclude

$$\nabla^{-1}_{k-2}\left(\left(\mathcal{E}^{\dagger\dagger}_{\infty,\underline{\delta}}\left(\vec{\mathfrak{x}}_{[\lambda]}\right)^{\pm}\ \mathrm{mod}\ P\right)\otimes\mathcal{X}_{k-2}\right)\ =\ \mathcal{C}_{\mathbf{f}_P}\times\mathcal{E}_{k,\epsilon,\gamma}\left(\vec{\mathfrak{x}}_{[\lambda]}\ \mathrm{mod}\ P\right)^{\pm} \qquad \textbf{(***)}$$

for all components $\vec{\mathfrak{x}}_{[\lambda]}$ living in the **f**-isotypic part $\mathbb{H}^1_{\mathrm{Eul},p}(\mathbb{T}_\infty)[\lambda]$.

Remark: In Steps 3 and 4, for any $\wp\in\Lambda^{\mathrm{wt}}$ satisfying $\wp\vec{\mathfrak{Z}}\in\mathbb{H}^1_{\mathrm{Eul},p}(\mathbb{T}_\infty)$ we have just shown $\mathcal{E}^{\dagger\dagger}_{\infty,\underline{\delta}}\left(\wp\vec{\mathfrak{Z}}_{[\lambda]}\right)$ belongs to the module $\mathcal{MS}^{\mathrm{ord}}(\mathbf{I})[\lambda]$, up to a power of p. Thus tensoring over the fraction field of the weight algebra, induces an injection

$$\mathcal{E}^{\dagger\dagger}_{\infty,\underline{\delta}}:\ \mathcal{L}'.\vec{\mathfrak{Z}}_{[\lambda]}\ \hookrightarrow\ \mathcal{MS}^{\mathrm{ord}}(\mathcal{L}')[\lambda].$$

Under either of (Rk1) or (UF), there must then exist non-zero elements $\ell^+_\lambda,\ell^-_\lambda\in\mathcal{L}'$ such that $\mathcal{E}^{\dagger\dagger}_{\infty,\underline{\delta}}\left(\vec{\mathfrak{Z}}_{[\lambda]}\right)^{\pm}\ =\ \ell^{\pm}_\lambda\ \times\ \Xi^{\pm}_\lambda$.

Courtesy of Equation(*******), we know that $\nabla^{-1}_{k-2}\left(\left(\mathcal{E}^{\dagger\dagger}_{\infty,\underline{\delta}}\left(\vec{\mathfrak{Z}}_{[\lambda]}\right)^{\pm}\ \mathrm{mod}\ P\right)\otimes\mathcal{X}_{k-2}\right)$ equals the symbol $\mathcal{C}_{\mathbf{f}_P}\times\mathcal{E}_{k,\epsilon,\gamma}\left(\vec{\mathfrak{Z}}_{[\lambda]}\ \mathrm{mod}\ P\right)^{\pm}$, so provided we can compute both quantities we will be done.

Focussing on the former term,

$$\begin{aligned}\nabla^{-1}_{k-2}\left(\left(\mathcal{E}^{\dagger\dagger}_{\infty,\underline{\delta}}\left(\vec{\mathfrak{Z}}_{[\lambda]}\right)^{\pm}\ \mathrm{mod}\ P\right)\otimes\mathcal{X}_{k-2}\right)\ &=\ \nabla^{-1}_{k-2}\left(\left(\ell^{\pm}_\lambda\times\Xi^{\pm}_\lambda\ \mathrm{mod}\ P\right)\otimes\mathcal{X}_{k-2}\right)\\ &=\ \left(P\circ\ell^{\pm}_\lambda\right)\times\mathrm{Per}^{\pm}_{\mathbf{I},\lambda_P}\ \times\ \eta^{\pm}_{\mathbf{f}_P}\end{aligned}$$

by the properties of the universal **I**-adic modular symbol at P.

For the latter term, in tandem Theorems 6.2(b) and 3.10(ii) imply

$$\mathcal{C}_{\mathbf{f}_P} \times \mathcal{E}_{k,\epsilon,\gamma}\left(\vec{\mathfrak{Z}}_{[\lambda]} \bmod P\right)^{\pm} = \mathcal{C}_{\mathbf{f}_P} \times \left\langle \gamma^{\pm},\, \nabla_{k-2}^{-1} \otimes \mathcal{E}_{k,\epsilon}\left(\vec{\mathfrak{Z}}_{[\lambda]} \bmod P\right)\right\rangle$$
$$= \mathcal{C}_{\mathbf{f}_P} \times \left\langle \gamma^{\pm},\, \gamma^{*} \otimes \mathcal{E}_{k,\epsilon}(\vec{\mathbf{z}}_{k,\epsilon})\right\rangle = \mathcal{C}_{\mathbf{f}_P} \times \eta_{\mathbf{f}_P}^{\pm}.$$

Combining these two equations together, and observing that the modular symbols $\eta_{\mathbf{f}_P}^{\pm}$ are both non-zero, we have an equality between **I**-adic periods

$$\mathcal{C}_{\mathbf{f}_P} = \left(P \circ \ell_\lambda^{\pm}\right) \times \mathrm{Per}^{\pm}_{\mathbf{I},\lambda_P} \quad \text{where } \ell_\lambda^{+}, \ell_\lambda^{-} \text{ satisfied } \mathcal{E}^{\dagger\dagger}_{\infty,\underline{\delta}}\left(\vec{\mathfrak{Z}}_{[\lambda]}\right)^{\pm} = \ell_\lambda^{\pm} \times \Xi_\lambda^{\pm}.$$

Fundamentally, we view this calculation as stating that the ambiguity inherent in the de Rham periods $\mathcal{C}_{\mathbf{f}_P}$ is exactly counterbalanced by the ambiguities present in the $\ell_\lambda^{\pm}$'s modulo P. These ambiguities are then related via $\mathcal{E}^{\dagger}_{\infty}$, to the structure of the Betti cohomologies

$$\mathrm{Hom}_{\Lambda_{\mathcal{K}}^{\mathrm{wt}}}\left(\mathbf{e}.\varprojlim_{r} H_1\Big(X_1(Np^r), \Gamma_1(Np^r)\backslash\mathbb{P}^1;\, \mathcal{O}_{\mathcal{K}}\Big),\ \mathbf{I}\right)^{\pm}[\lambda].$$

This is the crux of the preceding 60 pages' or so work: the arithmetic specialisations of $\mathbb{D}^{\mathbf{I}}_{\mathrm{cris}}(\mathbb{T}^*_\infty)[\lambda]$ are described – using our two-variable symbol interpolations – in terms of the behaviour of $\mathcal{MS}^{\mathrm{ord}}(\mathbf{I})[\lambda]$.

Definition 6.5. *We now renormalise our prototype symbol maps* $\mathcal{E}^{\dagger}_{\infty}$, $\mathcal{E}^{\dagger\dagger}_{\infty}$, $\mathcal{E}^{\dagger\dagger}_{\infty,\underline{\delta}}$ *defining* $\mathcal{E}_\infty : \mathbb{H}^1_{\mathrm{Eul},p}(\mathbb{T}_\infty)[\lambda] \longrightarrow \mathrm{Hom}_{\mathbb{Q}\mathrm{cts}}\left(\mathrm{div}^0\mathbb{P}^1(\mathbb{Q}), \mathcal{L}'\right)$ *via the formula*

$$\mathcal{E}_\infty\left(\vec{\mathfrak{X}}_{[\lambda]}\right) := \frac{1}{\ell_\lambda^{+}} \times \mathcal{E}^{\dagger\dagger}_{\infty,\underline{\delta}}\left(\vec{\mathfrak{X}}_{[\lambda]}\right)^{+} + \frac{1}{\ell_\lambda^{-}} \times \mathcal{E}^{\dagger\dagger}_{\infty,\underline{\delta}}\left(\vec{\mathfrak{X}}_{[\lambda]}\right)^{-}.$$

This modified map has the beautiful property, that

$$\nabla_{k-2}^{-1}\left(\left(\mathcal{E}_\infty\left(\vec{\mathfrak{X}}_{[\lambda]}\right)^{\pm} \bmod P\right) \otimes \mathcal{X}_{k-2}\right) = \left(P \circ \ell_\lambda^{\pm}\right)^{-1} \times \mathcal{C}_{\mathbf{f}_P} \times \mathcal{E}_{k,\epsilon,\gamma}\left(\vec{\mathfrak{X}}_{[\lambda]} \bmod P\right)^{\pm}$$
$$= \mathrm{Per}^{\pm}_{\mathbf{I},\lambda_P} \times \mathcal{E}_{k,\epsilon,\gamma}\left(\vec{\mathfrak{X}}_{[\lambda]} \bmod P\right)^{\pm} \quad \text{as required.}$$

It is also independent of the initial choice of crystalline period $\mathbf{v}_{\mathbf{I},\lambda}$.

The proof of our main result is finished. □

Remark: It is interesting to ask how much of this theorem works if neither of the hypotheses (Rk1) or (UF) actually hold. In this unhappy situation, the torsion submodule of the space of **I**-adic modular symbols might well be non-trivial, thus there are no longer distinguished generators $\Xi_\lambda^{\pm}$.

Nevertheless, the prototype homomorphism $\mathcal{E}^{\dagger}_{\infty}$ clearly exists unconditionally. When evaluated at $\vec{\mathfrak{Z}}_{[\lambda]}$, what two-variable p-adic L-function does $\mathcal{E}^{\dagger}_{\infty}$ interpolate?

6.3 Applications to Iwasawa theory

The most technical and difficult parts of the construction have been accomplished in the previous two sections. We are almost ready to apply them in the real world. What is essentially a statement about the deformation theory of modular symbols, has striking consequences in the Iwasawa theory of these big Galois representations. However we cannot quite use the result in its present form, as it is written in a language alien to Selmer groups, Tate-Shafarevich groups, and so on.

Let us start by reinterpreting the preceding section in the more familiar context of power series rings. Fix an integer M coprime to $p \geq 5$, and define $\mathbf{F} = \mathbb{Q}(\mu_M)$. We write $G_{\infty,\mathbf{F}}$ for the topological group $\mathrm{Gal}\big(\mathbf{F}(\mu_{p^\infty})/\mathbb{Q}\big)$, which is isomorphic to $\mathbb{Z}_{p,M}^\times$ via the cyclotomic character.

Corollary 6.6. *There exists a unique* $\Lambda^{\mathrm{wt}}[[G_{\infty,\mathbf{F}}]]$*-homomorphism*

$$\mathrm{PR}_{\infty,\mathbf{F}} : \overline{\mathcal{MS}}^{\mathrm{ord}} \otimes_{\mathbf{h}^{\mathrm{ord}}} \varprojlim_n H^1\Big(\mathbf{F}(\mu_{p^n}) \otimes \mathbb{Z}_p, \mathbb{T}_\infty[\lambda]\Big) \longrightarrow \mathbf{I}[[G_{\infty,\mathbf{F}}]] \otimes_{\mathbf{I}} \mathcal{L}'$$

for each triple (λ, k_0, ϵ)*, satisfying the interpolation rule*

$$\psi\chi_{\mathrm{cy}}^j \circ P\Big(\mathrm{PR}_{\infty,\mathbf{F}}\left(\mathfrak{X}_{[\lambda],p^n}\right)_n\Big) = \frac{j!M^j p^{nj}\left(1 - \psi^{-1}(p)p^j/\lambda_P(U_p)\right)}{\lambda_P(U_p)^n} G(\psi^{-1}, \zeta_{p^nM})$$
$$\times \ \mathrm{Per}^{\pm}_{\mathbf{I},\lambda_P} \sum_{\sigma \in \mathrm{Gal}\left(\mathbf{F}(\mu_{p^n})/\mathbb{Q}\right)} \psi(\sigma)\left\langle \gamma^{\pm}_{\mathbf{f}^*_P}, \exp^*_j\left(\mathfrak{X}_{[\lambda],p^n} \otimes \zeta_{p^n}^{\otimes -j} \bmod P\right)^\sigma \right\rangle$$

at almost all points $\left(P, \psi\chi_{\mathrm{cy}}^j\right) \in \mathrm{Spec}\, \mathbf{I}(\mathcal{O}_\mathcal{K})^{\mathrm{alg}} \times G^*_{\infty,\mathbf{F}}$*, with sign* $\pm = \psi(-1)(-1)^j$*. Here* P *is algebraic of type* (k, ϵ) *and* $k \in \mathbb{U}_{k_0}$*, the character* ψ *is viewed* $\bmod\ p^nM$*, and the integer* $j \in \{0, \ldots, k-2\}$*.*

Proof: This follows almost immediately from previous arguments, so we sketch it. First lift $\left(\mathfrak{X}_{[\lambda],p^n}\right)_n$ to an element $\vec{\mathfrak{X}}_{[\lambda]} \in \overline{\mathcal{MS}}^{\mathrm{ord}} \otimes_{\mathbf{h}^{\mathrm{ord}}} \mathbb{H}^1_{\mathrm{Eul},p}\Big(\mathbb{T}_\infty[\lambda]\Big)$, then map it via the interpolation $\mathcal{E}_\infty$ to the space $\mathrm{Hom}_{\mathbb{Q}\mathrm{cts}}\left(\mathrm{div}^0\mathbb{P}^1(\mathbb{Q}), \mathcal{L}'\right)$.

Applying Proposition 3.6(ii), we form the measure $\mu^\bullet_{\mathcal{E}_\infty(\vec{\mathfrak{X}}_{[\lambda]})} \in \mathrm{Meas}\left(\mathbb{Z}_{p,M}^\times, \mathcal{L}'\right)$ which corresponds to a two-variable power series in the algebra $\mathbf{I}[[\mathbb{Z}_{p,M}^\times]] \otimes_{\mathbf{I}} \mathcal{L}'$. Then Theorem 6.4 tells us how $\mathcal{E}_\infty\left(\vec{\mathfrak{X}}_{[\lambda]}\right)$ and thus $\mu^\bullet_{\mathcal{E}_\infty(\vec{\mathfrak{X}}_{[\lambda]})}$ behave modulo P, which explains the reappearance of those ubiquitous $\mathbf{I}$-adic periods.

Finally, the special value formulae are easily checked from Theorem 3.10. □

Remark: It is fairly cumbersome to keep writing '$\overline{\mathcal{MS}}^{\mathrm{ord}} \otimes_{\mathbf{h}^{\mathrm{ord}}} -$' behind every single cohomology group, and we shall sometimes neglect to mention it hereon in. However the reader should be aware that the $\mathcal{R}$-adic Euler system is not canonically defined, unless one first tensors $\mathbb{H}^1_{\mathrm{Eul}}$ by this universal space.

Question. *What does the super Euler system* $\vec{\mathfrak{Z}} \in \overline{\mathcal{MS}}^{\mathrm{ord}} \otimes_{\mathbf{h}^{\mathrm{ord}}} \mathbb{H}^1_{\mathrm{Eul}}(\mathbb{T}_\infty) \otimes_{\mathcal{R}} \widetilde{\mathcal{R}}$ *transform into under each homomorphism* $\mathrm{PR}_{\infty,\mathbf{F}}$*?*

Let $\left(\mathfrak{Z}_{[\lambda],p^n}\right)_n$ denote the piece of $\vec{\mathfrak{Z}}$ lying in $\varprojlim_n H^1\Big(\mathbf{F}(\mu_{p^n}) \otimes \mathbb{Z}_p, \mathbb{T}_\infty[\lambda]\Big) \otimes_{\mathbf{I}} \mathcal{L}'$. One calculates that $\psi\chi^j_{\mathrm{cy}} \circ P\Big(\mathrm{PR}_{\infty,\mathbf{F}}\left(\mathfrak{Z}_{[\lambda],p^n}\right)_n\Big)$ equals

$$\frac{j!M^j p^{nj}\left(1-\psi^{-1}(p)p^j/a_p(\mathbf{f}_P)\right)}{a_p(\mathbf{f}_P)^n} \times \mathrm{Per}^{\psi(-1)(-1)^j}_{\mathbf{I},\lambda_P} \times \frac{G(\psi^{-1},\zeta_{p^n M})L(\mathbf{f}_P,\psi,j+1)}{(2\pi i)^j\,\Omega^{\psi(-1)(-1)^j}_{\mathbf{f}_P}}$$

at critical points $(P,\psi\chi^j_{\mathrm{cy}}) \in \mathrm{Spec}\,\mathbf{I}(\mathcal{O}_{\mathcal{K}})^{\mathrm{alg}} \times G^{\,*}_{\infty,\mathbf{F}}$ as above.

The knowledgeable reader will recognise these L-values as precisely those of the two-variable analytic p-adic L-function $\mathbf{L}^{\mathrm{GS}}_{p,\mathbf{F}}(\mathbf{f}) \in \mathbf{I}[[G_{\infty,\mathbf{F}}]]$ introduced in §4.4. Because the special characters $\psi\chi^j_{\mathrm{cy}}$ are dense in $G^{\,*}_{\infty,\mathbf{F}}$, and under (Arith) the algebraic points P densely populate Spec $\mathbf{I}(\overline{\mathbb{Q}}_p)$, we have shown the following:

Corollary 6.7. $\mathrm{PR}_{\infty,\mathbf{F}}\Big(\mathfrak{Z}_{[\lambda],p^n}\Big)_n \;=\; \mathbf{L}^{\mathrm{GS}}_{p,\mathbf{F}}(\mathbf{f})\,.$

At this point it is appropriate to mention two closely related works, that have been formulated in the case of a p-ordinary deformation. The second, in particular, suggests the existence of a Λ-adic version of classical Eichler-Shimura theory.

The algebraic p-adic L-function of T. Ochiai.
The above corollary identifies the deformed Euler system with the p-adic L-function. A similar type of result was obtained in the case $\mathbf{F}=\mathbb{Q}$ in [Oc1, Thms 3.13, 3.17]. Ochiai has a two-variable parametrisation of $\varprojlim_n H^1\Big(\mathbb{Q}_p(\mu_{p^n}),\mathbb{T}_\infty\Big)$ which sends Kato's zeta-elements, to an algebraic p-adic L-function $\mathbf{L}^{\mathrm{Och}}_{p\text{-adic}}$ living in $\mathcal{R}[\![G_{\infty,\mathbb{Q}}]\!]$.

It is stated in Remark 3.18 of his paper, that it's unknown at present how to relate the λ-part of $\mathbf{L}^{\mathrm{Och}}_{p\text{-adic}}$ with the analytic function of Greenberg-Stevens and Kitagawa. Perhaps the reason is that the periods occurring in his algebraic constructions are all error terms on the de Rham side, whilst the periods of $\mathbf{L}^{\mathrm{GS}}_{p,\mathbb{Q}}(\mathbf{f})$ are error terms on the Betti side. By contrast, our theory actually links the structure of $\mathcal{MS}^{\mathrm{ord}}(\mathbf{I})^{\pm}$ with the structure of $\mathbb{D}^{\mathbf{I}}_{\mathrm{cris}}(\mathbb{T}^*_\infty)$. In fact one may define canonical elements in the crystalline deformation space, as follows.

Let's recall for each primitive homomorphism $\lambda : \mathbf{h}^{\mathrm{ord}}(Np^\infty;\mathcal{O}_{\mathcal{K}}) \otimes_{\Lambda^{\mathrm{wt}}_{\mathcal{K}}} \mathbf{I} \longrightarrow \mathbf{I}$, we made a choice of generator $\mathbf{v}_{\mathbf{I},\lambda} \in \mathbb{D}^{\mathbf{I}}_{\mathrm{cris}}(\mathbb{T}^*_\infty)[\lambda]$. This choice in turn normalised the dual exponentials $\mathrm{EXP}^*_{\mathbb{T}_\infty}$ in Definition 5.11, and thence the prototype $\mathcal{E}^{\dagger\dagger}_{\infty,\underline{\delta}}$. Furthermore, $\ell^+_\lambda, \ell^-_\lambda \in \mathcal{L}'$ denoted precisely those elements satisfying the identity

$$\mathcal{E}^{\dagger\dagger}_{\infty,\underline{\delta}}\left(\vec{\mathfrak{Z}}_{[\lambda]}\right)^{\pm} \;=\; \ell^{\pm}_{\lambda} \times \Xi^{\pm}_{\lambda} \qquad \text{of } \mathbf{I}\text{-adic modular symbols.}$$

Changing $\mathbf{v}_{\mathbf{I},\lambda}$ by a scalar in $\mathcal{L}'$, has the effect of multiplying $\mathrm{EXP}^*_{\mathbb{T}_\infty}$, $\mathcal{E}^{\dagger\dagger}_{\infty,\underline{\delta}}$ and thus $\ell^{\pm}_\lambda$ by the same scalar amount. However, it changes the quantities $\left(\ell^{\pm}_\lambda\right)^{-1}$ by the reciprocal amount.

Definition 6.8. *We define 'universal crystalline zeta-periods', lying inside of the one-dimensional space* $\mathbb{D}^{\mathbf{I}}_{\text{cris}}(\mathbb{T}^*_\infty)[\lambda] \otimes_{\mathbf{I}} \mathcal{L}'$, *by setting*

$$\mathbf{v}^{\text{zeta}}_{\lambda,+} \;:=\; \frac{1}{\ell^+_\lambda} \times \mathbf{v}_{\mathbf{I},\lambda} \qquad \textit{and} \qquad \mathbf{v}^{\text{zeta}}_{\lambda,-} \;:=\; \frac{1}{\ell^-_\lambda} \times \mathbf{v}_{\mathbf{I},\lambda}.$$

In particular, the $\mathbf{v}^{\text{zeta}}_{\lambda,\pm}$ *are canonical elements which depend only on the original choice of universal modular symbol* $\Xi_\lambda = \Xi^+_\lambda + \Xi^-_\lambda$ *in* $\mathcal{MS}^{\text{ord}}(\mathbf{I})[\lambda]$, *nothing else.*

Of course the space $\mathbb{D}^{\mathbf{I}}_{\text{cris}}(\mathbb{T}^*_\infty)[\lambda] \otimes_{\mathbf{I}} \mathcal{L}'$ itself cannot be decomposed into $\pm$-parts, this is simply a notational device.

One might envisage these periods $\mathbf{v}^{\text{zeta}}_{\lambda,\pm}$ as playing the same rôle in a Hida family, that the Néron differential

$$\omega_E \in \text{Fil}^0 H^1_{\text{dR}}(E) \otimes_{\mathbb{Q}} \mathbb{Q}_p \;\cong\; \text{Fil}^0 \mathbf{D}_{\text{dR}}\big(h^1(E)\big)$$

associated to a modular elliptic curve E over $\mathbb{Q}$, plays within p-adic Hodge theory. It is exactly the selection of these universal zeta-periods, which yielded the correct symbol maps

$$\mathcal{E}^{\pm}_\infty : \overline{\mathcal{MS}}^{\text{ord}} \otimes_{\mathbf{h}^{\text{ord}}} \mathbb{H}^1_{\text{Eul},p}(\mathbb{T}_\infty)[\lambda] \otimes_{\mathcal{R}} \widetilde{\mathcal{R}} \longrightarrow \text{Hom}_{\mathcal{Q}\text{cts}}\Big(\text{div}^0\mathbb{P}^1(\mathbb{Q}), \mathcal{L}'\Big)^{\pm}$$

and consequently, the correct interpolation $\text{PR}_{\infty,\mathbf{F}} : \big(\mathfrak{z}_{[\lambda],p^n}\big)_n \mapsto \mathbf{L}^{\text{GS}}_{p,\mathbf{F}}(\mathbf{f})$.

Remark: There is yet another interesting arithmetic application of these elements. For abelian extensions $\mathbf{F}/\mathbb{Q}$, both $t^{-1}.\mathbf{v}^{\text{zeta}}_{\lambda,+}$ and $t^{-1}.\mathbf{v}^{\text{zeta}}_{\lambda,-}$ generate lattices inside

$$H^0\Big(\mathbf{F} \otimes \mathbb{Z}_p,\; \mathbb{T}^*_\infty(1)[\lambda] \otimes_{\Lambda^{\text{wt}}} \mathbb{B}^{\text{wt}}_{\text{dR}}/\mathbb{B}^{+,\text{wt}}_{\text{dR}}\Big).$$

These $\mathbf{I}$-adic lattices induce Tamagawa measures, on the geometric part $H^1_{\mathcal{G}}$ of the semi-local cohomology. Such measures occur in nature when calculating Euler characteristics of Tate-Shafarevich groups (as is explained in upcoming chapters).

The universal zeta-modular form of T. Fukaya.
Fukaya [Fu] has a beautiful theory, converting Beilinson elements into two-variable p-adic L-functions. In her work she considers the K-groups $K_2\Big(Y(Np^m, p^m)\Big)$ and converts the corresponding zeta-elements directly, into something called the 'universal zeta-modular form'. Calculating the special values of this object $z^{\text{univ}}_{Np^\infty}$, she obtains what is essentially $\mathbf{L}^{\text{GS}}_{p,\mathbb{Q}}(\mathbf{f})$ up to some stray Euler factors.

More precisely, her Coleman map is a homomorphism

$$\mathcal{C}_N : \varprojlim_m K_2\Big(Y(Np^m, p^m)\Big) \;\longrightarrow\; \text{Frac}\Big(\mathcal{O}_H[[1 + p\mathbb{Z}_p, G_\infty]]\Big)$$

where $H = \Big(\varprojlim_m \mathbb{Z}/p^m\mathbb{Z}[\![q]\!][q^{-1}]\Big)[p^{-1}]$ is a discrete valuation field of q-expansions. The image of the ${}_{c,d}z_{Np^m,p^m}$'s is a convolution of two Λ^{wt}-adic Eisenstein series. Again the dependence on the integers c, d is purely an artefact of K-theory, it can be removed once inside the space $\mathcal{O}_H[\![1 + p\mathbb{Z}_p, G_\infty]\!]$.

Remark: When dealing with arithmetic objects such as Tate-Shafarevich groups, it is nice to have an interpolation factoring through étale cohomology (as $\mathcal{E}_\infty$ does). Probably the map $\mathcal{C}_N$ itself possesses a full description in terms of Bloch-Kato exponentials, though the building blocks of her K_2-homomorphism are themselves logarithmic differentials. How then can we compare these two maps $\mathcal{E}_\infty$ and $\mathcal{C}_N$?

We first observe that the zeta-elements constructed in §6.1 started life inside of the double projective limit $\varprojlim_r \varprojlim_m K_2\Big(Y\big(Mp^{n+m}, MNp^{n+r+m}\big)\Big)$, and clearly encode identical arithmetic to Fukaya's zeta-elements in $\varprojlim_m K_2\Big(Y\big(Np^m, p^m\big)\Big)$. One strongly suspects that the diagram

$$\begin{array}{ccc}
\text{Galois cohomology of } \mathbb{T}_\infty & \overset{\mathcal{E}_\infty}{\longrightarrow} & \mathcal{Q}\text{-continuous } \mathbf{I}\text{-adic symbols} \\
\mathfrak{c}^{\text{univ}}_{\infty,p}\Big\uparrow & & \Big\downarrow \mu^\bullet \\
K_2\text{-theory of } Y(-,-) & & \text{two-variable power series rings} \\
\mathcal{C}_N \Big\downarrow & & \Big\uparrow \text{special values} \\
\text{Frac}\Big(\mathcal{O}_H[[1+p\mathbb{Z}_p, G_\infty]]\Big) & \overset{\text{idempotents}}{\longrightarrow} & \Big(\text{ordinary } q\text{-expansions}\Big)[[G_\infty]]
\end{array}$$

commutes, in a suitable sense. Certainly it is true there are equalities

$$\mu^\bullet \circ \mathcal{E}_\infty\Big(\mathfrak{c}^{\text{univ}}_{\infty,p}(z_{-,-})[\lambda]\Big) \quad = \quad \mathbf{L}^{\text{GS}}_{p,\mathbb{Q}}(\mathbf{f}) \quad = \quad \text{the } \lambda\text{-part of } \mathcal{C}_N(z_{-,-})$$

so why not for non zeta-elements as well?

Let us speculate a little further. For any weight $k \geq 2$, Eichler-Shimura theory provides isomorphisms

$$\Big(\mathcal{S}_k \oplus \mathcal{S}^{\text{anti}}_k \oplus \text{Eis}_k\Big)(\Gamma_1(Np^r)) \;\cong\; H^1\Big(Y_1(Np^r), \text{Sym}^{k-2}(\mathbb{C})\Big)$$

of finite-dimensional $\mathbb{C}$-vector spaces, with action of the Hecke algebra at level Np^r. Poincaré duality identifies the right-hand side with $H^1_{\text{c}}\Big(Y_1(Np^r), \text{Sym}^{k-2}(\mathbb{C})\Big)^*$ and at weight two, one knows that $H^1_{\text{c}}\big(Y_1(Np^r), \mathbb{C}\big)^* \cong \text{Hom}_{\Gamma_1(Np^r)}\big(\text{div}^0\mathbb{P}^1(\mathbb{Q}), \mathbb{C}\big)^*$. Therefore classical q-expansions of tame level N, are eventually included in

$$\Big(\mathcal{S}_2 \oplus \mathcal{S}^{\text{anti}}_2 \oplus \text{Eis}_2\Big)(\Gamma_1(Np^r)) \;\overset{\sim}{\longrightarrow}\; \text{Hom}_{\mathbb{C}}\Big(\text{Hom}_{\Gamma_1(Np^r)}\big(\text{div}^0\mathbb{P}^1(\mathbb{Q}), \mathbb{C}\big), \,\mathbb{C}\Big)$$

if we allow r to increase through the positive integers.

Question. *Is there a deformation-theoretic version of these isomorphisms, which identifies the q-expansions* $\mathcal{O}_H[\![1+p\mathbb{Z}_p, G_\infty]\!]$ *with the* $\Lambda^{\text{wt}}[\![G_\infty]\!]$*-dual of* $\text{Hom}_{\mathcal{Q}\text{cts}}\Big(\text{div}^0\mathbb{P}^1(\mathbb{Q}), \,\mathbf{I}\Big)$, *at least on their ordinary components?*

If the answer is affirmative, under this identification we would expect to be able to relate the universal zeta-modular form with the universal $\mathbf{I}$-adic modular symbol. Without such a map, we are at a loss how to link Fukaya's work with Theorem 6.4.

6.4 The Coleman exact sequence

In his famous paper [Co] on division values of Lubin-Tate groups, Coleman showed there was an exact sequence of Λ^{cy}-modules, operating as a machine for converting the input $\left(\frac{1-\zeta^c_{p^m}}{1-\zeta_{p^m}}\right)_{m\geq 1}$ of cyclotomic units into the p-adic Riemann zeta-function. Casting our mind back to §3.2, we should also mention Coates and Wiles [CW] used this same sequence to relate elliptic units with the p-adic Hecke L-function, associated to an elliptic curve with complex multiplication [Kz,Ya].

This machine was then generalised extensively by Perrin-Riou [PR2] and others to convert any (hypothetical) Euler system attached to a pure motive over $\mathbb{Q}$ into a p-adic L-function – see Chapter II for more details, particularly Theorem 2.7. In order to construct two-variable versions of these exact sequences within the cohomology of $\mathbb{T}_\infty$, we simply have to answer the following:

Question. *What is the kernel and image of* $\mathrm{PR}_{\infty,\mathbf{F}}$ *for each* λ*-eigenspace?*

As before, we have written $\mathbf{F}$ for the cyclotomic field $\mathbb{Q}(\mu_M)$ with M coprime to p. Let $S_{\mathbf{F}}$ denote the set of finite primes in Spec $\mathbb{Z}[\mu_M]$, and $S_{\mathbf{F},p}$ denotes the subset consisting of those primes lying over p.

From Definition 5.11(iii), the big dual exponential map factorises through local Galois cohomology groups. It follows that $\mathrm{PR}_{\infty,\mathbf{F}}$ also factorises via

$$\varprojlim_n H^1\Big(\mathbf{F}(\mu_{p^n})\otimes\mathbb{Z}_p,\mathbb{T}_\infty[\lambda]\Big) \xrightarrow{\sim} \prod_{\mathfrak{P}\in S_{\mathbf{F},p}} H^1_{\text{Iw}}\Big(\mathbf{F}_{\mathfrak{P}},\mathbb{T}_\infty[\lambda]\Big) \overset{\prod \mathrm{PR}_{\infty,\mathbf{F}_{\mathfrak{P}}}}{\longrightarrow} \mathbf{I}[[G_{\infty,\mathbf{F}}]]\otimes_{\mathbf{I}}\mathcal{L}'$$

and we may then compute its kernel semi-locally, at each point of Spec $\mathbb{Z}[\mu_M]$. Notice we have neglected to write $\overline{\mathcal{MS}}^{\text{ord}}\otimes -$ above, and will continue in this vein.

Pick a prime $\mathfrak{P}\in S_{\mathbf{F},p}$, and let us denote by L the completed local field $\mathbf{F}_{\mathfrak{P}}$. By Corollary 6.6, we know that the λ-part of $\mathrm{PR}_{\infty,L}$ modulo $P_{k,\epsilon}$ coincides with $\mathrm{PR}_{k,\epsilon,L}$ (up to a scalar). Furthermore, since $\mathrm{Ker}\big(\mathrm{PR}_{k,\epsilon,L}\big)$ contains

$$\mathbb{Q}_p\otimes_{\mathbb{Z}_p}\varprojlim_n H^1_g\Big(L(\mu_{p^n}),\ \mathbb{T}_\infty[\lambda]\otimes_{\mathbf{I}}\mathbf{I}/P_{k,\epsilon}\Big)$$

it follows that the mapping $\mathrm{PR}_{\infty,L}$ must kill off the p-ordinary, N-primitive part of $\varprojlim_{n,r} H^1_g\big(L(\mu_{p^n}),\mathrm{Ta}_p(J_r)\big)$.

Remark: Thanks to the cohomological description of the Kummer map in [BK, §3], one may then identify $H^1_g\big(L(\mu_{p^n}),\mathrm{Ta}_p(J_r)\big)$ with the p-adic points $J_r\big(L(\mu_{p^n})\big)\widehat{\otimes}\,\mathbb{Z}_p$. Whenever the exponent $m\geq \mathrm{ord}_p\Big(\#J_r\big(L(\mu_{p^n})\big)_{\text{tors}}\Big)$, the natural map

$$J_r\big(L(\mu_{p^n})\big)[p^m] \longrightarrow J_r\big(L(\mu_{p^n})\big)/p^m$$

becomes injective. As an immediate consequence, $\varprojlim_{n,r} J_r\big(L(\mu_{p^n})\big)\widehat{\otimes}\,\mathbb{Z}_p$ properly contains $\varprojlim_{n,r,m} J_r\big(L(\mu_{p^n})\big)[p^m] = \mathrm{Ta}_p(J_\infty)^{G_{L(\mu_{p^\infty})}}$ as a sub-Iwasawa module.

Definition 6.9. *(i) For each $L = \mathbf{F}_{\mathfrak{P}}$, we define the $\mathcal{R}\big[\big[\mathrm{Gal}\big(L(\mu_{p^\infty})/\mathbb{Q}_p\big)\big]\big]$-module $\mathbf{H}^1_{\mathrm{kum},L}(\mathbb{T}_\infty)$ by saturating*

$$\mathbf{e}_{\mathrm{prim}}\left(\varprojlim_{n,r} J_r\big(L(\mu_{p^n})\big)\widehat{\otimes}_{\Lambda^{\mathrm{wt}}}\mathrm{Frac}\big(\Lambda^{\mathrm{wt}}\big)\right) \cap \left(\varprojlim_{n,r} J_r\big(L(\mu_{p^n})\big)\widehat{\otimes}\mathbb{Z}_p\right)^{\mathrm{ord}}$$

inside of $\varprojlim_n H^1\big(L(\mu_{p^n}), \mathbb{T}_\infty\big)$; in particular, $\mathbf{H}^1_{\mathrm{kum},L}(\mathbb{T}_\infty)$ must contain $\mathbb{T}_\infty^{G_{L(\mu_{p^\infty})}}$ by the above discussion.

(ii) Similarly, we define $\mathbf{H}^1_{\mathrm{kum},\mathbf{F}}(\mathbb{T}_\infty)$ to be the maximal $\mathcal{R}[\![G_{\infty,\mathbf{F}}]\!]$-submodule of $\varprojlim_n H^1\Big(\mathbf{F}(\mu_{p^n})\otimes\mathbb{Z}_p, \mathbb{T}_\infty\Big)$ with the property that at each finite prime $\mathfrak{P} \in S_{\mathbf{F},p}$, its restriction to $H^1_{\mathrm{Iw}}\Big(\mathbf{F}_{\mathfrak{P}}, \mathbb{T}_\infty\Big)$ lies in $\mathbf{H}^1_{\mathrm{kum},\mathbf{F}_{\mathfrak{P}}}(\mathbb{T}_\infty)$.

Let's try and understand what is happening in this definition, at each finite layer. As already mentioned in §5.4, the author and P. Smith studied $\mathbf{X}_{\mathbb{T}_\infty}\Big(\mathbf{F}(\mu_{p^n})\otimes\mathbb{Z}_p\Big)$ which was defined as the $\mathcal{R}$-saturation of the semi-local points

$$\mathbf{e}_{\mathrm{prim}}\left(\varprojlim_{r} J_r\big(\mathbf{F}(\mu_{p^n})\otimes\mathbb{Z}_p\big)\widehat{\otimes}_{\Lambda^{\mathrm{wt}}}\mathrm{Frac}\big(\Lambda^{\mathrm{wt}}\big)\right) \cap \left(\varprojlim_{r} J_r\big(\mathbf{F}(\mu_{p^n})\otimes\mathbb{Z}_p\big)\widehat{\otimes}\,\mathbb{Z}_p\right)^{\mathrm{ord}}.$$

Under the Kummer map to cohomology, $\mathbf{X}_{\mathbb{T}_\infty}(-)$ coincided exactly with

$$\mathrm{Ker}\left(H^1\Big(\mathbf{F}(\mu_{p^n})\otimes\mathbb{Z}_p, \mathbb{T}_\infty\Big) \stackrel{-\otimes 1}{\longrightarrow} H^1\Big(\mathbf{F}(\mu_{p^n})\otimes\mathbb{Z}_p, \mathbb{T}_\infty\otimes_{\Lambda^{\mathrm{wt}}}\mathbb{B}^{\mathrm{wt}}_{\mathrm{dR}}\Big)\otimes_{\mathcal{R}}\mathcal{F}\right).$$

Thus passing to the limit over the whole cyclotomic tower, there is an isomorphism between $\mathbf{H}^1_{\mathrm{kum},\mathbf{F}}(\mathbb{T}_\infty)$ and the $\mathcal{R}[\![G_{\infty,\mathbf{F}}]\!]$-saturation of $\varprojlim_n \mathbf{X}_{\mathbb{T}_\infty}\Big(\mathbf{F}(\mu_{p^n})\otimes\mathbb{Z}_p\Big)$.

Let us again remind the reader that Γ^{cy} was the Galois group of the cyclotomic $\mathbb{Z}_p$-extension of $\mathbb{Q}$, and is isomorphic to $\mathrm{Gal}\big(\mathbf{F}(\mu_{p^\infty})/\mathbf{F}(\mu_p)\big)$ topologically.

Lemma 6.10. *Provided that $\mathcal{R}$ is a Gorenstein ring, then*

$$\mathrm{Tors}_{\mathcal{R}[\![\Gamma^{\mathrm{cy}}]\!]}\left(\varprojlim_n H^1\big(\mathbf{F}(\mu_{p^n})\otimes\mathbb{Z}_p, \mathbb{T}_\infty\big)\right) \;\cong\; H^0\big(\mathbf{F}(\mu_{p^\infty})\otimes\mathbb{Z}_p, \mathbb{T}_\infty\big)$$

which is also the $\mathcal{R}[\![\Gamma^{\mathrm{cy}}]\!]$-torsion submodule of $\mathbf{H}^1_{\mathrm{kum},\mathbf{F}}(\mathbb{T}_\infty)$.

In order not to interrupt our calculation of the kernel/cokernel of the map $\mathrm{PR}_{\infty,\mathbf{F}}$, we postpone the proof of this result to the final section.

From now on we shall assume that the ring $\mathcal{R}$ is Gorenstein; we claim that the induced homorphism

$$\frac{\varprojlim_n H^1\big(\mathbf{F}(\mu_{p^n})\otimes\mathbb{Z}_p, \mathbb{T}_\infty[\lambda]\big)}{\mathbf{H}^1_{\mathrm{kum},\mathbf{F}}(\mathbb{T}_\infty)[\lambda]} \quad\stackrel{\mathrm{PR}_{\infty,\mathbf{F}}}{\longrightarrow}\quad \mathbf{I}[[G_{\infty,\mathbf{F}}]]\otimes_{\mathbf{I}}\mathcal{L}'$$

is an inclusion. By our lemma $\varprojlim_n H^1\big(\mathbf{F}(\mu_{p^n})\otimes\mathbb{Z}_p, \mathbb{T}_\infty[\lambda]\big)$ and $\mathbf{H}^1_{\mathrm{kum},\mathbf{F}}(\mathbb{T}_\infty)[\lambda]$ share the same torsion submodule and, as the latter is saturated in the former, their quotient must be $\mathbf{I}[\![G_{\infty,\mathbf{F}}]\!]$-free.

Remark: To establish injectivity, we simply need to calculate

(a) $\mathrm{rank}_{\mathbf{I}[\![\Gamma^{\mathrm{cy}}]\!]}\Big(\varprojlim_n H^1\big(\mathbf{F}(\mu_{p^n})\otimes\mathbb{Z}_p, \mathbb{T}_\infty[\lambda]\big)\Big)$

(b) $\mathrm{rank}_{\mathbf{I}[\![\Gamma^{\mathrm{cy}}]\!]}\Big(\mathbf{H}^1_{\mathrm{kum},\mathbf{F}}(\mathbb{T}_\infty)[\lambda]\Big)$

(c) $\mathrm{rank}_{\mathbf{I}[\![\Gamma^{\mathrm{cy}}]\!]}\Big(\mathrm{Im}\big(\mathrm{PR}_{\infty,\mathbf{F}}\big)\Big)$

then hope and pray that (a) = (b)+(c).

To compute the ranks predicted in (a) and (b), it's sensible to work semi-locally. We write $L = \mathbf{F}_{\mathfrak{P}}$ as the prime $\mathfrak{P}$ ranges over points of Spec $\mathbb{Z}[\mu_M]$ above p. Assuming for simplicity that $\mathbf{I}$ is regular (so its height one primes are all principal) at any point $P \in \mathrm{Spec}\,\mathbf{I}$, the sequence

$$0 \longrightarrow \frac{\varprojlim_n H^1\big(L(\mu_{p^n}), \mathbb{T}_\infty[\lambda]\big)}{P.\varprojlim_n H^1\big(L(\mu_{p^n}), \mathbb{T}_\infty[\lambda]\big)} \longrightarrow \varprojlim_n H^1\Big(L(\mu_{p^n}), \mathbb{T}_\infty[\lambda]\otimes_{\mathbf{I}}\mathbf{I}/P\Big) \longrightarrow \varprojlim_n H^2\Big(L(\mu_{p^n}), \mathbb{T}_\infty[\lambda]\Big)[P]$$

will certainly be exact.

Our strategy is to relate the $\mathbf{I}[\![\Gamma^{\mathrm{cy}}]\!]$-rank with the generic rank after specialisation. However, we must be careful to ditch primes that cause the rank to jump around. We define $\Sigma_{\mathrm{bad},\lambda} \subset \mathrm{Spec}\,\mathbf{I}[\![\Gamma^{\mathrm{cy}}]\!]$ to be the set of associated primes ideals of the Iwasawa modules $\mathrm{Tors}_{\mathbf{I}[\![\Gamma^{\mathrm{cy}}]\!]}\Big(\varprojlim_n H^j\big(L(\mu_{p^n}), \mathbb{T}_\infty[\lambda]\big)\Big)$ for $j = 0, 1, 2$.

Let $\mathcal{J}_{\mathrm{cy}}$ denote the augmentation ideal of $\mathbf{I}[\![\Gamma^{\mathrm{cy}}]\!]$.

Warning: For each $\mathbf{P} \in \Sigma_{\mathrm{bad},\lambda}$, its reduction $\widetilde{\mathbf{P}}$ modulo $\mathcal{J}_{\mathrm{cy}}$ has divisor

$$\mathrm{div}_{\mathbf{I}}(\mathbf{I}/\widetilde{\mathbf{P}}) \;=\; \sum_{P\in\mathrm{Spec}\,\mathbf{I}} \mathrm{ord}_P(\widetilde{\mathbf{P}}).(P) \qquad \text{inside the free group on Spec } \mathbf{I}\backslash\{\mathcal{J}_{\mathrm{cy}}\}.$$

In our rank calculations, we must avoid at all costs height one primes P occurring in the above divisors, i.e. one should never stray outside the set

$$\widetilde{\Sigma}_{\mathrm{good},\lambda} \;:=\; \mathrm{Spec}\,\mathbf{I}(\mathcal{O}_{\mathcal{K}})^{\mathrm{alg}}\Big\backslash \bigcup_{\mathbf{P}\in\Sigma_{\mathrm{bad},\lambda}} \Big\{P\in\mathrm{Spec}\,\mathbf{I}\quad \text{such that } \mathrm{ord}_P(\widetilde{\mathbf{P}})>0\Big\}.$$

Because $\#\Sigma_{\mathrm{bad},\lambda}$ is clearly finite, there exist infinitely many primes $P \in \mathrm{Spec}\,\mathbf{I}^{\mathrm{alg}}$ for which we can estimate $\mathbf{I}[\![\Gamma^{\mathrm{cy}}]\!]$-ranks, just by specialising modulo P.

Making a judicious choice of arithmetic prime $P \in \widetilde{\Sigma}_{\mathrm{good},\lambda}$, we calculate

$$\begin{aligned}\mathrm{rank}_{\mathbf{I}[\![\Gamma]\!]}\Big(\varprojlim_n H^1\big(L(\mu_{p^n}), \mathbb{T}_\infty[\lambda]\big)\Big) \;&=\; \mathrm{rank}_{\mathbf{I}/P[\![\Gamma]\!]}\Big(\varprojlim_n H^1\big(L(\mu_{p^n}), \mathbb{T}_\infty[\lambda]\otimes_{\mathbf{I}}\mathbf{I}/P\big)\Big)\\ &=\; [L(\mu_p):\mathbb{Q}_p]\times \mathrm{rank}_{\mathbf{I}/P}\big(\mathbb{T}_\infty[\lambda]\otimes_{\mathbf{I}}\mathbf{I}/P\big)\\ &=\; 2\,[L(\mu_p):\mathbb{Q}_p].\end{aligned}$$

Furthermore, taking the sum over $S_{\mathbf{F},p}$ yields the equality

$$\mathrm{rank}_{\mathbf{I}[\![\Gamma]\!]}\Big(\varprojlim_n H^1\big(\mathbf{F}(\mu_{p^n})\otimes\mathbb{Z}_p,\mathbb{T}_\infty[\lambda]\big)\Big) = \sum_{\mathfrak{P}\in S_{\mathbf{F},p}} 2[\mathbf{F}_{\mathfrak{P}}(\mu_p):\mathbb{Q}_p] \;=\; 2(p-1)[\mathbf{F}:\mathbb{Q}]. \qquad \textbf{(a)}$$

Similarly for each $L=\mathbf{F}_{\mathfrak{P}}$, the rank of the saturated image of the Kummer map is

$$\begin{aligned}\mathrm{rank}_{\mathbf{I}[\![\Gamma]\!]}\Big(\mathbf{H}^1_{\mathrm{kum},L}(\mathbb{T}_\infty)[\lambda]\Big) &= \mathrm{rank}_{\mathbf{I}/P[\![\Gamma]\!]}\Big(\varprojlim_n H^1_g\big(L(\mu_{p^n}),\mathbb{T}_\infty[\lambda]\otimes_{\mathbf{I}}\mathbf{I}/P\big)\Big)\\ &= [L(\mu_p):\mathbb{Q}_p]\times\mathrm{rank}_{\mathbf{I}/P}\Big(\mathrm{F}^+\mathbb{T}_\infty[\lambda]\otimes_{\mathbf{I}}\mathbf{I}/P\Big)\\ &= [L(\mu_p):\mathbb{Q}_p].\end{aligned}$$

as the rank of $\varprojlim_n H^1_g$ is the dimension of the tangent space. Summing as before,

$$\mathrm{rank}_{\mathbf{I}[\![\Gamma]\!]}\Big(\mathbf{H}^1_{\mathrm{kum},\mathbf{F}}(\mathbb{T}_\infty)[\lambda]\Big) \;=\; \sum_{\mathfrak{P}\in S_{\mathbf{F},p}}[\mathbf{F}_{\mathfrak{P}}(\mu_p):\mathbb{Q}_p] \;=\; (p-1)\,[\mathbf{F}:\mathbb{Q}]. \qquad \textbf{(b)}$$

This leaves us with the task of working out (c).

Remark: By construction $\mathrm{PR}_{\infty,\mathbf{F}}\left(\mathfrak{X}_{[\lambda],p^n}\right)_n$ is the Mellin transform of the bounded measure $\mu^\bullet_{\mathcal{E}_\infty(\vec{\mathfrak{X}}_{[\lambda]})}\in\mathrm{Meas}\left(\mathbb{Z}^\times_{p,M},\mathcal{L}'\right)$, whose special values on compact opens are evaluations of $\mathcal{E}_\infty\left(\vec{\mathfrak{X}}_{[\lambda]}\right)$ at divisors of the form $D=\left(\frac{a}{p^nM}\right)-(\infty)$, $\gcd(a,M)=1$. Hence to compute the quantity in (c), it is sufficient to calculate the rank of

$$\mathrm{Im}\left(\mathbb{H}^1_{\mathrm{Eul},p}(\mathbb{T}_\infty)[\lambda]\ \xrightarrow{\mathcal{E}_\infty}\ \mathrm{Hom}_{\mathcal{Q}\mathrm{cts}}\left(\mathrm{div}^0\mathbb{P}^1(\mathbb{Q}),\mathcal{L}'\right)\right)$$

restricted to D^0_M, the set of degree zero divisors on $\bigcup_{\gcd(c,M)=1}\Sigma_{M,c}\cup\{\infty\}$.

The rank can thus be broken up over $(\mathbb{Z}/M\mathbb{Z})^\times$ into

$$\begin{aligned}\mathrm{rank}_{\mathbf{I}[\![\Gamma]\!]}\Big(\mathrm{Im}(\mathrm{PR}_{\infty,\mathbf{F}})\Big) &= \sum_{\gcd(c,M)=1}\mathrm{rank}_{\mathbf{I}[\![\Gamma]\!]}\Big(\mathrm{Im}(\mathcal{E}_\infty)\Big|_{\mathrm{div}^0\left(\Sigma_{M,c}\cup\{\infty\}\right)}\Big)\\ &= (p-1)\times\sum_{\gcd(c,M)=1}\mathrm{rank}_{\mathbf{I}[\![G_\infty]\!]}\Big(\mathrm{Im}(\mathcal{E}_\infty)\Big|_{\mathrm{div}^0\left(\Sigma_{M,c}\cup\{\infty\}\right)}\Big).\end{aligned}$$

We already know the image of $\mathcal{E}_\infty$ is dominated by the $\mathbf{I}$-adic modular symbol, i.e.

$$\mathrm{Frac}\big(\mathbf{I}[\![G_\infty]\!]\big).\,\Xi_\lambda\Big|_{\mathrm{div}^0\left(\Sigma_{M,c}\cup\{\infty\}\right)} = \mathrm{Frac}\big(\mathbf{I}[\![G_\infty]\!]\big)\otimes_{\mathbf{I}[\![G_\infty]\!]}\mathrm{Im}(\mathcal{E}_\infty)\Big|_{\mathrm{div}^0\left(\Sigma_{M,c}\cup\{\infty\}\right)}$$

because the corresponding statements modulo $P_{k,\epsilon}$ were proved in Theorem 3.10(iii) at every algebraic point of type (k,ϵ).

It follows from the one-dimensionality of these $\mathrm{Frac}\big(\mathbf{I}[\![G_\infty]\!]\big)$-vector spaces, that

$$\begin{aligned}\mathrm{rank}_{\mathbf{I}[\![\Gamma]\!]}\Big(\mathrm{Im}(\mathrm{PR}_{\infty,\mathbf{F}})\Big) &= (p-1)\times\sum_{\gcd(c,M)=1}1\\ &= (p-1)\times\#(\mathbb{Z}/M\mathbb{Z})^\times \;=\; (p-1)\,[\mathbf{F}:\mathbb{Q}]. \qquad \textbf{(c)}\end{aligned}$$

As a consequence of (a), (b) and (c), we conclude that the $\mathbf{I}[\![\Gamma]\!]$-rank of both $\mathbf{H}^1_{\mathrm{kum},\mathbf{F}}$ and $\mathrm{Im}(\mathrm{PR}_{\infty,\mathbf{F}})$, is exactly half that of the projective limit of the semi-local H^1's. This finally establishes the injectivity of $\mathrm{PR}_{\infty,\mathbf{F}}$ modulo the image of Kummer.

In conclusion, we have shown the following:

Theorem 6.11. *Assume $\mathcal{R}$ is a Gorenstein ring, and the hypotheses of §4.3 hold. Then for each primitive $\mathbf{I}$-algebra homomorphism $\lambda : \mathbf{h}^{\mathrm{ord}}(Np^{\infty};\mathcal{O}_{\mathcal{K}}) \otimes_{\Lambda^{\mathrm{wt}}_{\mathcal{K}}} \mathbf{I} \to \mathbf{I}$ the corresponding 4-term sequence*

$$0 \longrightarrow H^0\big(\mathbf{F}(\mu_{p^\infty}) \otimes \mathbb{Z}_p, \mathbb{T}_\infty\big)[\lambda] \longrightarrow \mathbf{H}^1_{\mathrm{kum},\mathbf{F}}(\mathbb{T}_\infty)[\lambda] \longrightarrow$$
$$\varprojlim_n H^1\big(\mathbf{F}(\mu_{p^n}) \otimes \mathbb{Z}_p, \mathbb{T}_\infty[\lambda]\big) \Big/_{\mathbf{I}[\![G_{\infty,\mathbf{F}}]\!]\text{-tors}} \overset{\mathrm{PR}_{\infty,\mathbf{F}}}{\longrightarrow} \mathbf{I}[\![G_{\infty,\mathbf{F}}]\!] \otimes_{\mathbf{I}} \mathcal{L}'$$

is exact, and the image of $\mathrm{PR}_{\infty,\mathbf{F}}$ is free of rank one over $\mathbf{I}[\![G_{\infty,\mathbf{F}}]\!]$.

It is interesting to ask whether the image of $\mathrm{PR}_{\infty,\mathbf{F}}$ has any poles in the weight variable (in Hida's terminology if it is *genuine*).

Certainly, we know the image of the zeta-elements yields an analytic L-function. If there are any poles, they must therefore occur in the divisor of a $\mathbf{P}_i \mod \mathcal{J}_{\mathrm{cy}}$, where the $\mathbf{P}_i$'s are associated primes of the characteristic ideal

$$\mathrm{char}_{\mathcal{O}_{\mathcal{K}}[\![X,Y]\!]}\left(\frac{\varprojlim_n H^1\big(\mathbf{F}(\mu_{p^n}) \otimes \mathbb{Z}_p, \mathbb{T}_\infty[\lambda]\big)}{\varprojlim_n H^1(\dots) \cap \mathbf{I}[\![G_{\infty,\mathbf{F}}]\!].\big(\mathfrak{Z}_{[\lambda],p^n}\big)_n}\right).$$

This quotient module measures the failure of the super Euler system to generate the full semi-local cohomology. In this situation, the best inference we can make is that there are probably no poles, provided the indices of the p-adic zeta-elements inside étale cohomology are controlled by a fixed absolute bound.

If $\mathcal{R}$ is not Gorenstein, then our sequence almost certainly requires modification. Indeed $\mathbb{T}_\infty/\mathrm{F}^+\mathbb{T}_\infty$ will no longer be $\mathcal{R}$-free, meaning the $\mathbf{I}[\![\Gamma^{\mathrm{cy}}]\!]$-torsion submodules of $\mathbf{H}^1_{\mathrm{kum},\mathbf{F}}$ and $\varprojlim_n H^1$ must differ.

Example 6.12 Assume the ground field $\mathbf{F} = \mathbb{Q}$, and that ϵ is the trivial character. It follows that the q-coefficient $a_p(\mathbf{f}_k) \neq 1$ at some weight $k \geq 2$, in which case

$$H^0\Big(\mathbb{Q}(\mu_{p^\infty}) \otimes \mathbb{Z}_p, \mathbb{T}_\infty\Big)[\lambda] \;=\; \mathbb{T}_\infty[\lambda]^{G_{\mathbb{Q}_p(\mu_{p^\infty})}} \quad \text{is the zero subspace.}$$

As a corollary $\mathbf{H}^1_{\mathrm{kum},\mathbb{Q}}(\mathbb{T}_\infty)[\lambda]$ and $\varprojlim_n H^1\big(\mathbb{Q}_p(\mu_{p^n}), \mathbb{T}_\infty[\lambda]\big)$ will be $\mathbf{I}[\![G_{\infty,\mathbb{Q}}]\!]$-free, with $\mathrm{PR}_{\infty,\mathbb{Q}}$ mapping their quotient injectively into $\mathbf{I}[\![G_{\infty,\mathbb{Q}}]\!] \otimes_{\mathbf{I}} \mathcal{L}'$.

In a forthcoming paper [Db2] we study in some detail the specialisation of this 4-term exact sequence to weight one. Although considerable degeneration occurs (e.g. $\mathbb{T}_\infty$ at weight one isn't even Hodge-Tate), enough of the sequence remains intact to obtain results highly reminiscent of the p-adic analogue of Stark's formula. The Euler systems manifest themselves as S-units, the $\mathbf{I}$-adic L-function becomes an Artin L-function and, assuming the non-vanishing of certain $\mathcal{L}$-invariants, the big dual exponential map degenerates into a standard p-adic logarithm.

6.5 Computing the $\mathcal{R}[\![\Gamma]\!]$-torsion

Before supplying the missing proof of Lemma 6.10, it is necessary to review some notation from Chapter V. If M is a finitely-generated $\mathcal{R}$-module, we defined

$$A_M := \mathrm{Hom}_{\mathrm{cont}}\big(M, \mu_{p^\infty}\big), \quad \text{and wrote} \quad M^\vee := \mathrm{Hom}_{\mathrm{cont}}\big(M, \mathbb{Q}_p/\mathbb{Z}_p\big)$$

for the Pontrjagin dual of M. Moreover, if M possesses a continuous $\mathcal{R}$-linear G_L-action for some finite algebraic extension $L/\mathbb{Q}_p$, there are perfect dualities

$$H^i(L, M) \times H^{2-i}(L, A_M) \longrightarrow \mathbb{Q}_p/\mathbb{Z}_p \qquad \text{for } i = 0, 1 \text{ or } 2$$

induced by the invariant map of local class field theory.

The semi-local result 6.10 follows easily from the following local version.

Lemma 6.10'. *For every prime $\mathfrak{P}$ of* Spec $\mathbb{Z}[\mu_M]$ *lying over p,*

$$\mathrm{Tors}_{\mathcal{R}[\![\Gamma^{\mathrm{cy}}]\!]}\left(\varprojlim_n H^1\big(\mathbf{F}_{\mathfrak{P}}(\mu_{p^n}), \mathbb{T}_\infty\big)\right) \cong H^0\big(\mathbf{F}_{\mathfrak{P}}(\mu_{p^\infty}), \mathbb{T}_\infty\big) = \mathbb{T}_\infty^{G_{\mathbf{F}_{\mathfrak{P}}(\mu_{p^\infty})}}$$

assuming that the ring $\mathcal{R}$ is Gorenstein.

Deferring its demonstration for the moment, let's see how to deduce 6.10 from 6.10'. To simplify matters somewhat, we replace $\Gamma^{\mathrm{cy}} = \mathrm{Gal}\big(\mathbf{F}(\mu_{p^\infty})/\mathbf{F}(\mu_p)\big)$ just with Γ. We need to show that the $\mathcal{R}[\![\Gamma]\!]$-torsion submodule of $\varprojlim_n H^1\big(\mathbf{F}(\mu_{p^n}) \otimes \mathbb{Z}_p, \mathbb{T}_\infty\big)$ coincides with the image of $H^0\big(\mathbf{F}(\mu_{p^\infty}) \otimes \mathbb{Z}_p, \mathbb{T}_\infty\big)$ under the Kummer map.

If $(\mathfrak{X}_{p^n})_n \in \mathrm{Tors}_{\mathcal{R}[\![\Gamma]\!]}\Big(\varprojlim_n H^1\big(\mathbf{F}(\mu_{p^n}) \otimes \mathbb{Z}_p, \mathbb{T}_\infty\big)\Big)$, then $(\mathfrak{X}_{p^n})_n$ restricts at each prime ideal $\mathfrak{P} \in S_{\mathbf{F},p}$ to the $\mathcal{R}[\![\Gamma]\!]$-torsion submodule of $\varprojlim_n H^1\big(\mathbf{F}_{\mathfrak{P}}(\mu_{p^n}), \mathbb{T}_\infty\big)$. The latter object equals $H^0\big(\mathbf{F}_{\mathfrak{P}}(\mu_{p^\infty}), \mathbb{T}_\infty\big)$ by Lemma 6.10', thus $(\mathfrak{X}_{p^n})_n$ must have originated in the image of the H^0.

Conversely, the group $H^0\big(\mathbf{F}(\mu_{p^\infty}) \otimes \mathbb{Z}_p, \mathbb{T}_\infty\big) = \varprojlim_n H^0\big(\mathbf{F}(\mu_{p^n}) \otimes \mathbb{Z}_p, \mathbb{T}_\infty\big)$ is clearly torsion over $\mathcal{R}[\![\Gamma]\!] = \varprojlim_n \mathcal{R}[\Gamma/\Gamma^{p^{n-1}}]$, which means the first statement of Lemma 6.10 is true. To deduce the second, by 6.9(i) we know that $\mathbf{H}^1_{\mathrm{kum},\mathbf{F}_{\mathfrak{P}}}(\mathbb{T}_\infty)$ contains $H^0\big(\mathbf{F}_{\mathfrak{P}}(\mu_{p^\infty}), \mathbb{T}_\infty\big)$ for all $\mathfrak{P} \in S_{\mathbf{F},p}$. In particular $\mathbf{H}^1_{\mathrm{kum},\mathbf{F}}(\mathbb{T}_\infty)$ contains the entire $\mathcal{R}[\![\Gamma]\!]$-torsion module $H^0\big(\mathbf{F}(\mu_{p^\infty}) \otimes \mathbb{Z}_p, \mathbb{T}_\infty\big)$, and we are done.

Proof of Lemma 6.10': Let us denote by L the completion $\mathbf{F}_{\mathfrak{P}}$ at $\mathfrak{P} \in S_{\mathbf{F},p}$. We shall also write L_∞ for the field union of the $L(\mu_{p^n})$'s, which is an infinite Lie extension of dimension one.

The argument has a couple of steps in total. We initially use a standard trick of Iwasawa's, to construct a map from the inverse limit $\varprojlim_n H^1\big(L(\mu_{p^n}), \mathbb{T}_\infty\big)$ to an $\mathcal{R}[\![\Gamma]\!]$-free module. The second step is to show the kernel of this map is precisely the 0th-cohomology group $H^0\big(L_\infty, \mathbb{T}_\infty\big)$.

Step 1: Constructing the map $\alpha_3 \circ \alpha_2 \circ \alpha_1$.
The perfect duality between $H^1\big(L(\mu_{p^n}), \mathbb{T}_\infty\big)$ and $H^1\big(L(\mu_{p^n}), A_{\mathbb{T}_\infty}\big)$ furnishes us (upon passage to the limit) with an isomorphism

$$\alpha_1 : \varprojlim_n H^1\big(L(\mu_{p^n}), \mathbb{T}_\infty\big) \xrightarrow{\sim} H^1\big(L_\infty, A_{\mathbb{T}_\infty}\big)^\vee = H^1\big(L_\infty, \mathbb{T}^*_\infty(1) \otimes_{\mathcal{R}} \mathcal{R}^\vee\big)^\vee.$$

The right-hand equality follows because $A_M \cong M^*(1) \otimes_{\mathcal{R}} \mathcal{R}^\vee$ whenever M is $\mathcal{R}$-free. Now for any $\mathcal{R}[G]$-module M, there is a canonical homomorphism

$$\begin{aligned} \mathrm{Hom}_{\mathrm{cont}}\Big(H^1(G, M \otimes_{\mathcal{R}} \mathcal{R}^\vee),\ \mathbb{Q}_p/\mathbb{Z}_p\Big) &\longrightarrow \mathrm{Hom}_{\mathcal{R}}\Big(H^1(G, M),\ \mathcal{R}^{\vee\vee}\Big) \\ \vartheta \quad &\mapsto \quad \big\{\beta \mapsto \{\sigma \mapsto \vartheta(\beta \otimes \sigma)\}\big\}. \end{aligned}$$

In our situation $G = \mathrm{Gal}\big(\overline{L}/L_\infty\big)$ and $M = \mathbb{T}^*_\infty(1)$, so the above map becomes

$$\alpha_2 : \mathrm{Hom}_{\mathrm{cont}}\Big(H^1\big(L_\infty, \mathbb{T}^*_\infty(1) \otimes_{\mathcal{R}} \mathcal{R}^\vee\big),\ \mathbb{Q}_p/\mathbb{Z}_p\Big) \longrightarrow \mathrm{Hom}_{\mathcal{R}}\Big(H^1\big(L_\infty, \mathbb{T}^*_\infty(1)\big), \mathcal{R}\Big)$$

as $\mathcal{R}^{\vee\vee} \cong \mathcal{R}$ by Pontrjagin duality. The composition $\alpha_2 \circ \alpha_1$ just fails to be an isomorphism, because α_2 has a non-trivial kernel. In fact, there is a factorisation

$$\varprojlim_n \left(H^1\big(L(\mu_{p^n}), \mathbb{T}_\infty\big)_{/\mathcal{R}\text{-tors}} \right) \overset{\alpha_2 \circ \alpha_1}{\longrightarrow} \mathrm{Hom}_{\mathcal{R}}\Big(H^1\big(L_\infty, \mathbb{T}^*_\infty(1)\big),\ \mathcal{R}\Big)$$

since $\mathrm{Hom}_{\mathcal{R}}\Big(H^1\big(L_\infty, \mathbb{T}^*_\infty(1)\big), \mathcal{R}\Big) \cong \varprojlim_n \mathrm{Hom}_{\mathcal{R}}\Big(H^1\big(L(\mu_{p^n}), \mathbb{T}^*_\infty(1)\big), \mathcal{R}\Big)$ and the functor $\mathrm{Hom}_{\mathcal{R}}(\,-, \mathcal{R})$ kills off all $\mathcal{R}$-torsion.

Remark: Let us now write $A^{G_{L_\infty}}_{\mathbb{T}^*_\infty(1)}$ as a shorthand for the group $H^0\big(L_\infty, A_{\mathbb{T}^*_\infty(1)}\big)$. If $\Gamma_n = \mathrm{Gal}\big(L_\infty/L(\mu_{p^n})\big)$, the inflation-restriction sequence

$$0 \to H^1\left(\Gamma_n, A^{G_{L_\infty}}_{\mathbb{T}^*_\infty(1)}\right) \overset{\mathrm{infl}_n}{\to} H^1\left(L(\mu_{p^n}), A_{\mathbb{T}^*_\infty(1)}\right) \overset{\mathrm{rest}_n}{\to} H^1\left(L_\infty, A_{\mathbb{T}^*_\infty(1)}\right)^{\Gamma_n} \to 0$$

is right-exact provided $n \geq 1$, as Γ_n has cohomological dimension one.

Taking Pontrjagin duals of the latter, yields another exact sequence

$$0 \to \left(\varprojlim_m H^1\big(L(\mu_{p^m}), \mathbb{T}^*_\infty(1)\big)\right)_{\Gamma_n} \overset{\mathrm{rest}_n^\vee}{\to} H^1\big(L(\mu_{p^n}), \mathbb{T}^*_\infty(1)\big) \overset{\mathrm{infl}_n^\vee}{\to} H^1\Big(\Gamma_n, A^{G_{L_\infty}}_{\mathbb{T}^*_\infty(1)}\Big)^\vee \to 0$$

due to the isomorphism $\left(H^1\left(L_\infty, A_{\mathbb{T}^*_\infty(1)}\right)^{\Gamma_n}\right)^\vee \cong \left(\varprojlim_m H^1\big(L(\mu_{p^m}), \mathbb{T}^*_\infty(1)\big)\right)_{\Gamma_n}$. Applying the functor $\mathrm{Hom}_{\mathcal{R}}(\,-, \mathcal{R})$ once more, then passing to the limit over n, we obtain a third map

$$\begin{aligned} \alpha_3 : \mathrm{Hom}_{\mathcal{R}}\Big(H^1\big(L_\infty, \mathbb{T}^*_\infty(1)\big), \mathcal{R}\Big) &\cong \varprojlim_n \mathrm{Hom}_{\mathcal{R}}\Big(H^1\big(L(\mu_{p^n}), \mathbb{T}^*_\infty(1)\big), \mathcal{R}\Big) \\ &\overset{(\mathrm{rest}_n^\vee)^*_n}{\longrightarrow} \varprojlim_n \mathrm{Hom}_{\mathcal{R}}\left(\left(\varprojlim_m H^1\big(L(\mu_{p^m}), \mathbb{T}^*_\infty(1)\big)\right)_{\Gamma_n}, \mathcal{R}\right) \end{aligned}$$

whose kernel is precisely $\left(\varprojlim_n H^1\left(\Gamma_n, A^{G_{L_\infty}}_{\mathbb{T}^*_\infty(1)}\right)^\vee\right)^*$. Finally, the right-most space can be identified with $\varprojlim_n \mathrm{Hom}_{\mathcal{R}[\Gamma/\Gamma_n]}\left(\left(\varprojlim_m H^1(L(\mu_{p^m}), \mathbb{T}^*_\infty(1))\right)_{\Gamma_n}, \mathcal{R}[\Gamma/\Gamma_n]\right)^\bullet$,

i.e. the compact module $\mathrm{Hom}_{\mathcal{R}[\![\Gamma]\!]}\left(\varprojlim_m H^1(L(\mu_{p^m}), \mathbb{T}^*_\infty(1)),\ \mathcal{R}[\![\Gamma]\!]\right)^\bullet$

where as usual the superscript $^\bullet$ indicates that the Γ-action has been inverted.

Remark: We now have our desired map

$$\varprojlim_n H^1\left(L(\mu_{p^n}), \mathbb{T}_\infty\right) \overset{\alpha_3\circ\alpha_2\circ\alpha_1}{\longrightarrow} \mathrm{Hom}_{\mathcal{R}[\![\Gamma]\!]}\left(\varprojlim_n H^1(L(\mu_{p^n}), \mathbb{T}^*_\infty(1)),\ \mathcal{R}[\![\Gamma]\!]\right)^\bullet$$

and it remains to compute its kernel. In particular $\varprojlim_n H^1(L(\mu_{p^n}), \mathbb{T}^*_\infty(1))$ is of finite type over $\mathcal{R}[\![\Gamma]\!]$ by [Gr1]. Its dual $\mathrm{Hom}_{\mathcal{R}[\![\Gamma]\!]}(\ \dots\ )^\bullet$ must then be free, thus $\mathrm{Ker}(\alpha_3 \circ \alpha_2 \circ \alpha_1)$ identifies naturally with the $\mathcal{R}[\![\Gamma]\!]$-torsion submodule.

Step 2: Calculation of the kernel of $\alpha_3 \circ \alpha_2 \circ \alpha_1$.

Let us now exploit the condition that the ring $\mathcal{R}$ is Gorenstein; in particular, the dual module $\mathrm{Hom}_{\mathbb{Z}_p}(\mathcal{R}, \mathbb{Z}_p)$ will be free of rank one over $\mathcal{R}$. Using Lemma 5.5 we have the identification $\mathbb{T}_\infty \otimes_{\mathbb{Z}_p} \mathbb{Q}_p/\mathbb{Z}_p \cong A_{\mathbb{T}^*_\infty(1)}$ due to the fact that $\mathbb{T}_\infty$ is $\mathcal{R}$-free. Examining the Γ_n-cohomology of its $\mathrm{Gal}(\overline{L}/L_\infty)$-invariant subspace, there exists a canonical homomorphism

$$\beta:\ \mathbb{Q}_p/\mathbb{Z}_p \otimes_{\mathbb{Z}_p} H^1\left(\Gamma_n,\ \mathbb{T}^{G_{L_\infty}}_\infty\right) \longrightarrow H^1\left(\Gamma_n,\ A^{G_{L_\infty}}_{\mathbb{T}^*_\infty(1)}\right)$$

whose kernel and cokernel are both $\mathcal{R}$-cotorsion. Applying the Pontrjagin dual to both sides, yields a dual map

$$\begin{aligned} H^1\left(\Gamma_n, A^{G_{L_\infty}}_{\mathbb{T}^*_\infty(1)}\right)^\vee &\overset{\beta^\vee}{\longrightarrow} \mathrm{Hom}_{\mathbb{Z}_p}\left(H^1\left(\Gamma_n, \mathbb{T}^{G_{L_\infty}}_\infty\right), \mathbb{Z}_p\right) \\ &= \mathrm{Hom}_{\mathcal{R}}\left(H^1\left(\Gamma_n, \mathbb{T}^{G_{L_\infty}}_\infty\right), \mathcal{R}\right). \end{aligned}$$

The right-hand equality follows from the general fact that for Gorenstein rings $\mathcal{R}$, both the functors $\mathrm{Hom}_{\mathbb{Z}_p}(\ -, \mathbb{Z}_p)$ and $\mathrm{Hom}_{\mathcal{R}}(\ -, \mathcal{R})$ agree on the category of finitely-generated projective $\mathcal{R}$-modules. This time taking the $\mathcal{R}$-dual of both sides then the projective limit over n, we can switch between $\left(\varprojlim_n H^1\left(\Gamma_n, A^{G_{L_\infty}}_{\mathbb{T}^*_\infty(1)}\right)^\vee\right)^*$ and $\varprojlim_n \left(H^1\left(\Gamma_n, \mathbb{T}^{G_{L_\infty}}_\infty\right)_{/\mathcal{R}\text{-tors}}\right)$ as we please.

Let us now return to our calculation of $\mathrm{Ker}(\alpha_3 \circ \alpha_2 \circ \alpha_1)$. From the discussion about the kernels of $\alpha_2 \circ \alpha_1$ and α_3, there is an embedding

$$\varprojlim_n \left(\frac{H^1(L(\mu_{p^n}), \mathbb{T}_\infty)}{\mathcal{R}\text{-tors} + H^1\left(\Gamma_n, \mathbb{T}^{G_{L_\infty}}_\infty\right)}\right) \hookrightarrow \mathrm{Hom}_{\mathcal{R}[\![\Gamma]\!]}\left(\varprojlim_n H^1(L(\mu_{p^n}), \mathbb{T}^*_\infty(1)),\ \mathcal{R}[\![\Gamma]\!]\right)^\bullet$$

(****)

induced by the composition $\alpha_3 \circ \alpha_2 \circ \alpha_1$. On the other hand, the restriction map clearly injects $H^1\Big(L(\mu_{p^n}), \mathbb{T}_\infty\Big)_{\mathcal{R}\text{-tors}} + H^1\left(\Gamma_n, \mathbb{T}_\infty^{G_{L_\infty}}\right) \Big/ H^1\left(\Gamma_n, \mathbb{T}_\infty^{G_{L_\infty}}\right)$ into the Γ_n-invariants of $\mathrm{Tors}_{\mathcal{R}}\Big(H^1(L_\infty, \mathbb{T}_\infty)\Big)$.

Claim: $\varprojlim_n \mathrm{Tors}_{\mathcal{R}}\Big(H^1(L_\infty, \mathbb{T}_\infty)\Big)^{\Gamma_n} \;=\; 0.$

To see why this is so, we remark that one can replace $\mathrm{Tors}_{\mathcal{R}}(\ \dots\ )$ by $\mathrm{Tors}_{\Lambda^{\mathrm{wt}}}(\ \dots\ )$ because any element that is killed by $r \in \mathcal{R}$ will also be killed by $r^{\sigma_1} \dots r^{\sigma_d} \in \Lambda^{\mathrm{wt}}$ – here $\mathrm{Gal}(\mathcal{R}'/\Lambda^{\mathrm{wt}}) = \{\sigma_1, \dots, \sigma_d\}$ where $\mathcal{R}'$ is the Galois closure of $\mathcal{R}$ over Λ^{wt}.

Picking a suitable element $\varrho \in p\mathbb{Z}_p$ such that $\theta = u_0 + \varrho - 1 \in \Lambda^{\mathrm{wt}}$ does not divide the characteristic ideal of $\mathrm{Tors}_{\Lambda^{\mathrm{wt}}}\Big(H^1(L_\infty, \mathbb{T}_\infty)\Big)$, we have the equality

$$\varprojlim_n \mathrm{Tors}_{\Lambda^{\mathrm{wt}}}\Big(H^1(L_\infty, \mathbb{T}_\infty)\Big)^{\Gamma_n} \;=\; \varprojlim_n \mathrm{Tors}_{\mathbb{Z}_p}\left(\varprojlim_r H^1(L_\infty, \mathbb{T}_\infty/\theta^r\mathbb{T}_\infty)\right)^{\Gamma_n}.$$

One can interchange the order of n and r in the projective limits, since the Galois and diamond actions on $\mathbb{T}_\infty$ commute with each other. For a fixed integer $r \geq 1$, the $\mathbb{Z}_p$-torsion submodule of $H^1(L_\infty, \mathbb{T}_\infty/\theta^r\mathbb{T}_\infty)$ will be a finite group. It follows that $\varprojlim_n \mathrm{Tors}_{\mathbb{Z}_p}\Big(H^1(L_\infty, \mathbb{T}_\infty/\theta^r\mathbb{T}_\infty)\Big)^{\Gamma_n} = 0$, and our Claim is therefore true.

One useful consequence is that in passing to the projective limit over n, we trivialise the quotient $H^1\Big(L(\mu_{p^n}), \mathbb{T}_\infty\Big)_{\mathcal{R}\text{-tors}} + H^1\left(\Gamma_n, \mathbb{T}_\infty^{G_{L_\infty}}\right) \Big/ H^1\left(\Gamma_n, \mathbb{T}_\infty^{G_{L_\infty}}\right)$. In terms of Equation **(****)**, this becomes an injection

$$\varprojlim_n \left(\frac{H^1(L(\mu_{p^n}), \mathbb{T}_\infty)}{H^1\left(\Gamma_n, \mathbb{T}_\infty^{G_{L_\infty}}\right)}\right) \overset{\alpha_3\circ\alpha_2\circ\alpha_1}{\hookrightarrow} \mathrm{Hom}_{\mathcal{R}[\![\Gamma]\!]}\left(\varprojlim_n H^1(L(\mu_{p^n}), \mathbb{T}_\infty^*(1)),\ \mathcal{R}[\![\Gamma]\!]\right)^{\bullet}.$$

Finally, the left-hand space is none other than $\varprojlim_n H^1(L(\mu_{p^n}), \mathbb{T}_\infty)\Big/\mathbb{T}_\infty^{G_{L_\infty}}$, since $H^1\left(\Gamma_n, \mathbb{T}_\infty^{G_{L_\infty}}\right) \cong \left(\mathbb{T}_\infty^{G_{L_\infty}}\right)_{\Gamma_n}$ as the group Γ_n is pro-cyclic.

The proof is finished. □

Remark: Whilst we do not need it here, one can even prove the exactness of

$$0 \longrightarrow H^0(L_\infty, \mathbb{T}_\infty) \longrightarrow \varprojlim_n H^1(L(\mu_{p^n}), \mathbb{T}_\infty) \overset{\alpha_3\circ\alpha_2\circ\alpha_1}{\longrightarrow}$$

$$\mathrm{Hom}_{\mathcal{R}[\![\Gamma]\!]}\left(\varprojlim_n H^1(L(\mu_{p^n}), \mathbb{T}_\infty^*(1)),\ \mathcal{R}[\![\Gamma]\!]\right)^{\bullet} \longrightarrow \mathrm{Ext}^1_{\mathcal{R}}\Big(H^1(L_\infty, \mathbb{T}_\infty)^{\vee}_{\mathcal{R}\text{-tors}}, \mathcal{R}\Big).$$

At arithmetic points of $\mathrm{Spec}(\mathcal{R})$, this long exact sequence will specialise to the p-adic versions constructed by Perrin-Riou in [PR1, Prop 2.1.6].

CHAPTER VII

Vertical and Half-Twisted Arithmetic

This connection we have just found, between the Greenberg-Stevens L-function and the super zeta-elements, is very powerful indeed. However it is only half the story. In order to fully appreciate the applications to number theory, we need to establish a secondary link between these zeta-elements, and arithmetic objects such as **Ш**. Inevitably we are led to consider Selmer groups over deformation rings.

Fortunately, we already encountered these ideas in the context of elliptic curves. For an elliptic curve E defined over a number field $\mathbf{F}$, its Selmer group was the smallest cohomologically defined subgroup containing the Mordell-Weil group $E(\mathbf{F})$. More precisely, it is the set of one-cocycles in

$$H^1\Big(\mathrm{Gal}\big(\mathbf{F}_\Sigma/\mathbf{F}\big),\ E[\mathrm{tors}]\Big)$$

which restrict at all places ν of $\mathbf{F}$, to lie in the image of $E(\mathbf{F}_\nu) \otimes \mathbb{Q}/\mathbb{Z}$ under the Kummer map. Here the field $\mathbf{F}_\Sigma$ denotes the maximal algebraic extension of $\mathbf{F}$, unramified outside Σ and the infinite places. As usual, the finite set Σ is assumed to contain all the primes where E has bad reduction.

To mimic these constructs in the situation of an ordinary deformation, we should consider cocycles with coefficients inside $\mathbb{T}_\infty$ (or better still, its Pontrjagin dual). One then demands that they restrict everywhere locally, to produce families of points living on pro-Jacobians such as J_∞ (see §5.4 for a short discussion).

Once we have defined these big Selmer groups, it is worth asking:

Question 1. *Are they finitely-generated over the deformation ring?*

Question 2. *What are their ranks over the weight algebra?*

Question 3. *At weight two, how are their invariants related to abelian varieties?*

The third question seems rather arbitrary, and has been deliberately badly worded. Clearly there are infinitely many ways one may approach the point $(s, k) = (1, 2)$ within the (s, k)-plane. For example, along the line $k = 2$ we are essentially studying the Iwasawa theory of abelian varieties, and so never see the weight deformation.

Likewise, à priori we need not have made $k = 2$ our chosen weight and are free to specialise at any other integral weights, as we please – it is only the author's unhealthy obsession with the elliptic curve theory that has motivated this choice. One advantage of studying elliptic curves is that their Tamagawa numbers are easily computed (thanks to Tate's algorithm), whereas we know far less about the Tamagawa factors of arbitrary modular forms.

7.1 Big Selmer groups

Essentially there are three principal lines in the (s,k)-plane that we wish to examine: the vertical line $s=1$, the central line $s=k/2$, and the boundary line $s=k-1$. Luckily the boundary line is reflected into the vertical, courtesy of the two-variable functional equation. Consequently, it is enough to consider only $s=1$ and $s=k/2$. Our first task is to write down the correct Galois representations corresponding to each of these two lines.

We start with the vertical line $s=1$. Let $p\neq 2$ be a prime number, and as usual write $N\geq 1$ for the tame level. Recall from Chapter IV, we defined the universal p-ordinary Galois representation

$$\rho_\infty : G_{\mathbb{Q}} \longrightarrow \mathrm{Aut}_{\mathcal{R}}(\mathbb{T}_\infty) \;=\; \mathrm{GL}_2(\mathcal{R})$$

which was cut out of the Tate module of $J_\infty = \varprojlim_r \mathrm{Jac}X_1(Np^r)$ using idempotents in the abstract algebra $\mathbf{H}_{Np^\infty}\otimes_{\Lambda^{\mathrm{wt}}}\mathrm{Frac}(\Lambda^{\mathrm{wt}})$.

Definition 7.1. *For each primitive triple $\underline{v}=(\lambda,k_0,\epsilon)$, one defines $\mathbb{T}_{\infty,\underline{v}}$ to be the image of the Hecke eigenspace $\mathbb{T}_\infty[\lambda]$ under the analytic transform $\widetilde{\mathrm{Mell}}_{k_0,\epsilon}$.*

By its very construction the $G_{\mathbb{Q}}$-representation $\mathbb{T}_{\infty,\underline{v}}$ is free of rank two, over the affinoid $\mathcal{K}$-algebra of the closed disk $\mathbb{U}_{k_0}$. If $k\geq 2$ is an integral weight lying in $\mathbb{U}_{k_0}$, then each specialisation $\mathbb{T}_{\infty,\underline{v}}\Big|_{w=k}$ will be equivalent to the Deligne representation associated to the p-stabilisation $\mathbf{f}_{P_{k,\epsilon}}$.

We shall now explain how to attach a Selmer group to big Galois representations such as $\mathbb{T}_{\infty,\underline{v}}$. More generally, the techniques we describe here work equally well for pseudo-geometric representations in the sense of §5.4. Let $\mathbf{F}$ be any number field. As always we write $S_{\mathbf{F}}$ for the finite primes in $\mathrm{Spec}(\mathcal{O}_{\mathbf{F}})$, and also $S_{\mathbf{F},p}$ for those places lying above the prime p. We ensure that $\Sigma\subset S_{\mathbf{F}}$ is chosen to be a finite subset which includes all primes dividing Np.

For a finite place $\nu\nmid p$ of $\mathbf{F}$, let $H^1_{\mathrm{nr}}(\mathbf{F}_\nu,\mathbb{T}_{\infty,\underline{v}})$ denote the $\Lambda^{\mathrm{wt}}_{\mathcal{K}}$-saturation of

$$H^1\Big(\mathrm{Frob}_\nu,\,\mathbb{T}_{\infty,\underline{v}}^{I_{\mathbf{F}_\nu}}\Big) \;\cong\; \mathrm{Ker}\Big(H^1(\mathbf{F}_\nu,\mathbb{T}_{\infty,\underline{v}}) \longrightarrow H^1(I_{\mathbf{F}_\nu},\mathbb{T}_{\infty,\underline{v}})\Big)$$

inside the ambient cohomology group $H^1(\mathbf{F}_\nu,\mathbb{T}_{\infty,\underline{v}})$.

Definition 7.2. *(i) The compact Selmer group $\mathfrak{S}_{\mathbf{F}}(\mathbb{T}_{\infty,\underline{v}})$ is defined to be the kernel of the restriction maps*

$$H^1(\mathbf{F}_\Sigma/\mathbf{F},\mathbb{T}_{\infty,\underline{v}}) \overset{\oplus\mathrm{res}_v}{\longrightarrow} \bigoplus_{\nu\in\Sigma-S_{\mathbf{F},p}} \frac{H^1(\mathbf{F}_\nu,\mathbb{T}_{\infty,\underline{v}})}{H^1_{\mathrm{nr}}(\mathbf{F}_\nu,\mathbb{T}_{\infty,\underline{v}})} \oplus \bigoplus_{\nu\in S_{\mathbf{F},p}} \frac{H^1(\mathbf{F}_\nu,\mathbb{T}_{\infty,\underline{v}})}{\mathbf{X}_{\mathbb{T}_{\infty,\underline{v}}}(\mathbf{F}_\nu)}$$

where $\mathbf{X}_{\mathbb{T}_{\infty,\underline{v}}}(\mathbf{F}_\nu)$ denotes the image under $\widetilde{\mathrm{Mell}}_{k_0,\epsilon}$ of the local points $\mathbf{X}_{\mathbb{T}_\infty[\lambda]}(\mathbf{F}_\nu)$, arising from the pro-Jacobian variety J_∞ in Definition 5.17.

The cohomology groups occurring above, together with the subgroup $\mathbf{X}_{\mathbb{T}_{\infty,\underline{v}}}(\mathbf{F}_\nu)$, naturally possess the structure of compact, finitely-generated $\mathcal{O}_{\mathcal{K}}[\![\Gamma^{\mathrm{wt}}]\!]$-modules. More precisely, an element of the weight algebra $\Lambda_{\mathcal{K}}^{\mathrm{wt}}$ acts on them through the Mellin transform, centred on the base weight k_0 and finite order character ϵ.

Remark: In fact, the local condition $\mathbf{X}_{\mathbb{T}_{\infty,\underline{v}}}(\mathbf{F}_\nu)$ is automatically $\Lambda_{\mathcal{K}}^{\mathrm{wt}}$-saturated inside $H^1(\mathbf{F}_\nu, \mathbb{T}_{\infty,\underline{v}})$, because $\mathbf{X}_{\mathbb{T}_\infty}(-)$ was $\mathcal{R}$-saturated in its ambient cohomology. This observation will become very important to us later on, when we attempt to relate the Euler characteristic with the covolume of our big dual exponential maps.

When connecting arithmetic objects with analytic objects such as L-functions, it turns out that the natural Selmer groups to examine are their discrete versions, rather than the compact $\mathfrak{S}_{\mathbf{F}}(\mathbb{T}_{\infty,\underline{v}})$'s above. Recall first that the χ_{cy}-twisted dual of $\mathbb{T}_{\infty,\underline{v}}$ is the group $A_{\mathbb{T}_{\infty,\underline{v}}}$ of continuous homomorphisms $\mathrm{Hom}_{\mathrm{cont}}(\mathbb{T}_{\infty,\underline{v}}, \mu_{p^\infty})$. Under Pontrjagin duality, for a finite place $\nu \in \mathrm{Spec}(\mathcal{O}_{\mathbf{F}})$ there is a perfect pairing

$$H^1(\mathbf{F}_\nu, \mathbb{T}_{\infty,\underline{v}}) \times H^1(\mathbf{F}_\nu, A_{\mathbb{T}_{\infty,\underline{v}}}) \longrightarrow H^2(\mathbf{F}_\nu, \mu_{p^\infty}) \cong \mathbb{Q}_p/\mathbb{Z}_p.$$

If $\nu \nmid p$ then $H^1_{\mathrm{nr}}(\mathbf{F}_\nu, A_{\mathbb{T}_{\infty,\underline{v}}})$ denotes the orthogonal complement $H^1_{\mathrm{nr}}(\mathbf{F}_\nu, \mathbb{T}_{\infty,\underline{v}})^\perp$. Alternatively if $\nu | p$, we define the dual group $\mathbf{X}^D_{\mathbb{T}_{\infty,\underline{v}}}(\mathbf{F}_\nu)$ as the complement

$$\left\{\mathbf{x} \in H^1(\mathbf{F}_\nu, A_{\mathbb{T}_{\infty,\underline{v}}}) \quad \text{such that } \mathrm{inv}_{\mathbf{F}_\nu}\left(\mathbf{X}_{\mathbb{T}_{\infty,\underline{v}}}(\mathbf{F}_\nu) \cup \mathbf{x}\right) = 0\right\}.$$

It is hard to visualise $\mathbf{X}^D_{\mathbb{T}_{\infty,\underline{v}}}$ concretely, but it plays an analogous rôle to the group of p-primary torsion points on an abelian variety over a local field.

Definition 7.2. *(ii) The big Selmer group* $\mathrm{Sel}_{\mathbf{F}}(\mathbb{T}_{\infty,\underline{v}})$ *is defined to be the kernel of the restriction maps*

$$H^1(\mathbf{F}_\Sigma/\mathbf{F}, A_{\mathbb{T}_{\infty,\underline{v}}}) \overset{\oplus \mathrm{res}_\nu}{\longrightarrow} \bigoplus_{\nu \in \Sigma - S_{\mathbf{F},p}} \frac{H^1(\mathbf{F}_\nu, A_{\mathbb{T}_{\infty,\underline{v}}})}{H^1_{\mathrm{nr}}(\mathbf{F}_\nu, A_{\mathbb{T}_{\infty,\underline{v}}})} \oplus \bigoplus_{\nu \in S_{\mathbf{F},p}} \frac{H^1(\mathbf{F}_\nu, A_{\mathbb{T}_{\infty,\underline{v}}})}{\mathbf{X}^D_{\mathbb{T}_{\infty,\underline{v}}}(\mathbf{F}_\nu)}$$

induced on the discrete cohomology.

In particular, this Selmer group has the structure of a discrete module over the weight algebra. One of our initial chores will be to establish that $\mathrm{Sel}_{\mathbf{F}}(\mathbb{T}_{\infty,\underline{v}})$ is indeed cofinitely generated over $\Lambda_{\mathcal{K}}^{\mathrm{wt}}$.

For each arithmetic weight $k \in \mathbb{U}_{k_0} \cap \mathbb{N}_{\geq 2}$, the specialisation $\mathrm{Sel}_{\mathbf{F}}(\mathbb{T}_{\infty,\underline{v}})\Big|_{w=k}$ is interpolating the Bloch-Kato Selmer group which is attached to the p-adic Galois representation $\mathbb{T}_{\infty,\underline{v}}\Big|_{w=k}$. Given that the latter object (conjecturally) encodes those formulae related to the critical L-values of the eigenform $\mathbf{f}_{P_{k,\epsilon}}$, the big Selmer group should encapsulate the arithmetic of the whole Hida family $\mathbf{f} \in \mathbb{S}^{\mathrm{ord}}_{\mathbf{I}}$.

The main example: modular elliptic curves.
Let E be an elliptic curve defined over the rationals. In particular, by the work of Breuil, Conrad, Diamond and Taylor [BCDT], the curve E is necessarily modular. Without loss of generality, we shall henceforth assume that E is a strong Weil curve of conductor N_E, and denote by $\pm\phi$ the non-constant morphism of curves $\phi : X_0(N_E) \twoheadrightarrow E$ which is minimal amongst all $X_0(N_E)$-parametrisations.

In the following discussion, we shall assume that the prime number $p > 3$, although $p = 3$ should work provided the analogue of Theorem 4.3 holds true. Let's write $N = p^{-\mathrm{ord}_p N_E} N_E$ for the tame level again.

Hypothesis(p-Ord). *E has good ordinary or bad multiplicative reduction at p.*

If f_E is the newform of weight two and level N_E associated to $\phi : X_0(N_E) \twoheadrightarrow E$, then its p-stabilisation is

$$\mathbf{f}_2 \;:=\; \begin{cases} f_E(q) - \beta_p f_E(q^p) & \text{if } p \nmid N_E \\ f_E(q) & \text{if } p \| N_E \end{cases}$$

where β_p denotes the non-p-unit root of the Hecke polynomial $X^2 - a_p(E)X + p$. As the notation itself suggests, the p-stabilisation of f_E represents the weight two term in an ordinary family $\mathbf{f} \in \mathbb{Q}_{p,\mathbb{U}_{k_0}}\langle\langle w\rangle\rangle[\![q]\!]$ centred on $k_0 = 2$.

In this special situation we interchange $\mathbb{T}_{\infty,\underline{v}}$ with the notation $\mathbb{T}_{\infty,E}$, and the disk $\mathbb{U}_{k_0}$ above with $\mathbb{U}_E$. Finally, one writes $\mathbb{Z}_p\langle\langle w\rangle\rangle$ for the rigid analytic functions convergent on $\mathbb{U}_E$. The following result is simply a restatement of Theorem 4.3.

Corollary 7.3. *If $\rho_{\infty,E} : G_{\mathbb{Q}} \to \mathrm{GL}_2\big(\mathbb{Z}_p\langle\langle w\rangle\rangle\big)$ is the Galois representation lifting $\rho_{E,p} : G_{\mathbb{Q}} \to \mathrm{Aut}\big(\mathrm{Ta}_p(E)\big)$ to its modular deformation $\mathbb{T}_{\infty,E}$, then*

(i) The module $\mathbb{T}_{\infty,E}$ is free of rank 2 over $\mathbb{Z}_p\langle\langle w\rangle\rangle$;

(ii) As a $G_{\mathbb{Q}}$-representation $\mathbb{T}_{\infty,E}$ is unramified outside of Np, and for $l \nmid Np$

$$\det\big(1 - \rho_{\infty,E}(Frob_l).X\big)\Big|_{w=k} \;=\; 1 - a_l(\mathbf{f}_k)X + \omega(l)^{2-k} l^{k-1} X^2$$

at any integral weight $k \in \mathbb{U}_E$ with $k \geq 2$;

(iii) If ϕ_w is the unramified character sending $Frob_p$ to the image of U_p in $\mathbb{Z}_p\langle\langle w\rangle\rangle^\times$, then at a decomposition group above p

$$\rho_{\infty,E}\Big|_{G_{\mathbb{Q}_p}} \;\sim\; \begin{pmatrix} \chi_{\mathrm{cy}} < \chi_{\mathrm{cy}} >^{w-2} \phi_w^{-1} & * \\ 0 & \phi_w \end{pmatrix}.$$

The behaviour of the big Selmer group $\mathrm{Sel}_{\mathbf{F}}(\mathbb{T}_{\infty,E})$ at weight $k = 2$, is intimately connected with the p-primary portion of the classical Selmer group of E over $\mathbf{F}$. Instead one could consider the general situation of abelian varieties occurring as subquotients of $\mathrm{Jac}X_1(Np)$, by allowing non-trivial characters of $\mathbb{Z}_{p,N}^\times$. However the case of elliptic curves is more than enough for this author to cope with!

The half-twisted Galois representation.
Having dealt with the vertical line $s = 1$, we now turn our attention to the line of symmetry $s = k/2$ in the functional equation. Here we had better assume $p \geq 5$. Let $\underline{v} = (\lambda, k_0, \epsilon)$ be a primitive triple, so $\mathbb{T}_{\infty,\underline{v}}$ is the image of $\mathbb{T}_\infty[\lambda]$ under $\widetilde{\mathrm{Mell}}_{k_0,\epsilon}$.

The representation $\mathbb{T}_{\infty,\underline{v}}$ is not self-dual. However, there exists a skew-symmetric bilinear form on it

$$\mathbb{T}_{\infty,\underline{v}} \times \mathbb{T}_{\infty,\underline{v}} \quad \longrightarrow \quad \mathcal{O}_{\mathcal{K}}\langle\langle w\rangle\rangle(1) \otimes \big\langle \chi_{\mathrm{cy}} \circ \mathrm{Res}_{\mathbb{Q}^{\mathrm{cyc}}} \big\rangle$$

induced from the Poincaré pairings on Tate modules of each of the Jac $X_1(Np^r)$'s. N.B. The action of an element $g \in \mathrm{Gal}\big(\mathbb{Q}(\mu_{p^\infty})/\mathbb{Q}\big)$ on $\mathbb{T}_{\infty,\underline{v}}$ is given by $\omega^{k_0-2}(g)\langle \widetilde{g}\rangle$, where $\widetilde{g}$ denotes the projection of $\chi_{\mathrm{cy}}(g)$ to the principal units $1 + p\mathbb{Z}_p$.

Remark: In order to make the Galois representation 'nearly self-dual', there is a neat trick which was devised by Nekovář [NP, §3.2.3]. Firstly if $k_0 \equiv 2 \bmod p-1$, we suppose that $\rho_\infty[\lambda] : G_{\mathbb{Q}} \to \mathrm{GL}_2(\mathbf{I})$ is a residually irreducible representation. Then as a Hecke module

$$\mathbb{T}_{\infty,\underline{v}} \otimes \Psi^{-1/2} \quad \text{means exactly the same object as } \mathbb{T}_{\infty,\underline{v}},$$

except that its Galois action is now twisted by $\omega(-)^{1-k_0/2}\big\langle \chi_{\mathrm{cy}}^{-1/2} \circ \mathrm{Res}_{\mathbb{Q}^{\mathrm{cyc}}}(-)\big\rangle$. One therefore obtains a *nearly* self-dual, skew-symmetric pairing

$$\mathbb{T}_{\infty,\underline{v}} \otimes \Psi^{-1/2} \; \times \; \mathbb{T}_{\infty,\underline{v}} \otimes \Psi^{-1/2} \quad \longrightarrow \quad \mathcal{O}_{\mathcal{K}}\langle\langle w\rangle\rangle(1);$$

the latter produces an isomorphism between $\mathbb{T}_{\infty,\underline{v}} \otimes \Psi^{-1/2}$ and its Kummer dual, at those arithmetic weights $k \geq 2$ which are congruent to k_0 modulo $2(p-1)$.

To write down the p-adic representations $\mathbb{T}_{\infty,\underline{v}} \otimes \Psi^{-1/2}$ encodes, let us recall that at an arithmetic prime $P_{k,\epsilon} \in \mathrm{Spec}\,\mathbf{I}(\mathcal{O}_{\mathcal{K}})^{\mathrm{alg}}$ of type (k,ϵ), the specialisation $\mathbb{T}_{\infty,\underline{v}} \otimes_{\mathbf{I}} \mathbf{I}/P_{k,\epsilon}$ is a $G_{\mathbb{Q}}$-lattice inside of the dual space $V^*_{\mathbf{f}_{P_{k,\epsilon}}}$ for each $k \in \mathbb{U}_{k_0} \cap \mathbb{N}_{\geq 2}$. It follows in the case of trivial character ϵ, that

$$\mathbb{Q} \otimes \left(\mathbb{T}_{\infty,\underline{v}} \otimes \Psi^{-1/2}\right)\Big|_{w=k} \cong V^*_{\mathbf{f}_k} \otimes \omega^{1-k_0/2} < \chi_{\mathrm{cy}} >^{1-k/2}$$
$$\cong V_{\mathbf{f}_k}(k/2) \otimes \left(\frac{\cdot}{p}\right)^{(k-k_0)/(p-1)}.$$

Moreover if $k \equiv k_0 \bmod 2(p-1)$ and $k \in \mathbb{U}_{k_0}$ is even, then at $w = k$ the half-twisted representation $\mathbb{T}_{\infty,\underline{v}} \otimes \Psi^{-1/2}$ specialises to give a lattice in $V_{\mathbf{f}_k}(k/2)$.

Unfortunately, we don't have the local conditions $\mathbf{X}_{\mathbb{T}_{\infty,\underline{v}}}(-)$ any more, as the half-twisted representation is certainly not pseudo-geometric in the sense of §5.4. Nevertheless, we can still use the ordinary filtration at the primes above p.

Definition 7.4. *(i) The compact Selmer group* $\mathfrak{S}_{\mathbf{F}}\left(\mathbb{T}_{\infty,\underline{v}}\otimes\Psi^{-1/2}\right)$ *consists of the cocycles inside* $H^1\left(\mathbf{F}_\Sigma/\mathbf{F},\mathbb{T}_{\infty,\underline{v}}\otimes\Psi^{-1/2}\right)$ *whose restrictions are zero under*

$$\stackrel{\oplus\mathrm{res}_v}{\longrightarrow}\quad\bigoplus_{\nu\in\Sigma-S_{\mathbf{F},p}}\frac{H^1\big(\mathbf{F}_\nu,\mathbb{T}_{\infty,\underline{v}}\otimes\Psi^{-1/2}\big)}{H^1_{\mathrm{nr}}\big(\mathbf{F}_\nu,\mathbb{T}_{\infty,\underline{v}}\otimes\Psi^{-1/2}\big)}\oplus\bigoplus_{\nu\in S_{\mathbf{F},p}}\frac{H^1\big(\mathbf{F}_\nu,\mathbb{T}_{\infty,\underline{v}}\otimes\Psi^{-1/2}\big)}{H^1_+\big(\mathbf{F}_\nu,\mathbb{T}_{\infty,\underline{v}}\otimes\Psi^{-1/2}\big)};$$

note that here $H^1_+(\mathbf{F}_\nu)$ *refers to the* $\Lambda^{\mathrm{wt}}_{\mathcal{K}}$*-saturation of the local condition*

$$H^1_{+,0}\ :=\ \mathrm{Im}\Big(H^1\big(\mathbf{F}_\nu,\mathrm{F}^+\mathbb{T}_{\infty,\underline{v}}\otimes\Psi^{-1/2}\big)\to H^1\big(\mathbf{F}_\nu,\mathbb{T}_{\infty,\underline{v}}\otimes\Psi^{-1/2}\big)\Big).$$

(ii) The big Selmer group $\mathrm{Sel}_{\mathbf{F}}\left(\mathbb{T}_{\infty,\underline{v}}\otimes\Psi^{-1/2}\right)$ *is defined to be the kernel of*

$$H^1\Big(\mathbf{F}_\Sigma/\mathbf{F},A^{-1/2}_{\mathbb{T}_{\infty,\underline{v}}}\Big)\ \stackrel{\oplus\mathrm{res}_v}{\longrightarrow}\quad\bigoplus_{\nu\in\Sigma-S_{\mathbf{F},p}}\frac{H^1\Big(\mathbf{F}_\nu,A^{-1/2}_{\mathbb{T}_{\infty,\underline{v}}}\Big)}{H^1_{\mathrm{nr}}\Big(\mathbf{F}_\nu,A^{-1/2}_{\mathbb{T}_{\infty,\underline{v}}}\Big)}\oplus\bigoplus_{\nu\in S_{\mathbf{F},p}}\frac{H^1\Big(\mathbf{F}_\nu,A^{-1/2}_{\mathbb{T}_{\infty,\underline{v}}}\Big)}{H^1_+(\mathbf{F}_\nu)^\perp}$$

where the notation $A^{-1/2}_{\mathbb{T}_{\infty,\underline{v}}}$ *indicates the discrete dual* $\mathrm{Hom}_{\mathrm{cont}}\left(\mathbb{T}_{\infty,\underline{v}}\otimes\Psi^{-1/2},\mu_{p^\infty}\right)$.

The orthogonal complement $H^1_+(\mathbf{F}_\nu)^\perp$ above is with respect to the perfect pairing

$$H^1\Big(\mathbf{F}_\nu,\mathbb{T}_{\infty,\underline{v}}\otimes\Psi^{-1/2}\Big)\ \times\ H^1\Big(\mathbf{F}_\nu,A^{-1/2}_{\mathbb{T}_{\infty,\underline{v}}}\Big)\ \longrightarrow\ \mathbb{Q}_p/\mathbb{Z}_p$$

induced by the cup-product, which is between compact and discrete cohomologies. Needless to say, in the special situation where $\mathbf{f}_2$ is the p-stabilisation of the newform associated to a strong Weil curve E satisfying the condition (p-Ord), we shall replace $\mathbb{T}_{\infty,\underline{v}}\otimes\Psi^{-1/2}$ with $\mathbb{T}_{\infty,E}\otimes\Psi^{-1/2}$, and likewise $A^{-1/2}_{\mathbb{T}_{\infty,\underline{v}}}$ with $A^{-1/2}_{\mathbb{T}_{\infty,E}}$.

The behaviour of the vertical Selmer groups $\mathrm{Sel}_{\mathbf{F}}\left(\mathbb{T}_{\infty,\underline{v}}\right)$ and their half-twisted versions $\mathrm{Sel}_{\mathbf{F}}\left(\mathbb{T}_{\infty,\underline{v}}\otimes\Psi^{-1/2}\right)$ is strikingly different. As a useful illustration, the specialisations of the former at arithmetic weights $k\geq 3$ should be finite groups, whilst specialisations of the latter can be infinite all of the time, conjecturally. However there is the following prediction of Greenberg, which suggests that the number of cogenerators of the divisible part is generically fixed:

Conjecture 7.5. *(Greenberg) The order of vanishing of the two-variable p-adic L-function* $\mathbf{L}^{\mathrm{GS}}_{p,\mathbb{Q}}$ *along the line* $s=k/2$ *should be either always zero or always one, at all bar finitely many bad weights* $k\in\mathbb{U}_{k_0}$.

Translating back in terms of our half-twisted Selmer groups over $\mathbb{Q}$, this indicates their $\mathcal{O}_{\mathcal{K}}$-coranks at arithmetic primes should be either zero or one, up to finitely many exceptional weights. One of the important achievements in Nekovář's work on Selmer complexes was to verify that this is very frequently the case 'modulo 2', i.e. the parity expected in these coranks is almost always either even or odd.

7.2 The fundamental commutative diagrams

We begin our study of these Selmer groups with some fairly general comments. For the purposes of discussion G will denote either the group $\mathrm{Gal}(\mathbf{F}_\Sigma/\mathbf{F})$, or one of the decomposition groups $\mathrm{Gal}(\overline{\mathbf{F}_\nu}/\mathbf{F}_\nu)$. In particular, both examples satisfy the p-finiteness condition mentioned in Definition 5.1.

Let $\mathbf{T}$ denote a free $\mathcal{O}_\mathcal{K}\langle\langle w\rangle\rangle$-module of finite rank with a continuous G-action, which is also a Λ^{wt}-module via the Mellin transform. Let $P = (\mathfrak{g}_P) \in \mathrm{Spec}(\Lambda^{\mathrm{wt}})$ be a principal height one prime ideal. The tautological exact sequence

$$0 = \mathrm{Ker}\big(\times(\mathfrak{g}_P)\big) \longrightarrow \mathbf{T} \overset{\times(\mathfrak{g}_P)}{\longrightarrow} \mathbf{T} \longrightarrow \mathbf{T}\otimes_{\Lambda^{\mathrm{wt}}} \Lambda^{\mathrm{wt}}/P \longrightarrow 0$$

induces long exact sequences: in continuous G-cohomology,

$$\ldots \longrightarrow H^j(G,\mathbf{T}) \overset{\times(\mathfrak{g}_P)}{\longrightarrow} H^j(G,\mathbf{T}) \longrightarrow H^j\left(G,\, \mathbf{T}\otimes_{\Lambda^{\mathrm{wt}}}\Lambda^{\mathrm{wt}}/P\right) \longrightarrow \ldots$$

and in discrete G-cohomology,

$$\ldots \longrightarrow H^j\left(G,\, A_{\mathbf{T},P}\right) \longrightarrow H^j(G,A_\mathbf{T}) \longrightarrow H^j(G,A_\mathbf{T}) \longrightarrow \ldots.$$

Note we've written $A_{\mathbf{T},P}$ for the χ_{cy}-twisted dual $\mathrm{Hom}_{\mathrm{cont}}\Big(\mathbf{T}\otimes_{\Lambda^{\mathrm{wt}}}\Lambda^{\mathrm{wt}}/P,\ \mu_{p^\infty}\Big)$. For example, if $\mathfrak{g}_P = u_0 - (1+p)^{k-k_0}$ then clearly $\mathbf{T}\otimes_{\Lambda^{\mathrm{wt}}}\Lambda^{\mathrm{wt}}/P \;=\; \mathbf{T}\Big|_{w=k-k_0}$. In this manner, we can relate G-cohomologies over deformations rings with their standard p-adic counterparts.

Question. *What do these long exact sequences reveal in the specific situation of a modular elliptic curve E satisfying assumption (p-Ord)?*

Lemma 7.6. *If $\mathbf{T}$ equals either $\mathbb{T}_{\infty,E}$ or $\mathbb{T}_{\infty,E}\otimes\Psi^{-1/2}$, then putting $k = k_0 = 2$ we have equalities*

$$\mathbf{T}\Big|_{w=0} = \mathbf{T}\otimes_{\Lambda^{\mathrm{wt}}}\mathbb{Z}_p \cong \mathrm{Ta}_p(\mathcal{C}^{\min}) \qquad \textit{and} \qquad H^0\left(\Gamma^{\mathrm{wt}}, A_\mathbf{T}\right) \cong \mathcal{C}^{\min}[p^\infty]$$

where $\mathcal{C}^{\min}$ denotes the $\mathbb{Q}$-isogenous elliptic curve to E which occurs as a subvariety inside the Jacobian of $X_1(Np)$.

As an application, suppose $\mathbf{T}$ is either the vertical or the half-twisted lifting of $\rho_{E,p} : G_\mathbb{Q} \to \mathrm{Aut}\big(\mathrm{Ta}_p(E)\big)$. The multiplication by $\big(u_0 - (1+p)^{k-k_0}\big)$-map is just multiplication by $u_0 - 1$ for a topological generator u_0 of the diamond group Γ^{wt}. Thus the specialisation $(-)_{w=0}$ must correspond to taking the Γ^{wt}-coinvariants. Also, the kernel of multiplication by $u_0 - (1+p)^{k-k_0}$ is precisely the submodule of Γ^{wt}-invariants.

Truncating the continuous long exact sequence above, we obtain

$$0 \longrightarrow H^j(G, \mathbf{T}) \otimes_{\Lambda^{\mathrm{wt}}} \mathbb{Z}_p \longrightarrow H^j\big(G, \mathrm{Ta}_p(\mathcal{C}^{\min})\big) \longrightarrow H^{j+1}(G, \mathbf{T})^{\Gamma^{\mathrm{wt}}} \longrightarrow 0.$$

Similarly, in terms of discrete cohomology

$$0 \longrightarrow H^j(G, A_{\mathbf{T}}) \otimes_{\Lambda^{\mathrm{wt}}} \mathbb{Z}_p \longrightarrow H^{j+1}\left(G, \mathcal{C}^{\min}[p^\infty]\right) \longrightarrow H^{j+1}(G, A_{\mathbf{T}})^{\Gamma^{\mathrm{wt}}} \longrightarrow 0$$

is also exact.

Proof of Lemma 7.6: Consider the arithmetic pro-variety $\widehat{X} = \varprojlim_{r \geq 1} X_1(Np^r)$ endowed with its canonical $\mathbb{Q}$-structure. Let $\mathbf{T}$ be the vertical representation $\mathbb{T}_{\infty,E}$. The specialisation map

$$\mathbb{T}_{\infty,E} \twoheadrightarrow \mathbb{T}_{\infty,E} \otimes_{\Lambda^{\mathrm{wt}}} \mathbb{Z}_p \hookrightarrow \mathrm{Ta}_p\Big(\mathrm{Jac}\, X_1(Np)\Big)$$

is clearly induced from the projection $\widehat{X} \overset{\mathrm{proj}}{\longrightarrow} X_1(Np)$.

Now the Galois representation on $\mathbb{T}_{\infty,E} \otimes_{\Lambda^{\mathrm{wt}}} \mathbb{Z}_p$ is equivalent to $\rho_{E,p}$, in fact the former is a $G_{\mathbb{Q}}$-stable lattice in the vector space $\mathrm{Ta}_p(E) \otimes_{\mathbb{Z}_p} \mathbb{Q}_p$. By construction it must be the p-adic Tate module of the elliptic curve $\mathcal{C}^{\min}$, which is a subvariety of Jac $X_1(Np)$; the identification $\mathbb{T}_{\infty,E} \otimes_{\Lambda^{\mathrm{wt}}} \mathbb{Z}_p \cong \mathrm{Ta}_p(\mathcal{C}^{\min})$ therefore occurs on a p-integral level. Furthermore, there are naturally isomorphisms

$$\begin{aligned} H^0\left(\Gamma^{\mathrm{wt}}, A_{\mathbb{T}_{\infty,E}}\right) &\cong \mathrm{Hom}_{\mathrm{cont}}\Big(\mathbb{T}_{\infty,E} \otimes_{\Lambda^{\mathrm{wt}}} \mathbb{Z}_p,\ \mu_{p^\infty}\Big) \\ &\cong \mathrm{Hom}_{\mathrm{cont}}\Big(\mathrm{Ta}_p(\mathcal{C}^{\min}), \mu_{p^\infty}\Big) \cong \mathcal{C}^{\min}[p^\infty] \quad \text{by the Weil pairing.} \end{aligned}$$

Finally, the argument for $\mathbf{T} = \mathbb{T}_{\infty,E} \otimes \Psi^{-1/2}$ is identical because at weight $k = 2$ it looks like $\mathbb{T}_{\infty,E}$ as a Galois module (they differ only at weight $k > 2$).

□

Remarks: (i) We should point out that if $\mathrm{Ta}_p(E) \ncong \mathrm{Ta}_p(\mathcal{C}^{\min})$, there must exist a $\mathbb{Q}$-rational cyclic p-isogeny between these two elliptic curves. This situation can only happen if our prime is very small, i.e. $p \leq 163$. Equivalently, when $p > 163$ we automatically have an isomorphism $\mathrm{Ta}_p(E) \cong \mathrm{Ta}_p(\mathcal{C}^{\min})$ on an integral level.

(ii) The curve $\mathcal{C}^{\min}$ is the optimal $X_1(Np)$-curve lying in the $\mathbb{Q}$-isogeny class for E. In other words, it is the elliptic curve which admits the smallest degree modular parametrisation by $X_1(Np)$. Conjecturally, the Manin constant associated to this morphism $X_1(Np) \twoheadrightarrow \mathcal{C}^{\min}$ should be ± 1, according to Stevens [St, Conj. I, II].

(iii) It is possible to adjust the universal representation $\mathbb{T}_\infty$ by an element of $\mathrm{GL}_2(\mathcal{R})$ so that it specialises to give $\mathrm{Ta}_p(E)$ p-integrally. However in so doing, we pick up an extra term in our Tamagawa number calculations (via the archimedean periods) hence there is no net gain in the long term.

The vertical Selmer diagram.
The obvious strategy is to compare the vertical Selmer group with the p-primary part of the classical Selmer group for E over $\mathbf{F}$. However we cannot always achieve this directly, rather the optimal curve $\mathcal{C}^{\min}$ seems to be the natural integral object to consider. Nevertheless, it is easy to switch between $\mathcal{C}^{\min}$ and E courtesy of the Isogeny Theorem 1.19.

Let's begin by writing down maps between the various discrete cohomologies, both local and global. Assume first that G is the global Galois group $\mathrm{Gal}(\mathbf{F}_\Sigma/\mathbf{F})$. Putting $j = 0$ in our truncated exact sequence, we obtain

$$0 \longrightarrow H^0\left(\mathbf{F}_\Sigma/\mathbf{F},\ A_{\mathbb{T}_{\infty,E}}\right) \otimes_{\Lambda^{\mathrm{wt}}} \mathbb{Z}_p \longrightarrow H^1\left(\mathbf{F}_\Sigma/\mathbf{F}, \mathcal{C}^{\min}[p^\infty]\right)$$
$$\overset{\beta_{\infty,E}}{\longrightarrow} H^1\left(\mathbf{F}_\Sigma/\mathbf{F}, A_{\mathbb{T}_{\infty,E}}\right)^{\Gamma^{\mathrm{wt}}} \longrightarrow 0.$$

In particular, observe that the right-hand homomorphism $\beta_{\infty,E}$ is clearly surjective. The coinvariants $H^0\left(\mathbf{F}_\Sigma/\mathbf{F},\ A_{\mathbb{T}_{\infty,E}}\right) \otimes_{\Lambda^{\mathrm{wt}}} \mathbb{Z}_p$ turn up in our Euler characteristic computations – we abbreviate them with the notation $A_{\mathbb{T}_{\infty,E}}(\mathbf{F})_{\Gamma^{\mathrm{wt}}}$.

Turning our attention locally, assume $\nu \in \Sigma$ is any place of $\mathbf{F}$ not lying above p and set $G = \mathrm{Gal}(\overline{\mathbf{F}_\nu}/\mathbf{F}_\nu)$. We claim there exists a map

$$\delta_{\infty,E,\nu} : \frac{H^1(\mathbf{F}_\nu, \mathcal{C}^{\min}[p^\infty])}{H^1_{\mathrm{nr}}(\mathbf{F}_\nu, \mathcal{C}^{\min}[p^\infty])} \longrightarrow \left(\frac{H^1(\mathbf{F}_\nu, A_{\mathbb{T}_{\infty,E}})}{H^1_{\mathrm{nr}}(\mathbf{F}_\nu, A_{\mathbb{T}_{\infty,E}})}\right)^{\Gamma^{\mathrm{wt}}}$$

where $H^1_{\mathrm{nr}}(\mathbf{F}_\nu, \mathcal{C}^{\min}[p^\infty])$ denotes the orthogonal complement, to the p-saturation of the unramified cocycles $H^1(\mathrm{Frob}_\nu, \mathrm{Ta}_p(\mathcal{C}^{\min})^{I_{\mathbf{F}_\nu}})$ inside of $H^1(\mathbf{F}_\nu, \mathrm{Ta}_p(\mathcal{C}^{\min}))$. Since $H^1_{\mathrm{nr}}(\mathbf{F}_\nu, \mathbb{T}_{\infty,E})$ is Λ^{wt}-torsion, it follows easily that $H^1_{\mathrm{nr}}(\mathbf{F}_\nu, \mathbb{T}_{\infty,E}) \otimes_{\Lambda^{\mathrm{wt}}} \mathbb{Z}_p$ is p^∞-torsion, and must lie in any p-saturated subgroup of $H^1(\mathbf{F}_\nu, \mathrm{Ta}_p(\mathcal{C}^{\min}))$. Consequently, the Γ^{wt}-coinvariants inject via

$$H^1_{\mathrm{nr}}(\mathbf{F}_\nu, \mathbb{T}_{\infty,E})_{\Gamma^{\mathrm{wt}}} \hookrightarrow \text{ the } p\text{-saturation of } H^1(\mathrm{Frob}_\nu, \mathrm{Ta}_p(\mathcal{C}^{\min})^{I_{\mathbf{F}_\nu}})$$

and then dualising we obtain $\delta_{\infty,E,\nu}$.

On the other hand, let $\mathfrak{p} \in S_{\mathbf{F},p}$ be a finite place of $\mathbf{F}$ lying above the prime p. It is a direct corollary of Theorem 5.18, that the geometric part $H^1_{\mathcal{G}}(\mathbf{F}_\mathfrak{p}, \mathbb{T}_{\infty,E})$ is identified with the family of local points $\mathbf{X}_{\mathbb{T}_{\infty,E}}(\mathbf{F}_\mathfrak{p})$. By exploiting the finiteness of the invariants $H^1(\mathbf{F}_\mathfrak{p}, \mathbb{T}_{\infty,E})^{\Gamma^{\mathrm{wt}}}$, it is a simple exercise to show

$$\mathbf{X}_{\mathbb{T}_{\infty,E}}(\mathbf{F}_\mathfrak{p})_{\Gamma^{\mathrm{wt}}} = H^1_{\mathcal{G}}(\mathbf{F}_\mathfrak{p}, \mathbb{T}_{\infty,E}) \otimes_{\Lambda^{\mathrm{wt}}} \mathbb{Z}_p$$
$$\hookrightarrow H^1_g(\mathbf{F}_\mathfrak{p}, \mathrm{Ta}_p(\mathcal{C}^{\min})) \cong \mathcal{C}^{\min}(\mathbf{F}_\mathfrak{p}) \widehat{\otimes} \mathbb{Z}_p$$

the latter isomorphism arising from [BK, Section 3]. Dualising the above yields

$$\delta_{\infty,E,\mathfrak{p}} : \frac{H^1(\mathbf{F}_\mathfrak{p}, \mathcal{C}^{\min}[p^\infty])}{H^1_e(\mathbf{F}_\mathfrak{p}, \mathcal{C}^{\min}[p^\infty])} \longrightarrow \left(\frac{H^1(\mathbf{F}_\mathfrak{p}, A_{\mathbb{T}_{\infty,E}})}{\mathbf{X}^D_{\mathbb{T}_{\infty,E}}(\mathbf{F}_\mathfrak{p})}\right)^{\Gamma^{\mathrm{wt}}}$$

because $H^1_e(\mathbf{F}_\mathfrak{p}, \mathcal{C}^{\min}[p^\infty]) \cong \mathcal{C}^{\min}(\mathbf{F}_\mathfrak{p}) \otimes \mathbb{Q}_p/\mathbb{Z}_p$ and $\mathbf{X}_{\mathbb{T}_{\infty,E}}(\mathbf{F}_\mathfrak{p})^\perp = \mathbf{X}^D_{\mathbb{T}_{\infty,E}}(\mathbf{F}_\mathfrak{p})$.

These facts are neatly summarised in the following commutative diagram:

$$
\begin{array}{ccc}
0 & & 0 \\
\downarrow & & \downarrow \\
\mathrm{Sel}_{\mathbf{F}}\left(\mathcal{C}^{\min}\right)[p^\infty] & \overset{\alpha_{\infty,E}}{\longrightarrow} & \mathrm{Sel}_{\mathbf{F}}\left(\mathbb{T}_{\infty,E}\right)^{\Gamma^{\mathrm{wt}}} \\
\downarrow & & \downarrow \\
H^1\left(\mathbf{F}_\Sigma/\mathbf{F},\, \mathcal{C}^{\min}[p^\infty]\right) & \overset{\beta_{\infty,E}}{\longrightarrow} & H^1\left(\mathbf{F}_\Sigma/\mathbf{F},\, A_{\mathbb{T}_{\infty,E}}\right)^{\Gamma^{\mathrm{wt}}} \\
\downarrow \lambda_{\mathcal{C}^{\min}} & & \downarrow \lambda_{\infty,E} \\
\bigoplus_{\nu\in\Sigma-S_{\mathbf{F},p}} \dfrac{H^1\left(\mathbf{F}_\nu, \mathcal{C}^{\min}[p^\infty]\right)}{H^1_{\mathrm{nr}}\left(\mathbf{F}_\nu, \mathcal{C}^{\min}[p^\infty]\right)} & \overset{\oplus\delta_{\infty,E,\nu}}{\longrightarrow} & \bigoplus_{\nu\in\Sigma-S_{\mathbf{F},p}} \left(\dfrac{H^1\left(\mathbf{F}_\nu, A_{\mathbb{T}_{\infty,E}}\right)}{H^1_{\mathrm{nr}}\left(\mathbf{F}_\nu, A_{\mathbb{T}_{\infty,E}}\right)}\right)^{\Gamma^{\mathrm{wt}}} \\
\oplus \bigoplus_{\nu\in S_{\mathbf{F},p}} \dfrac{H^1\left(\mathbf{F}_\nu, \mathcal{C}^{\min}[p^\infty]\right)}{H^1_{e}\left(\mathbf{F}_\nu, \mathcal{C}^{\min}[p^\infty]\right)} & & \oplus \bigoplus_{\nu\in S_{\mathbf{F},p}} \left(\dfrac{H^1\left(\mathbf{F}_\nu, A_{\mathbb{T}_{\infty,E}}\right)}{\mathbf{X}^D_{\mathbb{T}_{\infty,E}}\left(\mathbf{F}_\nu\right)}\right)^{\Gamma^{\mathrm{wt}}}
\end{array}
$$

Figure 7.1

Remark: Both of the columns appearing in Figure 7.1 are left-exact sequences. The diagram itself is strongly reminiscent of the fundamental diagram involved in the cyclotomic Iwasawa theory of elliptic curves. However, in the latter theory there are no extra complications concerning E and the optimal curve $\mathcal{C}^{\min}$, as the underlying deformation space is canonically $\mathrm{Ta}_p(E)\otimes_{\mathbb{Z}_p}\mathbb{Z}_p\left[\left[\mathrm{Gal}\left(\mathbf{F}(\mu_{p^\infty})/\mathbf{F}\right)\right]\right]$.

Applying the Snake lemma to Figure 7.1, we obtain a long exact sequence

$$
\begin{aligned}
0 \longrightarrow \mathrm{Ker}(\alpha_{\infty,E}) \longrightarrow \mathrm{Ker}(\beta_{\infty,E}) \longrightarrow& \\
\mathrm{Im}\left(\lambda_{\mathcal{C}^{\min}}\right) \cap \oplus\mathrm{Ker}(\delta_{\infty,E,\nu}) &\longrightarrow \mathrm{Coker}(\alpha_{\infty,E}) \rightarrow 0
\end{aligned}
$$

since the map $\beta_{\infty,E}$ is surjective. Analysing the various terms above, will be our principal objective during the next few pages.

Lemma 7.7. *The kernel of $\beta_{\infty,E}$ is a finite p-group of size $\#\mathcal{C}^{\min}(\mathbf{F})[p^\infty]$.*

Proof: By definition, the kernel is the module of Γ^{wt}-coinvariants

$$A_{\mathbb{T}_{\infty,E}}(\mathbf{F})_{\Gamma^{\mathrm{wt}}} \;=\; H^1\Big(\Gamma^{\mathrm{wt}},\, H^0\big(\mathbf{F}_\Sigma/\mathbf{F}, A_{\mathbb{T}_{\infty,E}}\big)\Big).$$

As Γ^{wt} is a pro-cyclic group and $A_{\mathbb{T}_{\infty,E}}$ is a discrete module, the size of the H^1 is the same as that of the H^0.

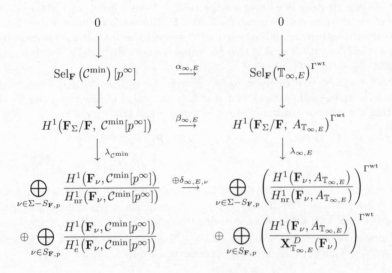

It follows that

$$\begin{aligned}
\#H^1\big(\Gamma^{\rm wt},\ H^0(\mathbf{F}_\Sigma/\mathbf{F}, A_{\mathbb{T}_{\infty,E}})\big) &= \#H^0\big(\Gamma^{\rm wt},\ H^0(\mathbf{F}_\Sigma/\mathbf{F}, A_{\mathbb{T}_{\infty,E}})\big)\\
&= \#H^0\big(\mathbf{F}_\Sigma/\mathbf{F},\ H^0(\Gamma^{\rm wt}, A_{\mathbb{T}_{\infty,E}})\big)\\
&= \#H^0\Big(\mathbf{F}_\Sigma/\mathbf{F},\ \mathrm{Hom}_{\rm cont}\big(\mathbb{T}_{\infty,E}\otimes_{\Lambda^{\rm wt}}\mathbb{Z}_p, \mu_{p^\infty}\big)\Big)\\
&= \#H^0\Big(\mathbf{F}_\Sigma/\mathbf{F},\ \mathrm{Hom}_{\rm cont}\big(\mathrm{Ta}_p(\mathcal{C}^{\rm min}), \mu_{p^\infty}\big)\Big).
\end{aligned}$$

By the Weil pairing, the interior term is none other than the torsion part $\mathcal{C}^{\rm min}[p^\infty]$. We conclude that the size of the $\Gamma^{\rm wt}$-coinvariants equals $\#H^0\big(\mathbf{F}_\Sigma/\mathbf{F}, \mathcal{C}^{\rm min}[p^\infty]\big)$ which is well known to be finite.

□

Lemma 7.8. *For each place $\nu\in\Sigma$, the kernel of $\delta_{\infty,E,\nu}$ is a finite p-group.*

Proof: It makes good sense to subdivide the calculation into two separate parts. To begin with, if $\nu\in\Sigma$ does not lie over p then $\#\mathrm{Ker}(\delta_{\infty,E,\nu})$ equals the size of the cokernel of

$$\widehat{\delta}_{\infty,E,\nu} : H^1_{\rm nr}\big(\mathbf{F}_\nu, \mathbb{T}_{\infty,E}\big)_{\Gamma^{\rm wt}} \hookrightarrow H^1_{\rm nr}\big(\mathbf{F}_\nu, \mathrm{Ta}_p(\mathcal{C}^{\rm min})\big)$$

where the right-hand side above denotes the p-saturation of the unramified cocycles. However $H^1_{\rm nr}\big(\mathbf{F}_\nu, \mathrm{Ta}_p(\mathcal{C}^{\rm min})\big)$ is itself a finite p-group as $\nu\nmid p$, whence the size of the kernel of our map $\delta_{\infty,E,\nu}$ is bounded by $\#H^1_{\rm nr}\big(\mathbf{F}_\nu, \mathrm{Ta}_p(\mathcal{C}^{\rm min})\big)$.

Alternatively, if $\mathfrak{p}\in\Sigma$ does lie over p then $\#\mathrm{Ker}(\delta_{\infty,E,\mathfrak{p}})$ coincides with the size of the cokernel of

$$\widehat{\delta}_{\infty,E,\mathfrak{p}} : H^1_{\mathcal{G}}\big(\mathbf{F}_\mathfrak{p}, \mathbb{T}_{\infty,E}\big)\otimes_{\Lambda^{\rm wt}}\mathbb{Z}_p \hookrightarrow H^1_g\big(\mathbf{F}_\mathfrak{p}, \mathrm{Ta}_p(\mathcal{C}^{\rm min})\big).$$

Clearly the $\mathbb{Z}_p$-rank of $H^1_{\mathcal{G}}\big(\mathbf{F}_\mathfrak{p}, \mathbb{T}_{\infty,E}\big)\otimes_{\Lambda^{\rm wt}}\mathbb{Z}_p$ is bounded below by the $\mathbb{Z}_p\langle\langle w\rangle\rangle$-rank of $H^1_{\mathcal{G}}\big(\mathbf{F}_\mathfrak{p}, \mathbb{T}_{\infty,E}\big)$, which is $[\mathbf{F}_\mathfrak{p}:\mathbb{Q}_p]$ by Theorem 5.18. On the other hand,

$$\mathrm{rank}_{\mathbb{Z}_p} H^1_g\big(\mathbf{F}_\mathfrak{p}, \mathrm{Ta}_p(\mathcal{C}^{\rm min})\big) = \mathrm{rank}_{\mathbb{Z}_p}\Big(E(\mathbf{F}_\mathfrak{p})\widehat{\otimes}\mathbb{Z}_p\Big) = [\mathbf{F}_\mathfrak{p}:\mathbb{Q}_p]$$

since the formal group of E over $\mathcal{O}_{\mathbf{F}_\mathfrak{p}}$ has height one whenever (p-Ord) holds true. Consequently, both $H^1_{\mathcal{G}}\big(\mathbf{F}_\mathfrak{p}, \mathbb{T}_{\infty,E}\big)\otimes_{\Lambda^{\rm wt}}\mathbb{Z}_p$ and $H^1_g\big(\mathbf{F}_\mathfrak{p}, \mathrm{Ta}_p(\mathcal{C}^{\rm min})\big)$ are $\mathbb{Z}_p$-modules of finite type, sharing the same rank. Thus they are commensurate, and the kernel of $\delta_{\infty,E,\mathfrak{p}}$ is therefore finite.

□

Even though the kernels of the maps $\delta_{\infty,E,\nu}$ are finite, to actually calculate their exact sizes is a highly non-trivial problem. In fact, for the primes lying above p one needs to use all of the Super Euler system machinery from Chapters V and VI. However, the formulae we obtain for their orders are both surprising, and suggest strong links to the improved p-adic L-function discussed in §4.4.

Before discussing these notions, we had better check the discrete Selmer group is well behaved as a bundle over weight-space.

Proposition 7.9. *The Selmer group* $\mathrm{Sel}_{\mathbf{F}}(\mathbb{T}_{\infty,E})$ *is cofinitely generated over* Λ^{wt}, *i.e. its Pontrjagin dual*

$$\mathrm{Sel}_{\mathbf{F}}(\mathbb{T}_{\infty,E})^{\vee} \;=\; \mathrm{Hom}_{\mathrm{cont}}\Big(\mathrm{Sel}_{\mathbf{F}}(\mathbb{T}_{\infty,E}),\, \mathbb{Q}_p/\mathbb{Z}_p\Big)$$

is a compact Λ^{wt}*-module of finite type.*

These types of finiteness statement have well known analogues in the cyclotomic Iwasawa theory of elliptic curves. In the latter situation, the cofinite generation of the Selmer group over the extension $\mathbf{F}(\mu_{p^\infty})$ was first established by Mazur [Mz2], and then generalised to GL_2-extensions in the work of Harris [Ha].

Proof: Applying Lemmas 7.7 and 7.8 to the long exact sequence

$$0 \longrightarrow \mathrm{Ker}(\alpha_{\infty,E}) \longrightarrow \mathrm{Ker}(\beta_{\infty,E}) \longrightarrow$$
$$\mathrm{Im}\,(\lambda_{\mathcal{C}^{\min}}) \,\cap\, \oplus\mathrm{Ker}(\delta_{\infty,E,\nu}) \longrightarrow \mathrm{Coker}(\alpha_{\infty,E}) \to 0$$

we deduce that both the kernel and the cokernel of the map $\alpha_{\infty,E}$ are finite groups. It follows that $\mathrm{Sel}_{\mathbf{F}}\left(\mathcal{C}^{\min}\right)[p^\infty] \overset{\alpha_{\infty,E}}{\longrightarrow} \mathrm{Sel}_{\mathbf{F}}(\mathbb{T}_{\infty,E})^{\Gamma^{\mathrm{wt}}}$ is a quasi-isomorphism.

Because the p-primary part of the Selmer group of $\mathcal{C}^{\min}$ over the field $\mathbf{F}$ is cofinitely generated over the ring $\mathbb{Z}_p$, so are the Γ^{wt}-invariants of $\mathrm{Sel}_{\mathbf{F}}(\mathbb{T}_{\infty,E})$. Equivalently, the number of generators of the compact $\mathbb{Z}_p$-module of coinvariants $\left(\mathrm{Sel}_{\mathbf{F}}(\mathbb{T}_{\infty,E})^{\vee}\right)_{\Gamma^{\mathrm{wt}}}$ must then be finite. By Nakayama's lemma, the number of generators of $\mathrm{Sel}_{\mathbf{F}}(\mathbb{T}_{\infty,E})^{\vee}$ over Λ^{wt} is also finite and bounded.

□

The half-twisted Selmer diagram.
We once more focus our attention on the line of symmetry $s = k/2$ in the functional equation, corresponding to the half-twisted big $G_{\mathbb{Q}}$-representation $\mathbb{T}_{\infty,E} \otimes \Psi^{-1/2}$. However it should be pointed out that à priori, the half-twisted Selmer group might not specialise (at weight two) to produce the exact p-primary part of $\mathrm{Sel}_{\mathbf{F}}\left(\mathcal{C}^{\min}\right)$. Following [NP, §2.3.6] one defines $\mathrm{Sel}^{\dagger}_{\mathbf{F}}\left(\mathcal{C}^{\min}\right)$ as the kernel of

$$H^1\Big(\mathbf{F}_\Sigma/\mathbf{F},\, \mathcal{C}^{\min}[p^\infty]\Big) \overset{\oplus\mathrm{res}_\nu}{\longrightarrow} \bigoplus_{\nu\in\Sigma - S_{\mathbf{F},p}} \frac{H^1(\mathbf{F}_\nu, \mathcal{C}^{\min}[p^\infty])}{\mathcal{C}^{\min}(\mathbf{F}_\nu)\otimes\mathbb{Q}_p/\mathbb{Z}_p} \oplus \bigoplus_{\nu\in S_{\mathbf{F},p}} \frac{H^1(\mathbf{F}_\nu, \mathcal{C}^{\min}[p^\infty])}{\mathcal{W}_{\mathcal{C}^{\min},+}(\mathbf{F}_\nu)^{\perp}}$$

where $\mathcal{W}_{\mathcal{C}^{\min},+}(\mathbf{F}_\nu)$ denotes the p-saturation of the subgroup

$$\mathcal{W}_{\mathcal{C}^{\min},+,0} \;:=\; \mathrm{Im}\Big(H^1(\mathbf{F}_\nu, \mathrm{Fil}^+\mathrm{Ta}_p(\mathcal{C}^{\min})) \to H^1(\mathbf{F}_\nu, \mathrm{Ta}_p(\mathcal{C}^{\min}))\Big).$$

To cut a long story short, we claim the commutativity of the following diagram with exact columns:

$$
\begin{array}{ccc}
0 & & 0 \\
\downarrow & & \downarrow \\
\mathrm{Sel}^{\dagger}_{\mathbf{F}}\left(\mathcal{C}^{\min}\right) & \overset{\alpha^{-1/2}_{\infty,E}}{\longrightarrow} & \mathrm{Sel}_{\mathbf{F}}\left(\mathbb{T}_{\infty,E}\otimes\Psi^{-1/2}\right)^{\Gamma^{\mathrm{wt}}} \\
\downarrow & & \downarrow \\
H^1\left(\mathbf{F}_\Sigma/\mathbf{F},\, \mathcal{C}^{\min}[p^\infty]\right) & \overset{\beta^{-1/2}_{\infty,E}}{\longrightarrow} & H^1\left(\mathbf{F}_\Sigma/\mathbf{F},\, A^{-1/2}_{\mathbb{T}_{\infty,E}}\right)^{\Gamma^{\mathrm{wt}}} \\
\big\downarrow \lambda^{-1/2}_{\mathcal{C}^{\min}} & & \big\downarrow \lambda^{-1/2}_{\infty,E} \\
\displaystyle\bigoplus_{\nu\in\Sigma-S_{\mathbf{F},p}} \frac{H^1\left(\mathbf{F}_\nu,\mathcal{C}^{\min}[p^\infty]\right)}{H^1_{\mathrm{nr}}\left(\mathbf{F}_\nu,\mathcal{C}^{\min}[p^\infty]\right)} & \overset{\oplus\delta^{-1/2}_{\infty,E,\nu}}{\longrightarrow} & \displaystyle\bigoplus_{\nu\in\Sigma-S_{\mathbf{F},p}} \left(\frac{H^1\left(\mathbf{F}_\nu, A^{-1/2}_{\mathbb{T}_{\infty,E}}\right)}{H^1_{\mathrm{nr}}\left(\mathbf{F}_\nu, A^{-1/2}_{\mathbb{T}_{\infty,E}}\right)}\right)^{\Gamma^{\mathrm{wt}}} \\
\displaystyle\oplus\bigoplus_{\nu\in S_{\mathbf{F},p}} \frac{H^1\left(\mathbf{F}_\nu,\mathcal{C}^{\min}[p^\infty]\right)}{\mathcal{W}_{\mathcal{C}^{\min},+}\left(\mathbf{F}_\nu\right)^{\perp}} & & \displaystyle\oplus\bigoplus_{\nu\in S_{\mathbf{F},p}} \left(\frac{H^1\left(\mathbf{F}_\nu, A^{-1/2}_{\mathbb{T}_{\infty,E}}\right)}{H^1_{+}\left(\mathbf{F}_\nu\right)^{\perp}}\right)^{\Gamma^{\mathrm{wt}}}
\end{array}
$$

Figure 7.2

Let's explain both its existence and properties over the next couple of paragraphs. Firstly, the global map $\beta^{-1/2}_{\infty,E}$ arises from the specialisation sequence

$$
0 \longrightarrow H^0\left(\mathbf{F}_\Sigma/\mathbf{F},\, A^{-1/2}_{\mathbb{T}_{\infty,E}}\right)\otimes_{\Lambda^{\mathrm{wt}}}\mathbb{Z}_p \longrightarrow H^1\left(\mathbf{F}_\Sigma/\mathbf{F},\mathcal{C}^{\min}[p^\infty]\right)
$$
$$
\overset{\beta^{-1/2}_{\infty,E}}{\longrightarrow} H^1\left(\mathbf{F}_\Sigma/\mathbf{F}, A^{-1/2}_{\mathbb{T}_{\infty,E}}\right)^{\Gamma^{\mathrm{wt}}} \longrightarrow 0.
$$

If $\nu\in\Sigma$ is a place not lying over p, then $\delta^{-1/2}_{\infty,E,\nu}$ is the dual of the injection

$$
H^1_{\mathrm{nr}}\left(\mathbf{F}_\nu, \mathbb{T}_{\infty,E}\otimes\Psi^{-1/2}\right)_{\Gamma^{\mathrm{wt}}} \hookrightarrow \text{ the } p\text{-saturation of } H^1\left(\mathrm{Frob}_\nu, \mathrm{Ta}_p(\mathcal{C}^{\min})^{I_{\mathbf{F}_\nu}}\right)
$$

N.B. this method is identical to the construction of its vertical analogue $\delta_{\infty,E,\nu}$.

Remark: To deal with those primes of $\mathcal{O}_{\mathbf{F}}$ which lie above p, observe that there exists a filtered exact sequence of local $G_{\mathbb{Q}_p}$-modules

$$
0 \longrightarrow \mathrm{Fil}^{+}\mathrm{Ta}_p(\mathcal{C}^{\min}) \longrightarrow \mathrm{Ta}_p(\mathcal{C}^{\min}) \longrightarrow \mathrm{Ta}_p(\mathcal{C}^{\min})/\mathrm{Fil}^{+} \longrightarrow 0
$$

where Fil^{+} is identified with the Tate module of the formal group of $\mathcal{C}^{\min}$ over $\mathbb{Z}_p$. In particular, we may consider the Γ^{wt}-coinvariants $\left(\mathrm{F}^{+}\mathbb{T}_{\infty,E}\otimes\Psi^{-1/2}\right)\otimes_{\Lambda^{\mathrm{wt}}}\mathbb{Z}_p$ as a $G_{\mathbb{Q}_p}$-stable $\mathbb{Z}_p$-submodule of the filtered term $\mathrm{Fil}^{+}\mathrm{Ta}_p(\mathcal{C}^{\min})$ above.

Therefore, one naturally obtains a commutative descent diagram

$$\begin{array}{ccccc} H^1\big(G,\mathrm{F}^+\mathbb{T}_{\infty,E}\otimes\Psi^{-1/2}\big) & \to & H^1\big(G,\mathbb{T}_{\infty,E}\otimes\Psi^{-1/2}\big) & \to & H^1\big(G,\mathbb{T}_{\infty,E}/\mathrm{F}^+\otimes\Psi^{-1/2}\big) \\ \downarrow{\scriptstyle\otimes_{\Lambda^{\mathrm{wt}}}\mathbb{Z}_p} & & \downarrow{\scriptstyle\otimes_{\Lambda^{\mathrm{wt}}}\mathbb{Z}_p} & & \downarrow{\scriptstyle\otimes_{\Lambda^{\mathrm{wt}}}\mathbb{Z}_p} \\ H^1\big(G,\mathrm{Fil}^+\mathrm{Ta}_p(\mathcal{C}^{\min})\big) & \longrightarrow & H^1\big(G,\mathrm{Ta}_p(\mathcal{C}^{\min})\big) & \longrightarrow & H^1\big(G,\mathrm{Ta}_p(\mathcal{C}^{\min})/\mathrm{Fil}^+\big) \end{array}$$

which works well for any open subgroup G of the decomposition group above p. Consequently, if $\mathfrak{p}$ divides p then $\delta_{\infty,E,\mathfrak{p}}^{-1/2}$ is taken as the dual homomorphism to

$$H^1_+\Big(\mathbf{F}_\mathfrak{p},\mathbb{T}_{\infty,E}\otimes\Psi^{-1/2}\Big)\otimes_{\Lambda^{\mathrm{wt}}}\mathbb{Z}_p \quad\hookrightarrow\quad \mathcal{W}_{\mathcal{C}^{\min},+}(\mathbf{F}_\mathfrak{p}).$$

In this manner, the definition and also the commutativity of the bottom square occurring in Figure 7.2, should now become evident.

Question. *What is the precise difference between our filtration version* $\mathrm{Sel}^\dagger_{\mathbf{F}}\big(\mathcal{C}^{\min}\big)$, *and the classical p^∞-Selmer group of $\mathcal{C}^{\min}$ over $\mathbf{F}$?*

Lemma 7.10. *The p-group* $\mathrm{Sel}^\dagger_{\mathbf{F}}\big(\mathcal{C}^{\min}\big)$ *coincides with* $\mathrm{Sel}_{\mathbf{F}}\big(\mathcal{C}^{\min}\big)[p^\infty]$.

Proof: The only conceivable places where the local conditions for $\mathrm{Sel}^\dagger_{\mathbf{F}}\big(\mathcal{C}^{\min}\big)$ and $\mathrm{Sel}_{\mathbf{F}}\big(\mathcal{C}^{\min}\big)[p^\infty]$ might differ, are at the subset of primes $S_{\mathbf{F},p}$ which lie above p. Let us write V to denote the p-adic representation $\mathrm{Ta}_p(\mathcal{C}^{\min})\otimes_{\mathbb{Z}_p}\mathbb{Q}_p$.

If $\mathfrak{p}$ divides p and $\mathcal{C}^{\min}$ does not possess split multiplicative reduction over $\mathbf{F}_\mathfrak{p}$, then we have a strict equality

$$H^1_g(\mathbf{F}_\mathfrak{p},V) \;=\; \mathrm{Im}\Big(H^1(\mathbf{F}_\mathfrak{p},\mathrm{Fil}^+V)\to H^1(\mathbf{F}_\mathfrak{p},V)\Big).$$

On the other hand, if the curve $\mathcal{C}^{\min}$ admits split multiplicative reduction over $\mathbf{F}_\mathfrak{p}$, then Tate's rigid analytic parametrisation $0\to\mathbb{Q}_p(1)\to V\to\mathbb{Q}_p\to 0$ yields the long exact sequence

$$\begin{array}{l} 0\longrightarrow H^0(\mathbf{F}_\mathfrak{p},\mathbb{Q}_p)\longrightarrow \mathbf{F}_\mathfrak{p}^\times\widehat{\otimes}\,\mathbb{Q}_p\longrightarrow H^1_g(\mathbf{F}_\mathfrak{p},V) \\ \qquad\qquad\longrightarrow H^1_g(\mathbf{F}_\mathfrak{p},\mathbb{Q}_p)\overset{\partial}{\longrightarrow} H^2_g(\mathbf{F}_\mathfrak{p},\mathbb{Q}_p(1))\longrightarrow H^2_g(\mathbf{F}_\mathfrak{p},V) \end{array}$$

where $H^2_g(\mathbf{F}_\mathfrak{p},W)$ is the linear dual of the $\mathbb{Q}_p$-vector space $\mathbf{D}_{\mathrm{cris},\mathbf{F}_\mathfrak{p}}\big(W^*(1)\big)^{\varphi=1}$. When $W=V$ the H^2_g is zero, and ∂ is an isomorphism; the above degenerates into

$$\begin{array}{ccccccc} & & \mathbf{F}_\mathfrak{p}^\times\widehat{\otimes}\,\mathbb{Q}_p & & & & \\ & & \| & & & & \\ 0\longrightarrow\mathbb{Q}_p & \longrightarrow & H^1(\mathbf{F}_\mathfrak{p},\mathrm{Fil}^+V) & \longrightarrow & H^1_g(\mathbf{F}_\mathfrak{p},V) & \longrightarrow & 0. \end{array}$$

In the exceptional case, the corank of H^1_g is one less than that of the $H^1\big(\mathbf{F}_\mathfrak{p},\mathrm{Fil}^+V\big)$. However, we may still identify $H^1_g\big(\mathbf{F}_\mathfrak{p},V\big)$ with the image of $H^1\big(\mathbf{F}_\mathfrak{p},\mathrm{Fil}^+V\big)$.

The result follows because the local conditions are identical.

□

What this result reveals is that one may replace $\mathrm{Sel}^{\dagger}_{\mathbf{F}}\left(\mathcal{C}^{\min}\right)$ with $\mathrm{Sel}_{\mathbf{F}}\left(\mathcal{C}^{\min}\right)[p^{\infty}]$ in Figure 7.2, without compromising either its commutativity nor its exactness.

Let us now turn our attention to the restriction maps. Recall that the abstract Hecke algebra $\mathbf{H}_{Np^\infty}$ operates on both $\mathbb{T}_{\infty,E}$, and the half-twist $\mathbb{T}_{\infty,E}\otimes\Psi^{-1/2}$. Henceforth, we shall write $\mathcal{R}_E$ for the local component through which it acts. To make the most accurate statement, we are forced to work under the following assumption (which is closely related to Greenberg's Conjecture 7.5).

Conjecture 7.11. *If at each place $\mathfrak{p}\in\Sigma$ lying over p, we write*

$$\mathcal{Z}_{\mathfrak{p}} \;=\; \mathrm{Im}\Big(H^1\big(\mathbf{F}_{\mathfrak{p}},\mathbb{T}_{\infty,E}\otimes\Psi^{-1/2}\big)\to H^1\big(\mathbf{F}_{\mathfrak{p}},\mathbb{T}_{\infty,E}/\mathrm{F}^{+}\otimes\Psi^{-1/2}\big)\Big)$$

then the indices $\left[H^1\big(\mathbf{F}_{\mathfrak{p}},\mathbb{T}_{\infty,E}/\mathrm{F}^{+}\otimes\Psi^{-1/2}\big)^{\Gamma^{\mathrm{wt}}}:\mathcal{Z}_{\mathfrak{p}}^{\Gamma^{\mathrm{wt}}}\right]$ *are always finite.*

As we shall see in due course, this conjecture can be easily proven in the case where the curve E does not have split multiplicative reduction at any prime above p. However, we believe that it holds true in general.

Lemma 7.12. *(a) The kernel of $\beta^{-1/2}_{\infty,E}$ is finite of size $\#\mathcal{C}^{\min}(\mathbf{F})[p^\infty]$;*

(b) For each place $\nu\in\Sigma-S_{\mathbf{F},p}$, the kernel of $\delta^{-1/2}_{\infty,E,\nu}$ is finite;

(c) If $\mathcal{R}_E$ is Gorenstein, then the kernel of $\delta^{-1/2}_{\infty,E,\mathfrak{p}}$ is cofinitely-generated over $\mathbb{Z}_p$ at every place $\mathfrak{p}\in S_{\mathbf{F},p}$. Furthermore, under Conjecture 7.11

$$\mathrm{corank}_{\mathbb{Z}_p}\mathrm{Ker}\left(\delta^{-1/2}_{\infty,E,\mathfrak{p}}\right) \;=\; \begin{cases} 1 & \textit{if } E \textit{ is split multiplicative at } \mathfrak{p}\\ 0 & \textit{otherwise.}\end{cases}$$

Proof: The demonstration of (a) is almost identical to the proof of Lemma 7.7. Likewise, the argument for part (b) is very similar to the first half of Lemma 7.8. This leaves us with the task of proving (c).

Assume $\mathfrak{p}\in\Sigma$ is a place over p, and let's abbreviate $\mathbb{T}_{\infty,E}\otimes\Psi^{-1/2}$ by $\mathbb{T}^{-1/2}_{\infty,E}$. Firstly, the invariants/coinvariants sequence for the filtration on $\mathbb{T}_{\infty,E}$ yields

$$\cdots\to\left(\mathbb{T}_{\infty,E}/\mathrm{F}^{+}\right)^{\Gamma^{\mathrm{wt}}}\to\left(\mathrm{F}^{+}\mathbb{T}_{\infty,E}\right)_{\Gamma^{\mathrm{wt}}}\to\left(\mathbb{T}_{\infty,E}\right)_{\Gamma^{\mathrm{wt}}}\to\left(\mathbb{T}_{\infty,E}/\mathrm{F}^{+}\right)_{\Gamma^{\mathrm{wt}}}\to 0$$

and the left-hand term is zero, because $\mathbb{T}_{\infty,E}/\mathrm{F}^{+}$ is $\mathcal{R}_E$-free in the Gorenstein case. It follows that one has an identification

$$\begin{array}{ccccccc} 0\to\left(\mathrm{F}^{+}\mathbb{T}^{-1/2}_{\infty,E}\right)_{\Gamma^{\mathrm{wt}}} & \to & \left(\mathbb{T}^{-1/2}_{\infty,E}\right)_{\Gamma^{\mathrm{wt}}} & \to & \left(\mathbb{T}^{-1/2}_{\infty,E}/\mathrm{F}^{+}\right)_{\Gamma^{\mathrm{wt}}} & \to 0\\ \downarrow\cong & & \downarrow\cong & & \downarrow\cong & \\ 0\to\mathrm{Fil}^{+}\mathrm{Ta}_p(\mathcal{C}^{\min}) & \to & \mathrm{Ta}_p(\mathcal{C}^{\min}) & \to & \mathrm{Ta}_p(\mathcal{C}^{\min})/\mathrm{Fil}^{+} & \to 0\end{array}$$

since at weight two only, both $\mathbb{T}_{\infty,E}$ and $\mathbb{T}^{-1/2}_{\infty,E}$ look the same as Galois modules.

Taking its $\mathrm{Gal}(\overline{\mathbf{F}_p}/\mathbf{F}_\mathfrak{p})$-cohomology, one deduces the exactness of

$$\begin{aligned} 0 = H^0\Big(\mathbf{F}_\mathfrak{p}, \mathbb{T}^{-1/2}_{\infty,E}/\mathrm{F}^+\Big) &\longrightarrow H^1\Big(\mathbf{F}_\mathfrak{p}, \mathrm{F}^+\mathbb{T}^{-1/2}_{\infty,E}\Big) \longrightarrow H^1\Big(\mathbf{F}_\mathfrak{p}, \mathbb{T}^{-1/2}_{\infty,E}\Big) \\ &\longrightarrow H^1\Big(\mathbf{F}_\mathfrak{p}, \mathbb{T}^{-1/2}_{\infty,E}/\mathrm{F}^+\Big) \longrightarrow H^2\Big(\mathbf{F}_\mathfrak{p}, \mathrm{F}^+\mathbb{T}^{-1/2}_{\infty,E}\Big) \end{aligned}$$

which shortens to become

$$0 \longrightarrow H^1\Big(\mathbf{F}_\mathfrak{p}, \mathrm{F}^+\mathbb{T}^{-1/2}_{\infty,E}\Big) \longrightarrow H^1\Big(\mathbf{F}_\mathfrak{p}, \mathbb{T}^{-1/2}_{\infty,E}\Big) \longrightarrow \mathcal{Z}_\mathfrak{p} \longrightarrow 0.$$

In particular, the quotient $H^1\big(\mathbf{F}_\mathfrak{p}, \mathbb{T}^{-1/2}_{\infty,E}/\mathrm{F}^+\big)/\mathcal{Z}_\mathfrak{p}$ must be a Λ^{wt}-torsion module because the $H^2\Big(\mathbf{F}_\mathfrak{p}, \mathrm{F}^+\mathbb{T}^{-1/2}_{\infty,E}\Big)$ clearly is.

This time forming the Γ^{wt}-cohomology, we obtain a long exact sequence

$$\begin{aligned} \ldots \longrightarrow H^0\Big(\Gamma^{\mathrm{wt}}, H^1\Big(\mathbf{F}_\mathfrak{p}, \mathbb{T}^{-1/2}_{\infty,E}\Big)\Big) &\longrightarrow H^0\big(\Gamma^{\mathrm{wt}}, \mathcal{Z}_\mathfrak{p}\big) \longrightarrow \\ H^1\Big(\mathbf{F}_\mathfrak{p}, \mathrm{F}^+\mathbb{T}^{-1/2}_{\infty,E}\Big)_{\Gamma^{\mathrm{wt}}} &\longrightarrow H^1\Big(\mathbf{F}_\mathfrak{p}, \mathbb{T}^{-1/2}_{\infty,E}\Big)_{\Gamma^{\mathrm{wt}}} \longrightarrow \Big(\mathcal{Z}_\mathfrak{p}\Big)_{\Gamma^{\mathrm{wt}}} \longrightarrow 0. \end{aligned}$$

The left-hand term is isomorphic to $H^0\big(\mathbf{F}_\mathfrak{p}, \mathrm{Ta}_p(\mathcal{C}^{\min})\big)$, which is obviously finite. We claim there exists a commutative diagram, with exact rows and columns:

$$\begin{array}{ccccccc} & \text{finite} & & \text{finite} & & & \\ & \downarrow & & \downarrow & & & \\ & H^0\big(\Gamma^{\mathrm{wt}}, \mathcal{Z}_\mathfrak{p}\big) & & H^0\Big(\mathbf{F}_\mathfrak{p}, \mathrm{Ta}_p(\mathcal{C}^{\min})/\mathrm{Fil}^+\Big) & & & \\ & \downarrow & & \downarrow & & & \\ 0 \rightarrow & H^1\Big(\mathbf{F}_\mathfrak{p}, \mathrm{F}^+\mathbb{T}^{-1/2}_{\infty,E}\Big)_{\Gamma^{\mathrm{wt}}} & \overset{f_1}{\rightarrow} & H^1\Big(\mathbf{F}_\mathfrak{p}, \mathrm{Fil}^+\mathrm{Ta}_p(\mathcal{C}^{\min})\Big) & \rightarrow & H^2\Big(\mathbf{F}_\mathfrak{p}, \mathrm{F}^+\mathbb{T}^{-1/2}_{\infty,E}\Big)^{\Gamma^{\mathrm{wt}}} & \rightarrow 0 \\ & \downarrow g_1 & & \downarrow g_2 & & \downarrow & \\ 0 \rightarrow & H^1\Big(\mathbf{F}_\mathfrak{p}, \mathbb{T}^{-1/2}_{\infty,E}\Big)_{\Gamma^{\mathrm{wt}}} & \overset{f_2}{\longrightarrow} & H^1\Big(\mathbf{F}_\mathfrak{p}, \mathrm{Ta}_p(\mathcal{C}^{\min})\Big) & \longrightarrow & \text{finite group} & \longrightarrow 0 \\ & \downarrow & & \downarrow & & & \\ & H^1\Big(\mathbf{F}_\mathfrak{p}, \mathbb{T}^{-1/2}_{\infty,E}/\mathrm{F}^+\Big)_{\Gamma^{\mathrm{wt}}} & \overset{f_3}{\rightarrow} & H^1\Big(\mathbf{F}_\mathfrak{p}, \mathrm{Ta}_p(\mathcal{C}^{\min})/\mathrm{Fil}^+\Big). & & & \end{array}$$

To see why the map f_2 has a finite cokernel, it is equivalent to demonstrate that the Γ^{wt}-invariants of $H^2\Big(\mathbf{F}_\mathfrak{p}, \mathbb{T}_{\infty,E} \otimes \Psi^{-1/2}\Big)$ are always finite. By local duality, they correspond to the Γ^{wt}-coinvariants of its dual $H^0\Big(\mathbf{F}_\mathfrak{p},\ A^{-1/2}_{\mathbb{T}_{\infty,E}}\Big)$. However

$$\begin{aligned} \#H^1\Big(\Gamma^{\mathrm{wt}},\ H^0\Big(\mathbf{F}_\mathfrak{p},\ A^{-1/2}_{\mathbb{T}_{\infty,E}}\Big)\Big) &= \#H^0\Big(\Gamma^{\mathrm{wt}},\ H^0\Big(\mathbf{F}_\mathfrak{p},\ A^{-1/2}_{\mathbb{T}_{\infty,E}}\Big)\Big) \\ &= \#H^0\Big(\mathbf{F}_\mathfrak{p},\ \mathrm{Hom}_{\mathrm{cont}}\Big(\mathbb{T}^{-1/2}_{\infty,E} \otimes_{\Lambda^{\mathrm{wt}}} \mathbb{Z}_p,\ \mu_{p^\infty}\Big)\Big) \\ &= \#H^0\Big(\mathbf{F}_\mathfrak{p},\ \mathcal{C}^{\min}[p^\infty]\Big) = \#\mathcal{C}^{\min}(\mathbf{F}_\mathfrak{p})[p^\infty] < \infty. \end{aligned}$$

As an immediate corollary,

$$\#H^2\Big(\mathbf{F}_{\mathfrak{p}}, \mathbb{T}_{\infty,E}^{-1/2}\Big)^{\Gamma^{\mathrm{wt}}} \;=\; \Big[H^1\Big(\mathbf{F}_{\mathfrak{p}}, \mathrm{Ta}_p(\mathcal{C}^{\min})\Big) : H^1\Big(\mathbf{F}_{\mathfrak{p}}, \mathbb{T}_{\infty,E}^{-1/2}\Big)_{\Gamma^{\mathrm{wt}}}\Big] < \infty$$

and our diagram must therefore be correct.

Remark: Returning to the proof of (c), we are required to establish that

$$\mathrm{Coker}\left(\widehat{\delta}_{\infty,E,\mathfrak{p}}^{-1/2} : H^1_+\Big(\mathbf{F}_{\mathfrak{p}}, \mathbb{T}_{\infty,E}^{-1/2}\Big)_{\Gamma^{\mathrm{wt}}} \longrightarrow \mathcal{W}_{\mathcal{C}^{\min},+}(\mathbf{F}_{\mathfrak{p}})\right)$$

is finitely-generated over $\mathbb{Z}_p$, with the rank required from the statement of 7.12(c). It is enough to prove this for the quotient module

$$\frac{g_2\Big(H^1\Big(\mathbf{F}_{\mathfrak{p}}, \mathrm{Fil}^+\mathrm{Ta}_p(\mathcal{C}^{\min})\Big)\Big)}{f_2\circ g_1\Big(H^1\Big(\mathbf{F}_{\mathfrak{p}}, \mathrm{F}^+\mathbb{T}_{\infty,E}^{-1/2}\Big)_{\Gamma^{\mathrm{wt}}}\Big)}.$$

We proceed by examining the $\mathbb{Z}_p$-ranks of $g_2(\dots)$ and $f_2\circ g_1(\dots)$, and then show that they differ by at most one.

Chasing round our diagram and noting the injectivity of f_2, yields the equality

$$\mathrm{rank}_{\mathbb{Z}_p}\mathrm{Im}\big(f_2\circ g_1\big) \;+\; \mathrm{rank}_{\mathbb{Z}_p}H^0\big(\Gamma^{\mathrm{wt}}, \mathcal{Z}_{\mathfrak{p}}\big) \;=\; \mathrm{rank}_{\mathbb{Z}_p}H^1\Big(\mathbf{F}_{\mathfrak{p}}, \mathrm{F}^+\mathbb{T}_{\infty,E}^{-1/2}\Big)_{\Gamma^{\mathrm{wt}}}$$

– we shall refer to this quantity as $r^+(\mathbf{F}_{\mathfrak{p}})$, say. On the other hand,

$$\begin{aligned}\mathrm{rank}_{\mathbb{Z}_p}\mathrm{Im}(g_2) &= \mathrm{rank}_{\mathbb{Z}_p}H^1\Big(\mathbf{F}_{\mathfrak{p}}, \mathrm{Fil}^+\mathrm{Ta}_p(\mathcal{C}^{\min})\Big) - \mathrm{rank}_{\mathbb{Z}_p}H^0\Big(\mathbf{F}_{\mathfrak{p}}, \mathrm{Ta}_p(\mathcal{C}^{\min})/\mathrm{Fil}^+\Big)\\ &= r^+(\mathbf{F}_{\mathfrak{p}}) \;+\; \mathrm{rank}_{\mathbb{Z}_p}H^2\Big(\mathbf{F}_{\mathfrak{p}}, \mathrm{F}^+\mathbb{T}_{\infty,E}^{-1/2}\Big)^{\Gamma^{\mathrm{wt}}} \;-\; \dim_{\mathbb{Q}_p}\big(V/\mathrm{Fil}^+\big)^{G_{\mathbf{F}_{\mathfrak{p}}}}\end{aligned}$$

where $V = \mathrm{Ta}_p(\mathcal{C}^{\min})\otimes_{\mathbb{Z}_p}\mathbb{Q}_p$. Furthermore, by local duality

$$\begin{aligned}\mathrm{rank}_{\mathbb{Z}_p}H^2\Big(\mathbf{F}_{\mathfrak{p}}, \mathrm{F}^+\mathbb{T}_{\infty,E}^{-1/2}\Big)^{\Gamma^{\mathrm{wt}}} &= \mathrm{corank}_{\mathbb{Z}_p}H^0\Big(\mathbf{F}_{\mathfrak{p}}, A_{\mathrm{F}^+\mathbb{T}_{\infty,E}^{-1/2}}\Big)_{\Gamma^{\mathrm{wt}}}\\ &= \dim_{\mathbb{Q}_p}H^0\Big(\mathbf{F}_{\mathfrak{p}}, \mathbb{Q}_p\otimes_{\mathbb{Z}_p}\mathrm{Fil}^+\mathrm{Ta}_p(\mathcal{C}^{\min})^*(1)\Big).\end{aligned}$$

Remark: To simplify matters we now suppose Conjecture 7.11 holds, whence

$$\begin{aligned}\mathrm{rank}_{\mathbb{Z}_p}H^0(\Gamma^{\mathrm{wt}}, \mathcal{Z}_{\mathfrak{p}}) &= \mathrm{rank}_{\mathbb{Z}_p}H^0\Big(\Gamma^{\mathrm{wt}},\, H^1\Big(\mathbf{F}_{\mathfrak{p}}, \mathbb{T}_{\infty,E}^{-1/2}/\mathrm{F}^+\Big)\Big)\\ &= \mathrm{rank}_{\mathbb{Z}_p}H^0\Big(\mathbf{F}_{\mathfrak{p}}, \mathrm{Ta}_p(\mathcal{C}^{\min})/\mathrm{Fil}^+\Big) = \dim_{\mathbb{Q}_p}\big(V/\mathrm{Fil}^+\big)^{G_{\mathbf{F}_{\mathfrak{p}}}}.\end{aligned}$$

As an immediate consequence,

$$\begin{aligned}\mathrm{rank}_{\mathbb{Z}_p}\mathrm{Im}(g_2) &= r^+(\mathbf{F}_{\mathfrak{p}}) \;-\; \mathrm{rank}_{\mathbb{Z}_p}H^0(\Gamma^{\mathrm{wt}}, \mathcal{Z}_{\mathfrak{p}}) \;+\; \mathrm{rank}_{\mathbb{Z}_p}H^2\Big(\mathbf{F}_{\mathfrak{p}}, \mathrm{F}^+\mathbb{T}_{\infty,E}^{-1/2}\Big)^{\Gamma^{\mathrm{wt}}}\\ &= \mathrm{rank}_{\mathbb{Z}_p}\mathrm{Im}\big(f_2\circ g_1\big) + \dim_{\mathbb{Q}_p}\Big(\mathbb{Q}_p\otimes_{\mathbb{Z}_p}\mathrm{Fil}^+\mathrm{Ta}_p(\mathcal{C}^{\min})^*(1)\Big)^{G_{\mathbf{F}_{\mathfrak{p}}}}.\end{aligned}$$

In the non-exceptional case, the $\mathbb{Q}_p$-dimension of $\left(\mathbb{Q}_p \otimes_{\mathbb{Z}_p} \mathrm{Fil}^+\mathrm{Ta}_p(\mathcal{C}^{\min})^*(1)\right)^{G_{\mathbf{F}_\mathfrak{p}}}$ is zero, which means that

$$\mathrm{rank}_{\mathbb{Z}_p}\mathrm{Im}(g_2) \;=\; \mathrm{rank}_{\mathbb{Z}_p}\mathrm{Im}\big(f_2 \circ g_1\big) \;=\; r^+(\mathbf{F}_\mathfrak{p}).$$

In the exceptional case, the $\mathbb{Q}_p$-dimensions of $\left(\mathbb{Q}_p \otimes_{\mathbb{Z}_p} \mathrm{Fil}^+\mathrm{Ta}_p(\mathcal{C}^{\min})^*(1)\right)^{G_{\mathbf{F}_\mathfrak{p}}}$ and also of $H^0\big(\mathbf{F}_\mathfrak{p}, V/\mathrm{Fil}^+\big)$ are both one, hence

$$\mathrm{rank}_{\mathbb{Z}_p}\mathrm{Im}(g_2) \;=\; \mathrm{rank}_{\mathbb{Z}_p}\mathrm{Im}\big(f_2 \circ g_1\big) + 1 \;=\; r^+(\mathbf{F}_\mathfrak{p}).$$

Finally, if Conjecture 7.11 is false(!) then the whole argument works fine, except we are unable to eliminate $\mathrm{rank}_{\mathbb{Z}_p}H^0\big(\Gamma^{\mathrm{wt}}, \mathcal{Z}_\mathfrak{p}\big)$ from our formulae.

□

Arguing as we did in the vertical case, the Snake lemma applied to Figure 7.2 produces the long exact sequence

$$0 \longrightarrow \mathrm{Ker}\left(\alpha_{\infty,E}^{-1/2}\right) \longrightarrow \mathrm{Ker}\left(\beta_{\infty,E}^{-1/2}\right) \longrightarrow$$
$$\mathrm{Im}\left(\lambda_{\mathcal{C}^{\min}}^{-1/2}\right) \cap \oplus\mathrm{Ker}\left(\delta_{\infty,E,\nu}^{-1/2}\right) \longrightarrow \mathrm{Coker}\left(\alpha_{\infty,E}^{-1/2}\right) \to 0.$$

In the Gorenstein situation, by Lemma 7.12 each term is of cofinite-type over $\mathbb{Z}_p$, and the first two terms are certainly finite p-groups. Consequently

$$\mathrm{Sel}_{\mathbf{F}}\big(\mathcal{C}^{\min}\big)[p^\infty] \overset{\alpha_{\infty,E}^{-1/2}}{\longrightarrow} \mathrm{Sel}_{\mathbf{F}}\big(\mathbb{T}_{\infty,E} \otimes \Psi^{-1/2}\big)^{\Gamma^{\mathrm{wt}}}$$

is a quasi-injective map, and its cokernel is cofinitely-generated. The module of coinvariants $\left(\mathrm{Sel}_{\mathbf{F}}\big(\mathbb{T}_{\infty,E} \otimes \Psi^{-1/2}\big)^\vee\right)_{\Gamma^{\mathrm{wt}}}$ must then be finitely-generated over $\mathbb{Z}_p$, in which case Nakayama's lemma implies

Corollary 7.13. *The half-twisted Selmer group* $\mathrm{Sel}_{\mathbf{F}}\big(\mathbb{T}_{\infty,E} \otimes \Psi^{-1/2}\big)$ *is cofinitely generated over the weight algebra* Λ^{wt}, *provided the local factor* $\mathcal{R}_E$ *is Gorenstein.*

If the local component is not Gorenstein, we still expect the half-twisted Selmer group to be of cofinite-type over Λ^{wt}. However, we are unable to prove that the kernel of $\delta_{\infty,E,\mathfrak{p}}^{-1/2}$ is cofinitely-generated over $\mathbb{Z}_p$. This apparent anomaly should not worry us too much, since these universal deformations tend to yield complete intersection rings, anyway.

It should be pointed out that the proof of cofinite-generation for both the vertical and the half-twisted Selmer groups, is very much inspired by analogous techniques in the cyclotomic Iwasawa theory of elliptic curves. The reader who is familiar with the latter theory will probably spot many connections between the two approaches. Indeed many of the theorems in the next section rely heavily on finiteness results, which so far have only been proved using p-adic Euler systems.

7.3 Control theory for Selmer coranks

Let us begin with some general comments on the structure theory of Λ-modules. Assume that M is some finitely-generated, compact Λ-module where $\Lambda \cong \mathbb{Z}_p[\![X]\!]$, e.g. Λ might denote the weight algebra Λ^{wt}, or the cyclotomic Iwasawa algebra. The first invariant is its Λ-rank, i.e. the number of copies of Λ contained inside M. For instance if $\mathrm{rank}_\Lambda(M) = 0$, we say that M is Λ-torsion.

Remark: Given $M = \mathrm{Hom}_{\mathrm{cont}}(S, \mathbb{Q}_p/\mathbb{Z}_p)$ with S of cofinite-type over Λ, then

$$\mathrm{Tors}_\Lambda(M) \;\cong\; \mathrm{Hom}_{\mathrm{cont}}\left(S_{/\Lambda\text{-div}},\, \mathbb{Q}_p/\mathbb{Z}_p\right)$$

will automatically be Λ-torsion (here $S_{/\Lambda\text{-div}}$ means we have quotiented S by its maximal (p, X)-divisible submodule). In particular, this construction is a neat way of extracting a Λ-torsion component from a non-Λ-torsion module.

We shall now assume that M is Λ-torsion. By the structure theory, it sits inside

$$0 \longrightarrow \text{finite group} \longrightarrow M \longrightarrow \bigoplus_{i=1}^{r} \frac{\Lambda}{\mathcal{P}_i^{e_i}} \longrightarrow \text{finite group} \longrightarrow 0$$

which is an exact sequence of Λ-modules. The $\mathcal{P}_i$'s are those height one prime ideals in $\mathrm{Spec}(\Lambda)$, occurring in the support of M. Moreover, the product

$$\mathrm{char}_\Lambda(M) \;=\; \prod_{i=1}^{r} \mathcal{P}_i^{e_i}$$

constitutes the characteristic ideal of M.

Since Λ is a principal ideal domain, $\mathrm{char}_\Lambda(M)$ has a single generator called the characteristic power series (which is well-defined modulo $\Lambda^\times$). Rather confusingly, we sometimes refer to this generator of the characteristic ideal, also as $\mathrm{char}_\Lambda(M)$. Lastly, under the isomorphism $\Lambda \cong \mathbb{Z}_p[\![X]\!]$ there is a factorisation

$$\mathrm{char}_\Lambda(M) \;=\; p^{\mu(M)} \times F(X) \times U(X)$$

where $F(X)$ is a distinguished polynomial, and $U(X)$ is an invertible polynomial. The quantity $\mu(M) \geq 0$ is commonly known as the μ-invariant of M.

Definition 7.14. *Suppose* $\mathbf{T}$ *denotes either* $\mathbb{T}_{\infty,E}$ *or its half-twist* $\mathbb{T}_{\infty,E} \otimes \Psi^{-1/2}$. *We define the Tate-Shafarevich series*

$$Ш_{\mathbf{F}}(\mathbf{T}) \;:=\; \mathrm{char}_{\Lambda^{\mathrm{wt}}}\left(\mathrm{Hom}_{\mathrm{cont}}\left(\mathrm{Sel}_{\mathbf{F}}(\mathbf{T})_{/\Lambda^{\mathrm{wt}}\text{-div}},\, \mathbb{Q}_p/\mathbb{Z}_p\right)\right)$$

or equivalently, it's the characteristic power series of $\mathrm{Tors}_{\Lambda^{\mathrm{wt}}}\left(\mathrm{Sel}_{\mathbf{F}}(\mathbf{T})^\vee\right)$.

As the notation suggests, the power series $\text{Ш}_{\mathbf{F}}(\mathbf{T})$ should interpolate the individual Tate-Shafarevich groups attached to the eigenforms in the Hida family $\{\mathbf{f}_k\}_{k\in\mathbb{U}_E}$. More precisely, for appropriate values of $n \geq 0$ we expect connections between

$$\left.\frac{\mathrm{d}^n\text{Ш}_{\mathbf{F}}(\mathbf{T})}{\mathrm{d}X^n}\right|_{X=(1+p)^{k-2}-1} \quad \text{and} \quad \#\text{Ш}_{\mathbf{F}}\left(\mathbf{T}|_{X=(1+p)^{k-2}-1}\right)[p^\infty].$$

At weight $k = 2$, we may therefore ask the following:

Question. *What is the precise relationship between the leading term of* $\text{Ш}_{\mathbf{F}}(\mathbf{T})$ *and the p-primary part of the Tate-Shafarevich group of E over* $\mathbf{F}$*?*

The actual answer is quite involved, and will occupy the next couple of chapters. For the moment, let us focus exclusively on the vertical representation $\mathbf{T} = \mathbb{T}_{\infty,E}$. We want to prove the Λ^{wt}-cotorsion of the vertical Selmer group under various hypotheses, so the first result we will need is a Control Theorem.

Let V be a p-adic representation equipped with a continuous $\mathrm{Gal}(\mathbf{F}_\Sigma/\mathbf{F})$-action. Following [BK, §5.1], one defines $H^1_{f,\mathrm{Spec}\mathcal{O}_{\mathbf{F}}}(\mathbf{F}_\Sigma/\mathbf{F},\, V)$ as the kernel of

$$H^1(\mathbf{F}_\Sigma/\mathbf{F},\, V) \xrightarrow{\oplus\mathrm{res}_\nu} \bigoplus_{\nu\in\Sigma - S_{\mathbf{F},p}} \frac{H^1(\mathbf{F}_\nu, V)}{H^1_{\mathrm{nr}}(\mathbf{F}_\nu, V)} \oplus \bigoplus_{\nu\in S_{\mathbf{F},p}} \frac{H^1(\mathbf{F}_\nu, V)}{H^1_f(\mathbf{F}_\nu, V)}.$$

Moreover, if $T \subset V$ is a Galois-stable $\mathbb{Z}_p$-lattice and $A = V/T$, we define its discrete version $H^1_{f,\mathrm{Spec}\mathcal{O}_{\mathbf{F}}}(\mathbf{F}_\Sigma/\mathbf{F},\, A)$ to be the kernel of

$$H^1(\mathbf{F}_\Sigma/\mathbf{F},\, A) \xrightarrow{\oplus\mathrm{res}_\nu} \bigoplus_{\nu\in\Sigma - S_{\mathbf{F},p}} \frac{H^1(\mathbf{F}_\nu, A)}{H^1_{\mathrm{nr}}(\mathbf{F}_\nu, A)} \oplus \bigoplus_{\nu\in S_{\mathbf{F},p}} \frac{H^1(\mathbf{F}_\nu, A)}{\mathrm{pr}_* H^1_f(\mathbf{F}_\nu, V)}.$$

One may then replace $H^1_f(\mathbf{F}_\nu, -)$ with the possibly larger local condition $H^1_g(\mathbf{F}_\nu, -)$, and obtain analogous objects $H^1_{g,\mathrm{Spec}\mathcal{O}_{\mathbf{F}}}(\mathbf{F}_\Sigma/\mathbf{F}, -)$.

Theorem 7.15. *(Smith) Assume that the prime p does not ramify in the field* $\mathbf{F}$*. Then for all bar finitely many arithmetic weights* $k \in \mathbb{U}_E$*, the natural map*

$$H^1_{g,\mathrm{Spec}\mathcal{O}_{\mathbf{F}}}\left(\mathbf{F}_\Sigma/\mathbf{F},\, A_{\mathbb{T}_{\infty,E}}[\times\mathfrak{g}_k]\right) \longrightarrow \mathrm{Sel}_{\mathbf{F}}(\mathbb{T}_{\infty,E})[\times\mathfrak{g}_k]$$

has finite kernel and cokernel, bounded independently of $\mathfrak{g}_k = u_0 - (1+p)^{k-2}$.

Note that we have written $[\times\mathfrak{g}]$ above, for the kernel of multiplication by $\mathfrak{g} \in \Lambda^{\mathrm{wt}}$. The control theorem itself allows us to relate the specialisation of $\mathrm{Sel}_{\mathbf{F}}(\mathbb{T}_{\infty,E})$ at weight $k \in \mathbb{U}_E$, with the Bloch-Kato Selmer group attached to each eigenform $\mathbf{f}_k$. Provided the latter are generically small, then so is the Λ^{wt}-corank of $\mathrm{Sel}_{\mathbf{F}}(\mathbb{T}_{\infty,E})$.

A proof of Theorem 7.15 is supplied in Appendix C by P. Smith.

Theorem 7.16. *(Kato) Assume that* $\mathbf{F}$ *is an abelian number field, and that* $f = f_k$ *denotes an eigenform whose p-adic Galois representation is residually irreducible. Then the vector space Selmer group*

$$H^1_{f,\mathrm{Spec}\mathcal{O}_\mathbf{F}}\left(\mathbf{F}_\Sigma/\mathbf{F},\ V^*_{f_k}\right) \ = \ 0$$

whenever the weight k *of* f_k *is at least* 3.

In fact, if the weight $k = 2$ then $H^1_{f,\mathrm{Spec}\mathcal{O}_\mathbf{F}}\left(\mathbf{F}_\Sigma/\mathbf{F},\ V^*_{f_E}\right)$ must contain $E(\mathbf{F})\widehat{\otimes}\,\mathbb{Q}_p$. It follows when the Mordell-Weil group has an $\mathbf{F}$-rational point of infinite order, the Selmer group $H^1_{f,\mathrm{Spec}\mathcal{O}_\mathbf{F}}$ cannot be zero.

Splicing together Theorems 7.15 and 7.16, we can at last make an interesting statement about the arithmetic of the vertical deformation.

Corollary 7.17. *Assume (i)* $\mathbf{F}$ *is an abelian number field, (ii) the prime* p *does not ramify in* $\mathbf{F}$, *and (iii) that* $\rho_{\infty,E}$ *is residually irreducible. Then*

$$\mathrm{rank}_{\Lambda^{\mathrm{wt}}}\left(\mathrm{Sel}_\mathbf{F}\left(\mathbb{T}_{\infty,E}\right)^\vee\right) \ = \ 0$$

i.e. the discrete Selmer group $\mathrm{Sel}_\mathbf{F}\left(\mathbb{T}_{\infty,E}\right)$ *is* Λ^{wt}*-cotorsion.*

Proof: We first verify $H^1_{f,\mathrm{Spec}\mathcal{O}_\mathbf{F}}\left(\mathbf{F}_\Sigma/\mathbf{F}, V^*_{\mathbf{f}_k}\right)$ is equal to $H^1_{g,\mathrm{Spec}\mathcal{O}_\mathbf{F}}\left(\mathbf{F}_\Sigma/\mathbf{F}, V^*_{\mathbf{f}_k}\right)$. This claim certainly follows, provided the local conditions at the primes $\mathfrak{p} \in S_{\mathbf{F},p}$ are the same for each Selmer group. By [BK, Corr 3.8.4], we have an equality

$$\dim_{\mathbb{Q}_p}\left(H^1_{g/f}\left(\mathbf{F}_\mathfrak{p}, V^*_{\mathbf{f}_k}\right)\right) \ = \ \dim_{\mathbb{Q}_p}\left(\mathbf{D}_{\mathrm{cris},\mathbf{F}_\mathfrak{p}}\left(V_{\mathbf{f}_k}(1)\right)/(\varphi - 1)\right).$$

However, the Frobenius cannot have eigenvalue one as the slopes would be wrong. We conclude that $H^1_{g/f}$ is indeed zero, whence our claim.

By condition (iii), there exists an infinite subset $\mathcal{U}' \subset \mathbb{U}_E \cap \mathbb{N}$ of integral weights such that if $k \geq 2$ belongs to $\mathcal{U}'$, then $\rho_{\infty,E} \bmod \mathfrak{g}_k \cong \rho^\vee_{\mathbf{f}_k}$ is residually irreducible as a $G_\mathbb{Q}$-representation. Specifically at weight $k \geq 3$, by applying Theorem 7.16 one discovers that $H^1_{g,\mathrm{Spec}\mathcal{O}_\mathbf{F}}\left(\mathbf{F}_\Sigma/\mathbf{F}, V^*_{\mathbf{f}_k}\right) = 0$. Consequently, the discrete version

$$H^1_{g,\mathrm{Spec}\mathcal{O}_\mathbf{F}}\left(\mathbf{F}_\Sigma/\mathbf{F},\ A_{\mathbb{T}_{\infty,E}}\left[\times\, \mathfrak{g}_k\right]\right)$$

corresponding to the lattice $\mathbb{T}_{\infty,E} \otimes_{\Lambda^{\mathrm{wt}},\mathfrak{g}_k} \mathbb{Z}_p \subset V^*_{\mathbf{f}_k}$, must then be a finite p-group. The Control Theorem 7.15 implies that $\mathrm{Sel}_\mathbf{F}\left(\mathbb{T}_{\infty,E}\right)\left[\times\, \mathfrak{g}_k\right]$ is also a finite group, after possibly jettisoning finitely many bad weights from $\mathcal{U}'$.

To finish off the argument, we show $\mathrm{Sel}_\mathbf{F}\left(\mathbb{T}_{\infty,E}\right)$ contains no copies of $\left(\Lambda^{\mathrm{wt}}\right)^\vee$. Courtesy of Proposition 7.9, we know $\mathrm{Sel}_\mathbf{F}\left(\mathbb{T}_{\infty,E}\right)$ is cofinitely-generated over Λ^{wt}. In particular, its (p, X)-divisible part must be trivial, because the specialisations at infinitely many distinct weights $k \in \mathcal{U}'$ are finite.

Thus the Pontrjagin dual of $\mathrm{Sel}_\mathbf{F}\left(\mathbb{T}_{\infty,E}\right)$ is Λ^{wt}-torsion, as required.

$\square$

The connection with Greenberg's conjecture.
We now shift discussion to the half-twisted Galois representation $\mathbf{T} = \mathbb{T}_{\infty,E} \otimes \Psi^{-1/2}$ which deforms $\rho_{E,p} : G_{\mathbb{Q}} \longrightarrow \mathrm{Aut}_{\mathbb{Z}_p}(\mathrm{Ta}_p(E))$ along the line of symmetry $s = k/2$. To simplify matters, for the moment assume that $\mathbf{F}$ is an abelian extension of $\mathbb{Q}$ such that the prime p does not ramify in $\mathcal{O}_{\mathbf{F}}$.

For every eigenform $\mathbf{f}_k$ in our Hida family, there is a product decomposition

$$L(\mathbf{f}_k/\mathbf{F}, s_\infty) \quad = \quad \prod_{\psi \in X_{\mathbf{F}}} L(\mathbf{f}_k \otimes \psi, s_\infty), \quad s_\infty \in \mathbb{C}$$

where $X_{\mathbf{F}} = \mathrm{Hom}\big(\mathrm{Gal}(\mathbf{F}/\mathbb{Q}), \mathbb{G}_{\mathrm{mult}}\big)$ is the character group associated to $\mathbf{F}/\mathbb{Q}$. Because p is unramified, the conductor of each ψ is coprime to the level of $\mathbf{f}_k$. Consequently these twists $\mathbf{f}_k \otimes \psi$ are all p-stabilised newforms.

Recall that Greenberg's Conjecture 7.5 predicts for each given ψ, the order of vanishing of the p-adic L-function $\mathbf{L}_p(\mathbf{f}_k \otimes \psi, s)$ at the point $s = k/2$ is either always zero or always one (except at finitely many bad weights).

Remarks: Assume that $V_{\mathbf{f}_k}$ is residually irreducible as a $G_{\mathbb{Q}}$-representation, and the weight $k \in \mathbb{U}_E$ is an even integer:

(a) If $k \geq 3$ and $\mathrm{ord}_{s=k/2}\mathbf{L}_p(\mathbf{f}_k, \psi, s) = 1$, then

$$\dim_{\mathbb{Q}_p} H^1_{f,\mathrm{Spec}\mathbb{Z}}\Big(\mathbb{Q}, V_{\mathbf{f}_k} \otimes \psi\chi_{\mathrm{cy}}^{k/2}\Big) \;=\; 1 \qquad \text{by [Ne2, Thms C, D];}$$

(b) If $k \geq 2$ and $\mathrm{ord}_{s=k/2}\mathbf{L}_p(\mathbf{f}_k, \psi, s) = 0$, then

$$H^1_{f,\mathrm{Spec}\mathbb{Z}}\Big(\mathbb{Q}, V_{\mathbf{f}_k} \otimes \psi\chi_{\mathrm{cy}}^{k/2}\Big) \;=\; 0 \qquad \text{by [Ka1, Thm 14.2].}$$

It clearly follows that the ψ^{-1}-component of $H^1_{f,\mathrm{Spec}\mathcal{O}_{\mathbf{F}}}\Big(\mathbf{F}_\Sigma/\mathbf{F}, V_{\mathbf{f}_k} \otimes \chi_{\mathrm{cy}}^{k/2}\Big)$ will have dimension equal to $\mathrm{ord}_{s=k/2}\mathbf{L}_p(\mathbf{f}_k, \psi, s)$, again discarding finitely many weights. However in the non-exceptional situation, it is also widely conjectured

$$\mathrm{ord}_{s=k/2}\, \mathbf{L}_p(\mathbf{f}_k, \psi, s) \quad = \quad \mathrm{ord}_{s_\infty = k/2}\, L(\mathbf{f}_k, \psi, s_\infty)$$

because there is no trivial p-adic zero to contend with.

Therefore Greenberg's conjecture implies the order of vanishing of $L(\mathbf{f}_k/\mathbf{F}, s_\infty)$ along the central line $s_\infty = k/2$, should coincide **generically** with

$$\mathrm{corank}_{\mathbb{Z}_p}\, H^1_{f,\mathrm{Spec}\mathcal{O}_{\mathbf{F}}}\Big(\mathbf{F}_\Sigma/\mathbf{F},\ \mathbb{T}_{\infty,E}^{-1/2} \otimes_{\Lambda^{\mathrm{wt}},\mathfrak{g}_k} \mathbb{Z}_p\Big).$$

We can now relax our assumption that the number field $\mathbf{F}$ is abelian.

Conjecture 7.18. *Providing the right-hand side exists,*

$$\mathrm{rank}_{\Lambda^{\mathrm{wt}}}\Big(\mathrm{Sel}_{\mathbf{F}}\big(\mathbb{T}_{\infty,E} \otimes \Psi^{-1/2}\big)^{\vee}\Big) \;=\; \textit{the generic order}_{s_\infty = k/2}\, L(\mathbf{f}_k/\mathbf{F}, s_\infty).$$

This prediction is a direct translation of Conjecture 7.5, in terms of the behaviour of the half-twisted Selmer group interpolating $\{\mathbf{f}_k\}_{k \in \mathbb{U}_E}$.

Let's try to understand what this statement means about the Selmer rank of E. However we might need to make a few assumptions, in order to obtain a formula with some meat on it.

Throughout we suppose $\mathcal{R}_E$ is a Gorenstein ring, and that Conjecture 7.11 holds. Let $r_{\mathrm{gen}}(\mathbf{F})$ denote the generic order of vanishing of $L(\mathbf{f}_k/\mathbf{F}, s_\infty)$ along $s_\infty = k/2$. Writing $\widetilde{\text{Ш}}^{-1/2}$ for the Λ^{wt}-torsion submodule of $\mathrm{Sel}_{\mathbf{F}}\left(\mathbb{T}_{\infty,E} \otimes \Psi^{-1/2}\right)^\vee$, and with the proviso that Conjecture 7.18 is true:

$$\mathrm{rank}_{\mathbb{Z}_p}\left(\widetilde{\text{Ш}}^{-1/2} \otimes_{\Lambda^{\mathrm{wt}}} \mathbb{Z}_p\right) \;=\; \mathrm{corank}_{\mathbb{Z}_p}\mathrm{Sel}_{\mathbf{F}}\Big(\mathbb{T}_{\infty,E} \otimes \Psi^{-1/2}\Big)^{\Gamma^{\mathrm{wt}}} \;-\; r_{\mathrm{gen}}(\mathbf{F}).$$

Furthermore, diagram chasing around Figure 7.2 then employing Lemma 7.12(c), we quickly discover

$$\mathrm{corank}_{\mathbb{Z}_p}\mathrm{Sel}_{\mathbf{F}}\Big(\mathbb{T}_{\infty,E} \otimes \Psi^{-1/2}\Big)^{\Gamma^{\mathrm{wt}}} \;=\; \mathrm{corank}_{\mathbb{Z}_p}\mathrm{Sel}_{\mathbf{F}}\left(\mathcal{C}^{\mathrm{min}}\right)[p^\infty] \;+\; \#S^{\mathrm{split}}_{\mathbf{F},p}$$

where $S^{\mathrm{split}}_{\mathbf{F},p} \subset S_{\mathbf{F},p}$ is the set of primes at which E has split multiplicative reduction. Since E and $\mathcal{C}^{\mathrm{min}}$ are isogenous elliptic curves over $\mathbf{F}$ (in fact over the rationals), it is acceptable to swap the coranks of their respective p^∞-Selmer groups over $\mathbf{F}$. One concludes that

$$\mathrm{rank}_{\mathbb{Z}_p}\left(\widetilde{\text{Ш}}^{-1/2}\right)_{\Gamma^{\mathrm{wt}}} \;=\; \mathrm{corank}_{\mathbb{Z}_p}\mathrm{Sel}_{\mathbf{F}}(E)[p^\infty] \;+\; \#S^{\mathrm{split}}_{\mathbf{F},p} \;-\; r_{\mathrm{gen}}(\mathbf{F}).$$

Remark: To make further progress, we will assume $\widetilde{\text{Ш}}^{-1/2}$ exhibits a semi-simple Λ^{wt}-structure at the prime ideal (X). In other words $\widetilde{\text{Ш}}^{-1/2}$ should contain no pseudo-summands of the form $\mathbb{Z}_p[\![X]\!]/X^j\mathbb{Z}_p[\![X]\!]$ with $j > 1$.

In this situation, both $\mathbb{Z}_p$-ranks of the Γ^{wt}-coinvariants and the Γ^{wt}-invariants of $\widetilde{\text{Ш}}^{-1/2}$ are exactly the same. Moreover, this quantity will coincide with the order of vanishing of the characteristic power series $\text{Ш}_{\mathbf{F}}\left(\mathbb{T}^{-1/2}_{\infty,E}\right)$ at the point $X = 0$. As a corollary,

$$\mathrm{ord}_{X=0}\text{Ш}_{\mathbf{F}}\left(\mathbb{T}^{-1/2}_{\infty,E}\right) = \mathrm{rank}_{\mathbb{Z}}E(\mathbf{F}) + \mathrm{corank}_{\mathbb{Z}_p}\text{Ш}_{\mathbf{F}}(E)[p^\infty] + \#S^{\mathrm{split}}_{\mathbf{F},p} - r_{\mathrm{gen}}(\mathbf{F}).$$

Last of all, we'll now take the field $\mathbf{F}$ to be an abelian extension of the rationals. The p-adic Birch and Swinnerton-Dyer Conjecture predicts the triple summation

$$\mathrm{rank}_{\mathbb{Z}}E(\mathbf{F}) \;+\; \mathrm{corank}_{\mathbb{Z}_p}\text{Ш}_{\mathbf{F}}(E)[p^\infty] \;+\; \#S^{\mathrm{split}}_{\mathbf{F},p}$$

is equal to the order of vanishing of the p-adic L-function $\mathbf{L}_p(E/\mathbf{F}, s)$ at $s = 1$. However the latter function coincides with $\mathbf{L}_p^{\mathrm{GS}}\left(\mathbf{f}/\mathbf{F}, 2, s\right)$ up to an $\mathbf{I}$-adic period, i.e. up to a non-zero scalar.

Conjecture 7.19. *The order of vanishing of* $\text{Ш}_{\mathbf{F}}\left(\mathbb{T}_{\infty,E}^{-1/2}\right)$ *at* $X=0$ *equals*

$$\text{order}_{s=1}\mathbf{L}_p^{\text{GS}}\Big(\mathbf{f}/\mathbf{F},2,s\Big) \quad - \quad \textit{the generic order}_{s=k/2}\mathbf{L}_p^{\text{GS}}\Big(\mathbf{f}/\mathbf{F},k,s\Big)$$

under the various hypotheses made above.

The Conjectures 7.18 and 7.19 appear very plausible when considered in tandem.

The former predicts that the number of copies of $(\Lambda^{\text{wt}})^{\vee} = \text{Hom}_{\text{cont}}(\Lambda^{\text{wt}}, \mathbb{Q}/\mathbb{Z})$ lying inside the half-twisted Selmer group, is governed by the generic behaviour of $\mathbf{L}_p^{\text{GS}}\Big(\mathbf{f}/\mathbf{F},k,s\Big)$. The latter conjecture then corrects by the p-adic L-function of E which plays the rôle of a 'weight two error term'.

Example 7.20. Consider the modular elliptic curve $E = X_0(11)$ and put $\mathbf{F} = \mathbb{Q}$. As we already saw in 4.14, if $p = 11$ then $\mathcal{R}_E \cong \mathbb{Z}_{11}[\![X]\!]$ which is clearly Gorenstein. Writing $\mathbf{f} \in \Lambda^{\text{wt}}[\![q]\!]$ for the Hida family lifting $X_0(11)$ at p, then $\mathbf{f}_2 = f_{X_0(11)}$ and $\mathbf{f}_{12}$ is the 11-stabilisation of the Ramanujan Δ-function.

The generic rank may be computed via specialisation at weight 12:

$$r_{\text{gen}}(\mathbb{Q}) \;=\; \text{order}_{s=6}\,\mathbf{L}_p(\mathbf{f}_{12},s) \;=\; \text{order}_{s=6}\,L(\Delta,s) \;=\; 1.$$

The Greenberg-Stevens formula implies that $\mathbf{L}_{11}(\mathbf{f}_2,s)$ has a simple zero at $s=1$ because $L\big(X_0(11),1\big) \neq 0$. Finally, we expect from Conjecture 7.19 that

$$\text{ord}_{X=0}\,\text{Ш}_{\mathbf{F}}\left(\mathbb{T}_{\infty,E}^{-1/2}\right) \;=\; \text{order}_{s=1}\,\mathbf{L}_{11}(\mathbf{f}_2,s) \;-\; r_{\text{gen}}(\mathbb{Q}) \;=\; 0.$$

In other words, $\text{Ш}_{\mathbf{F}}\left(\mathbb{T}_{\infty,E}^{-1/2}\right) = a_0 + a_1X + a_2X^2 + \cdots$ where $a_0 \neq 0$.

Question. *What is the* 11*-adic valuation of the constant term* a_0*?*

In the next two chapters, we will find complete solutions to questions of this type. More generally, for any strong Weil curve E which is ordinary at a prime $p \geq 3$, we prove formulae for the leading term of the characteristic power series $\text{Ш}_{\mathbf{F}}(\mathbf{T})$ where $\mathbf{T}$ denotes either $\mathbb{T}_{\infty,E}$, or its half-twist $\mathbb{T}_{\infty,E} \otimes \Psi^{-1/2}$.

CHAPTER VIII

Diamond-Euler Characteristics: the Local Case

We now have a pretty good idea of what Λ^{wt}-coranks to expect for both the vertical and the half-twisted Selmer groups, the latter subject to Greenberg's conjecture. However there are several other invariants attached to these objects, which have deep arithmetic significance. Those occurring at weight two are connected to the Birch and Swinnerton-Dyer formula, whilst at weight greater than two they relate to the Bloch-Kato conjectures.

Pioneering the cyclotomic Iwasawa theory of elliptic curves, Mazur et al [Mz2] analysed the structure of the Selmer group over the $\mathbb{Z}_p$-extension, and essentially recovered the BSD formulae. This corresponds to the line $k = 2$ in the (s, k)-plane. One of our tasks will be to compute invariants along the vertical line $s = 1$, and along the line of symmetry $s = k/2$. These quantities equate with the power of p occurring in the leading term of $\text{Ш}_{\mathbf{F}}(\mathbb{T}_{\infty,E})$ and $\text{Ш}_{\mathbf{F}}(\mathbb{T}_{\infty,E}^{-1/2})$, respectively.

One of the principal reasons why we wish to know the leading terms, relates to the so-called two-variable 'Main Conjecture of Iwasawa theory' for the Hida family. This statement predicts that the Greenberg-Stevens p-adic L-function generates the characteristic ideal of the Selmer group, deformed as a sheaf over the (s, k)-plane. To make the connection precise, it is necessary to match up the leading terms of the analytic and algebraic L-functions. The former is estimated using p-adic BSD, whilst the latter arises from determinants of certain Galois cohomology sequences.

To work out these quantities, we need to study the Γ^{wt}-Euler characteristic of both vertical and half-twisted versions – the calculation breaks up into a local problem and a global problem. The global issues are resolved in Chapters IX and X, and involve introducing p-adic weight pairings, and likewise p-adic height pairings. These in turn allow us to define regulators, which (conjecturally) should not vanish.

The local side of the problem will be dealt with exclusively here in this chapter. The main difficulty is to evaluate the size of the kernels of the local restriction maps $\delta_{\infty,E,\nu}$ and $\delta_{\infty,E,\nu}^{-1/2}$ at the finite primes ν of $\mathbf{F}$. In general, this is way too difficult! To make the calculation tractable, there are two assumptions we are forced into. The first hypothesis requires that the number field $\mathbf{F}$ be an abelian extension of $\mathbb{Q}$ (this allows us to exploit properties of the super zeta-elements). The second one assumes that the local component through which the Hecke algebra acts on $\mathbb{T}_{\infty,E}$ is a complete intersection ring.

As motivation for what is to come, let's begin by explaining the simplest case.

8.1 Analytic rank zero

Throughout this chapter, we suppose that E is a modular elliptic curve over $\mathbb{Q}$ satisfying the condition (p-Ord). In addition, we also assume

Hypothesis(F-ab). *The number field* $\mathbf{F}$ *is an abelian extension of the rationals such that the prime* p *does not ramify as an ideal of* $\mathcal{O}_{\mathbf{F}}$.

Rather than rush headlong into tackling the most general situation, in this section we outline a very special case of our calculations. The reason for this simplification is due to the following deep result [Kv,Ka1].

Theorem 8.1. *If* $L(E/\mathbf{F},1)$ *is non-zero and* $\rho_{E,p}: G_{\mathbb{Q}} \to \mathrm{GL}_2(\mathbb{Z}_p)$ *is a residually irreducible Galois representation, then the Selmer group of* E *over* $\mathbf{F}$ *is finite.*

In particular, whenever E has analytic rank zero over $\mathbf{F}$ then $\#\mathrm{Sel}_{\mathbf{F}}(E)[p^\infty] < \infty$. Because E and $\mathcal{C}^{\min}$ are $\mathbb{Q}$-isogenous, the p-primary part of $\mathrm{Sel}_{\mathbf{F}}(\mathcal{C}^{\min})$ is also finite.

In this section only, we will work under the assumption that $L(E/\mathbf{F},1) \neq 0$. Let's focus first on the vertical deformation of $\mathrm{Ta}_p(E)$, i.e. the straight line $s=1$. Recalling the existence of the quasi-isomorphism

$$\mathrm{Sel}_{\mathbf{F}}\left(\mathcal{C}^{\min}\right)[p^\infty] \overset{\alpha_{\infty,E}}{\longrightarrow} \mathrm{Sel}_{\mathbf{F}}\left(\mathbb{T}_{\infty,E}\right)^{\Gamma^{\mathrm{wt}}}$$

one easily deduces the finiteness of the Γ^{wt}-invariants of the vertical Selmer group. Furthermore, the characteristic power series of $\mathrm{Sel}_{\mathbf{F}}(\mathbb{T}_{\infty,E})$ is non-zero at $X=0$, and has leading term

$$Ш_{\mathbf{F}}\left(\mathbb{T}_{\infty,E}\right)\Big|_{X=0} \quad = \quad \chi\Big(\Gamma^{\mathrm{wt}},\ \mathrm{Sel}_{\mathbf{F}}\left(\mathbb{T}_{\infty,E}\right)\Big)$$

up to p-adic units, of course. Note that the Γ^{wt}-Euler characteristic is defined by

$$\chi\Big(\Gamma^{\mathrm{wt}},\ \mathrm{Sel}_{\mathbf{F}}\left(\mathbb{T}_{\infty,E}\right)\Big) \ := \ \prod_{j=0}^{\infty}\Big(\#H^j\left(\Gamma^{\mathrm{wt}},\ \mathrm{Sel}_{\mathbf{F}}(\mathbb{T}_{\infty,E})\right)\Big)^{(-1)^j}$$
$$= \ \frac{\#H^0\Big(\Gamma^{\mathrm{wt}},\ \mathrm{Sel}_{\mathbf{F}}\left(\mathbb{T}_{\infty,E}\right)\Big)}{\#H^1\Big(\Gamma^{\mathrm{wt}},\ \mathrm{Sel}_{\mathbf{F}}\left(\mathbb{T}_{\infty,E}\right)\Big)}$$

since $\Gamma^{\mathrm{wt}} \cong \mathbb{Z}_p$ has cohomological dimension one.

Question. *How should we compute alternating products of this form?*

The answer lies hidden in the fundamental commutative diagram in Figure 7.1. Chasing around the kernel/cokernels of the three horizontal maps, we find

$$\chi\Big(\Gamma^{\mathrm{wt}}, \mathrm{Sel}_{\mathbf{F}}(\mathbb{T}_{\infty,E})\Big) = \frac{\#\mathrm{Sel}_{\mathbf{F}}\left(\mathcal{C}^{\min}\right)[p^\infty] \times \#\Big(\mathrm{Im}(\lambda_{\mathcal{C}^{\min}}) \cap \bigoplus_{\nu\in\Sigma} \#\mathrm{Ker}\left(\delta_{\infty,E,\nu}\right)\Big)}{\#\mathrm{Ker}\left(\beta_{\infty,E}\right) \ \times \ \#H^1\Big(\Gamma^{\mathrm{wt}},\ \mathrm{Sel}_{\mathbf{F}}\left(\mathbb{T}_{\infty,E}\right)\Big)}.$$

In particular, by Lemmas 7.7 and 7.8 each of these terms above is a finite p-group.

Corollary 8.2. *If $L(E/\mathbf{F},1)\neq 0$ and $\rho_{E,p}$ is residually irreducible, then*

$$\text{Ш}_{\mathbf{F}}(\mathbb{T}_{\infty,E})\Big|_{X=0} \;=\; \frac{\#\text{Ш}_{\mathbf{F}}(\mathcal{C}^{\min})[p^\infty]}{\#\mathcal{C}^{\min}(\mathbf{F})[p^\infty]}\times \#H^1\Big(\Gamma^{\text{wt}},\,\text{Sel}_{\mathbf{F}}(\mathbb{T}_{\infty,E})\Big)^{-1}$$
$$\times\; \frac{\prod_{\nu\in\Sigma}\#\text{Ker}(\delta_{\infty,E,\nu})}{\prod_{\nu\in\Sigma}\Big[\text{Ker}(\delta_{\infty,E,\nu}):\text{Im}(\lambda_{\mathcal{C}^{\min}})\cap\text{Ker}(\delta_{\infty,E,\nu})\Big]}$$

up to an element in $\mathbb{Z}_p^\times$.

Proof: Observe that $\mathcal{C}^{\min}(\mathbf{F})$ is finite because the Selmer group containing it is, whence $\text{Sel}_{\mathbf{F}}(\mathcal{C}^{\min})$ is isomorphic to $\text{Ш}_{\mathbf{F}}(\mathcal{C}^{\min})$. Moreover, the kernel of $\beta_{\infty,E}$ has size $\#\mathcal{C}^{\min}(\mathbf{F})[p^\infty]$ by Lemma 7.7 again.

□

Remark: In the analytic rank zero situation, this corollary reduces the evaluation of $\text{Ш}_{\mathbf{F}}(\mathbb{T}_{\infty,E})$ at $X=0$ into the computation of two separate quantities:

(A) $\prod_{\nu\in\Sigma}\#\text{Ker}(\delta_{\infty,E,\nu})$; and

(B) $\#H^1\Big(\Gamma^{\text{wt}},\,\text{Sel}_{\mathbf{F}}(\mathbb{T}_{\infty,E})\Big)\times\prod_{\nu\in\Sigma}\Big[\text{Ker}(\delta_{\infty,E,\nu}):\text{Im}(\lambda_{\mathcal{C}^{\min}})\cap\text{Ker}(\delta_{\infty,E,\nu})\Big]$.

The first calculation is basically local, and occupies the remainder of this chapter. The second part (B) is global, and constitutes the majority of Chapter IX.

Eliminating Corollary 7.11.
Let us now shift our attention to the half-twisted representation $\mathbf{T}=\mathbb{T}_{\infty,E}\otimes\Psi^{-1/2}$, i.e. to the central line $s=k/2$. What causes something of a nuisance is the failure of the quasi-injective map

$$\text{Sel}_{\mathbf{F}}(\mathcal{C}^{\min})[p^\infty]\;\overset{\alpha_{\infty,E}^{-1/2}}{\longrightarrow}\;\text{Sel}_{\mathbf{F}}\Big(\mathbb{T}_{\infty,E}\otimes\Psi^{-1/2}\Big)^{\Gamma^{\text{wt}}}$$

to be quasi-surjective, if there are primes above p where E is split multiplicative. More precisely, if $\mathcal{R}_E$ is Gorenstein and under Conjecture 7.11

$$\begin{aligned}\text{corank}_{\mathbb{Z}_p}\text{Sel}_{\mathbf{F}}\Big(\mathbb{T}_{\infty,E}\otimes\Psi^{-1/2}\Big)^{\Gamma^{\text{wt}}} &= \text{corank}_{\mathbb{Z}_p}\text{Sel}_{\mathbf{F}}(\mathcal{C}^{\min})[p^\infty]\;+\;\#S_{\mathbf{F},p}^{\text{split}}\\ &= 0\;+\;\#S_{\mathbf{F},p}^{\text{split}}\end{aligned}$$

where the notation $S_{\mathbf{F},p}^{\text{split}}$ denotes the split multiplicative primes for E lying over p. The following result works whether or not $L(E/\mathbf{F},s)$ vanishes at $s=1$.

Lemma 8.3. *If $S_{\mathbf{F},p}^{\text{split}}=\emptyset$ and $\mathcal{R}_E$ is Gorenstein, then Conjecture 7.11 holds true.*

For example, if E admits complex multiplication then its j-invariant is p-integral, in which case there are no primes of split multiplicative reduction, that is $S_{\mathbf{F},p}^{\text{split}}=\emptyset$. It follows that 7.11 is true for all CM elliptic curves over $\mathbb{Q}$.

Proof of 8.3: We need to establish for all $\mathfrak{p} \in S_{\mathbf{F},p}$ that the index

$$\left[H^1\big(\mathbf{F}_{\mathfrak{p}}, \mathbb{T}_{\infty,E}/\mathrm{F}^+ \otimes \Psi^{-1/2}\big)^{\Gamma^{\mathrm{wt}}} : \mathcal{Z}_{\mathfrak{p}}^{\Gamma^{\mathrm{wt}}}\right]$$

is finite, where $\mathcal{Z}_{\mathfrak{p}} = \mathrm{Im}\Big(H^1\big(\mathbf{F}_{\mathfrak{p}}, \mathbb{T}_{\infty,E} \otimes \Psi^{-1/2}\big) \to H^1\big(\mathbf{F}_{\mathfrak{p}}, \mathbb{T}_{\infty,E}/\mathrm{F}^+ \otimes \Psi^{-1/2}\big)\Big)$. If we set $\mathbf{T} = \mathbb{T}_{\infty,E} \otimes \Psi^{-1/2}$ then there is a short exact sequence

$$0 \longrightarrow H^0\big(\mathbf{F}_{\mathfrak{p}}, \mathbf{T}/\mathrm{F}^+\big) \otimes_{\Lambda^{\mathrm{wt}}} \mathbb{Z}_p \longrightarrow H^0\big(\mathbf{F}_{\mathfrak{p}}, \mathrm{Ta}_p(\mathcal{C}^{\min})/\mathrm{Fil}^+\big) \longrightarrow H^1\big(\mathbf{F}_{\mathfrak{p}}, \mathbf{T}/\mathrm{F}^+\big)^{\Gamma^{\mathrm{wt}}} \longrightarrow 0.$$

The middle group is finite, because as a $\mathrm{Gal}\big(\overline{\mathbf{F}_{\mathfrak{p}}}/\mathbf{F}_{\mathfrak{p}}\big)$-module

$$\mathbb{Q}_p \otimes_{\mathbb{Z}_p} \Big(\mathrm{Ta}_p(\mathcal{C}^{\min})/\mathrm{Fil}^+\Big) \ncong \mathbb{Q}_p$$

when E is not split multiplicative at $\mathfrak{p}$. The finiteness of $H^1\big(\mathbf{F}_{\mathfrak{p}}, \mathbf{T}/\mathrm{F}^+\big)^{\Gamma^{\mathrm{wt}}}$ follows. □

We are understandably cautious in relying on too many hypotheses and conjectures. Thus our calculation of the Euler characteristic for the half-twisted Selmer group will only be in the situation where E does *not* have split multiplicative reduction at any of the primes above p.

Corollary 8.4. *Assume that $\mathcal{R}_E$ is Gorenstein, and the subset $S_{\mathbf{F},p}^{\mathrm{split}} \subset \Sigma$ is empty. If $L(E/\mathbf{F},1) \neq 0$ and $\rho_{E,p}$ is residually irreducible,*

$$Ш_{\mathbf{F}}\Big(\mathbb{T}_{\infty,E} \otimes \Psi^{-1/2}\Big)\Big|_{X=0} = \frac{\#Ш_{\mathbf{F}}\big(\mathcal{C}^{\min}\big)[p^\infty]}{\#\mathcal{C}^{\min}\big(\mathbf{F}\big)[p^\infty]} \times \#H^1\Big(\Gamma^{\mathrm{wt}}, \mathrm{Sel}_{\mathbf{F}}\big(\mathbb{T}_{\infty,E}^{-1/2}\big)\Big)^{-1}$$
$$\times \frac{\prod_{\nu\in\Sigma} \#\mathrm{Ker}\big(\delta_{\infty,E,\nu}^{-1/2}\big)}{\prod_{\nu\in\Sigma}\Big[\mathrm{Ker}\big(\delta_{\infty,E,\nu}^{-1/2}\big) : \mathrm{Im}\big(\lambda_{\mathcal{C}^{\min}}^{-1/2}\big) \cap \mathrm{Ker}\big(\delta_{\infty,E,\nu}^{-1/2}\big)\Big]}$$

up to elements of $\mathbb{Z}_p^{\times}$.

Proof: To establish the finiteness of $\mathrm{Sel}_{\mathbf{F}}\big(\mathbb{T}_{\infty,E}^{-1/2}\big)^{\Gamma^{\mathrm{wt}}}$ we just apply Lemma 7.12(c). The argument reduces to a familiar diagram chase, this time around Figure 7.2 rather than Figure 7.1 (otherwise, the details are identical to the previous corollary). □

Let us now return to the general case, and ask what happens if $L(E/\mathbf{F},1)$ is zero. We can no longer say that the Γ^{wt}-invariants of our big Selmer groups are finite, and therefore expect the associated Tate-Shafarevich series to vanish at $X = 0$.

In Chapter IX, we obtain formulae for the leading term of these characteristic power series $Ш_{\mathbf{F}}(-)$, even when the analytic rank of E over $\mathbf{F}$ is strictly positive. However, we need to work under a certain semi-simplicity criterion governing the underlying $\mathbb{Z}_p[\![X]\!]$-modules.

In the sequel, we drop all hypotheses concerning the order of $L(E/\mathbf{F},s)$ at $s = 1$.

8.2 The Tamagawa factors away from p

In order to work out the kernels of the restriction maps on local cohomology, it is sensible to distinguish between primes that lie over p, and the primes that don't. In this section we complete the latter analysis, which is far easier than the former. Let $\nu \in \Sigma$ such that $\nu \nmid p$; we want to calculate the size of

$$\mathrm{Ker}\left(\delta_{\infty,E,\nu} : \frac{H^1(\mathbf{F}_\nu, \mathcal{C}^{\min}[p^\infty])}{H^1_{\mathrm{nr}}(\mathbf{F}_\nu, \mathcal{C}^{\min}[p^\infty])} \longrightarrow \left(\frac{H^1(\mathbf{F}_\nu, A_{\mathbb{T}_{\infty,E}})}{H^1_{\mathrm{nr}}(\mathbf{F}_\nu, A_{\mathbb{T}_{\infty,E}})}\right)^{\Gamma^{\mathrm{wt}}}\right)$$

and in the half-twisted scenario, the size of

$$\mathrm{Ker}\left(\delta^{-1/2}_{\infty,E,\nu} : \frac{H^1(\mathbf{F}_\nu, \mathcal{C}^{\min}[p^\infty])}{H^1_{\mathrm{nr}}(\mathbf{F}_\nu, \mathcal{C}^{\min}[p^\infty])} \longrightarrow \left(\frac{H^1\left(\mathbf{F}_\nu, A^{-1/2}_{\mathbb{T}_{\infty,E}}\right)}{H^1_{\mathrm{nr}}\left(\mathbf{F}_\nu, A^{-1/2}_{\mathbb{T}_{\infty,E}}\right)}\right)^{\Gamma^{\mathrm{wt}}}\right).$$

This leads us nicely onto the concept of a Tamagawa factor for a p-adic deformation.

Definition 8.5. *Let $\mathbf{T}$ denote either $\mathbb{T}_{\infty,E}$, or its half-twisted form $\mathbb{T}_{\infty,E} \otimes \Psi^{-1/2}$. For a positive integral weight $k \in \mathbb{U}_E$, we define*

$$\mathrm{Tam}^{(k)}_{\mathbf{F}_\nu}(\mathbf{T}) \quad := \quad \#\left(\mathbb{Z}_p \otimes_{\Lambda^{\mathrm{wt}},\mathfrak{g}_k} \mathrm{Tors}_{\Lambda^{\mathrm{wt}}} H^1(I_{\mathbf{F}_\nu}, \mathbf{T})^{\mathrm{Frob}_\nu = 1}\right)$$

which is an integral power of p, called **the νth-Tamagawa number** *of $\mathbf{T}$ at k.*

The following compiles a list of some basic properties for these numbers.

Lemma 8.6. *(i) If the prime $\nu \nmid Np$, then $\mathrm{Tam}^{(k)}_{\mathbf{F}_\nu}(\mathbf{T}) = 1$ for all k as above;*

(ii) If the prime $\nu | N$, then $\mathrm{Tam}^{(k)}_{\mathbf{F}_\nu}(\mathbf{T})$ divides the Tamagawa number associated to the lattice $\mathbf{T} \otimes_{\Lambda^{\mathrm{wt}},\mathfrak{g}_k} \mathbb{Z}_p$ by Fontaine and Perrin-Riou;

(iii) At weight two, $\mathrm{Tam}^{(2)}_{\mathbf{F}_\nu}(\mathbf{T})$ divides the p-primary part of $[\mathcal{C}^{\min}(\mathbf{F}_\nu) : \mathcal{C}^{\min}_0(\mathbf{F}_\nu)]$.

Proof: To show (i), we simply remark that $H^1(I_{\mathbf{F}_\nu}, \mathbf{T}) \cong \mathbf{T}(-1)$ since the lattice $\mathbf{T}$ is unramified at the prime ν. In particular, its Λ^{wt}-torsion submodule is trivial.

To prove that part (ii) is true, one observes any cocycle $x \in H^1(I_{\mathbf{F}_\nu}, \mathbf{T})$ which is $\mathfrak{g}_k$-torsion, is also killed by the power p^n with $n = \mathrm{ord}_p\left(\#\frac{(\mathbf{T} \otimes_{\Lambda^{\mathrm{wt}},\mathfrak{g}_k} \mathbb{Z}_p)^{I_{\mathbf{F}_\nu}}}{\mathbf{T}^{I_{\mathbf{F}_\nu}} \otimes_{\Lambda^{\mathrm{wt}},\mathfrak{g}_k} \mathbb{Z}_p}\right)$. Clearly there is an injection

$$\mathrm{Tors}_{\Lambda^{\mathrm{wt}}} H^1(I_{\mathbf{F}_\nu}, \mathbf{T}) \otimes_{\Lambda^{\mathrm{wt}},\mathfrak{g}_k} \mathbb{Z}_p \quad \hookrightarrow \quad \mathrm{Tors}_{\mathbb{Z}_p} H^1\left(I_{\mathbf{F}_\nu}, \mathbf{T} \otimes_{\Lambda^{\mathrm{wt}},\mathfrak{g}_k} \mathbb{Z}_p\right)$$

which respects taking the Frobenius invariants at ν. It follows that $\mathrm{Tam}^{(k)}_{\mathbf{F}_\nu}(\mathbf{T})$ divides into $\#\mathrm{Tors}_{\mathbb{Z}_p} H^1\left(I_{\mathbf{F}_\nu}, \mathbf{T} \otimes_{\Lambda^{\mathrm{wt}},\mathfrak{g}_k} \mathbb{Z}_p\right)^{\mathrm{Frob}_\nu = 1}$. The latter is precisely the definition of the p-adic Tamagawa factors given in [FPR].

Finally, part (iii) is just a special case of (ii) since $\mathbf{T} \otimes_{\Lambda^{\mathrm{wt}},\mathfrak{g}_2} \mathbb{Z}_p \cong \mathrm{Ta}_p(\mathcal{C}^{\min})$.

□

Proposition 8.7. *(i)* $\#\mathrm{Ker}\big(\delta_{\infty,E,\nu}\big) = \dfrac{\Big|\big[\mathcal{C}^{\min}(\mathbf{F}_\nu):\mathcal{C}_0^{\min}(\mathbf{F}_\nu)\big]\Big|_p^{-1}}{\mathrm{Tam}^{(2)}_{\mathbf{F}_\nu}\big(\mathbb{T}_{\infty,E}\big)}$ *for all* $\nu \nmid p$*;*

(ii) $\#\mathrm{Ker}\Big(\delta^{-1/2}_{\infty,E,\nu}\Big) = \dfrac{\Big|\big[\mathcal{C}^{\min}(\mathbf{F}_\nu):\mathcal{C}_0^{\min}(\mathbf{F}_\nu)\big]\Big|_p^{-1}}{\mathrm{Tam}^{(2)}_{\mathbf{F}_\nu}\big(\mathbb{T}_{\infty,E\otimes\Psi^{-1/2}}\big)}$ *for all* $\nu \nmid p$.

Proof: We start with the vertical case (i). By definition $\delta_{\infty,E,\nu}$ is the dual of

$$\widehat{\delta}_{\infty,E,\nu} \;:\; H^1_{\mathrm{nr}}\big(\mathbf{F}_\nu,\mathbb{T}_{\infty,E}\big)\otimes_{\Lambda^{\mathrm{wt}}}\mathbb{Z}_p \;\hookrightarrow\; H^1_{\mathrm{nr}}\big(\mathbf{F}_\nu,\mathrm{Ta}_p(\mathcal{C}^{\min})\big)$$

where the right-hand term was the p-saturation of

$$H^1\Big(\mathrm{Frob}_\nu,\ \mathrm{Ta}_p(\mathcal{C}^{\min})^{I_{\mathbf{F}_\nu}}\Big) \;\cong\; \frac{\mathrm{Ta}_p(\mathcal{C}^{\min})^{I_{\mathbf{F}_\nu}}}{(\mathrm{Frob}_\nu-1)}.$$

In fact $H^1_{\mathrm{nr}}\big(\mathbf{F}_\nu,\mathbb{T}_{\infty,E}\big)$ is all of $H^1\big(\mathbf{F}_\nu,\mathbb{T}_{\infty,E}\big)$ when $\nu\nmid p$, as the Λ^{wt}-rank is zero.

Remark: The key quantity we need to calculate is

$$\#\mathrm{Coker}\Big(\widehat{\delta}_{\infty,E,\nu}\Big) = \Big[H^1_{\mathrm{nr}}\big(\mathbf{F}_\nu,\mathrm{Ta}_p(\mathcal{C}^{\min})\big) : H^1\big(\mathbf{F}_\nu,\mathbb{T}_{\infty,E}\big)\otimes_{\Lambda^{\mathrm{wt}}}\mathbb{Z}_p\Big].$$

To begin with, there is an exact sequence

$$0 \longrightarrow \mathbb{T}_{\infty,E}^{I_{\mathbf{F}_\nu}}\otimes_{\Lambda^{\mathrm{wt}}}\mathbb{Z}_p \longrightarrow \mathrm{Ta}_p(\mathcal{C}^{\min})^{I_{\mathbf{F}_\nu}} \longrightarrow H^1\big(I_{\mathbf{F}_\nu},\mathbb{T}_{\infty,E}\big)^{\Gamma^{\mathrm{wt}}} \longrightarrow 0.$$

Further, $\mathbb{T}_{\infty,E}^{I_{\mathbf{F}_\nu}}\otimes_{\Lambda^{\mathrm{wt}}}\mathbb{Z}_p$ coincides with $\big(\mathbb{T}_{\infty,E}\otimes_{\Lambda^{\mathrm{wt}}}\mathbb{Z}_p\big)^{I_{\mathbf{F}_\nu}} = \mathrm{Ta}_p(\mathcal{C}^{\min})^{I_{\mathbf{F}_\nu}}$ because inertia must act trivially on $\mathbb{Z}_p$, through the weight two specialisation $\mathfrak{g}_2 = u_0 - 1$. As a corollary $H^1\big(I_{\mathbf{F}_\nu},\mathbb{T}_{\infty,E}\big)^{\Gamma^{\mathrm{wt}}}$ must be zero.

The group $\mathrm{Gal}\big(\mathbf{F}_\nu^{\mathrm{unr}}/\mathbf{F}_\nu\big)$ is topologically generated by the Frobenius element, and has cohomological dimension ≤ 1. There is an identification

$$\begin{aligned} H^1\Big(\mathrm{Frob}_\nu,\mathbb{T}_{\infty,E}^{I_{\mathbf{F}_\nu}}\Big)_{\Gamma^{\mathrm{wt}}} &\cong \left(\frac{\mathbb{T}_{\infty,E}^{I_{\mathbf{F}_\nu}}}{(\mathrm{Frob}_\nu-1).\mathbb{T}_{\infty,E}^{I_{\mathbf{F}_\nu}}}\right)\otimes_{\Lambda^{\mathrm{wt}}}\mathbb{Z}_p \\ &= \left(\frac{\big(\mathbb{T}_{\infty,E}\otimes_{\Lambda^{\mathrm{wt}}}\mathbb{Z}_p\big)^{I_{\mathbf{F}_\nu}}}{(\mathrm{Frob}_\nu-1).\big(\mathbb{T}_{\infty,E}\otimes_{\Lambda^{\mathrm{wt}}}\mathbb{Z}_p\big)^{I_{\mathbf{F}_\nu}}}\right) \cong H^1\Big(\mathrm{Frob}_\nu,\ \mathrm{Ta}_p(\mathcal{C}^{\min})^{I_{\mathbf{F}_\nu}}\Big). \end{aligned}$$

Since the local cohomology $H^1\big(\mathbf{F}_\nu,\mathbb{T}_{\infty,E}\big)$ is always Λ^{wt}-torsion whenever $\nu\nmid p$, inflation-restriction provides us with a short exact sequence

$$0 \to H^1\Big(\mathrm{Frob}_\nu,\mathbb{T}_{\infty,E}^{I_{\mathbf{F}_\nu}}\Big) \overset{\mathrm{infl}}{\to} H^1\big(\mathbf{F}_\nu,\mathbb{T}_{\infty,E}\big) \overset{\mathrm{rest}}{\to} \mathrm{Tors}_{\Lambda^{\mathrm{wt}}}H^1\big(I_{\mathbf{F}_\nu},\mathbb{T}_{\infty,E}\big)^{\mathrm{Frob}_\nu} \to 0.$$

The boundary map $\mathrm{Tors}_{\Lambda^{\mathrm{wt}}}\Big(H^1\big(I_{\mathbf{F}_\nu},\mathbb{T}_{\infty,E}\big)^{\mathrm{Frob}_\nu}\Big)^{\Gamma^{\mathrm{wt}}} \longrightarrow H^1\Big(\mathrm{Frob}_\nu,\mathbb{T}_{\infty,E}^{I_{\mathbf{F}_\nu}}\Big)_{\Gamma^{\mathrm{wt}}}$ in Γ^{wt}-cohomology trivialises because $H^1\big(I_{\mathbf{F}_\nu},\mathbb{T}_{\infty,E}\big)^{\Gamma^{\mathrm{wt}}} = 0$. It therefore follows that the coinvariants $H^1\Big(\mathrm{Frob}_\nu,\mathbb{T}_{\infty,E}^{I_{\mathbf{F}_\nu}}\Big)_{\Gamma^{\mathrm{wt}}}$ inject into $H^1\big(\mathbf{F}_\nu,\mathbb{T}_{\infty,E}\big)_{\Gamma^{\mathrm{wt}}}$ under inflation.

We deduce that there is a commutative diagram, with exact rows and columns:

$$\begin{array}{ccccc}
 & & 0 & & 0 \\
 & & \downarrow & & \downarrow \\
H^1\Big(\mathrm{Frob}_\nu, \mathbb{T}_{\infty,E}^{I_{\mathbf{F}_\nu}}\Big)_{\Gamma^{\mathrm{wt}}} & \overset{\mathrm{infl}}{\hookrightarrow} & H^1(\mathbf{F}_\nu, \mathbb{T}_{\infty,E})_{\Gamma^{wt}} & \overset{\mathrm{rest}}{\twoheadrightarrow} & \mathrm{Tors}_{\Lambda^{\mathrm{wt}}}\Big(H^1(I_{\mathbf{F}_\nu}, \mathbb{T}_{\infty,E})^{\mathrm{Frob}_\nu}\Big)_{\Gamma^{\mathrm{wt}}} \\
\| & & \downarrow & & \downarrow \theta_\nu \\
H^1\Big(\mathrm{Frob}_\nu, \mathrm{Ta}_p(\mathcal{C}^{\min})^{I_{\mathbf{F}_\nu}}\Big) & \overset{\mathrm{infl}}{\hookrightarrow} & H^1_{\mathrm{nr}}(\mathbf{F}_\nu, \mathrm{Ta}_p(\mathcal{C}^{\min})) & \overset{\mathrm{rest}}{\twoheadrightarrow} & H^1(I_{\mathbf{F}_\nu}, \mathrm{Ta}_p(\mathcal{C}^{\min}))^{\mathrm{Frob}_\nu}[p^\infty] \\
 & & \downarrow & & \downarrow \\
 & & H^2(\mathbf{F}_\nu, \mathbb{T}_{\infty,E})^{\Gamma^{\mathrm{wt}}} & \cong & \mathrm{Coker}(\theta_\nu) \\
 & & \downarrow & & \downarrow \\
 & & 0 & & 0
\end{array}$$

Exploiting the above to compute indices, general nonsense informs us that

$$\begin{aligned}
\#\mathrm{Coker}\left(\widehat{\delta}_{\infty,E,\nu}\right) &= \Big[H^1_{\mathrm{nr}}(\mathbf{F}_\nu, \mathrm{Ta}_p(\mathcal{C}^{\min})) : H^1(\mathbf{F}_\nu, \mathbb{T}_{\infty,E})_{\Gamma^{\mathrm{wt}}}\Big] = \#\mathrm{Coker}(\theta_\nu) \\
&= \frac{\#H^1\Big(I_{\mathbf{F}_\nu}, \mathrm{Ta}_p(\mathcal{C}^{\min})\Big)^{\mathrm{Frob}_\nu}[p^\infty]}{\#\mathrm{Tors}_{\Lambda^{\mathrm{wt}}}\Big(H^1(I_{\mathbf{F}_\nu}, \mathbb{T}_{\infty,E})^{\mathrm{Frob}_\nu}\Big)_{\Gamma^{\mathrm{wt}}}} = \frac{\Big|\big[\mathcal{C}^{\min}(\mathbf{F}_\nu) : \mathcal{C}_0^{\min}(\mathbf{F}_\nu)\big]\Big|_p^{-1}}{\mathrm{Tam}^{(2)}_{\mathbf{F}_\nu}(\mathbb{T}_{\infty,E})}.
\end{aligned}$$

However this is also the size of the kernel of $\delta_{\infty,E,\nu}$, and part (i) follows.

Finally to prove (ii), one employs an absolutely identical argument to before except $\mathbb{T}_{\infty,E}$ is replaced with $\mathbb{T}_{\infty,E} \otimes \Psi^{-1/2}$, and $\delta_{\infty,E,\nu}$ is replaced with $\delta^{-1/2}_{\infty,E,\nu}$. □

We now have a suitable working definition in place, for Tamagawa numbers inside a p-adic deformation. It is natural to ask what relation (if any) they bear to the standard Tamagawa numbers of the individual modular forms living in the family. In particular, for our original elliptic curve $E_{/\mathbb{Q}}$ we may ask the following

Question. *Can the* $\mathrm{Tam}^{(2)}_{\mathbf{F}_\nu}(\mathbf{T})$*'s ever differ from the p-part of* $\big[E(\mathbf{F}_\nu) : E_0(\mathbf{F}_\nu)\big]$*?*

Let's look for an answer in the special situation where the ground field $\mathbf{F} = \mathbb{Q}$. Certainly one way of spotting when there is a difference occurs if the p-primary part of $\big[\mathcal{C}^{\min}(\mathbb{Q}_l) : \mathcal{C}_0^{\min}(\mathbb{Q}_l)\big]$ is strictly smaller than p-part of $\big[E(\mathbb{Q}_l) : E_0(\mathbb{Q}_l)\big]$, for some rational prime number $l \neq p$.

Clearly for this to happen, there must exist a rational cyclic p-isogeny between the elliptic curves E and $\mathcal{C}^{\min}$, otherwise their two p-primary factors are the same. When E has split multiplicative reduction at l, then $\Big|\big[E(\mathbb{Q}_l) : E_0(\mathbb{Q}_l)\big]\Big|_p^{-1}$ equals the p-part of $\mathrm{ord}_l(\mathbf{q}_E)$, which can be arbitrarily large. It follows that the likeliest candidates to consider are the primes at which E and $\mathcal{C}^{\min}$ become Tate curves.

Table of differing Tamagawa factors up to conductor $\leq 10,000$

E	$[a_1, a_2, a_3, a_4, a_6]$	p	l	$c_l(E)$	$\mathcal{R}_E = \Lambda^{\mathrm{wt}}$?
11A1	[0,–1,1,–10,–20]	5	11	5	Yes
14A1	[1,0,1,4,–6]	3	7	3	Yes
19A1	[0,1,1,–9,–15]	3	19	3	Yes
26A1	[1,0,1,–5,–8]	3	13	3	Yes
35A1	[0,1,1,9,1]	3	7	3	Yes
37B1	[0,1,1,–23,–50]	3	37	3	Yes
38A1	[1,0,1,9,90]	3	19	3	Yes
77B1	[0,1,1,–49,600]	3	7	6	Yes
158D1	[1,0,1,–82,–92]	3	79	3	Yes
278B1	[1,0,1,–537,6908]	3	139	3	Yes
326C1	[1,0,1,–355,1182]	3	163	3	Yes
370C1	[1,0,1,–19,342]	3	37	3	No
485A1	[0,1,1,–121,–64]	3	97	3	Yes
2771A1	[0,1,1,–737,10177]	3	163	3	Yes
4385B1	[0,1,1,–12931,539706]	3	877	3	Yes
6070C1	[1,0,1,–5934,163232]	3	607	3	Yes
6886A1	[1,0,1,–2630,58512]	3	313	3	No
7094C1	[1,0,1,–288181,59747904]	3	3547	3	No
8027A1	[0,1,1,–3243,77986]	3	349	3	No

Figure 8.1

Let us quickly explain how to read this table. The first column refers to Cremona's notation [Cr] for the strong Weil curve E (rather than the Antwerp nomenclature) and the second column supplies its coefficients in Weierstrass form, i.e.

$$E \;:\; y^2 \;+\; a_1xy \;+\; a_3y \;\;=\;\; x^3 \;+\; a_2x^2 \;+\; a_4x \;+\; a_6.$$

The third and fourth columns indicate the choice of $p \neq l$, whilst the fifth column is the Tamagawa factor $c_l(E) = \left[E(\mathbb{Q}_l) : E_0(\mathbb{Q}_l)\right]$ at the prime l.

In all cases the Tamagawa factor $\mathrm{Tam}^{(2)}_{\mathbb{Q}_l}(\mathbf{T})$ of both deformations equals one, because $c_l(\mathcal{C}^{\min})$ is coprime to p. Since we stopped searching at conductor 10,000 and found only 19 examples, it appears the Λ^{wt}-adic Tamagawa number at l differs from its p-adic counterpart very infrequently (see [Sm, App. A] for the algorithm).

8.3 The Tamagawa factors above p (the vertical case)

Having disposed of the primes away from p, we now deal with those above p. Throughout we are forced to make:

Hypothesis(Gor). *The deformation ring $\mathcal{R}_E$ is Gorenstein.*

Let us recall that for all places $\mathfrak{p}$ above p, we are hoping to calculate the size of

$$\mathrm{Ker}\left(\delta_{\infty,E,\mathfrak{p}} : \frac{H^1(\mathbf{F}_\mathfrak{p}, \mathcal{C}^{\min}[p^\infty])}{H^1_e(\mathbf{F}_\mathfrak{p}, \mathcal{C}^{\min}[p^\infty])} \longrightarrow \left(\frac{H^1(\mathbf{F}_\mathfrak{p}, A_{\mathbb{T}_{\infty,E}})}{\mathbf{X}^D_{\mathbb{T}_{\infty,E}}(\mathbf{F}_\mathfrak{p})}\right)^{\Gamma^{\mathrm{wt}}}\right)$$

and for the half-twisted lifting, the size of

$$\mathrm{Ker}\left(\delta^{-1/2}_{\infty,E,\mathfrak{p}} : \frac{H^1(\mathbf{F}_\mathfrak{p}, \mathcal{C}^{\min}[p^\infty])}{\mathcal{W}_{\mathcal{C}^{\min},+}(\mathbf{F}_\mathfrak{p})^\perp} \longrightarrow \left(\frac{H^1\Big(\mathbf{F}_\mathfrak{p}, A^{-1/2}_{\mathbb{T}_{\infty,E}}\Big)}{H^1_+(\mathbf{F}_\mathfrak{p})^\perp}\right)^{\Gamma^{\mathrm{wt}}}\right).$$

In order not to overload the reader with too much exposition, we defer the analysis in the half-twisted case to the next section. From now on consider $\mathbf{T} = \mathbb{T}_{\infty,E}$.

Definition 8.8. *For a positive integral weight $k \in \mathbb{U}_E$, one defines*

$$\mathrm{Tam}^{(k)}_{\mathbf{F}_\mathfrak{p}}(\mathbb{T}_{\infty,E}) := \frac{\Big[H^1_f\Big(\mathbf{F}_\mathfrak{p}, \mathbb{T}_{\infty,E} \otimes_{\Lambda^{\mathrm{wt}},\mathfrak{g}_k} \mathbb{Z}_p\Big) : H^1_f\Big(\mathbf{F}_\mathfrak{p}, \mathrm{F}^+\mathbb{T}_{\infty,E} \otimes_{\Lambda^{\mathrm{wt}},\mathfrak{g}_k} \mathbb{Z}_p\Big)\Big]}{\#H^0\Big(\mathbf{F}_\mathfrak{p},\ A_{\mathbb{T}_{\infty,E}}\Big) \otimes_{\Lambda^{\mathrm{wt}},\mathfrak{g}_k} \mathbb{Z}_p}$$

which is an integral power of p, *called* **the $\mathfrak{p}^{\mathrm{th}}$-Tamagawa number** *of* $\mathbb{T}_{\infty,E}$ *at* k.

As we shall shortly see, at weight two these Tamagawa numbers are always one; however this appears to be a fluke, in general we expect them to be non-trivial. The rest of this section is devoted to the proof of the following

Proposition 8.9. *(a) If E has good ordinary reduction over $\mathbb{Q}_p$, then*

$$\#\mathrm{Ker}\Big(\oplus_{\mathfrak{p}|p}\,\delta_{\infty,E,\mathfrak{p}}\Big) = \prod_{\mathfrak{p}|p} \#\widetilde{E}(f_\mathfrak{p})[p^\infty] \times \frac{\Big|\big[\mathcal{C}^{\min}(\mathbf{F}_\mathfrak{p}) : \mathcal{C}^{\min}_0(\mathbf{F}_\mathfrak{p})\big]\Big|_p^{-1}}{\mathrm{covol}^{(2)}_{\mathbb{T}_{\infty,E}}(\mathbf{F}_\mathfrak{p}) \times \mathrm{Tam}^{(2)}_{\mathbf{F}_\mathfrak{p}}(\mathbb{T}_{\infty,E})};$$

(b) If E has bad multiplicative reduction over $\mathbb{Q}_p$, then

$$\#\mathrm{Ker}\Big(\oplus_{\mathfrak{p}|p}\,\delta_{\infty,E,\mathfrak{p}}\Big) = \prod_{\mathfrak{p}|p} \frac{\Big|\big[\mathcal{C}^{\min}(\mathbf{F}_\mathfrak{p}) : \mathcal{C}^{\min}_0(\mathbf{F}_\mathfrak{p})\big]\Big|_p^{-1}}{\mathrm{covol}^{(2)}_{\mathbb{T}_{\infty,E}}(\mathbf{F}_\mathfrak{p}) \times \mathrm{Tam}^{(2)}_{\mathbf{F}_\mathfrak{p}}(\mathbb{T}_{\infty,E})};$$

(c) At weight $k = 2$, the factor $\mathrm{Tam}^{(2)}_{\mathbf{F}_\mathfrak{p}}(\mathbb{T}_{\infty,E}) = 1$ *for every place* $\mathfrak{p}|p$.

N.B. At those primes $\mathfrak{p}$ lying over p, we have defined $\mathrm{covol}^{(2)}_{\mathbb{T}_{\infty,E}}(\mathbf{F}_\mathfrak{p})$ to be the index

$$\Big[\exp^*_\omega\Big(H^1(\mathbf{F}_\mathfrak{p}, \mathrm{Ta}_p(\mathcal{C}^{\min}))\Big) : \exp^*_\omega\Big(H^1(\mathbf{F}_\mathfrak{p}, \mathbb{T}_{\infty,E}) \otimes_{\Lambda^{\mathrm{wt}}} \mathbb{Z}_p\Big)\Big].$$

Note that $\exp^*_\omega$ is the homomorphism obtained by cupping the dual exponential map with a Néron differential $\omega \in H^1_{\mathrm{dR}}$, living on the optimal elliptic curve $\mathcal{C}^{\min}$.

Remarks: (i) In the statement of the above proposition, we have grouped together the kernels at all the primes above p simultaneously. The underlying reason is that the calculation is best carried out semi-locally rather than locally, as the connection with L-functions becomes more apparent.

(ii) When E has good ordinary reduction over $\mathbb{Q}_p$, the extra terms $\#\widetilde{E}(f_{\mathfrak{p}})[p^\infty]$ enter into our formulae – by definition, they are the number of p-power torsion points on the reduction of E over $f_{\mathfrak{p}}$, the residue field of $\mathbf{F}$ at $\mathfrak{p}$.

There are two main technical tools we need: the super zeta-elements of §6.1, and the big dual exponential map (constructed in §5.3 via Iovita-Stevens' theory). Employing Definition 5.11(ii), since $\mathbf{F}_{\mathfrak{p}}$ is a normal extension of the p-adic numbers there exists a unique Λ^{wt}-homomorphism EXP^*, which makes

$$\begin{array}{ccc} H^1(\mathbf{F}_{\mathfrak{p}}, \mathbb{T}_{\infty,E}) & \xrightarrow{\mathrm{EXP}^*_{\mathbb{T}_\infty,\mathbf{F}_{\mathfrak{p}}}} & \mathbf{F}_{\mathfrak{p}} \otimes_{\mathbb{Z}_p} \mathbb{Z}_p\langle\langle w\rangle\rangle \\ \Big\downarrow{\scriptstyle w=k} & & \Big\downarrow{\scriptstyle w=k} \\ H^1\left(\mathbf{F}_{\mathfrak{p}}, V^*_{\mathbf{f}_k}\right) \xrightarrow{\exp^*_{V^*_{\mathbf{f}_k}}} \mathrm{cotang}\left(V^*_{\mathbf{f}_k}/\mathbf{F}_{\mathfrak{p}}\right) & \xrightarrow{\cup(1-\varphi)^{-1}\left(t^{-1}\cdot\mathbf{v}_{\mathbf{I},\lambda_E}\right)_{w=k}} & \mathbf{D}_{\mathrm{dR},\mathbf{F}_{\mathfrak{p}}}(\mathcal{K}_{\mathbf{f}_k}) \end{array}$$

commute at all arithmetic weights $k \in \mathbb{U}_E \cap \mathbb{N}_{\geq 2}$. It is at this point we require $\mathcal{R}_E$ to be Gorenstein, otherwise we might have to throw away a dense set of weights.

Warning: One must be very careful here precisely what is meant by *uniqueness*, i.e. only after a generator $\mathbf{v}_{\mathbf{I},\lambda_E} \in \mathbb{D}^{\mathbf{I}}_{\mathrm{cris}}(\mathbb{T}^*_{\infty,E})$ over the affinoid algebra $\mathbb{Q}_p\langle\langle w\rangle\rangle$ has actually been chosen, can the homomorphism EXP^* be considered unique.

Now we indicated in 5.11(iii) that these maps can be glued along primes $\mathfrak{p} \in S_{\mathbf{F},p}$; together they yielded a homomorphism on semi-local cohomology

$$\mathrm{EXP}^*_{\mathbb{T}_\infty,E} : H^1\Big(\mathbf{F}\otimes\mathbb{Z}_p, \mathbb{T}_{\infty,E}\Big) \xrightarrow{\sim} \prod_{\mathfrak{p}|p} H^1\Big(D_{\mathfrak{p}}, \mathbb{T}_{\infty,E}\Big) \xrightarrow{\mathrm{EXP}^*_{\mathbb{T}_\infty,\mathbf{F}_{\mathfrak{p}}}} \mathbf{F}\otimes\mathbb{Z}_p\langle\langle w\rangle\rangle.$$

Let us begin the proof of Proposition 8.9 by checking how large the image is.

Lemma 8.10. $\mathrm{rank}_{\Lambda^{\mathrm{wt}}}\Big(\mathrm{Im}\big(\mathrm{EXP}^*_{\mathbb{T}_\infty,E}\big)\Big) = [\mathbf{F}:\mathbb{Q}]$.

Proof: Because $\mathbf{F}$ is an abelian extension of $\mathbb{Q}$, we can split up the calculation over the $X_{\mathbf{F}}$-eigenspaces, i.e.

$$\begin{aligned} \mathrm{rank}_{\Lambda^{\mathrm{wt}}}\Big(\mathrm{Im}\big(\mathrm{EXP}^*_{\mathbb{T}_\infty,E}\big)\Big) &= \sum_{\psi\in X_{\mathbf{F}}} \mathrm{rank}_{\Lambda^{\mathrm{wt}}_\psi}\Big(\mathrm{EXP}^*_{\mathbb{T}_\infty,E}\Big(H^1(\mathbf{F}\otimes\mathbb{Z}_p, \mathbb{T}_{\infty,E})\Big)^{(\psi)}\Big) \\ &= \sum_{\psi\in X_{\mathbf{F}}} \mathrm{rank}_{\mathbb{Z}_{p,\psi}}\left(\mathrm{EXP}^*_{\mathbb{T}_\infty,E}\Big(H^1(\mathbf{F}\otimes\mathbb{Z}_p, \mathbb{T}_{\infty,E}) \otimes_{\Lambda^{\mathrm{wt}},\mathfrak{g}_k} \mathbb{Z}_p\Big)^{(\psi)}\right) \end{aligned}$$

at all bar finitely many weights $k \in \mathbb{U}_E \cap \mathbb{N}_{\geq 2}$. In fact, the weights we must discard correspond precisely to those primes occurring in the support of the semi-local H^1.

As a consequence of Theorem 6.2, there exist zeta-elements $\mathfrak{Z} = \mathfrak{Z}_{\mathbf{F}}$ belonging to

$$\mathcal{MS}^{\mathrm{ord}} \otimes_{\mathbf{h}^{\mathrm{ord}}(Np^\infty)} H^1\Big(\mathbf{F}, \mathbb{T}_{\infty,E}\Big) \otimes_{\mathcal{R}} \widetilde{\mathcal{R}};$$

we don't care about their exact normalisation, and just drop the $\mathcal{MS}^{\mathrm{ord}}$ hereon in.

Exploiting the interpolation properties of EXP^*, at each character $\psi \in X_{\mathbf{F}}$

$$\begin{aligned}\mathrm{EXP}^*_{\mathbb{T}_{\infty,E}}\Big(\mathrm{loc}_p(\mathfrak{Z}_{\mathbf{F}})^{(\psi)}\Big)_{w=k} &= \exp^*_{V^*_{\mathbf{f}_k}}\Big(\mathrm{loc}_p(\mathfrak{Z}_{\mathbf{F}})^{(\psi)}_{w=k}\Big) \cup (1-\varphi)^{-1}.(t^{-1}.\mathbf{v}_{\mathbf{I},\lambda_E})_{w=k} \\ &= (\text{a non-zero number}) \times \frac{L(\mathbf{f}_k \otimes \psi^{-1}, 1)}{\Omega^{\mathrm{sign}(\psi)}_{\mathbf{f}_k}}\end{aligned}$$

which does not vanish provided $k \geq 3$. It follows directly that the $\mathbb{Z}_{p,\psi}$-rank of the ψ-part of $\mathrm{EXP}^*_{\mathbb{T}_{\infty,E}}(-)_{w=k}$ is at least one, in which case

$$\mathrm{rank}_{\Lambda^{\mathrm{wt}}}\Big(\mathrm{Im}\big(\mathrm{EXP}^*_{\mathbb{T}_{\infty,E}}\big)\Big) \geq \sum_{\psi \in X_{\mathbf{F}}} 1 = \# X_{\mathbf{F}} = [\mathbf{F} : \mathbb{Q}].$$

Conversely, the image of $\mathrm{EXP}^*_{\mathbb{T}_{\infty,E}}$ is a compact subset of $\mathbf{F} \otimes \mathbb{Z}_p\langle\langle w\rangle\rangle$, which means its Λ^{wt}-rank is also bounded above by $[\mathbf{F} : \mathbb{Q}]$.

□

Corollary 8.11. *(i) There is a left-exact sequence of Λ^{wt}-modules*

$$0 \longrightarrow \mathbf{X}_{\mathbb{T}_{\infty,E}}(\mathbf{F} \otimes \mathbb{Z}_p) \longrightarrow H^1\Big(\mathbf{F} \otimes \mathbb{Z}_p,\ \mathbb{T}_{\infty,E}\Big) \overset{\mathrm{EXP}^*_{\mathbb{T}_{\infty,E}}}{\longrightarrow} \mathbf{F} \otimes \mathbb{Z}_p\langle\langle w\rangle\rangle.$$

(ii) At weight two, this specialises to a left-exact sequence of $\mathbb{Z}_p$-modules

$$0 \longrightarrow \mathbf{X}_{\mathbb{T}_{\infty,E}}(\mathbf{F} \otimes \mathbb{Z}_p) \otimes_{\Lambda^{\mathrm{wt}}} \mathbb{Z}_p \longrightarrow H^1\Big(\mathbf{F} \otimes \mathbb{Z}_p,\ \mathbb{T}_{\infty,E}\Big) \otimes_{\Lambda^{\mathrm{wt}}} \mathbb{Z}_p \overset{\exp^*_\omega}{\longrightarrow} \mathbf{F} \otimes \mathbb{Z}_p.$$

Proof: We start with part (i). If $\mathfrak{X} \in \mathbf{X}_{\mathbb{T}_{\infty,E}}(\mathbf{F} \otimes \mathbb{Z}_p)$, then for all $k \in \mathbb{U}_E \cap \mathbb{N}_{\geq 2}$

$$\mathrm{EXP}^*_{\mathbb{T}_{\infty,E}}(\mathfrak{X})_{w=k} = \exp^*_{V^*_{\mathbf{f}_k}}(\mathfrak{X}_{w=k}) \cup (1-\varphi)^{-1}.(t^{-1}.\mathbf{v}_{\mathbf{I},\lambda_E})_{w=k} = 0$$

as the specialisation $\mathfrak{X}_{w=k}$ lies in $H^1_g\Big(\mathbf{F} \otimes \mathbb{Z}_p,\ \mathbb{T}_{\infty,E} \otimes_{\Lambda^{\mathrm{wt}},\mathfrak{g}_k} \mathbb{Z}_p\Big) \subset \mathrm{Ker}\Big(\exp^*_{V^*_{\mathbf{f}_k}}\Big)$. It follows that $\mathrm{EXP}^*_{\mathbb{T}_{\infty,E}}(\mathfrak{X})$ vanishes at infinitely many distinct k, and so is zero. However $\mathbf{X}_{\mathbb{T}_{\infty,E}}$ is Λ^{wt}-saturated inside the semi-local H^1; using Lemma 8.10

$$\begin{aligned}\mathrm{rank}_{\Lambda^{\mathrm{wt}}} H^1\Big(\mathbf{F} \otimes \mathbb{Z}_p,\ \mathbb{T}_{\infty,E}\Big) &= 2[\mathbf{F} : \mathbb{Q}] = [\mathbf{F} : \mathbb{Q}] + [\mathbf{F} : \mathbb{Q}] \\ &= \mathrm{rank}_{\Lambda^{\mathrm{wt}}} \mathbf{X}_{\mathbb{T}_{\infty,E}}(\mathbf{F} \otimes \mathbb{Z}_p) + \mathrm{rank}_{\Lambda^{\mathrm{wt}}} \mathrm{Im}\big(\mathrm{EXP}^*_{\mathbb{T}_{\infty,E}}\big).\end{aligned}$$

Therefore the first sequence above must be left-exact.

To deduce part (ii), we will examine the $\Gamma^{\rm wt}$-invariants and $\Gamma^{\rm wt}$-coinvariants of (i). In particular, one obtains a long-exact sequence in cohomology

$$\cdots \longrightarrow \left(\frac{H^1\left(\mathbf{F}\otimes\mathbb{Z}_p, \mathbb{T}_{\infty,E}\right)}{\mathbf{X}_{\mathbb{T}_{\infty,E}}(\mathbf{F}\otimes\mathbb{Z}_p)}\right)^{\Gamma^{\rm wt}} \stackrel{\partial}{\longrightarrow} \mathbf{X}_{\mathbb{T}_{\infty,E}}\left(\mathbf{F}\otimes\mathbb{Z}_p\right)_{\Gamma^{\rm wt}}$$

$$\longrightarrow H^1\left(\mathbf{F}\otimes\mathbb{Z}_p, \mathbb{T}_{\infty,E}\right)_{\Gamma^{\rm wt}} \stackrel{\left(\mathrm{EXP}^*_{\mathbb{T}_{\infty,E}}\right)_{w=2}}{\longrightarrow} \mathbf{F}\otimes\mathbb{Z}_p\langle\langle w\rangle\rangle\Big|_{w=2}.$$

Key Claim: The image of the boundary homomorphism ∂ is always zero.

To justify this assertion, observe the quotient of H^1 by $\mathbf{X}_{\mathbb{T}_{\infty,E}}$ is $\Lambda^{\rm wt}$-torsion free, in which case the $\Gamma^{\rm wt}$-invariants of $\frac{H^1(\mathbf{F}\otimes\mathbb{Z}_p,\mathbb{T}_{\infty,E})}{\mathbf{X}_{\mathbb{T}_{\infty,E}}(\mathbf{F}\otimes\mathbb{Z}_p)}$ have no option but to be $\mathbb{Z}_p$-free. Thus it suffices to show the $\mathbb{Z}_p$-rank of $H^1\left(\mathbf{F}\otimes\mathbb{Z}_p,\ \mathbb{T}_{\infty,E}\right)_{\Gamma^{\rm wt}}$ is equal to

$$\mathrm{rank}_{\mathbb{Z}_p}\mathbf{X}_{\mathbb{T}_{\infty,E}}\left(\mathbf{F}\otimes\mathbb{Z}_p\right)_{\Gamma^{\rm wt}} + \mathrm{rank}_{\mathbb{Z}_p}\mathrm{Im}\left(\mathrm{EXP}^*_{\mathbb{T}_{\infty,E}}\right)_{w=2}.$$

Firstly, the former rank is $2[\mathbf{F}:\mathbb{Q}]$. The $\mathbb{Z}_p$-rank of $\mathbf{X}_{\mathbb{T}_{\infty,E}}(-)_{\Gamma^{\rm wt}}$ is bounded below by the $\Lambda^{\rm wt}$-rank of $\mathbf{X}_{\mathbb{T}_{\infty,E}}(-)$ which is $[\mathbf{F}:\mathbb{Q}]$ courtesy of Theorem 5.18; however, it is also bounded above by the $\mathbb{Z}_p$-rank of the $H^1_g(\mathbf{F}\otimes\mathbb{Z}_p,\ \mathrm{Ta}_p(\mathcal{C}^{\min}))$ which again equals the degree $[\mathbf{F}:\mathbb{Q}]$. Lastly, if $p^n = \#H^2(\mathbf{F}\otimes\mathbb{Z}_p,\mathbb{T}_{\infty,E})^{\Gamma^{\rm wt}}$ then $p^n.\left(\text{the pullback of }\mathrm{Im}\left(\exp^*_{V^*_{\mathbf{f}_2}}\right)\right)$ has $\mathbb{Z}_p$-rank $[\mathbf{F}:\mathbb{Q}]$ in $H^1_{/g}(\mathbf{F}\otimes\mathbb{Z}_p,\mathrm{Ta}_p(\mathcal{C}^{\min}))$; it also lies inside the pullback of $\mathrm{Im}\left(\mathrm{EXP}^*_{\mathbb{T}_{\infty,E}}\right)_{w=2}$ modulo $\mathbf{X}_{\mathbb{T}_{\infty,E}}\left(\mathbf{F}\otimes\mathbb{Z}_p\right)_{\Gamma^{\rm wt}}$ with the same maximal $\mathbb{Z}_p$-rank.

As a consequence of this Key Claim, we obtain the exactness of the sequence

$$0 \longrightarrow \mathbf{X}_{\mathbb{T}_{\infty,E}}\left(\mathbf{F}\otimes\mathbb{Z}_p\right)_{\Gamma^{\rm wt}} \longrightarrow H^1\left(\mathbf{F}\otimes\mathbb{Z}_p,\ \mathbb{T}_{\infty,E}\right)_{\Gamma^{\rm wt}} \stackrel{\left(\mathrm{EXP}^*_{\mathbb{T}_{\infty,E}}\right)_{w=2}}{\longrightarrow} \mathbf{F}\otimes\mathbb{Z}_p.$$

It remains to replace $\left(\mathrm{EXP}^*_{\mathbb{T}_{\infty,E}}\right)_{w=2}$ on the right with $\exp^*_\omega$, then we are done. Indeed, for any $\mathfrak{X}\in\mathbf{X}_{\mathbb{T}_{\infty,E}}(\mathbf{F}\otimes\mathbb{Z}_p)$

$$\begin{aligned}\mathrm{EXP}^*_{\mathbb{T}_{\infty,E}}(\mathfrak{X})_{w=2} &= \exp^*_{V^*_{\mathbf{f}_2}}(\mathfrak{X}_{w=2})\cup(1-\varphi)^{-1}.(t^{-1}.\mathbf{v}_{\mathbf{I},\lambda_E})_{w=2}\\ &= \frac{\mathcal{C}_{\mathbf{f}_2}}{(1-p^{-1}a_p(\mathbf{f}_2))}\times\exp^*_{V^*_{\mathbf{f}_2}}(\mathfrak{X}_{w=2})\cup(t^{-1}.\mathbf{v}_E)\end{aligned}$$

N.B. here $\mathbf{v}_E\in\mathbf{D}_{\mathrm{cris},\mathbb{Q}_p}(V_{\mathbf{f}_2})^{\varphi=U_p}$ satisfies $\mathbf{f}_2^*\cup\mathbf{v}_E = 1$ and $\left(\mathbf{v}_{\mathbf{I},\lambda_E}\right)_{w=2} = \mathcal{C}_{\mathbf{f}_2}\mathbf{v}_E$. However $\exp^*_{V^*_{\mathbf{f}_2}}(-)\cup(t^{-1}.\mathbf{v}_E)$ coincides with $\exp^*_\omega(-)$, up to a non-zero scalar.

Since the quotient $\dfrac{H^1\left(\mathbf{F}\otimes\mathbb{Z}_p,\ \mathbb{T}_{\infty,E}\right)_{\Gamma^{\rm wt}}}{\mathbf{X}_{\mathbb{T}_{\infty,E}}\left(\mathbf{F}\otimes\mathbb{Z}_p\right)_{\Gamma^{\rm wt}}}$ is free of any p^∞-torsion, the restriction of $\exp^*_\omega$ to $H^1\left(\mathbf{F}\otimes\mathbb{Z}_p,\ \mathbb{T}_{\infty,E}\right)_{\Gamma^{\rm wt}}$ must share the same kernel as $\mathrm{EXP}^*_{\mathbb{T}_{\infty,E}}(-)_{w=2}$.

The proof of part (ii) is completed.

□

We should bear in mind our original task was to compute the size of $\mathrm{Ker}(\delta_{\infty,E,\mathfrak{p}})$. By local duality, this corresponds to the size of the cokernel of

$$\widehat{\delta}_{\infty,E,\mathfrak{p}} : \mathbf{X}_{\mathbb{T}_{\infty,E}}(\mathbf{F}_{\mathfrak{p}})_{\Gamma^{\mathrm{wt}}} \hookrightarrow H^1_g(\mathbf{F}_{\mathfrak{p}}, \mathrm{Ta}_p(\mathcal{C}^{\min})) \cong \mathcal{C}^{\min}(\mathbf{F}_{\mathfrak{p}})\widehat{\otimes}\mathbb{Z}_p$$

i.e. to the index of the Γ^{wt}-coinvariants of $\mathbf{X}_{\mathbb{T}_{\infty,E}}$ inside the local H^1_g for $\mathcal{C}^{\min}$. Applying the Snake lemma to the commutative diagram

$$\begin{array}{ccccccc}
0 \longrightarrow & \mathbf{X}_{\mathbb{T}_{\infty,E}}(\mathbf{F}\otimes\mathbb{Z}_p)\otimes_{\Lambda^{\mathrm{wt}}}\mathbb{Z}_p & \longrightarrow & H^1\Big(\mathbf{F}\otimes\mathbb{Z}_p,\ \mathbb{T}_{\infty,E}\Big)\otimes_{\Lambda^{\mathrm{wt}}}\mathbb{Z}_p & \stackrel{\exp^*_\omega}{\longrightarrow} & \mathbf{F}\otimes\mathbb{Z}_p \\
 & \downarrow\mathrm{incl} & & \downarrow\mathrm{incl} & & \| \\
0 \longrightarrow & H^1_g\Big(\mathbf{F}\otimes\mathbb{Z}_p,\ \mathrm{Ta}_p(\mathcal{C}^{\min})\Big) & \longrightarrow & H^1\Big(\mathbf{F}\otimes\mathbb{Z}_p,\ \mathrm{Ta}_p(\mathcal{C}^{\min})\Big) & \stackrel{\exp^*_\omega}{\longrightarrow} & \mathbf{F}\otimes\mathbb{Z}_p
\end{array}$$

via the usual nonsense, one finds that

$$\#\mathrm{Ker}\Big(\oplus_{\mathfrak{p}|p}\,\delta_{\infty,E,\mathfrak{p}}\Big) = \Big[H^1_g\Big(\mathbf{F}\otimes\mathbb{Z}_p,\ \mathrm{Ta}_p(\mathcal{C}^{\min})\Big) : \mathbf{X}_{\mathbb{T}_{\infty,E}}(\mathbf{F}\otimes\mathbb{Z}_p)\otimes_{\Lambda^{\mathrm{wt}}}\mathbb{Z}_p\Big]$$

$$= \frac{\Big[H^1\Big(\mathbf{F}\otimes\mathbb{Z}_p,\ \mathrm{Ta}_p(\mathcal{C}^{\min})\Big) : H^1\Big(\mathbf{F}\otimes\mathbb{Z}_p,\ \mathbb{T}_{\infty,E}\Big)\otimes_{\Lambda^{\mathrm{wt}}}\mathbb{Z}_p\Big]}{\Big[\exp^*_\omega\Big(H^1\Big(\mathbf{F}\otimes\mathbb{Z}_p,\ \mathrm{Ta}_p(\mathcal{C}^{\min})\Big)\Big) : \exp^*_\omega\Big(H^1\Big(\mathbf{F}\otimes\mathbb{Z}_p,\ \mathbb{T}_{\infty,E}\Big)\otimes_{\Lambda^{\mathrm{wt}}}\mathbb{Z}_p\Big)\Big]}.$$

Significantly the latter quotient no longer involves either H^1_g nor $\mathbf{X}_{\mathbb{T}_{\infty,E}}$.

Remark: The denominator above is equal to $\prod_{\mathfrak{p}|p}\mathrm{covol}^{(2)}_{\mathbb{T}_{\infty,E}}(\mathbf{F}_{\mathfrak{p}})$ by definition. On the other hand, using the short exact sequence

$$\begin{array}{c}
\hfill \Big(H^0(\mathbf{F}_{\mathfrak{p}}, A_{\mathbb{T}_{\infty,E}})_{\Gamma^{\mathrm{wt}}}\Big)^\vee \\
\hfill \uparrow\cong \\
0 \to H^1\Big(\mathbf{F}_{\mathfrak{p}}, \mathbb{T}_{\infty,E}\Big)\otimes_{\Lambda^{\mathrm{wt}}}\mathbb{Z}_p \to H^1\Big(\mathbf{F}_{\mathfrak{p}}, \mathrm{Ta}_p(\mathcal{C}^{\min})\Big) \to H^2\Big(\mathbf{F}_{\mathfrak{p}}, \mathbb{T}_{\infty,E}\Big)^{\Gamma^{\mathrm{wt}}} \to 0
\end{array}$$

we immediately deduce the numerator is the product $\prod_{\mathfrak{p}|p}\#H^0(\mathbf{F}_{\mathfrak{p}}, A_{\mathbb{T}_{\infty,E}})_{\Gamma^{\mathrm{wt}}}$.

As a corollary, one obtains the closed formula

$$\#\mathrm{Ker}\Big(\oplus_{\mathfrak{p}|p}\,\delta_{\infty,E,\mathfrak{p}}\Big) = \prod_{\mathfrak{p}|p}\frac{\#H^0\Big(\mathbf{F}_{\mathfrak{p}}, A_{\mathbb{T}_{\infty,E}}\Big)_{\Gamma^{\mathrm{wt}}}}{\mathrm{covol}^{(2)}_{\mathbb{T}_{\infty,E}}(\mathbf{F}_{\mathfrak{p}})}.$$

If we elect to write out Definition 8.8 in full, then at weight two

$$\mathrm{Tam}^{(2)}_{\mathbf{F}_{\mathfrak{p}}}(\mathbb{T}_{\infty,E}) = \frac{\Big[H^1_f\Big(\mathbf{F}_{\mathfrak{p}}, \mathbb{T}_{\infty,E}\otimes_{\Lambda^{\mathrm{wt}}}\mathbb{Z}_p\Big) : H^1_f\Big(\mathbf{F}_{\mathfrak{p}}, \mathrm{F}^+\mathbb{T}_{\infty,E}\otimes_{\Lambda^{\mathrm{wt}}}\mathbb{Z}_p\Big)\Big]}{\#H^0\Big(\mathbf{F}_{\mathfrak{p}}, A_{\mathbb{T}_{\infty,E}}\Big)_{\Gamma^{\mathrm{wt}}}}. \qquad (*)$$

Clearly Proposition 8.9 will follow, and thence the triviality of these $\mathrm{Tam}^{(2)}_{\mathbf{F}_{\mathfrak{p}}}(-)$'s, provided we can establish:

Proposition 8.9'. *(a) If E has good ordinary reduction over $\mathbb{Q}_p$, then*

$$\#H^0\Big(\mathbf{F}_\mathfrak{p}, A_{\mathbb{T}_{\infty,E}}\Big)_{\Gamma^{\mathrm{wt}}} = \#\widetilde{E}(f_\mathfrak{p})[p^\infty] \times \Big|\big[\mathcal{C}^{\min}(\mathbf{F}_\mathfrak{p}) : \mathcal{C}_0^{\min}(\mathbf{F}_\mathfrak{p})\big]\Big|_p^{-1} \quad \textit{for all } \mathfrak{p}|p;$$

(b) If E has bad multiplicative reduction over $\mathbb{Q}_p$, then

$$\#H^0\Big(\mathbf{F}_\mathfrak{p}, A_{\mathbb{T}_{\infty,E}}\Big)_{\Gamma^{\mathrm{wt}}} = \Big|\big[\mathcal{C}^{\min}(\mathbf{F}_\mathfrak{p}) : \mathcal{C}_0^{\min}(\mathbf{F}_\mathfrak{p})\big]\Big|_p^{-1} \quad \textit{for all } \mathfrak{p}|p;$$

(c) In either case it equals $\Big[H^1_f\Big(\mathbf{F}_\mathfrak{p}, \mathbb{T}_{\infty,E}\otimes_{\Lambda^{\mathrm{wt}}}\mathbb{Z}_p\Big) : H^1_f\Big(\mathbf{F}_\mathfrak{p}, \mathrm{F}^+\mathbb{T}_{\infty,E}\otimes_{\Lambda^{\mathrm{wt}}}\mathbb{Z}_p\Big)\Big]$.

Proof: We start by calculating these coinvariant terms:

$$\begin{aligned}\#H^0\Big(\mathbf{F}_\mathfrak{p}, A_{\mathbb{T}_{\infty,E}}\Big)_{\Gamma^{\mathrm{wt}}} &= \#H^0\Big(\mathbf{F}_\mathfrak{p}, A_{\mathbb{T}_{\infty,E}}\Big)^{\Gamma^{\mathrm{wt}}} \quad \text{as } \Gamma^{\mathrm{wt}} \text{ is pro-cyclic}\\ &= \#\mathrm{Hom}_{\mathrm{cont}}\Big(\mathbb{T}_{\infty,E}\otimes_{\Lambda^{\mathrm{wt}}}\mathbb{Z}_p, \mu_{p^\infty}\Big)^{G_{\mathbf{F}_\mathfrak{p}}} = \#\mathcal{C}^{\min}(\mathbf{F}_\mathfrak{p})[p^\infty].\end{aligned}$$

It makes sense to consider the good ordinary and bad multiplicative cases separately. Recall that the index $\Big[H^1_f\Big(\mathbf{F}_\mathfrak{p}, \mathbb{T}_{\infty,E}\otimes_{\Lambda^{\mathrm{wt}}}\mathbb{Z}_p\Big) : H^1_f\Big(\mathbf{F}_\mathfrak{p}, \mathrm{F}^+\mathbb{T}_{\infty,E}\otimes_{\Lambda^{\mathrm{wt}}}\mathbb{Z}_p\Big)\Big]$ is the numerator of (*).

First case: E and $\mathcal{C}^{\min}$ have good ordinary reduction at p.
Write $\widehat{\mathcal{C}}^{\min}_{/\mathcal{O}_{\mathbf{F}_\mathfrak{p}}}$ for the height one formal group of $\mathcal{C}^{\min}$ over the ring of integers of $\mathbf{F}_\mathfrak{p}$. As $\mathrm{Gal}(\overline{\mathbf{F}_\mathfrak{p}}/\mathbf{F}_\mathfrak{p})$-modules, $\mathrm{F}^+\mathbb{T}_{\infty,E}\otimes_{\Lambda^{\mathrm{wt}}}\mathbb{Z}_p \cong \mathrm{Ta}_p\Big(\widehat{\mathcal{C}}^{\min}_{/\mathcal{O}_{\mathbf{F}_\mathfrak{p}}}\Big)$ on a p-integral level. Consequently the numerator in (*) transforms into

$$\Big[H^1_f\Big(\mathbf{F}_\mathfrak{p}, \mathrm{Ta}_p(\mathcal{C}^{\min})\Big) : H^1_f\Big(\mathbf{F}_\mathfrak{p}, \mathrm{Ta}_p\Big(\widehat{\mathcal{C}}^{\min}_{/\mathcal{O}_{\mathbf{F}_\mathfrak{p}}}\Big)\Big)\Big]$$

which is precisely the index $\Big[\mathcal{C}^{\min}(\mathbf{F}_\mathfrak{p})\widehat{\otimes}\mathbb{Z}_p : \widehat{\mathcal{C}}^{\min}_{/\mathcal{O}_{\mathbf{F}_\mathfrak{p}}}(\mathfrak{p})\widehat{\otimes}\mathbb{Z}_p\Big]$.

Because $p \geq 3$ does not ramify in the field $\mathbf{F}$, the formal group has no p-torsion. It follows that the quotient $\frac{\mathcal{C}^{\min}(\mathbf{F}_\mathfrak{p})\widehat{\otimes}\mathbb{Z}_p}{\widehat{\mathcal{C}}^{\min}_{/\mathcal{O}_{\mathbf{F}_\mathfrak{p}}}(\mathfrak{p})\widehat{\otimes}\mathbb{Z}_p} \cong \widetilde{\mathcal{C}^{\min}}(f_\mathfrak{p})\widehat{\otimes}\mathbb{Z}_p$, where $\widetilde{\mathcal{C}^{\min}}$ denotes the reduction of the optimal elliptic curve $\mathcal{C}^{\min}$ over the residue field $f_\mathfrak{p} = \mathcal{O}_{\mathbf{F}_\mathfrak{p}}/\mathfrak{p}$. Moreover, there are equalities

$$\#\widetilde{\mathcal{C}^{\min}}(f_\mathfrak{p})\widehat{\otimes}\mathbb{Z}_p = \#\widetilde{\mathcal{C}^{\min}}(f_\mathfrak{p})[p^\infty] = \#\widetilde{E}(f_\mathfrak{p})[p^\infty]$$

due to the fact that $\widetilde{\mathcal{C}^{\min}}$ and $\widetilde{E}$ are $\mathbb{F}_p$-isogenous elliptic curves, hence they have the same number of points over the residue field extension. As had been predicted

$$\begin{aligned}\text{the numerator of (*)} &= \#\widetilde{E}(f_\mathfrak{p})[p^\infty] \times \Big|\big[\mathcal{C}^{\min}(\mathbf{F}_\mathfrak{p}) : \mathcal{C}_0^{\min}(\mathbf{F}_\mathfrak{p})\big]\Big|_p^{-1}\\ &= \text{the denominator of (*)}\end{aligned}$$

since $\big[\mathcal{C}^{\min}(\mathbf{F}_\mathfrak{p}) : \mathcal{C}_0^{\min}(\mathbf{F}_\mathfrak{p})\big]$ is trivial in the good ordinary case.

Second case: E and $\mathcal{C}^{\min}$ have bad multiplicative reduction at p.
To make the calculation work, we'll examine the connected piece of $\mathcal{C}^{\min}$ over $\mathbf{F}_{\mathfrak{p}}$. From Chapter I, there is the well-known short exact sequence

$$0 \longrightarrow \widehat{\mathcal{C}}^{\min}_{/\mathcal{O}_{\mathbf{F}_{\mathfrak{p}}}}(\mathfrak{p}) \longrightarrow \mathcal{C}^{\min}_0(\mathbf{F}_{\mathfrak{p}}) \longrightarrow \mathbb{G}_{\mathrm{mult}}(f_{\mathfrak{p}}) \longrightarrow 0.$$

Now $\#\mathbb{G}_{\mathrm{mult}}(f_{\mathfrak{p}}) = \#f_{\mathfrak{p}} - 1$ is coprime to p, hence $\mathcal{C}^{\min}_0(\mathbf{F}_{\mathfrak{p}})/p^n \cong \widehat{\mathcal{C}}^{\min}_{/\mathcal{O}_{\mathbf{F}_{\mathfrak{p}}}}(\mathfrak{p})/p^n$ and similarly $\mathcal{C}^{\min}_0(\mathbf{F}_{\mathfrak{p}})[p^n] \cong \widehat{\mathcal{C}}^{\min}_{/\mathcal{O}_{\mathbf{F}_{\mathfrak{p}}}}(\mathfrak{p})[p^n] = \{O_{\mathcal{C}^{\min}}\}$.

Arguing in an identical fashion to the situation of good ordinary reduction at p, again the numerator in **(*)** coincides with

$$\left[\mathcal{C}^{\min}(\mathbf{F}_{\mathfrak{p}})\widehat{\otimes}\mathbb{Z}_p : \widehat{\mathcal{C}}^{\min}_{/\mathcal{O}_{\mathbf{F}_{\mathfrak{p}}}}(\mathfrak{p})\widehat{\otimes}\mathbb{Z}_p\right]$$

which is none other than the stable index $\left[\mathcal{C}^{\min}(\mathbf{F}_{\mathfrak{p}})/p^n : \widehat{\mathcal{C}}^{\min}_{/\mathcal{O}_{\mathbf{F}_{\mathfrak{p}}}}(\mathfrak{p})/p^n\right]$ for $n \gg 0$. But we already know $\widehat{\mathcal{C}}^{\min}_{/\mathcal{O}_{\mathbf{F}_{\mathfrak{p}}}}(\mathfrak{p})/p^n \cong \mathcal{C}^{\min}_0(\mathbf{F}_{\mathfrak{p}})/p^n$, in which case

$$\text{the numerator of } \textbf{(*)} \quad = \quad \left|\left[\mathcal{C}^{\min}(\mathbf{F}_{\mathfrak{p}}) : \mathcal{C}^{\min}_0(\mathbf{F}_{\mathfrak{p}})\right]\right|_p^{-1}.$$

Let $\Phi^{\min}_{\mathfrak{p}}$ denote the group of connected components of $\mathcal{C}^{\min}$ over the local field $\mathbf{F}_{\mathfrak{p}}$. If we now consider the effect of multiplication by p^n on the tautological short exact sequence $0 \to \mathcal{C}^{\min}_0(\mathbf{F}_{\mathfrak{p}}) \to \mathcal{C}^{\min}(\mathbf{F}_{\mathfrak{p}}) \to \Phi^{\min}_{\mathfrak{p}} \to 0$, we obtain a long exact sequence

$$\begin{aligned} 0 \longrightarrow \mathcal{C}^{\min}_0(\mathbf{F}_{\mathfrak{p}})[p^n] &\longrightarrow \mathcal{C}^{\min}(\mathbf{F}_{\mathfrak{p}})[p^n] \longrightarrow \Phi^{\min}_{\mathfrak{p}}[p^n] \\ \longrightarrow \mathcal{C}^{\min}_0(\mathbf{F}_{\mathfrak{p}})/p^n &\longrightarrow \mathcal{C}^{\min}(\mathbf{F}_{\mathfrak{p}})/p^n \longrightarrow \Phi^{\min}_{\mathfrak{p}}/p^n \longrightarrow 0. \end{aligned}$$

Because $\Phi^{\min}_{\mathfrak{p}}$ is known to be a finite abelian group, for all integers $n \geq \mathrm{ord}_p\left(\#\Phi^{\min}_{\mathfrak{p}}\right)$ this becomes

$$\begin{aligned} 0 = \mathcal{C}^{\min}_0(\mathbf{F}_{\mathfrak{p}})[p^\infty] &\longrightarrow \mathcal{C}^{\min}(\mathbf{F}_{\mathfrak{p}})[p^\infty] \longrightarrow \Phi^{\min}_{\mathfrak{p}}[p^\infty] \\ \longrightarrow \mathcal{C}^{\min}_0(\mathbf{F}_{\mathfrak{p}})/p^n &\longrightarrow \mathcal{C}^{\min}(\mathbf{F}_{\mathfrak{p}})/p^n \longrightarrow \Phi^{\min}_{\mathfrak{p}}/\mathcal{A}'_{\mathfrak{p}} \longrightarrow 0 \end{aligned}$$

where $\mathcal{A}'_{\mathfrak{p}}$ denotes the non-p-primary subgroup of the components $\Phi^{\min}_{\mathfrak{p}}$.

Computing orders along this sequence, we may conclude that $\#\mathcal{C}^{\min}(\mathbf{F}_{\mathfrak{p}})[p^\infty]$ coincides with $\#\Phi^{\min}_{\mathfrak{p}}[p^\infty] = \left|\left[\mathcal{C}^{\min}(\mathbf{F}_{\mathfrak{p}}) : \mathcal{C}^{\min}_0(\mathbf{F}_{\mathfrak{p}})\right]\right|_p^{-1}$, i.e.

$$\text{the denominator of } \textbf{(*)} \quad = \quad \left|\left[\mathcal{C}^{\min}(\mathbf{F}_{\mathfrak{p}}) : \mathcal{C}^{\min}_0(\mathbf{F}_{\mathfrak{p}})\right]\right|_p^{-1}.$$

This is strong enough to imply the triviality of $\mathrm{Tam}^{(2)}_{\mathbf{F}_{\mathfrak{p}}}(\mathbb{T}_{\infty,E})$ at primes above p.

The proof of Propositions 8.9' and 8.9 is finished.

□

8.4 The Tamagawa factors above p (the half-twisted case)

Now that we have computed the local kernels above p in the vertical situation, we switch to the line $s = k/2$ instead. Throughout this section we shall consider $\mathbf{T} = \mathbb{T}_{\infty,E} \otimes \Psi^{-1/2}$ which was also abbreviated by $\mathbb{T}_{\infty,E}^{-1/2}$ earlier on in the text. The main objective is to calculate the size of

$$\mathrm{Ker}\left(\delta_{\infty,E,\mathfrak{p}}^{-1/2} : \frac{H^1(\mathbf{F}_\mathfrak{p}, \mathcal{C}^{\min}[p^\infty])}{\mathcal{W}_{\mathcal{C}^{\min},+}(\mathbf{F}_\mathfrak{p})^\perp} \longrightarrow \left(\frac{H^1\left(\mathbf{F}_\mathfrak{p}, A_{\mathbb{T}_{\infty,E}}^{-1/2}\right)}{H^1_+(\mathbf{F}_\mathfrak{p})^\perp}\right)^{\Gamma^{\mathrm{wt}}}\right)$$

where as before $A_{\mathbb{T}_{\infty,E}}^{-1/2}$ is the continuous χ_{cy}-twisted dual $\mathrm{Hom}_{\mathrm{cont}}\left(\mathbb{T}_{\infty,E}^{-1/2}, \mu_{p^\infty}\right)$. The local conditions $\mathcal{W}_{\mathcal{C}^{\min},+}$ and H^1_+ were themselves given in §7.1-§7.2.

Notation: For each prime $\mathfrak{p}|p$, let us employ the prefix

$$\mathcal{N}_\infty^{\mathrm{univ}} H^1(\mathbf{F}_\mathfrak{p}, -) \;:=\; \bigcap_{n\geq 0} \mathrm{cores}_{\mathbf{F}_\mathfrak{p}(\mu_{p^n})/\mathbf{F}_\mathfrak{p}} H^1\left(\mathbf{F}_\mathfrak{p}(\mu_{p^n}), -\right)$$

to indicate the group of universal norms over the cyclotomic $\mathbb{Z}_p^\times$-extension.

In the half-twisted scenario, the universal norms play a key rôle in the calculation of the local kernels. First we must redefine the covolume terms mentioned in §8.3; precisely, the label $\mathrm{covol}^{(2)}_{\mathbb{T}_{\infty,E}^{-1/2}}(\mathbf{F}_\mathfrak{p})$ denotes the regularised index

$$\frac{\left[\exp_\omega^*\left(\mathcal{N}_\infty^{\mathrm{univ}} H^1(\mathbf{F}_\mathfrak{p}, \mathrm{Ta}_p(\mathcal{C}^{\min}))\right) : \exp_\omega^*\left(\mathcal{N}_\infty^{\mathrm{univ}} H^1(\mathbf{F}_\mathfrak{p}, \mathbb{T}_{\infty,E}^{-1/2})_{\Gamma^{\mathrm{wt}}}\right)\right]}{\left[\mathcal{N}_\infty^{\mathrm{univ}} H^1(\mathbf{F}_\mathfrak{p}, \mathrm{Ta}_p(\mathcal{C}^{\min})) : \mathcal{N}_\infty^{\mathrm{univ}} H^1(\mathbf{F}_\mathfrak{p}, \mathbb{T}_{\infty,E}^{-1/2})_{\Gamma^{\mathrm{wt}}}\right]}$$

at all primes $\mathfrak{p}$ lying above p. This formula is very ugly, but thankfully temporary!

Definition 8.12. *For a positive weight $k \in \mathbb{U}_E$, one defines* $\mathrm{Tam}^{(k)}_{\mathbf{F}_\mathfrak{p}}\left(\mathbb{T}_{\infty,E}^{-1/2}\right)$ *as*

$$\left[H^1\left(\mathbf{F}_\mathfrak{p}, \mathrm{F}^+\mathbb{T}_{\infty,E}^{-1/2}\right) \otimes_{\Lambda^{\mathrm{wt}},\mathfrak{g}_k} \mathbb{Z}_p : \mathcal{N}_\infty^{\mathrm{univ}} H^1\left(\mathbf{F}_\mathfrak{p}, \mathrm{F}^+\mathbb{T}_{\infty,E}^{-1/2}\right) \otimes_{\Lambda^{\mathrm{wt}},\mathfrak{g}_k} \mathbb{Z}_p\right]$$

which is called **the $\mathfrak{p}^{\mathrm{th}}$-Tamagawa number** *of $\mathbb{T}_{\infty,E}^{-1/2}$ at k.*

The following result is essentially the half-twisted analogue of 8.9(a) and 8.9(b).

Proposition 8.13. *(a) If E has good ordinary reduction over $\mathbb{Q}_p$, then*

$$\#\mathrm{Ker}\left(\oplus_{\mathfrak{p}|p}\, \delta_{\infty,E,\mathfrak{p}}^{-1/2}\right) = \prod_{\mathfrak{p}|p} \left(\#\widetilde{E}(f_\mathfrak{p})[p^\infty]\right)^2 \times \frac{\left|\left[\mathcal{C}^{\min}(\mathbf{F}_\mathfrak{p}) : \mathcal{C}_0^{\min}(\mathbf{F}_\mathfrak{p})\right]\right|_p^{-1}}{\mathrm{covol}^{(2)}_{\mathbb{T}_{\infty,E}^{-1/2}}(\mathbf{F}_\mathfrak{p}) \times \mathrm{Tam}^{(2)}_{\mathbf{F}_\mathfrak{p}}\left(\mathbb{T}_{\infty,E}^{-1/2}\right)};$$

(b) If E has non-split multiplicative reduction over $\mathbb{Q}_p$, then

$$\#\mathrm{Ker}\left(\oplus_{\mathfrak{p}|p}\, \delta_{\infty,E,\mathfrak{p}}^{-1/2}\right) = \prod_{\mathfrak{p}|p} \frac{\left|\left[\mathcal{C}^{\min}(\mathbf{F}_\mathfrak{p}) : \mathcal{C}_0^{\min}(\mathbf{F}_\mathfrak{p})\right]\right|_p^{-1}}{\mathrm{covol}^{(2)}_{\mathbb{T}_{\infty,E}^{-1/2}}(\mathbf{F}_\mathfrak{p}) \times \mathrm{Tam}^{(2)}_{\mathbf{F}_\mathfrak{p}}\left(\mathbb{T}_{\infty,E}^{-1/2}\right)}.$$

Remarks: (i) When the elliptic curve E has non-split multiplicative reduction, the $\mathrm{Tam}^{(2)}_{\mathbf{F}_{\mathfrak{p}}}(-)$'s are always trivial. However, if E has good ordinary reduction then they divide into the square of $\#\widetilde{E}(f_{\mathfrak{p}})[p^\infty]$, and presumably can be non-trivial.

(ii) The reader will notice we have carefully avoided the case of split multiplicative reduction in our proposition. If Conjecture 7.11 is to be believed, in this case the size of the kernel of each $\delta^{-1/2}_{\infty,E,\mathfrak{p}}$ will be infinite, and the calculation stalls abruptly.

(iii) It is somewhat annoying that $\mathrm{covol}^{(2)}_{\mathbb{T}^{-1/2}_{\infty,E}}(\mathbf{F}_{\mathfrak{p}})$ incorporates the universal norms, rather than the whole local cohomology. Unfortunately this is something we shall just have to live with – there is no decent theory of the dual exponential map in the half-twisted case, which doesn't first rely on taking a cyclotomic deformation.

Sketch of the Proof of Proposition 8.13.
The argument is very similar to §8.3; we shall only briefly outline the main ideas. The missing ingredient is that there is no analogue of the homomorphism $\mathrm{EXP}^*_{\mathbb{T}_{\infty,E}}$ for the representation $\mathbb{T}_{\infty,E}\otimes\Psi^{-1/2}$, which is a major obstruction to working out the covolumes.

To rectify matters, we try and construct a large enough portion of $\mathrm{EXP}^*_{\mathbb{T}^{-1/2}_{\infty,E}}(?)$ using cyclotomic methods from §6.3, which will allow us to make the measurement. Let's first consider our interpolating homomorphism

$$PR_{\infty,\mathbf{F}}:\overline{\mathcal{MS}}^{\mathrm{ord}}\otimes_{\mathbf{h}^{\mathrm{ord}}}\varprojlim_n H^1\Big(\mathbf{F}(\mu_{p^n})\otimes\mathbb{Z}_p,\ \mathbb{T}_{\infty,E}\Big)\ \longrightarrow\ \mathbb{Z}_p\langle\langle w\rangle\rangle[[G_{\infty,\mathbf{F}}]]\otimes_{\mathbf{I}}\mathcal{L}'.$$

We drop the '$\overline{\mathcal{MS}}^{\mathrm{ord}}\otimes_{\mathbf{h}^{\mathrm{ord}}}-$' again as we don't care about its precise normalisation. Without loss of generality, we may also assume that there are no poles in the weight variable by clearing denominators in $\mathcal{L}'$.

The kernel will properly contain the submodule $\varprojlim_n H^1\Big(\mathbf{F}(\mu_{p^n})\otimes\mathbb{Z}_p,\ \mathrm{F}^+\mathbb{T}_{\infty,E}\Big)$; moreover, its $\Lambda^{\mathrm{wt}}[[G_{\infty,\mathbf{F}}]]$-saturation gives the whole kernel because the image of $PR_{\infty,\mathbf{F}}$ has $\Lambda^{\mathrm{wt}}[[G_\infty]]$-rank exactly equal to $[\mathbf{F}:\mathbb{Q}]$, courtesy of Theorem 6.11. One can twist these objects by the half-integral character $\langle\chi_{\mathrm{cy}}^{-1/2}\circ\mathrm{Res}_{\mathbb{Q}^{\mathrm{cyc}}}(-)\rangle$; over the cyclotomic tower, the twist absorbs into the coefficients $\mathbb{T}_{\infty,E}$ and $\mathrm{F}^+\mathbb{T}_{\infty,E}$. We therefore obtain the left-exact sequence

$$\begin{aligned}0\ \longrightarrow\ &\text{the saturation of }\varprojlim_n H^1\Big(\mathbf{F}(\mu_{p^n})\otimes\mathbb{Z}_p,\ \mathrm{F}^+\mathbb{T}^{-1/2}_{\infty,E}\Big)\\ &\longrightarrow\ \varprojlim_n H^1\Big(\mathbf{F}(\mu_{p^n})\otimes\mathbb{Z}_p,\ \mathbb{T}^{-1/2}_{\infty,E}\Big)\ \overset{PR_{\infty,\mathbf{F}}\otimes\mathrm{id}}{\longrightarrow}\ \mathbb{Z}_p\langle\langle w\rangle\rangle[[G_{\infty,\mathbf{F}}]]\otimes\Psi^{-1/2}.\end{aligned}$$

Taking its G_∞-coinvariants, the above degenerates into

$$\begin{aligned}0\ \longrightarrow\ \mathcal{N}^{\mathrm{univ}}_\infty H^1_+\Big(\mathbf{F}\otimes\mathbb{Z}_p,\ \mathbb{T}^{-1/2}_{\infty,E}\Big)\ &\longrightarrow\ \mathcal{N}^{\mathrm{univ}}_\infty H^1\Big(\mathbf{F}\otimes\mathbb{Z}_p,\ \mathbb{T}^{-1/2}_{\infty,E}\Big) \qquad (**)\\ &\overset{PR^\dagger_{\infty,\mathbf{F}}}{\longrightarrow}\ \mathbb{Z}_p\langle\langle w\rangle\rangle[\mathrm{Gal}(\mathbf{F}/\mathbb{Q})]\otimes\Big(\Psi^{-1/2}\Big|_{\mathrm{Gal}(\mathbf{F}/\mathbb{Q})}\Big)\end{aligned}$$

where $PR^\dagger_{\infty,\mathbf{F}}$ denotes the map '$PR_{\infty,\mathbf{F}}\otimes\mathrm{id}$' modulo $\mathcal{J}_{G_\infty}$, the augmentation ideal (in fact $\langle\chi_{\mathrm{cy}}^{-1/2}\circ\mathrm{Res}_{\mathbb{Q}^{\mathrm{cyc}}}(-)\rangle$ restricted to $\mathrm{Gal}(\mathbf{F}/\mathbb{Q})$ is just the trivial character).

Remark: We consider the number field $\mathbf{F}$ as a finite-dimensional $\mathbb{Q}$-vector space. For each choice of basis $\underline{\mathbf{e}} = \{\mathbf{e}_\psi\}_{\psi\in X_{\mathbf{F}}}$ of the integers $\mathcal{O}_{\mathbf{F}}$ with $\sigma(\mathbf{e}_\psi) = \psi(\sigma)\mathbf{e}_\psi$, one naturally obtains a map

$$\mathbb{Z}_p\langle\langle w\rangle\rangle[\mathrm{Gal}(\mathbf{F}/\mathbb{Q})] \stackrel{\mathrm{Tw}_{\underline{\mathbf{e}}}}{\longrightarrow} \mathbf{F}\otimes\mathbb{Z}_p\langle\langle w\rangle\rangle, \qquad \mathcal{G} \mapsto \mathrm{Tw}_{\underline{\mathbf{e}}}(\mathcal{G}) := \sum_{\psi\in X_{\mathbf{F}}} \psi(\mathcal{G}).\mathbf{e}_\psi.$$

The composition $\mathrm{Tw}_{\underline{\mathbf{e}}} \circ PR^{\dagger}_{\infty,\mathbf{F}} : \mathcal{N}^{\mathrm{univ}}_{\infty} H^1\Big(\mathbf{F}\otimes\mathbb{Z}_p, \mathbb{T}^{-1/2}_{\infty,E}\Big) \longrightarrow \mathbf{F}\otimes\mathbb{Z}_p\langle\langle w\rangle\rangle$ will now play an identical rôle to that of $\mathrm{EXP}^{*}_{\mathbb{T}_{\infty,E}}$ in the previous section.

Making a suitable choice of $\{\mathbf{e}_\psi\}_{\psi\in X_{\mathbf{F}}}$, the interpolation formulae in Corollary 6.6 imply at the bottom layer

$$\Big(\mathrm{Tw}_{\underline{\mathbf{e}}} \circ PR^{\dagger}_{\infty,\mathbf{F}}\Big)_{w=2} : \mathcal{N}^{\mathrm{univ}}_{\infty} H^1\Big(\mathbf{F}\otimes\mathbb{Z}_p,\ \mathbb{T}^{-1/2}_{\infty,E}\Big) \otimes_{\Lambda^{\mathrm{wt}}} \mathbb{Z}_p \longrightarrow \mathbf{F}\otimes\mathbb{Z}_p$$

coincides with the dual exponential map $\exp^{*}_{\omega}$, up to some non-zero scalar $\mathfrak{u} \in \mathbb{Q}_p^{\times}$. As a corollary, the sequence **(**)** at weight two specialises into

$$\cdots \longrightarrow \left(\frac{\mathcal{N}^{\mathrm{univ}}_{\infty} H^1\Big(\mathbf{F}\otimes\mathbb{Z}_p,\ \mathbb{T}^{-1/2}_{\infty,E}\Big)}{\mathcal{N}^{\mathrm{univ}}_{\infty} H^1_{+}\Big(\mathbf{F}\otimes\mathbb{Z}_p,\ \mathbb{T}^{-1/2}_{\infty,E}\Big)}\right)^{\Gamma^{\mathrm{wt}}} \stackrel{\partial}{\longrightarrow} \mathcal{N}^{\mathrm{univ}}_{\infty} H^1_{+}\Big(\mathbf{F}\otimes\mathbb{Z}_p,\ \mathbb{T}^{-1/2}_{\infty,E}\Big)_{\Gamma^{\mathrm{wt}}}$$
$$\longrightarrow \mathcal{N}^{\mathrm{univ}}_{\infty} H^1\Big(\mathbf{F}\otimes\mathbb{Z}_p,\ \mathbb{T}^{-1/2}_{\infty,E}\Big)_{\Gamma^{\mathrm{wt}}} \stackrel{\mathfrak{u}\times\exp^{*}_{\omega}}{\longrightarrow} \mathbf{F}\otimes\mathbb{Z}_p.$$

The quotient group on the left is $\mathbb{Z}_p$-torsion free, and a simple calculation of ranks shows that the image of ∂ must be trivial.

On the other hand, the kernel of $\mathfrak{u}\times\exp^{*}_{\omega} : H^1(\mathbf{F}\otimes\mathbb{Z}_p, \mathrm{Ta}_p(\mathcal{C}^{\min})) \longrightarrow \mathbf{F}\otimes\mathbb{Z}_p$ is the p-saturation of $H^1(\mathbf{F}\otimes\mathbb{Z}_p, \mathrm{Fil}^{+}\mathrm{Ta}_p(\mathcal{C}^{\min}))$, in which case

$$0 \longrightarrow \mathcal{W}_{\mathcal{C}^{\min},+}(\mathbf{F}\otimes\mathbb{Z}_p) \longrightarrow H^1\Big(\mathbf{F}\otimes\mathbb{Z}_p, \mathrm{Ta}_p(\mathcal{C}^{\min})\Big) \stackrel{\mathfrak{u}\times\exp^{*}_{\omega}}{\longrightarrow} \mathbf{F}\otimes\mathbb{Z}_p$$

is left-exact. Further, the sequence remains exact if one takes universal norms of the first two groups on the left-hand side.

Remark: As a direct consequence, we obtain a commutative diagram

$$\begin{array}{ccccc}
\mathcal{N}^{\mathrm{univ}}_{\infty} H^1_{+}\Big(\mathbf{F}\otimes\mathbb{Z}_p,\ \mathbb{T}^{-1/2}_{\infty,E}\Big)_{\Gamma^{\mathrm{wt}}} & \hookrightarrow & \mathcal{N}^{\mathrm{univ}}_{\infty} H^1\Big(\mathbf{F}\otimes\mathbb{Z}_p,\ \mathbb{T}^{-1/2}_{\infty,E}\Big)_{\Gamma^{\mathrm{wt}}} & \stackrel{\mathfrak{u}\times\exp^{*}_{\omega}}{\longrightarrow} & \mathbf{F}\otimes\mathbb{Z}_p \\
\downarrow{\scriptstyle \mathrm{incl}} & & \downarrow{\scriptstyle \mathrm{incl}} & & \| \\
\mathcal{N}^{\mathrm{univ}}_{\infty}\mathcal{W}_{\mathcal{C}^{\min},+}(\mathbf{F}\otimes\mathbb{Z}_p) & \hookrightarrow & \mathcal{N}^{\mathrm{univ}}_{\infty} H^1\Big(\mathbf{F}\otimes\mathbb{Z}_p, \mathrm{Ta}_p(\mathcal{C}^{\min})\Big) & \stackrel{\mathfrak{u}\times\exp^{*}_{\omega}}{\longrightarrow} & \mathbf{F}\otimes\mathbb{Z}_p
\end{array}$$

whose terms are (the duals of) the objects occurring in the kernel of $\oplus_{\mathfrak{p}|p}\ \delta^{-1/2}_{\infty,E,\mathfrak{p}}$, albeit their universal norms appear rather than the full cohomology.

Applying the Snake lemma to this diagram, one deduces that

$$\left[\mathcal{N}_\infty^{\mathrm{univ}}\mathcal{W}_{\mathcal{C}^{\min},+}(\mathbf{F}\otimes\mathbb{Z}_p):\mathcal{N}_\infty^{\mathrm{univ}}H^1_+\left(\mathbf{F}\otimes\mathbb{Z}_p,\ \mathbb{T}_{\infty,E}^{-1/2}\right)_{\Gamma^{\mathrm{wt}}}\right]\ =\ \mathrm{covol}^{(2)}_{\mathbb{T}_{\infty,E}^{-1/2}}(\mathbf{F}\otimes\mathbb{Z}_p)^{-1}.$$

Moreover the size of the kernel of $\oplus_{\mathfrak{p}|p}\,\delta_{\infty,E,\mathfrak{p}}^{-1/2}$ coincides with

$$\#\mathrm{Coker}\left(\oplus_{\mathfrak{p}|p}\,\widehat{\delta}_{\infty,E,\mathfrak{p}}^{-1/2}\right)\ =\ \left[\mathcal{W}_{\mathcal{C}^{\min},+}(\mathbf{F}\otimes\mathbb{Z}_p):H^1_+\left(\mathbf{F}\otimes\mathbb{Z}_p,\ \mathbb{T}_{\infty,E}^{-1/2}\right)_{\Gamma^{\mathrm{wt}}}\right]$$
$$=\ \frac{\left[\mathcal{W}_{\mathcal{C}^{\min},+}:\mathcal{N}_\infty^{\mathrm{univ}}\mathcal{W}_{\mathcal{C}^{\min},+}\right]\cdot\left[\mathcal{N}_\infty^{\mathrm{univ}}\mathcal{W}_{\mathcal{C}^{\min},+}:\mathcal{N}_\infty^{\mathrm{univ}}H^1_+\left(\mathbf{F}\otimes\mathbb{Z}_p,\ \mathbb{T}_{\infty,E}^{-1/2}\right)_{\Gamma^{\mathrm{wt}}}\right]}{\left[H^1_+\left(\mathbf{F}\otimes\mathbb{Z}_p,\ \mathbb{T}_{\infty,E}^{-1/2}\right)_{\Gamma^{\mathrm{wt}}}:\mathcal{N}_\infty^{\mathrm{univ}}H^1_+\left(\mathbf{F}\otimes\mathbb{Z}_p,\ \mathbb{T}_{\infty,E}^{-1/2}\right)_{\Gamma^{\mathrm{wt}}}\right]}$$

then plugging in the definition of Tamagawa numbers, the latter is none other than

$$\frac{\left[\mathcal{W}_{\mathcal{C}^{\min},+}(\mathbf{F}\otimes\mathbb{Z}_p):\mathcal{N}_\infty^{\mathrm{univ}}\mathcal{W}_{\mathcal{C}^{\min},+}(\mathbf{F}\otimes\mathbb{Z}_p)\right]}{\mathrm{covol}^{(2)}_{\mathbb{T}_{\infty,E}^{-1/2}}(\mathbf{F}\otimes\mathbb{Z}_p)\ \times\ \prod_{\mathfrak{p}|p}\mathrm{Tam}^{(2)}_{\mathbf{F}_\mathfrak{p}}\left(\mathbb{T}_{\infty,E}^{-1/2}\right)}.$$

Mazur showed in [Mz2] that the numerator is trivial if E has non-split multiplicative reduction, and equal to $\prod_{\mathfrak{p}|p}\left(\#\widetilde{\mathcal{C}^{\min}}(f_\mathfrak{p})[p^\infty]\right)^2$ if E has good ordinary reduction. Furthermore, in the good ordinary situation $\#\widetilde{\mathcal{C}^{\min}}(f_\mathfrak{p})$ actually equals $\#\widetilde{E}(f_\mathfrak{p})$, since the two reduced elliptic curves are $\mathbb{F}_p$-isogenous.

This completes our rather tiresome computation of $\#\mathrm{Ker}\left(\oplus_{\mathfrak{p}|p}\,\delta_{\infty,E,\mathfrak{p}}^{-1/2}\right)$.

8.5 Evaluating the covolumes

In this last section, we shall give an interpretation of the quantity $\mathrm{covol}^{(2)}_{\mathbb{T}_{\infty,E}}$ in terms of the $\mathbf{I}$-adic periods which manifest themselves in the two-variable L-function. This viewpoint arises naturally when one considers these covolume terms as being volumes on the dual space (the correction factors are then precisely the periods). In order to introduce the notion of a Λ^{wt}-adic volume, we shall first review the non-archimedean Haar measures used by Bloch and Kato in [BK, Sect 5].

Let $\mathcal{A}$ be an elliptic curve defined over $\mathbb{Q}$, and $\omega_\mathcal{A}\in\mathrm{Fil}^0 H^1_{\mathrm{dR}}$ a differential 1-form. For a local field K of residue characteristic $p>0$, the above fixes an isomorphism

$$\omega_\mathcal{A}:\det\nolimits_{\mathbb{Q}_p}\left(\mathbf{D}_{\mathrm{dR},K}(V)/\mathrm{Fil}^0\right)\ \longrightarrow\ \mathbb{Q}_p$$

where $V=\mathrm{Ta}_p(\mathcal{A})\otimes_{\mathbb{Z}_p}\mathbb{Q}_p$ denotes the p-adic $G_\mathbb{Q}$-representation associated to $\mathcal{A}$. This trivialisation of the determinant defines a Haar measure on the tangent space. On the other hand, the exponential map sends

$$\exp_V:\mathbf{D}_{\mathrm{dR},K}(V)/\mathrm{Fil}^0\ \longrightarrow\ H^1_e(K,V)\ \cong\ \mathcal{A}(K)\widehat{\otimes}\,\mathbb{Q}_p.$$

We write $\mathrm{vol}_{\mathcal{A},K}$ for the Haar measure induced on $H^1_e(K,V)$ via the exponential.

Remark: The product measure $\mu = \prod_{p\neq\infty} \mathrm{vol}_{\mathcal{A},\mathbb{Q}_p}(-)$ conjecturally converges, and should then yield

$$\mathrm{vol}_{\mathcal{A},\mathbb{R}}\Big(\mathcal{A}(\mathbb{R})/\mathcal{A}(\mathbb{Q})\Big) \times \mu\left(\prod_{p\neq\infty} \mathcal{A}(\mathbb{Q}_p)/\mathcal{A}(\mathbb{Q})\right) \stackrel{?}{=} \frac{\#H^0(\mathbb{Q},\ \mathcal{A}[\mathrm{tors}])}{\#\text{Ш}_{\mathbb{Q}}(\mathcal{A})_{/\mathrm{div}}}$$

provided that the Bloch-Kato conjecture over $\mathbb{Q}$ holds for the elliptic curve $\mathcal{A}$. Assuming finiteness of the Tate-Shafarevich group and other standard suppositions, this is equivalent to the formula predicted by Birch and Swinnerton-Dyer.

Question. *Is there some Λ^{wt}-adic analogue of the Haar measure?*

The answer is very straightforward: 'No'.

Whilst it is certainly possible to write down a weight deformation of the volume, we lose the necessary property of additivity thus we do not get a Haar measure. This will not be too serious a problem, provided one is ready and willing to sacrifice additivity (our 'measure' will at least be multiplicative along exact sequences).

Step 1: Background on the exponential.
Let $\mathbf{K}$ denote any number field – it need not be an abelian extension of the rationals. We now discuss how to measure volumes semi-locally over $\mathbf{K}\otimes\mathbb{Z}_p = \prod_{\mathfrak{p}|p}\mathbf{K}_{\mathfrak{p}}$. The constructions work in a wider context than the p-ordinary deformation of $\rho_{E,p}$. In fact, the scenario considered here will be identical to Chapters V and VI.

First fix a primitive $\mathbf{I}$-algebra homomorphism $\lambda : \mathbf{h}^{\mathrm{ord}}(Np^\infty;\mathcal{O}_{\mathcal{K}})\otimes_{\Lambda^{\mathrm{wt}}_{\mathcal{K}}}\mathbf{I} \to \mathbf{I}$ corresponding to $\mathbf{f}\in\mathbb{S}^{\mathrm{ord}}_{\mathbf{I}}$. In particular, either of the conditions (Rk1) or (UF) ensure that $\mathcal{MS}^{\mathrm{ord}}(\mathbf{I})^{\pm}[\lambda]$ is free of rank one, generated by the symbols $\Xi^{\pm}_{\lambda}$ say. Recall from §5.3 that there was a commutative square

$$\begin{array}{ccc}
\mathbf{I}.\mathbf{v}_{\mathbf{I},\lambda} & \stackrel{\mathrm{EXP}_{\mathbb{Q}_p,i}}{\longrightarrow} & H^1\Big(\mathbb{Q}_p,\mathbb{T}^*_\infty[\lambda]\otimes_{\mathbb{Z}_p}\mathbb{Q}_p(i)\Big) \\
{\scriptstyle .t^{-i} \bmod \mathbb{B}^{+,\mathrm{wt}}_{\mathrm{dR}}}\downarrow & & \downarrow{\scriptstyle \mathrm{res}_{\mathbf{K}_{\mathfrak{p}}/\mathbb{Q}_p}} \\
\Big(\mathbf{I}(\Phi\chi^i_{\mathrm{cy}})\otimes_{\Lambda^{\mathrm{wt}}}\mathbb{B}^{\mathrm{wt}}_{\mathrm{dR}}/\mathbb{B}^{+,\mathrm{wt}}_{\mathrm{dR}}\Big)^{G_{\mathbf{K}_{\mathfrak{p}}}} & \stackrel{(\mathcal{J}_-)_*\circ\mathrm{EXP}_{\Phi\chi^i_{\mathrm{cy}}}}{\longrightarrow} & H^1\Big(\mathbf{K}_{\mathfrak{p}},\ \mathbb{T}^*_\infty[\lambda]\otimes_{\mathbb{Z}_p}\mathbb{Q}_p(i)\Big)
\end{array}$$

for all integers $i\geq 1$.

If $i=1$ then $\Big(\mathbf{I}(\Phi\chi_{\mathrm{cy}})\otimes_{\Lambda^{\mathrm{wt}}}\mathbb{B}^{\mathrm{wt}}_{\mathrm{dR}}/\mathbb{B}^{+,\mathrm{wt}}_{\mathrm{dR}}\Big)^{G_{\mathbf{K}_{\mathfrak{p}}}}$ is free of rank $\big[\mathbf{K}_{\mathfrak{p}}:\mathbb{Q}_p\big]$ over $\mathbf{I}[1/p]$; further, the element $t^{-1}.\mathbf{v}_{\mathbf{I},\lambda}$ generates an $\mathcal{O}_{\mathbf{K}_{\mathfrak{p}}}\otimes_{\mathbb{Z}_p}\mathbf{I}$-lattice denoted by $\mathcal{D}_{\mathbf{K}_{\mathfrak{p}},\mathbf{v}_{\mathbf{I},\lambda}}$. We also abbreviate $(\mathcal{J}_-)_*\circ\mathrm{EXP}_{\Phi\chi_{\mathrm{cy}}}$ by the shorter $\mathrm{EXP}_{\infty,\lambda}$, in other words

$$\mathrm{EXP}_{\infty,\lambda} : \Big(\mathbb{T}^*_\infty[\lambda](1)\otimes_{\Lambda^{\mathrm{wt}}}\mathbb{B}^{\mathrm{wt}}_{\mathrm{dR}}/\mathbb{B}^{+,\mathrm{wt}}_{\mathrm{dR}}\Big)^{G_{\mathbf{K}_{\mathfrak{p}}}} \twoheadrightarrow H^1_{\mathcal{E}}\Big(\mathbf{K}_{\mathfrak{p}},\ \mathbb{T}^*_\infty[\lambda](1)\otimes_{\mathbb{Z}_p}\mathbb{Q}_p\Big).$$

It follows that $\mathrm{EXP}_{\infty,\lambda}\Big(\mathcal{D}_{\mathbf{K}_{\mathfrak{p}},\mathbf{v}_{\mathbf{I},\lambda}}\Big)$ is an $\mathbf{I}$-lattice of maximal rank inside of $H^1_{\mathcal{E}}$.

Remark: The precise connection between $\mathrm{EXP}_{\infty,\lambda}$ and our big dual exponential homomorphism $\mathrm{EXP}^*_{\mathbb{T}_\infty,\mathbf{K}_\mathfrak{p}} : H^1\big(\mathbf{K}_\mathfrak{p}, \mathbb{T}_\infty[\lambda]\big) \longrightarrow \mathbf{K}_\mathfrak{p} \otimes_{\mathbb{Z}_p} \mathbf{I}$ is easy to describe. Using Definition 5.11(ii),(iii) the two are related via the simple formula

$$\mathrm{EXP}^*_{\mathbb{T}_\infty,\mathbf{K}_\mathfrak{p}}(-) \;=\; \mathrm{EXP}^*_{\infty,\lambda}(-) \cup \big(1-\varphi\big)^{-1}.\big(t^{-1}.\mathbf{v}_{\mathbf{I},\lambda}\big) \qquad (***)$$

where $\mathrm{EXP}^*_{\infty,\lambda}$ denotes the $\mathbf{I}$-linear dual homomorphism to $\mathrm{EXP}_{\infty,\lambda}$.

There is one last bit of theory we need to review before we can define our volumes. For each $P_{k,\epsilon} \in \mathrm{Spec}\,\mathbf{I}(\mathcal{O}_\mathcal{K})^{\mathrm{alg}}$ of type (k,ϵ), there exist scalars $\mathcal{C}_{\mathbf{f}_{P_{k,\epsilon}}}$ such that

$$\mathbf{v}_{\mathbf{I},\lambda} \bmod P_{k,\epsilon} \;=\; \mathcal{C}_{\mathbf{f}_{P_{k,\epsilon}}} \times \mathbf{v}_{k,\epsilon} \qquad \text{with } \mathbf{f}^*_{P_{k,\epsilon}} \cup \mathbf{v}_{k,\epsilon} = 1.$$

We computed these $\mathbf{I}$-adic periods exactly in Theorem 6.4, and found that

$$\mathcal{C}_{\mathbf{f}_{P_{k,\epsilon}}} \;=\; \Big(P_{k,\epsilon}\circ \ell^{\pm}_{\lambda}\Big) \times \mathrm{Per}^{\pm}_{\mathbf{I},\lambda_P} \qquad (****)$$

where the elements $\ell^+_\lambda, \ell^-_\lambda \in \mathcal{L}'$ satisfy $\mathcal{E}^{\dagger\dagger}_{\infty,\underline{\delta}}\Big(\vec{\mathfrak{Z}}_{[\lambda]}\Big)^{\pm} = \ell^{\pm}_\lambda \times \Xi^{\pm}_\lambda$ in $\mathcal{MS}^{\mathrm{ord}}(\mathcal{L}')^{\pm}[\lambda]$. Under this identification, the crystalline zeta-periods given by

$$\mathbf{v}^{\mathrm{zeta}}_{\lambda,+} \;:=\; \frac{1}{\ell^+_\lambda} \times \mathbf{v}_{\mathbf{I},\lambda} \qquad \text{and} \qquad \mathbf{v}^{\mathrm{zeta}}_{\lambda,-} \;:=\; \frac{1}{\ell^-_\lambda} \times \mathbf{v}_{\mathbf{I},\lambda}$$

behaved like global invariants (c.f. Definition 6.8), depending only on the original choice of universal modular symbol $\Xi_\lambda = \Xi^+_\lambda + \Xi^-_\lambda$.

Step 2: $\mathbf{I}$*-adic Tamagawa measures of local points.*
For an $\mathbf{I}[1/p]$-module $\mathbb{M}$ of finite type, we use $\mathbb{M}_{\mathcal{L}'}$ to indicate the tensor product $\mathbb{M} \otimes_{\mathbf{I}[1/p]} \mathcal{L}'$. Let us fix some generator $\mathbf{v}_{\mathbf{I},\lambda}$ of the Hecke eigenspace $\mathbb{D}^{\mathbf{I}}_{\mathrm{cris}}\big(\mathbb{T}_\infty\big)[\lambda]$.

Notation: One writes $\mathcal{D}_{\mathbf{K}_\mathfrak{p},\mathbf{v}_{\mathbf{I},\lambda}}$ for the $\mathcal{O}_{\mathbf{K}_\mathfrak{p}} \otimes_{\mathbb{Z}_p} \mathbf{I}$-lattice generated by the element $t^{-1}.\mathbf{v}_{\mathbf{I},\lambda}$ inside the $\mathcal{L}'$-vector space $\Big(\mathbb{T}^*_\infty[\lambda](1) \otimes_{\Lambda^{\mathrm{wt}}} \mathbb{B}^{\mathrm{wt}}_{\mathrm{dR}}/\mathbb{B}^{+,\mathrm{wt}}_{\mathrm{dR}}\Big)^{G_{\mathbf{K}_\mathfrak{p}}}_{\mathcal{L}'}$.

By transport of structure $\mathrm{EXP}_{\infty,\lambda}\Big(\mathcal{D}_{\mathbf{K}_\mathfrak{p},\mathbf{v}_{\mathbf{I},\lambda}}\Big)$ will be an $\mathbf{I}$-lattice in the vector space $H^1_{\mathcal{E}}\Big(\mathbf{K}_\mathfrak{p},\ \mathbb{T}^*_\infty[\lambda](1) \otimes_{\mathbb{Z}_p} \mathbb{Q}_p\Big)_{\mathcal{L}'}$, which has dimension $\big[\mathbf{K}_\mathfrak{p} : \mathbb{Q}_p\big]$ over the field $\mathcal{L}'$.

Remark: Let $\mathbb{L}_\mathfrak{p}$ be an auxiliary $\mathbf{I}$-lattice inside of $H^1_{\mathcal{E}}\Big(\mathbf{K}_\mathfrak{p},\ \mathbb{T}^*_\infty[\lambda](1) \otimes_{\mathbb{Z}_p} \mathbb{Q}_p\Big)_{\mathcal{L}'}$. Then for a suitable choice of scalar element $\mathfrak{g} \in \mathbf{I}$, the lattice $\mathfrak{g}.\mathrm{EXP}_{\infty,\lambda}\Big(\mathcal{D}_{\mathbf{K}_\mathfrak{p},\mathbf{v}_{\mathbf{I},\lambda}}\Big)$ will be properly contained within $\mathbb{L}_\mathfrak{p}$. The divisor of the fractional ideal

$$\mathfrak{g}^{-[\mathbf{K}_\mathfrak{p}:\mathbb{Q}_p]} \times \mathrm{char}_{\mathbf{I}}\left(\frac{\mathbb{L}_\mathfrak{p}}{\mathfrak{g}.\mathrm{EXP}_{\infty,\lambda}\Big(\mathcal{D}_{\mathbf{K}_\mathfrak{p},\mathbf{v}_{\mathbf{I},\lambda}}\Big)}\right)$$

will be called $\mu_{\mathbf{K}_\mathfrak{p},\mathbf{v}_{\mathbf{I},\lambda}}\big(\mathbb{L}_\mathfrak{p}\big)$, and is clearly independent of this initial choice of $\mathfrak{g}$. Furthermore, if $\mathbf{I}$ is a regular local ring then we may view the divisor $\mu_{\mathbf{K}_\mathfrak{p},\mathbf{v}_{\mathbf{I},\lambda}}\big(\mathbb{L}_\mathfrak{p}\big)$ as being a well-defined element of $\mathcal{L}'^{\times}/\mathbf{I}^{\times}$.

The reader should regard $\mu_{\mathbf{K}_\mathfrak{p},\mathbf{v}_{\mathbf{I},\lambda}}$ as rudimentary prototypes of the volume forms $\mathrm{vol}_{\mathcal{A},\mathbf{K}_\mathfrak{p}}$ for the elliptic curve $\mathcal{A}$, however they depend heavily on the choice of $\mathbf{v}_{\mathbf{I},\lambda}$! Our next task, therefore, must be to remove this dependence.

Suppose that $\mathbb{L}$ is an $\mathbf{I}$-submodule of $H^1_{\mathcal{E}}\Big(\mathbf{K}\otimes\mathbb{Z}_p, \mathbb{T}^*_\infty[\lambda](1)\Big)_{\mathcal{L}'}$ of maximal rank. Moreover to simplify the exposition, we shall now assume that the ring $\mathbf{I}$ is regular.

Definition 8.14. *We define the* $\mathbf{I}$*-adic Tamagawa measure of* $\mathbb{L}$ *over* $\mathbf{K}\otimes\mathbb{Z}_p$ *by*

$$\mathrm{vol}_{\mathbb{T}_\infty[\lambda],\mathbf{K},p}(\mathbb{L}) \quad := \quad \left(\ell^+_\lambda\right)^{-r_1} \times \left(\ell^+_\lambda\ell^-_\lambda\right)^{-r_2} \times \prod_{\mathfrak{p}|p} \mu_{\mathbf{K}_\mathfrak{p},\mathbf{v}_{\mathbf{I},\lambda}}(\mathbb{L}_\mathfrak{p})$$

where r_1 *is the number of real embeddings of* $\mathbf{K}\hookrightarrow\mathbb{R}$*, and* r_2 *denotes the number of pairs of conjugate embeddings.*

Lemma 8.15. *The* $\mathbf{I}$*-adic volume form* $\mathrm{vol}_{\mathbb{T}_\infty[\lambda],\mathbf{K},p}(-)$ *is canonically defined.*

Proof: Let us first show the independence of $\mathrm{vol}_{\mathbb{T}_\infty[\lambda],\mathbf{K},p}$ from the choice of $\mathbf{v}_{\mathbf{I},\lambda}$. We study the effect of replacing $\mathbf{v}_{\mathbf{I},\lambda}$ by the multiple $\mathfrak{u}\times\mathbf{v}_{\mathbf{I},\lambda}$ with $\mathfrak{u}\in\mathcal{L}'$, $\mathfrak{u}\neq 0$. The product term in Definition 8.14 transforms into

$$\prod_{\mathfrak{p}|p} \mathfrak{u}^{\mathrm{rank}(\mathbb{L}_\mathfrak{p})} \times \mu_{\mathbf{K}_\mathfrak{p},\mathbf{v}_{\mathbf{I},\lambda}}(\mathbb{L}_\mathfrak{p}) \quad = \quad \mathfrak{u}^{\sum_{\mathfrak{p}|p}[\mathbf{K}_\mathfrak{p}:\mathbb{Q}_p]} \times \prod_{\mathfrak{p}|p} \mu_{\mathbf{K}_\mathfrak{p},\mathbf{v}_{\mathbf{I},\lambda}}(\mathbb{L}_\mathfrak{p}).$$

On the other hand, sending $\mathbf{v}_{\mathbf{I},\lambda}\mapsto\mathfrak{u}\times\mathbf{v}_{\mathbf{I},\lambda}$ multiplies $\mathrm{EXP}^*_{\mathbb{T}_\infty}$ by $\mathfrak{u}$, which in turn multiplies the interpolation $\mathcal{E}^{\dagger\dagger}_{\infty,\underline{\delta}}$ by $\mathfrak{u}$ again. However $\mathcal{E}^{\dagger\dagger}_{\infty,\underline{\delta}}\left(\vec{\mathfrak{Z}}_{[\lambda]}\right)^{\pm} = \ell^\pm_\lambda\times\Xi^\pm_\lambda$, which means both of the periods ℓ^+_λ and ℓ^-_λ will also be scaled by this factor $\mathfrak{u}$. Hence the first chunk of Definition 8.14 becomes

$$\mathfrak{u}^{-(r_1+2r_2)} \times \left(\ell^+_\lambda\right)^{-r_1} \times \left(\ell^+_\lambda\ell^-_\lambda\right)^{-r_2}.$$

But the discrepancy $-(r_1+2r_2)+\sum_{\mathfrak{p}|p}\left[\mathbf{K}_\mathfrak{p}:\mathbb{Q}_p\right]$ equals zero, so the volume itself is invariant under scaling.

Finally, the form $\mathrm{vol}_{\mathbb{T}_\infty[\lambda],\mathbf{K},p}$ cannot depend on how we chose the symbols $\Xi^\pm_\lambda$, because the volume only takes values modulo $\mathbf{I}^\times$.

□

An important feature is that we can now specialise at arithmetic weights, in order to obtain information about the standard Tamagawa measures in the analytic family. More precisely at every point $P_{k,\epsilon}\in\mathrm{Spec}\,\mathbf{I}(\mathcal{O}_\mathcal{K})^{\mathrm{alg}}$ of type (k,ϵ), the natural map $\mathcal{L}'^\times/\mathbf{I}^\times \xrightarrow{P_{k,\epsilon}} (\mathcal{K}^\times/\mathcal{O}^\times_\mathcal{K})\cup\{\infty\}$ produces an element

$$\mathrm{vol}^{(k,\epsilon)}_{\mathbb{T}_\infty[\lambda],\mathbf{K},p}(\mathbb{L}) \quad := \quad P_{k,\epsilon}\Big(\mathrm{vol}_{\mathbb{T}_\infty[\lambda],\mathbf{K},p}(\mathbb{L})\Big).$$

Warning: The poles (if there are any) occur with negative multiplicity in

$$\sum_{\mathfrak{p}|p}\mathrm{div}_\mathbf{I}\left(\mathfrak{g}_\mathfrak{p}^{-[\mathbf{K}_\mathfrak{p}:\mathbb{Q}_p]}\mathrm{char}_\mathbf{I}\left(\frac{\mathbb{L}_\mathfrak{p}}{\mathfrak{g}_\mathfrak{p}.\mathrm{EXP}_{\infty,\lambda}\left(\mathcal{D}_{\mathbf{K}_\mathfrak{p},\mathbf{v}_{\mathbf{I},\lambda}}\right)}\right)\right) - \mathrm{div}_\mathbf{I}\Big(\left(\ell^+_\lambda\right)^{r_1}\left(\ell^+_\lambda\ell^-_\lambda\right)^{r_2}\Big).$$

Step 3: Relation between the volume and covolume at weight two.
Let us return to the original situation where E is a modular elliptic curve over $\mathbb{Q}$, and $\mathbf{K} = \mathbf{F}$ is an abelian number field in which the prime $p \geq 3$ does not ramify. The hypotheses ($\mathbf{F}$-ab) and (p-Ord) are assumed to hold true, as ever.

We shall henceforth write $\mathrm{vol}^{(k)}_{\mathbb{T}_{\infty},E}(-)$ as shorthand in place of $\mathrm{vol}^{(k,\mathbf{1})}_{\mathbb{T}_{\infty}[\lambda_E],\mathbf{F},p}(-)$. In particular, one may interpret $\mathrm{vol}^{(k)}_{\mathbb{T}_{\infty},E}(-)$ as taking values inside $\mathbb{Q}_p^{\times}/\mathbb{Z}_p^{\times} \cong p^{\mathbb{Z}}$ at almost all positive integers $k \in \mathbb{U}_E$.

Question. *What is the connection between* $\mathrm{vol}^{(2)}_{\mathbb{T}_{\infty},E}$ *and* $\mathrm{covol}^{(2)}_{\mathbb{T}_{\infty},E}$*?*

More generally, one can ask a similar question at other arithmetic weights $k > 2$ but for the narrow purposes of this chapter, weight two will be more than sufficient. For each positive $k \in \mathbb{U}_E$, denote by $\mathrm{Per}^{(k)}_{\mathbf{I},\lambda_E}(\mathbf{F})$ the $\mathbf{I}$-adic periods

$$\left(\mathrm{Per}^{+}_{\mathbf{I},\lambda_{P_{k,1}}}\right)^{r_1} \times \left(\mathrm{Per}^{+}_{\mathbf{I},\lambda_{P_{k,1}}}\mathrm{Per}^{-}_{\mathbf{I},\lambda_{P_{k,1}}}\right)^{r_2}$$

which occur naturally in the improved p-adic L-function $\mathbf{L}_p^{\mathrm{imp}}(\mathbf{f}/\mathbf{F},k,1)$.

Proposition 8.16. *At weight $k = 2$, there is an equality*

$$\mathrm{Per}^{(2)}_{\mathbf{I},\lambda_E}(\mathbf{F}) \times \mathrm{covol}^{(2)}_{\mathbb{T}_{\infty},E}(\mathbf{F}\otimes\mathbb{Z}_p) \quad = \quad \frac{\mathrm{vol}^{(2)}_{\mathbb{T}_{\infty},E}\left(H^1_{\mathcal{E}}(\mathbf{F}\otimes\mathbb{Z}_p,\mathbb{T}^*_{\infty,E}(1))\right)}{\prod_{\mathfrak{p}|p}\mathrm{vol}_{\mathcal{C}^{\min},\mathbf{F}_{\mathfrak{p}}}\left(H^1_e(\mathbf{F}_{\mathfrak{p}},\mathrm{Ta}_p(\mathcal{C}^{\min}))\right)}.$$

Proof: The idea behind the demonstration is to break up the calculation into its constituent eigenspaces, under the action of the finite abelian group $G = \mathrm{Gal}(\mathbf{F}/\mathbb{Q})$. In fact one can subdivide it even further, by decomposing over the primes in $S_{\mathbf{F},p}$. Firstly, recall that $\mathrm{covol}^{(2)}_{\mathbb{T}_{\infty},E}(\mathbf{F}\otimes\mathbb{Z}_p)$ is by definition the index

$$\left[\exp^*_{\omega}\left(H^1(\mathbf{F}\otimes\mathbb{Z}_p,\mathrm{Ta}_p(\mathcal{C}^{\min}))\right) : \exp^*_{\omega}\left(H^1(\mathbf{F}\otimes\mathbb{Z}_p,\mathbb{T}_{\infty,E})\otimes_{\Lambda^{\mathrm{wt}}}\mathbb{Z}_p\right)\right]$$
$$= \prod_{\mathfrak{p}|p}\prod_{\psi\in X_{\mathbf{F}_{\mathfrak{p}}}}\left[\exp^*_{\omega}\left(H^1(\mathbf{F}_{\mathfrak{p}},\mathrm{Ta}_p(\mathcal{C}^{\min}))\right)^{(\psi)} : \exp^*_{\omega}\left(H^1(\mathbf{F}_{\mathfrak{p}},\mathbb{T}_{\infty,E})\otimes_{\Lambda^{\mathrm{wt}}}\mathbb{Z}_p\right)^{(\psi)}\right]$$

where $X_{\mathbf{F}_{\mathfrak{p}}}$ denotes the character group associated to $\mathrm{Gal}(\mathbf{F}_{\mathfrak{p}}/\mathbb{Q}_p)$ at every $\mathfrak{p} \in S_{\mathbf{F},p}$. Fortunately each ψ-eigenspace is one-dimensional, hence the problem reduces to estimating the p-adic distance between the respective $\exp^*_{\omega}(-)^{(\psi)}$'s.

Remark: Fix a prime $\mathfrak{p}$ and a character $\psi \in X_{\mathbf{F}_{\mathfrak{p}}}$. If $V = \mathrm{Ta}_p(\mathcal{C}^{\min})\otimes_{\mathbb{Z}_p}\mathbb{Q}_p$, then

$$\exp^*_{\omega}\left(H^1(\mathbf{F}_{\mathfrak{p}},\mathrm{Ta}_p(\mathcal{C}^{\min}))\right)^{(\psi)} \quad = \quad \exp^*_{V}\left(H^1(\mathbf{F}_{\mathfrak{p}},\mathrm{Ta}_p(\mathcal{C}^{\min}))\right)^{(\psi)} \cup t^{-1}.\mathbf{v}_{2,1}$$

since $\mathbf{v}_{2,1}$ is invariant under the $G_{\mathbb{Q}_p}$-action, and $\omega \cup t^{-1}.\mathbf{v}_{2,1} = 1 \in \mathbf{D}_{\mathrm{dR}}(\mathbb{Q}_p(1))$ by the properties of the elements $\mathbf{v}_{k,\epsilon}$ given in §2.3.

It follows directly that

$$\exp^*_\omega\Big(H^1\big(\mathbf{F}_\mathfrak{p},\mathrm{Ta}_p(\mathcal{C}^{\min})\big)\Big)^{(\psi)} = \exp^*_V\Big(H^1\big(\mathbf{F}_\mathfrak{p},\mathrm{Ta}_p(\mathcal{C}^{\min})\big)\Big)^{(\psi)} \cup t^{-1}\mathbf{v}_{2,1}.(\mathbf{e}_\psi^{-1}\mathbf{e}_\psi)$$

$$= \frac{1}{[\mathbf{F}_\mathfrak{p}:\mathbb{Q}_p]}\,\mathrm{Tr}_{\mathbf{F}_\mathfrak{p}/\mathbb{Q}_p}\Big(\exp^*_V\Big(H^1\big(\mathbf{F}_\mathfrak{p},\mathrm{Ta}_p(\mathcal{C}^{\min})\big)\Big)\cup\ \mathbb{Z}_{p,\psi}.\ \mathbf{e}_\psi^{-1}t^{-1}\mathbf{v}_{2,1}\Big).\mathbf{e}_\psi$$

$$= \frac{1}{[\mathbf{F}_\mathfrak{p}:\mathbb{Q}_p]}\,\mathrm{inv}_{\mathbf{F}_\mathfrak{p}}\Big(H^1\big(\mathbf{F}_\mathfrak{p},\mathrm{Ta}_p(\mathcal{C}^{\min})\big)\cup\ \exp_V\Big(\mathbb{Z}_{p,\psi}.\ \mathbf{e}_\psi^{-1}t^{-1}\mathbf{v}_{2,1}\Big)\Big).\mathbf{e}_\psi.$$

However, we can replace $\exp_V\left(\mathbb{Z}_{p,\psi}.\mathbf{e}_\psi^{-1}t^{-1}\mathbf{v}_{2,1}\right)$ by $\mathcal{V}_\mathfrak{p}^{(\psi)}.\ H^1_e\big(\mathbf{F}_\mathfrak{p},\mathrm{Ta}_p(\mathcal{C}^{\min})\big)^{(\psi^{-1})}$ where $\mathcal{V}_\mathfrak{p}^{(\psi)} := \mathrm{vol}_{\mathcal{C}^{\min},\mathbf{F}_\mathfrak{p}}\Big(H^1_e\big(\mathbf{F}_\mathfrak{p},\mathrm{Ta}_p(\mathcal{C}^{\min})\big)^{(\psi^{-1})}\Big)$. The reason is that the lattice $\exp_V\left(\mathbb{Z}_p.t^{-1}\mathbf{v}_{2,1}\right)$ is assigned Haar measure 1 under the volume form $\mathrm{vol}_{\mathcal{C}^{\min},\mathbf{F}_\mathfrak{p}}(-)$. As a basic corollary,

$$\exp^*_\omega\Big(H^1\big(\mathbf{F}_\mathfrak{p},\mathrm{Ta}_p(\mathcal{C}^{\min})\big)\Big)^{(\psi)} = \frac{\mathcal{V}_\mathfrak{p}^{(\psi)}}{[\mathbf{F}_\mathfrak{p}:\mathbb{Q}_p]}\,\mathrm{inv}_{\mathbf{F}_\mathfrak{p}}\left(H^1(\dots)\cup H^1_e(\dots)^{(\psi^{-1})}\right).\,\mathbf{e}_\psi$$

$$= \frac{\mathrm{vol}_{\mathcal{C}^{\min},\mathbf{F}_\mathfrak{p}}\Big(H^1_e\big(\mathbf{F}_\mathfrak{p},\mathrm{Ta}_p(\mathcal{C}^{\min})\big)^{(\psi^{-1})}\Big)}{[\mathbf{F}_\mathfrak{p}:\mathbb{Q}_p]}\,\mathbb{Z}_p.\ \mathbf{e}_\psi$$

since $H^1\big(\mathbf{F}_\mathfrak{p},\mathrm{Ta}_p(\mathcal{C}^{\min})\big)\cup H^1_e\big(\mathbf{F}_\mathfrak{p},\mathrm{Ta}_p(\mathcal{C}^{\min})\big)$ projects onto $\mathbb{Z}_p$ via the map $\mathrm{inv}_{\mathbf{F}_\mathfrak{p}}$.

Remark: To play the same game with the term $\exp^*_\omega\left(H^1\big(\mathbf{F}\otimes\mathbb{Z}_p,\mathbb{T}_{\infty,E}\big)\otimes_{\Lambda^{\mathrm{wt}}}\mathbb{Z}_p\right)$, we should now apply the p-adic interpolation properties of our big exponentials. The necessary background on these homomorphisms has been supplied in Step 1.

Once more we fix a prime $\mathfrak{p}$ dividing p, and a character $\psi\in X_{\mathbf{F}_\mathfrak{p}}$; then

$$\exp^*_\omega\Big(H^1\big(\mathbf{F}_\mathfrak{p},\mathbb{T}_{\infty,E}\big)\otimes_{\Lambda^{\mathrm{wt}}}\mathbb{Z}_p\Big)^{(\psi)} = \exp^*_V\Big(H^1\big(\mathbf{F}_\mathfrak{p},\mathbb{T}_{\infty,E}\big)\otimes_{\Lambda^{\mathrm{wt}}}\mathbb{Z}_p\Big)^{(\psi)}\cup t^{-1}.\mathbf{v}_{2,1}$$

$$= \mathcal{C}^{-1}_{\mathbf{f}_{P_{2,1}}}\times\exp^*_V\Big(H^1\big(\mathbf{F}_\mathfrak{p},\mathbb{T}_{\infty,E}\big)\otimes_{\Lambda^{\mathrm{wt}}}\mathbb{Z}_p\Big)^{(\psi)}\cup\Big(t^{-1}.\mathbf{v}_{\mathbf{I},\lambda_E}\bmod P_{2,1}\Big)$$

where as before $\mathbf{v}_{\mathbf{I},\lambda}\bmod P_{k,\epsilon} = \mathcal{C}_{\mathbf{f}_{P_{k,\epsilon}}}\times\mathbf{v}_{k,\epsilon}$ at an arithmetic point of type (k,ϵ). Applying Equation **(***)** one deduces that

$$\exp^*_\omega\Big(H^1\big(\mathbf{F}_\mathfrak{p},\mathbb{T}_{\infty,E}\big)\otimes_{\Lambda^{\mathrm{wt}}}\mathbb{Z}_p\Big)^{(\psi)}$$

$$= \mathcal{C}^{-1}_{\mathbf{f}_{P_{2,1}}}\times\Big(\mathrm{EXP}^*_{\infty,\lambda_E}\Big(H^1\big(\mathbf{F}_\mathfrak{p},\mathbb{T}_{\infty,E}\big)\Big)^{(\psi)}\cup t^{-1}.\mathbf{v}_{\mathbf{I},\lambda_E}\Big)\ \text{modulo}\ P_{2,1}.$$

Exploiting the same argument as in the previous case, this becomes

$$\exp^*_\omega\Big(H^1\big(\mathbf{F}_\mathfrak{p},\mathbb{T}_{\infty,E}\big)\otimes_{\Lambda^{\mathrm{wt}}}\mathbb{Z}_p\Big)^{(\psi)} = \mathcal{C}^{-1}_{\mathbf{f}_{P_{2,1}}}\times\frac{1}{[\mathbf{F}_\mathfrak{p}:\mathbb{Q}_p]}\,\mathrm{Tr}_{\mathbf{F}_\mathfrak{p}/\mathbb{Q}_p}\Big(\dots\dots\Big)$$

$$= \frac{\mathcal{C}^{-1}_{\mathbf{f}_{P_{2,1}}}}{[\mathbf{F}_\mathfrak{p}:\mathbb{Q}_p]}\,\mathrm{inv}_{\mathbf{F}_\mathfrak{p}}\left(H^1\big(\mathbf{F}_\mathfrak{p},\mathbb{T}_{\infty,E}\big)\cup\mathrm{EXP}_{\infty,\lambda_E}\left(\mathcal{D}_{\mathbf{F}_\mathfrak{p},\mathbf{v}_{\mathbf{I},\lambda_E}}\right)^{(\psi^{-1})}\bmod P_{2,1}\right).\,\mathbf{e}_\psi.$$

By our definition of the prototypes $\mu_{\mathbf{F}_\mathfrak{p},\mathbf{v}_{\mathbf{I},\lambda_E}}(-)$, the lattice $\mathcal{D}_{\mathbf{F}_\mathfrak{p},\mathbf{v}_{\mathbf{I},\lambda_E}}$ was assigned unit volume. Consequently, $\exp^*_\omega\Big(H^1(\dots)\otimes_{\Lambda^{\mathrm{wt}}}\mathbb{Z}_p\Big)^{(\psi)}$ coincides with

$$\frac{\mathcal{Y}^{(\psi)}_{\mathbf{I},\mathfrak{p}}\times\mathcal{C}^{-1}_{\mathbf{f}_{P_{2,1}}}}{[\mathbf{F}_\mathfrak{p}:\mathbb{Q}_p]}\ \mathrm{inv}_{\mathbf{F}_\mathfrak{p}}\Big(H^1(\mathbf{F}_\mathfrak{p},\mathbb{T}_{\infty,E})\cup H^1_{\mathcal{E}}(\mathbf{F}_\mathfrak{p},\mathbb{T}^*_{\infty,E}(1))^{(\psi^{-1})}\ \mathrm{mod}\ P_{2,\mathbf{1}}\Big)\,.\,\mathbf{e}_\psi$$

where we have defined $\mathcal{Y}^{(\psi)}_{\mathbf{I},\mathfrak{p}}\ :=\ \mu_{\mathbf{F}_\mathfrak{p},\mathbf{v}_{\mathbf{I},\lambda_E}}\Big(H^1_{\mathcal{E}}(\mathbf{F}_\mathfrak{p},\mathbb{T}^*_{\infty,E}(1))^{(\psi^{-1})}\Big)$ modulo $P_{2,\mathbf{1}}$. Fortunately $H^1_{\mathcal{E}}$ was already $\mathcal{R}_E$-saturated, which in turn implies

$$\begin{aligned}\mathrm{inv}_{\mathbf{F}_\mathfrak{p}}\Big(H^1(\dots)\cup H^1_{\mathcal{E}}(\dots)^{(\psi^{-1})}\ \mathrm{mod}\ P_{2,\mathbf{1}}\Big) &= \mathrm{inv}_{\mathbf{F}_\mathfrak{p}}\Big(H^2(\mathbf{F}_\mathfrak{p},\ \mathbf{I}(1))\ \mathrm{mod}\ P_{2,\mathbf{1}}\Big)\\ &= \mathrm{inv}_{\mathbf{F}_\mathfrak{p}}\Big(H^2(\mathbf{F}_\mathfrak{p},\ \mathbb{Z}_p(1))\Big)\ =\ \mathbb{Z}_p.\end{aligned}$$

One may therefore conclude

$$\exp^*_\omega\Big(H^1(\mathbf{F}_\mathfrak{p},\mathbb{T}_{\infty,E})\otimes_{\Lambda^{\mathrm{wt}}}\mathbb{Z}_p\Big)^{(\psi)}\ =\ \frac{\mathcal{Y}^{(\psi)}_{\mathbf{I},\mathfrak{p}}\times\mathcal{C}^{-1}_{\mathbf{f}_{P_{2,1}}}}{[\mathbf{F}_\mathfrak{p}:\mathbb{Q}_p]}\ \mathbb{Z}_p.\,\mathbf{e}_\psi.$$

Remark: Evaluating the p-adic distance between these two $\mathbb{Z}_p$-lines, we discover that each index $\left[\exp^*_\omega\Big(H^1(\mathbf{F}_\mathfrak{p},\mathrm{Ta}_p(\mathcal{C}^{\min}))\Big)^{(\psi)}:\exp^*_\omega\Big(H^1(\mathbf{F}_\mathfrak{p},\mathbb{T}_{\infty,E})\otimes_{\Lambda^{\mathrm{wt}}}\mathbb{Z}_p\Big)^{(\psi)}\right]$ must actually equal

$$\frac{\mathcal{Y}^{(\psi)}_{\mathbf{I},\mathfrak{p}}\times\mathcal{C}^{-1}_{\mathbf{f}_{P_{2,1}}}}{\mathcal{V}^{(\psi)}_\mathfrak{p}}\ =\ \frac{\mu_{\mathbf{F}_\mathfrak{p},\mathbf{v}_{\mathbf{I},\lambda_E}}\Big(H^1_{\mathcal{E}}(\mathbf{F}_\mathfrak{p},\mathbb{T}^*_{\infty,E}(1))^{(\psi^{-1})}\Big)\ \mathrm{mod}\ P_{2,\mathbf{1}}\times\mathcal{C}^{-1}_{\mathbf{f}_{P_{2,1}}}}{\mathrm{vol}_{\mathcal{C}^{\min},\mathbf{F}_\mathfrak{p}}\Big(H^1_e(\mathbf{F}_\mathfrak{p},\mathrm{Ta}_p(\mathcal{C}^{\min}))^{(\psi^{-1})}\Big)}.$$

Taking the product over all $\mathfrak{p}\in S_{\mathbf{F},p}$ and $\psi\in X_{\mathbf{F}_\mathfrak{p}}$, this becomes

$$\begin{aligned}\mathrm{covol}^{(2)}_{\mathbb{T}_{\infty,E}}(\mathbf{F}\otimes\mathbb{Z}_p) &= \prod_{\mathfrak{p}|p}\mathcal{C}^{-\#X_{\mathbf{F}_\mathfrak{p}}}_{\mathbf{f}_{P_{2,1}}}\prod_{\psi\in X_{\mathbf{F}_\mathfrak{p}}}\frac{\mathcal{Y}^{(\psi)}_{\mathbf{I},\mathfrak{p}}}{\mathcal{V}^{(\psi)}_\mathfrak{p}}\\ &= \mathcal{C}^{-[\mathbf{F}:\mathbb{Q}]}_{\mathbf{f}_{P_{2,1}}}\times\prod_{\mathfrak{p}|p}\frac{\mu_{\mathbf{F}_\mathfrak{p},\mathbf{v}_{\mathbf{I},\lambda_E}}\Big(H^1_{\mathcal{E}}(\mathbf{F}_\mathfrak{p},\mathbb{T}^*_{\infty,E}(1))\Big)\ \mathrm{mod}\ P_{2,1}}{\mathrm{vol}_{\mathcal{C}^{\min},\mathbf{F}_\mathfrak{p}}\Big(H^1_e(\mathbf{F}_\mathfrak{p},\mathrm{Ta}_p(\mathcal{C}^{\min}))\Big)}\\ = \ \mathcal{C}^{-[\mathbf{F}:\mathbb{Q}]}_{\mathbf{f}_{P_{2,1}}}\times&\ \frac{(\ell^+_\lambda)^{r_1}(\ell^+_\lambda\ell^-_\lambda)^{r_2}\times\mathrm{vol}_{\mathbb{T}_{\infty,E},\mathbf{F}}\Big(\prod_{\mathfrak{p}|p}H^1_{\mathcal{E}}(\mathbf{F}_\mathfrak{p},\mathbb{T}^*_{\infty,E}(1))\Big)\ \mathrm{mod}\ P_{2,1}}{\prod_{\mathfrak{p}|p}\mathrm{vol}_{\mathcal{C}^{\min},\mathbf{F}_\mathfrak{p}}\Big(H^1_e(\mathbf{F}_\mathfrak{p},\mathrm{Ta}_p(\mathcal{C}^{\min}))\Big)}.\end{aligned}$$

Finally, the error term $\mathcal{C}^{-[\mathbf{F}:\mathbb{Q}]}_{\mathbf{f}_{P_{2,1}}}\times(\ell^+_\lambda)^{r_1}(\ell^+_\lambda\ell^-_\lambda)^{r_2}$ mod $P_{2,\mathbf{1}}$ will equal the quantity

$$\Big(\mathrm{Per}^+_{\mathbf{I},\lambda_{P_{2,1}}}\Big)^{-r_1}\times\Big(\mathrm{Per}^+_{\mathbf{I},\lambda_{P_{2,1}}}\mathrm{Per}^-_{\mathbf{I},\lambda_{P_{2,1}}}\Big)^{-r_2}\ =\ 1/\mathrm{Per}^{(2)}_{\mathbf{I},\lambda_E}$$

as a consequence of Equation **(****)**.

The proof of Proposition 8.16 is finished.

□

This relationship between the covolume and volume is quite beautiful, and explains why the **I**-adic periods of $\Xi_{\lambda_E} = \Xi^+_{\lambda_E} + \Xi^-_{\lambda_E}$ enter the vertical arithmetic over **F**. Moreover, it means that we can replace the covolume term in $\#\mathrm{Ker}\Big(\oplus_{\mathfrak{p}|p}\,\delta_{\infty,E,\mathfrak{p}}\Big)$ with the ratio of two volumes, one **I**-adic and the other p-adic in nature.

Question. *Can one define an* **I***-adic volume for the half-twisted lifting* $\mathbb{T}^{-1/2}_{\infty,E}$ *?*

In particular, for any number field **K** we postulate the existence of a 'measure'

$$\mathrm{vol}_{\mathbb{T}^{-1/2}_{\infty,E},\mathbf{K},p}(-) \quad = \quad \left(\ell^+_\lambda\right)^{-r_1}\left(\ell^+_\lambda\ell^-_\lambda\right)^{-r_2} \times \prod_{\mathfrak{p}|p} \mu^{-1/2}_{\mathbf{K}_\mathfrak{p},\mathbf{v}_{\mathbf{I},\lambda}}(-)$$

taking values in $\mathcal{L}'^\times/\mathbf{I}^\times$, which plays an analogous rôle to that of $\mathrm{vol}_{\mathbb{T}_{\infty,E},\mathbf{K},p}(-)$. One would then demand a formula connecting

$$\mathrm{Per}^{(2)}_{\mathbf{I},\lambda_E}(\mathbf{K}) \times \mathrm{covol}^{(2)}_{\mathbb{T}^{-1/2}_{\infty,E}}(\mathbf{K}\otimes\mathbb{Z}_p) \quad \overset{??}{\longleftrightarrow} \quad \frac{\mathrm{vol}^{(2)}_{\mathbb{T}^{-1/2}_{\infty,E}}\Big(H^1_+\big(\mathbf{K}\otimes\mathbb{Z}_p, \mathbb{T}^{-1/2}_{\infty,E}\big)^*\Big)}{\prod_{\mathfrak{p}|p}\mathrm{vol}_{\mathcal{C}^{\min},\mathbf{K}_\mathfrak{p}}\Big(H^1_e\big(\mathbf{K}_\mathfrak{p},\mathrm{Ta}_p(\mathcal{C}^{\min})\big)\Big)}$$

very much in the spirit of Proposition 8.16.

We must confess that we have utterly failed to write down a suitable definition. Due to the lack of a good exponential map for these half-twisted cohomologies, one would first have to deform along the cyclotomic direction, define the 'measure' over the two-variable deformation ring, then specialise back down to the weight-axis. With tongue embedded in cheek, the reader is left with an unenviable task:

Exercise. *For a lattice* $\mathbb{L} \subset \Big(H^1_+\big(\mathbf{K}\otimes\mathbb{Z}_p, \mathbb{T}^{-1/2}_{\infty,E}\big)^*\Big)_{\mathcal{L}}$ *construct* $\mathrm{vol}_{\mathbb{T}^{-1/2}_{\infty,E},\mathbf{K},p}(\mathbb{L})$*!*

This concludes our discussion concerning $\mathrm{Ker}\Big(\oplus_{\mathfrak{p}|p}\,\delta_{\infty,E,\mathfrak{p}}\Big)$ and $\mathrm{Ker}\Big(\oplus_{\mathfrak{p}|p}\,\delta^{-1/2}_{\infty,E,\mathfrak{p}}\Big)$. It is now time to utilise Propositions 8.9, 8.13 and 8.16 in the systematic calculation of the global Γ^{wt}-Euler characteristic of E.

CHAPTER IX

Diamond-Euler Characteristics: the Global Case

To motivate the global calculations, we illustrate what happens in the situation of cyclotomic Iwasawa theory. In two well-known papers [PR1,Sn2] on this subject, Perrin-Riou and Schneider computed the leading term of an algebraic L-function attached to E over the cyclotomic $\mathbb{Z}_p$-extension of $\mathbf{F}$. In particular, they related the leading L-value with the determinant of a certain height pairing

$$\langle -,- \rangle_{\mathbf{F},p}^{\mathrm{cy}} : E(\mathbf{F}) \times E(\mathbf{F}) \longrightarrow \mathbb{Q}_p$$

which is a symmetric, bilinear form on $E(\mathbf{F}) \otimes \mathbb{Q}_p$. This pairing is known to be degenerate if E has split multiplicative reduction at some prime of $\mathbf{F}$ lying above p. In all cases, the discriminant of the p-adic height pairing should be closely related to the dominant term in $\mathbf{L}_p(E/\mathbf{F}, s)$ at the central point $s = 1$.

Surely one might then expect similar types of formulae, for both the vertical and the half-twisted deformation? Let us recall the main reason in evaluating

$$\chi\Big(\Gamma^{\mathrm{wt}}, S\Big) = \frac{\#H^0\big(\Gamma^{\mathrm{wt}}, S\big)}{\#H^1\big(\Gamma^{\mathrm{wt}}, S\big)} \quad \text{with } S = \mathrm{Sel}_{\mathbf{F}}\big(\mathbb{T}_{\infty,E}\big) \text{ or } \mathrm{Sel}_{\mathbf{F}}\big(\mathbb{T}_{\infty,E} \otimes \Psi^{-1/2}\big)$$

is that these invariants occur as leading terms, in the Taylor series expansions of $\text{Ш}_{\mathbf{F}}\big(\mathbb{T}_{\infty,E}\big)$ and $\text{Ш}_{\mathbf{F}}\big(\mathbb{T}_{\infty,E} \otimes \Psi^{-1/2}\big)$ respectively. Buoyed by the appearance of a p-adic regulator in the work of Perrin-Riou and Schneider, it is not unreasonable to expect analogous height pairings

$$\langle -,- \rangle_{\mathbf{F},p}^{\mathrm{wt}} \quad \text{and} \quad \langle -,- \rangle_{\mathbf{F},p}^{-1/2} : E(\mathbf{F}) \otimes \mathbb{Q}_p \times E(\mathbf{F}) \otimes \mathbb{Q}_p \longrightarrow \mathbb{Q}_p$$

to manifest themselves in the arithmetic of the two deformations we are studying. This is indeed the case.

Due to reasons of presentation, the calculations in the half-twisted scenario are postponed to Chapter X – here we deal exclusively with the vertical deformation. The major challenge facing us is the construction of a 'p-adic weight regulator' intrinsic to the big Galois representation $\mathbb{T}_{\infty,E}$.

The approach which seems to yield the most explicit description for our pairings involves using degeneration maps, connecting layers in the pro-variety J_∞ covering the elliptic curve $\mathcal{C}^{\min}$. Unfortunately, the weight pairing itself can only be written down on a subset of the (compact) Selmer group attached to $\mathcal{C}^{\min}$ over the field $\mathbf{F}$. Extending the pairing to the full Selmer group is only possible, *provided that the p-primary component of the Tate-Shafarevich group attached to E over $\mathbf{F}$ is finite.* Nobody has ever found an example where this condition doesn't hold.

9.1 The Poitou-Tate exact sequences

We now assume that $\mathbf{K}$ is a number field, not necessarily an abelian extension of $\mathbb{Q}$. Throughout $\mathcal{R}$ will be a complete local Noetherian ring, with finite residue field. The two examples we have in mind are when $\mathcal{R} = \mathbb{Z}_p$, or $\mathcal{R}$ is a finite flat extension of the weight algebra Λ^{wt}. Recall for a compact $\mathcal{R}$-module M of finite-type,

$$A_M \;=\; \mathrm{Hom}_{\mathrm{cont}}\Big(M,\ \mathbb{Q}_p/\mathbb{Z}_p(1)\Big)$$

denoted the χ_{cy}-twisted continuous dual of M.

For each place $\nu \in \Sigma$ of $\mathbf{K}$, fix local conditions X_ν which are finitely-generated $\mathcal{R}$-submodules of $H^1\big(\mathbf{K}_\nu, M\big)$. Let $Y_\nu = X_\nu^\perp$ be the orthogonal complement under

$$H^1\big(\mathbf{K}_\nu, A_M\big) \times H^1\big(\mathbf{K}_\nu, M\big) \overset{\mathrm{inv}_{\mathbf{K}_\nu}\circ\cup}{\longrightarrow} \mathbb{Q}_p/\mathbb{Z}_p.$$

We define $H^1_X\big(\mathbf{K}_\Sigma/\mathbf{K}, M\big)$ as the kernel of $H^1\big(\mathbf{K}_\Sigma/\mathbf{K}, M\big) \overset{\oplus\mathrm{res}_\nu}{\longrightarrow} \bigoplus_{\nu\in\Sigma} \frac{H^1(\mathbf{K}_\nu, M)}{X_\nu}$ which is finitely-generated over $\mathcal{R}$. Similarly, write $H^1_Y\big(\mathbf{K}_\Sigma/\mathbf{K}, A_M\big)$ for the kernel of $H^1\big(\mathbf{K}_\Sigma/\mathbf{K}, A_M\big) \overset{\oplus\mathrm{res}_\nu}{\longrightarrow} \bigoplus_{\nu\in\Sigma} \frac{H^1(\mathbf{K}_\nu, A_M)}{Y_\nu}$ which is cofinitely-generated over $\mathcal{R}$.

Proposition 9.1. *(Poitou-Tate) There exists a long exact sequence*

$$\begin{aligned} 0 \longrightarrow\ & H^1_Y\big(\mathbf{K}_\Sigma/\mathbf{K}, A_M\big) \longrightarrow H^1\big(\mathbf{K}_\Sigma/\mathbf{K}, A_M\big) \longrightarrow \bigoplus_{\nu\in\Sigma} \frac{H^1\big(\mathbf{K}_\nu, A_M\big)}{Y_\nu} \\ \longrightarrow\ & H^1_X\big(\mathbf{K}_\Sigma/\mathbf{K}, M\big)^\vee \longrightarrow H^2\big(\mathbf{K}_\Sigma/\mathbf{K}, A_M\big) \longrightarrow \bigoplus_{\nu\in\Sigma} H^2\big(\mathbf{K}_\nu, A_M\big) \\ \longrightarrow\ & H^0\big(\mathbf{K}_\Sigma/\mathbf{K}, M\big)^\vee \longrightarrow 0. \end{aligned}$$

The connecting homomorphisms in this sequence may be found in [FPR, Sect 1.2]. One important advantage of this proposition is that it is really the closest thing one has to a global duality theorem for Selmer groups. We shall apply it initially to the p-adic situation, and then afterwards to the Λ^{wt}-adic one.

Let $\mathcal{A}$ be an elliptic curve defined over the field $\mathbf{K}$, and p will be an odd prime. We choose M to be the p-adic Tate module of $\mathcal{A}$, in which case $A_M \cong \mathcal{A}[p^\infty]$ by the Weil pairing. The local conditions imposed are that X_ν is the p-saturation of the unramified cocycles if $\nu \nmid p$, and otherwise equals $H^1_f(\mathbf{K}_\nu, \mathrm{Ta}_p(\mathcal{A}))$ if $\nu | p$. Consequently, the Poitou-Tate sequence becomes

$$\begin{aligned} 0 \longrightarrow\ & \mathrm{Sel}_{\mathbf{K}}(\mathcal{A})[p^\infty] \longrightarrow H^1\Big(\mathbf{K}_\Sigma/\mathbf{K}, \mathcal{A}[p^\infty]\Big) \longrightarrow \bigoplus_{\nu\in\Sigma} \frac{H^1\big(\mathbf{K}_\nu, \mathcal{A}[p^\infty]\big)}{\mathcal{A}(\mathbf{K}_\nu)\otimes\mathbb{Q}_p/\mathbb{Z}_p} \\ \longrightarrow\ & H^1_f\Big(\mathbf{K}_\Sigma/\mathbf{K}, \mathrm{Ta}_p(\mathcal{A})\Big)^\vee \longrightarrow H^2\Big(\mathbf{K}_\Sigma/\mathbf{K}, \mathcal{A}[p^\infty]\Big) \longrightarrow \ \ldots\ . \end{aligned}$$

For instance, if the Selmer group of $\mathcal{A}$ over $\mathbf{K}$ is actually finite, it is a well-known corollary that $H^2\Big(\mathbf{K}_\Sigma/\mathbf{K}, \mathcal{A}[p^\infty]\Big)$ must be zero.

Remark: The elliptic curves we want to study are the strong Weil curve E and also the optimal $X_1(Np)$-curve $\mathcal{C}^{\min}$, which are both modular and defined over $\mathbb{Q} \subset \mathbf{K}$. Remember that the p^∞-Selmer group of $\mathcal{C}^{\min}$ coincides with the Γ^{wt}-invariant part of $\mathrm{Sel}_{\mathbf{K}}(\mathbb{T}_{\infty,E})$, up to a finite kernel and cokernel. Thus it makes good sense to write down the Poitou-Tate sequence for the deformed Selmer group.

We now set M equal to the vertical deformation $\mathbb{T}_{\infty,E}$, so vacuously $A_M = A_{\mathbb{T}_{\infty,E}}$. Let us further recall from Chapter VII:

Definition 7.2. *(i) The compact Selmer group $\mathfrak{S}_{\mathbf{K}}(\mathbb{T}_{\infty,E})$ is defined to be the kernel of the restriction maps*

$$H^1\Big(\mathbf{K}_\Sigma/\mathbf{K}, \mathbb{T}_{\infty,E}\Big) \overset{\oplus \mathrm{res}_\nu}{\longrightarrow} \bigoplus_{\nu \in \Sigma - S_{\mathbf{K},p}} \frac{H^1(\mathbf{K}_\nu, \mathbb{T}_{\infty,E})}{H^1_{\mathrm{nr}}(\mathbf{K}_\nu, \mathbb{T}_{\infty,E})} \oplus \bigoplus_{\nu \in S_{\mathbf{K},p}} \frac{H^1(\mathbf{K}_\nu, \mathbb{T}_{\infty,E})}{\mathbf{X}_{\mathbb{T}_{\infty,E}}(\mathbf{K}_\nu)}.$$

This time the local conditions are that X_ν is the p-saturation of the unramified cocycles if $\nu \nmid p$, and is otherwise equal to $\mathbf{X}_{\mathbb{T}_{\infty,E}}(\mathbf{K}_\nu)$ if ν is a place of $\mathbf{K}$ above p. Under local duality $Y_\nu = X_\nu^\perp$ is the family of local points $\mathbf{X}^D_{\mathbb{T}_{\infty,E}}(\mathbf{K}_\nu)$ at $\nu | p$. Therefore, upstairs the Poitou-Tate sequence reads as

$$\begin{aligned} 0 \longrightarrow\ & \mathrm{Sel}_{\mathbf{K}}(\mathbb{T}_{\infty,E}) \longrightarrow H^1\Big(\mathbf{K}_\Sigma/\mathbf{K}, A_{\mathbb{T}_{\infty,E}}\Big) \\ \overset{\widetilde{\lambda}_{\infty,E}}{\longrightarrow}\ & \bigoplus_{\nu \in \Sigma - S_{\mathbf{K},p}} \frac{H^1(\mathbf{K}_\nu, A_{\mathbb{T}_{\infty,E}})}{H^1_{\mathrm{nr}}(\mathbf{K}_\nu, A_{\mathbb{T}_{\infty,E}})} \oplus \bigoplus_{\nu \in S_{\mathbf{K},p}} \frac{H^1(\mathbf{K}_\nu, A_{\mathbb{T}_{\infty,E}})}{\mathbf{X}^D_{\mathbb{T}_{\infty,E}}(\mathbf{K}_\nu)} \\ \longrightarrow\ & \mathfrak{S}_{\mathbf{K}}(\mathbb{T}_{\infty,E})^\vee \longrightarrow H^2\Big(\mathbf{K}_\Sigma/\mathbf{K}, A_{\mathbb{T}_{\infty,E}}\Big) \longrightarrow \ \dots . \end{aligned}$$

One is particularly interested in the surjectivity or otherwise, of this restriction homomorphism $\widetilde{\lambda}_{\infty,E}$.

From now on we set $\mathbf{K} = \mathbf{F}$ to be an abelian extension of $\mathbb{Q}$ satisfying (**F**-ab).

Theorem 9.2. *If $\rho_{\infty,E}$ is residually irreducible, then $\mathfrak{S}_{\mathbf{F}}(\mathbb{T}_{\infty,E})$ is zero.*

The proof is deferred to the next section.

Assuming $\rho_{\infty,E}$ is residually absolutely irreducible, we deduce immediately that the map $\widetilde{\lambda}_{\infty,E}$ is surjective. The Γ^{wt}-cohomology induces a long exact sequence

$$\begin{aligned} 0 \longrightarrow\ & \mathrm{Sel}_{\mathbf{F}}(\mathbb{T}_{\infty,E})^{\Gamma^{\mathrm{wt}}} \longrightarrow H^1\Big(\mathbf{F}_\Sigma/\mathbf{F}, A_{\mathbb{T}_{\infty,E}}\Big)^{\Gamma^{\mathrm{wt}}} \\ \overset{\lambda_{\infty,E}}{\longrightarrow}\ & \bigoplus_{\nu \in \Sigma - S_{\mathbf{F},p}} \left(\frac{H^1(\mathbf{F}_\nu, A_{\mathbb{T}_{\infty,E}})}{H^1_{\mathrm{nr}}(\mathbf{F}_\nu, A_{\mathbb{T}_{\infty,E}})}\right)^{\Gamma^{\mathrm{wt}}} \oplus \bigoplus_{\nu \in S_{\mathbf{F},p}} \left(\frac{H^1(\mathbf{F}_\nu, A_{\mathbb{T}_{\infty,E}})}{\mathbf{X}^D_{\mathbb{T}_{\infty,E}}(\mathbf{F}_\nu)}\right)^{\Gamma^{\mathrm{wt}}} \\ \longrightarrow\ & \mathrm{Sel}_{\mathbf{F}}(\mathbb{T}_{\infty,E})_{\Gamma^{\mathrm{wt}}} \longrightarrow H^1\Big(\mathbf{F}_\Sigma/\mathbf{F}, A_{\mathbb{T}_{\infty,E}}\Big)_{\Gamma^{\mathrm{wt}}} \\ \overset{\lambda'_{\infty,E}}{\longrightarrow}\ & \bigoplus_{\nu \in \Sigma - S_{\mathbf{F},p}} \left(\frac{H^1(\mathbf{F}_\nu, A_{\mathbb{T}_{\infty,E}})}{H^1_{\mathrm{nr}}(\mathbf{F}_\nu, A_{\mathbb{T}_{\infty,E}})}\right)_{\Gamma^{\mathrm{wt}}} \oplus \bigoplus_{\nu \in S_{\mathbf{F},p}} \left(\frac{H^1(\mathbf{F}_\nu, A_{\mathbb{T}_{\infty,E}})}{\mathbf{X}^D_{\mathbb{T}_{\infty,E}}(\mathbf{F}_\nu)}\right)_{\Gamma^{\mathrm{wt}}} \longrightarrow 0. \end{aligned}$$

This notation is becoming rather cumbersome; to simplify things let's define

$$H^1_\star\big(\mathbf{F}_\nu, A_{\mathbb{T}_{\infty,E}}\big) \quad := \quad \begin{cases} H^1_{\mathrm{nr}}\big(\mathbf{F}_\nu, A_{\mathbb{T}_{\infty,E}}\big) & \text{if } \nu \nmid p \\ \mathbf{X}^D_{\mathbb{T}_{\infty,E}}(\mathbf{F}_\nu) & \text{if } \nu \mid p. \end{cases}$$

We claim that there is a commutative diagram naturally extending Figure 7.1:

$$\begin{array}{ccc}
0 & & 0 \\
\downarrow & & \downarrow \\
\mathrm{Sel}_{\mathbf{F}}(\mathcal{C}^{\min})[p^\infty] & \overset{\alpha_{\infty,E}}{\longrightarrow} & \mathrm{Sel}_{\mathbf{F}}(\mathbb{T}_{\infty,E})^{\Gamma^{\mathrm{wt}}} \\
\downarrow & & \downarrow \\
H^1\Big(\mathbf{F}_\Sigma/\mathbf{F},\, \mathcal{C}^{\min}[p^\infty]\Big) & \overset{\beta_{\infty,E}}{\longrightarrow} & H^1\Big(\mathbf{F}_\Sigma/\mathbf{F},\, A_{\mathbb{T}_{\infty,E}}\Big)^{\Gamma^{\mathrm{wt}}} \\
\downarrow \lambda_{\mathcal{C}^{\min}} & & \downarrow \lambda_{\infty,E} \\
\bigoplus_{\nu\in\Sigma} \dfrac{H^1(\mathbf{F}_\nu, \mathcal{C}^{\min}[p^\infty])}{\mathcal{C}^{\min}(\mathbf{F}_\nu)\otimes\mathbb{Q}_p/\mathbb{Z}_p} & \overset{\oplus\delta_{\infty,E,\nu}}{\longrightarrow} & \bigoplus_{\nu\in\Sigma} \left(\dfrac{H^1(\mathbf{F}_\nu, A_{\mathbb{T}_{\infty,E}})}{H^1_\star(\mathbf{F}_\nu, A_{\mathbb{T}_{\infty,E}})}\right)^{\Gamma^{\mathrm{wt}}} \\
\downarrow & & \downarrow \\
H^1_f\Big(\mathbf{F}, \mathrm{Ta}_p(\mathcal{C}^{\min})\Big)^\vee & \overset{\widehat{\theta}_{\infty,E}}{\longleftarrow} & \mathrm{Sel}_{\mathbf{F}}(\mathbb{T}_{\infty,E})_{\Gamma^{\mathrm{wt}}} \\
\downarrow & & \downarrow \\
H^2\Big(\mathbf{F}_\Sigma/\mathbf{F},\, \mathcal{C}^{\min}[p^\infty]\Big) & \overset{\tau_{\infty,E}}{\hookleftarrow} & H^1\Big(\mathbf{F}_\Sigma/\mathbf{F},\, A_{\mathbb{T}_{\infty,E}}\Big)_{\Gamma^{\mathrm{wt}}} \\
\downarrow & & \downarrow \\
\vdots & & \vdots
\end{array}$$

Figure 9.1

The columns are clearly exact from our discussion of both Poitou-Tate sequences. Similarly, the commutativity of the top two squares has already been established. The bottom horizontal map arises via the specialisation

$$0 \longrightarrow H^1\Big(\mathbf{F}_\Sigma/\mathbf{F},\, A_{\mathbb{T}_{\infty,E}}\Big)_{\Gamma^{\mathrm{wt}}} \overset{\tau_{\infty,E}}{\longrightarrow} H^2\Big(\mathbf{F}_\Sigma/\mathbf{F},\, \mathcal{C}^{\min}[p^\infty]\Big)$$
$$\longrightarrow H^2\Big(\mathbf{F}_\Sigma/\mathbf{F},\, A_{\mathbb{T}_{\infty,E}}\Big)^{\Gamma^{\mathrm{wt}}} \longrightarrow 0.$$

Finally, the construction of the homomorphism

$$\widehat{\theta}_{\infty,E} : \mathrm{Sel}_{\mathbf{F}}(\mathbb{T}_{\infty,E}) \otimes_{\Lambda^{\mathrm{wt}}} \mathbb{Z}_p \longrightarrow H^1_f\Big(\mathbf{F}, \mathrm{Ta}_p(\mathcal{C}^{\min})\Big)^\vee$$

is highly non-trivial. It is intimately connected to the existence of a p-adic pairing

$$H^1_f\Big(\mathbf{F}, \mathrm{Ta}_p(\mathcal{C}^{\min})\Big) \times H^1_f\Big(\mathbf{F}, \mathrm{Ta}_p(\mathcal{C}^{\min})\Big) \longrightarrow \mathbb{Q}_p.$$

These matters are considered at great length in §9.4 – for now one must be patient.

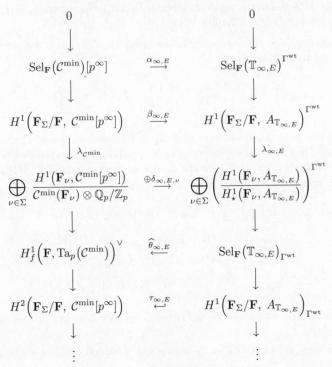

Analytic rank zero revisited.
As a useful illustration of the general approach, we return to the situation which was considered at the very start of the last chapter. In particular, for the moment we shall temporarily assume that the L-function of $E/\mathbf{F}$ does not vanish at $s = 1$.

Let us first define the local invariant $\mathcal{L}_p^{\mathrm{wt}}(E) := \begin{cases} \prod_{\mathfrak{p}|p} \#\widetilde{E}(f_{\mathfrak{p}})[p^\infty] & \text{if } p \nmid N_E \\ 1 & \text{if } p \,\|\, N_E. \end{cases}$

Theorem 9.3. *Assume $L(E/\mathbf{F},1) \neq 0$, that $\rho_{E,p}$ is residually irreducible, and there are no rational cyclic p-isogenies between $E_{/\mathbb{Q}}$ and any other elliptic curve. Then $\text{Ш}_{\mathbf{F}}(\mathbb{T}_{\infty,E})$ does not vanish at $X = 0$, in fact*

$$\text{Ш}_{\mathbf{F}}\big(\mathbb{T}_{\infty,E}\big)\Big|_{X=0} \times \prod_{\text{finite } \nu} \mathrm{Tam}^{(2)}_{\mathbf{F}_\nu}\big(\mathbb{T}_{\infty,E}\big) \times \mathrm{vol}^{(2)}_{\mathbb{T}_{\infty,E}}\big(H^1_{\mathcal{E}}\big)$$

$$= \mathcal{L}_p^{\mathrm{wt}}(E) \times \mathrm{Per}^{(2)}_{\mathbf{I},\lambda_E}(\mathbf{F}) \times \frac{\#\text{Ш}_{\mathbf{F}}(E)}{\#E(\mathbf{F})^2} \times \prod_{\text{finite } \nu} \big[E(\mathbf{F}_\nu) : E_0(\mathbf{F}_\nu)\big] \times \mathrm{vol}_E\big(H^1_e\big)$$

up to an element of $\mathbb{Z}_p^\times$.

A special case of this result was shown in [Db1, Thm 1.4] under certain restrictions. Before supplying the full demonstration, we should make a couple of observations. The left-hand side comprises the specialisations of Λ^{wt}-adic objects to weight two. Conversely, the right-hand side consists of the corresponding p-adic objects, which just so happen to be the terms occurring in the BSD conjecture.

Stretching the analogy a bit further, the right-hand side of this formula should *conjecturally* coincide with the special value

$$\mathcal{L}_p^{\mathrm{wt}}(E) \times \mathrm{Per}^{(2)}_{\mathbf{I},\lambda_E}(\mathbf{F}) \times \frac{L(E/\mathbf{F},1)}{\Omega_{E/\mathbf{F}}} \times \mathrm{vol}_E\big(H^1_e\big)$$

again up to a p-adic unit. However this is precisely the improved p-adic L-value $\mathbf{L}_p^{\mathrm{imp}}(\mathbf{f}/\mathbf{F},2,1)$, scaled by the non-archimedean volume of $H^1_e\big(\mathbf{F}\otimes\mathbb{Z}_p, \mathrm{Ta}_p(E)\big)$.

More on this strange coincidence later on …

Proof of 9.3: We begin by remarking that $\mathrm{Ta}_p(E) \cong \mathrm{Ta}_p(\mathcal{C}^{\min})$ as a $G_{\mathbb{Q}}$-module, otherwise there would have to exist a rational cyclic p-isogeny between E and $\mathcal{C}^{\min}$. N.B. All of the equalities in this proof are assumed to hold true only modulo $\mathbb{Z}_p^\times$.

It follows from Corollary 8.2 that

$$\text{Ш}_{\mathbf{F}}\big(\mathbb{T}_{\infty,E}\big)\Big|_{X=0} = \frac{\#\text{Ш}_{\mathbf{F}}(E)[p^\infty]}{\#E(\mathbf{F})[p^\infty]} \times \#H^1\Big(\Gamma^{\mathrm{wt}}, \mathrm{Sel}_{\mathbf{F}}\big(\mathbb{T}_{\infty,E}\big)\Big)^{-1}$$

$$\times \frac{\prod_{\nu\in\Sigma} \#\mathrm{Ker}\big(\delta_{\infty,E,\nu}\big)}{\prod_{\nu\in\Sigma}\Big[\mathrm{Ker}\big(\delta_{\infty,E,\nu}\big) : \mathrm{Im}\big(\lambda_E\big) \cap \mathrm{Ker}\big(\delta_{\infty,E,\nu}\big)\Big]}.$$

Now applying Propositions 8.7(i) and 8.9(a),(b) we have an equality

$$\prod_{\nu\in\Sigma} \#\mathrm{Ker}\big(\delta_{\infty,E,\nu}\big) = \frac{\mathcal{L}_p^{\mathrm{wt}}(E)}{\mathrm{covol}^{(2)}_{\mathbb{T}_{\infty,E}}\big(\mathbf{F}\otimes\mathbb{Z}_p\big)} \times \prod_{\nu\in\Sigma} \frac{\big[E(\mathbf{F}_\nu) : E_0(\mathbf{F}_\nu)\big]}{\mathrm{Tam}^{(2)}_{\mathbf{F}_\nu}\big(\mathbb{T}_{\infty,E}\big)}.$$

However Proposition 8.16 then informs us

$$\frac{1}{\mathrm{covol}^{(2)}_{\mathbb{T}_{\infty,E}}\left(\mathbf{F}\otimes\mathbb{Z}_p\right)} \quad = \quad \mathrm{Per}^{(2)}_{\mathbf{I},\lambda_E}(\mathbf{F}) \times \frac{\prod_{\mathfrak{p}|p} \mathrm{vol}_{E,\mathbf{F}_\mathfrak{p}}\Big(H^1_e\big(\mathbf{F}_\mathfrak{p},\mathrm{Ta}_p(E)\big)\Big)}{\mathrm{vol}^{(2)}_{\mathbb{T}_{\infty,E}}\Big(H^1_{\mathcal{E}}\big(\mathbf{F}\otimes\mathbb{Z}_p,\mathbb{T}^*_{\infty,E}(1)\big)\Big)}$$

which explains why the volume and **I**-adic period enter the formula.

Remark: The theorem will follow, provided one can establish that

$$\#H^1\Big(\Gamma^{\mathrm{wt}},\ \mathrm{Sel}_{\mathbf{F}}(\mathbb{T}_{\infty,E})\Big) \ \times \prod_{\nu\in\Sigma}\Big[\mathrm{Ker}(\delta_{\infty,E,\nu}) : \mathrm{Im}(\lambda_E)\cap\mathrm{Ker}(\delta_{\infty,E,\nu})\Big]$$

is equal to the size of the p^∞-Sylow subgroup of the Mordell-Weil group $E(\mathbf{F})$.

Under the circumstances we are considering, the Selmer group of E over $\mathbf{F}$ is finite, in which case both $H^2(\mathbf{F}_\Sigma/\mathbf{F},E[p^\infty])$ and $H^1(\mathbf{F}_\Sigma/\mathbf{F},A_{\mathbb{T}_{\infty,E}})_{\Gamma^{\mathrm{wt}}}$ must be trivial. Consequently the bottom half of Figure 9.1 looks like

$$\begin{array}{ccc}
\vdots & & \vdots \\
\downarrow & & \downarrow \\
H^1\Big(\mathbf{F}_\Sigma/\mathbf{F},\ E[p^\infty]\Big) & \xrightarrow{\beta_{\infty,E}} & H^1\Big(\mathbf{F}_\Sigma/\mathbf{F},\ A_{\mathbb{T}_{\infty,E}}\Big)^{\Gamma^{\mathrm{wt}}} \\
\downarrow \lambda_E & & \downarrow \lambda_{\infty,E} \\
\bigoplus_{\nu\in\Sigma}\dfrac{H^1(\mathbf{F}_\nu,E[p^\infty])}{E(\mathbf{F}_\nu)\otimes\mathbb{Q}_p/\mathbb{Z}_p} & \xrightarrow{\oplus\delta_{\infty,E,\nu}} & \bigoplus_{\nu\in\Sigma}\left(\dfrac{H^1(\mathbf{F}_\nu,A_{\mathbb{T}_{\infty,E}})}{H^1_\star(\mathbf{F}_\nu,A_{\mathbb{T}_{\infty,E}})}\right)^{\Gamma^{\mathrm{wt}}} \\
\downarrow & & \downarrow \\
H^1_f\Big(\mathbf{F},\mathrm{Ta}_p(E)\Big)^\vee & \dashrightarrow & H^1\Big(\Gamma^{\mathrm{wt}},\ \mathrm{Sel}_{\mathbf{F}}(\mathbb{T}_{\infty,E})\Big) \\
\downarrow & & \downarrow \\
0 & & 0
\end{array}$$

Via the standard nonsense, one readily deduces

$$\Big[\mathrm{Ker}(\oplus\delta_{\infty,E,\nu}) : \mathrm{Im}(\lambda_E)\cap\mathrm{Ker}(\oplus\delta_{\infty,E,\nu})\Big] = \frac{\Big[\bigoplus_{\nu\in\Sigma}H^1(\mathbf{F}_\nu,E)[p^\infty] : \mathrm{Im}(\lambda_E)\Big]}{\Big[\mathrm{Im}(\oplus\,\delta_{\infty,E,\nu}) : \oplus\delta_{\infty,E,\nu}(\mathrm{Im}(\lambda_E))\Big]}.$$

The numerator equals $\#H^1_f(\mathbf{F},\mathrm{Ta}_p(E))^\vee = \#E(\mathbf{F})[p^\infty]$ because $\text{Ш}_{\mathbf{F}}(E)$ is finite. Exploiting the surjectivity of $\beta_{\infty,E}$ and $\delta_{\infty,E,\nu}$, the denominator is $\#\mathrm{Coker}(\lambda_{\infty,E})$.

The demonstration is finished.

□

9.2 Triviality of the compact Selmer group

Let us now supply the missing proof of Theorem 9.2.

Its statement asserted that the compact Selmer group $\mathfrak{S}_{\mathbf{F}}(\mathbb{T}_{\infty,E})$ was trivial, subject to the hypothesis that $\rho_{\infty,E}: G_{\mathbb{Q}} \longrightarrow \mathrm{GL}_2(\mathcal{R}_E)$ is residually irreducible. At this stage we should point out that the local conditions $H^1_{\mathrm{nr}}(\mathbf{F}_\nu, \mathbb{T}_{\infty,E})$ defining $\mathfrak{S}_{\mathbf{F}}(\mathbb{T}_{\infty,E})$ at the finite places $\nu \nmid p$ must be the whole of the local cohomology, because the Λ^{wt}-ranks of the ambient groups are identically zero.

As a corollary, the compact Selmer group coincides with

$$\mathrm{Ker}\left(H^1\Big(\mathbf{F}_\Sigma/\mathbf{F},\ \mathbb{T}_{\infty,E}\Big) \overset{\oplus \mathrm{res}_{\mathfrak{p}}}{\longrightarrow} \bigoplus_{\mathfrak{p}\in S_{\mathbf{F},p}} \frac{H^1(\mathbf{F}_{\mathfrak{p}}, \mathbb{T}_{\infty,E})}{\mathbf{X}_{\mathbb{T}_{\infty,E}}(\mathbf{F}_{\mathfrak{p}})} \right).$$

There are three stages. We first show that the compact Selmer group is Λ^{wt}-torsion. Using a version of Nekovář's control theory along the critical line $(s,k) \in \{1\} \times \mathbb{U}_E$, we next establish it has finite order. Lastly one embeds $\mathfrak{S}_{\mathbf{F}}$ inside a pro-tower of $\mathbf{F}$-rational points, whose narrow structure implies the Selmer group is zero.

Step 1: The Λ^{wt}-rank is zero.

Examining the behaviour of our dual exponential $\mathrm{EXP}^*_{\mathbb{T}_{\infty,E}}$ from Definition 5.11, there is a tautological sequence of Λ^{wt}-homomorphisms

$$\begin{array}{ccccccc} 0 \longrightarrow \mathfrak{S}_{\mathbf{F}}(\mathbb{T}_{\infty,E}) \longrightarrow H^1\Big(\mathbf{F}_\Sigma/\mathbf{F},\ \mathbb{T}_{\infty,E}\Big) & \overset{\mathrm{loc}_p(-) \bmod \mathbf{X}_{\mathbb{T}_{\infty,E}}}{\longrightarrow} & \dfrac{H^1(\mathbf{F}\otimes\mathbb{Z}_p, \mathbb{T}_{\infty,E})}{\mathbf{X}_{\mathbb{T}_{\infty,E}}(\mathbf{F}\otimes\mathbb{Z}_p)} \\ & & \Big\downarrow \mathrm{EXP}^*_{\mathbb{T}_{\infty,E}} \\ & & \mathbf{F}\otimes_{\mathbb{Z}_p}\mathbb{Z}_p\langle\langle w\rangle\rangle \end{array}$$

which is exact along the row. By Lemma 8.10, the Λ^{wt}-rank of the image of the downwards arrow $\mathrm{EXP}^*_{\mathbb{T}_{\infty,E}}$ is equal to $[\mathbf{F}:\mathbb{Q}]$.

Remark: A global Euler characteristic calculation reveals that

$$\mathrm{rank}_{\Lambda^{\mathrm{wt}}}\Big(H^1(\mathbf{F}_\Sigma/\mathbf{F}, \mathbb{T}_{\infty,E})\Big) = \mathrm{rank}_{\Lambda^{\mathrm{wt}}}\Big(H^2(\mathbf{F}_\Sigma/\mathbf{F}, \mathbb{T}_{\infty,E})\Big) + d^+(\rho_{\infty,E})$$

where $d^+(\rho_{\infty,E})$ denotes the rank for the eigenspace of the induced representation $\mathrm{Ind}_{\mathbf{F}/\mathbb{Q}}(\mathbb{T}_{\infty,E})$ fixed by complex conjugation. In our situation, this will be precisely half the total rank, i.e. $[\mathbf{F}:\mathbb{Q}]$.

For all bar finitely many bad weights $k \in \mathbb{U}_E \cap \mathbb{N}_{\geq 2}$, the rank of the H^2 coincides with that of its specialisations, namely

$$\mathrm{rank}_{\Lambda^{\mathrm{wt}}}\Big(H^2(\mathbf{F}_\Sigma/\mathbf{F}, \mathbb{T}_{\infty,E})\Big) = \dim_{\mathbb{Q}_p}\Big(H^2(\mathbf{F}_\Sigma/\mathbf{F}, V^*_{\mathbf{f}_k})\Big).$$

Under our assumptions on $\rho_{\infty,E}$, at infinitely many such k the (contragredient) p-adic Galois representation attached to the eigenform $\mathbf{f}_k$ is residually irreducible; in this case, the $\mathbb{Q}_p$-dimension of $H^2(\mathbf{F}_\Sigma/\mathbf{F}, V^*_{\mathbf{f}_k})$ is zero due to [Ka1, Thm 14.5(1)].

On the other hand, the bottom layer of the super zeta-elements $\mathfrak{Z}_{\mathbf{F}}^{(\psi)}$ with $\psi \in X_{\mathbf{F}}$ generate a Λ^{wt}-submodule $\mathbb{M}_{\mathrm{glob}}$ of maximal rank inside $H^1(\mathbf{F}_\Sigma/\mathbf{F},\ \mathbb{T}_{\infty,E})$, and also a Λ^{wt}-submodule $\mathbb{M}_{\mathrm{loc}}$ inside the quotient $H^1(\mathbf{F}\otimes\mathbb{Z}_p, \mathbb{T}_{\infty,E})\big/\mathbf{X}_{\mathbb{T}_{\infty,E}}(\mathbf{F}\otimes\mathbb{Z}_p)$. In fact, $\mathbb{M}_{\mathrm{loc}}$ is injected into $\mathbf{F}\otimes_{\mathbb{Z}_p}\mathbb{Z}_p\langle\langle w\rangle\rangle$ under the big dual exponential map. Moreover the former submodule $\mathbb{M}_{\mathrm{glob}}$ is mapped onto the latter one $\mathbb{M}_{\mathrm{loc}}$ under the homomorphism $\mathrm{loc}_p(-)$ modulo $\mathbf{X}_{\mathbb{T}_{\infty,E}}$, and the kernel must then be Λ^{wt}-torsion since none of the individual $\mathfrak{Z}_{\mathbf{F}}^{(\psi)}$'s is killed by the composition $\mathrm{EXP}^*_{\mathbb{T}_\infty,E}\circ\mathrm{loc}_p(-)$. Consequently

$$\mathrm{rank}_{\Lambda^{\mathrm{wt}}}\Big(\mathfrak{S}_{\mathbf{F}}(\mathbb{T}_{\infty,E})\Big) = \mathrm{rank}_{\Lambda^{\mathrm{wt}}}\Big(H^1(\mathbf{F}_\Sigma/\mathbf{F},\mathbb{T}_{\infty,E})\Big) - \mathrm{rank}_{\Lambda^{\mathrm{wt}}}\Big(\mathbb{M}_{\mathrm{loc}}\Big)$$
$$\overset{\text{by Kato}}{=} 0 + d^+(\rho_{\infty,E}) - \mathrm{rank}_{\Lambda^{\mathrm{wt}}}\mathrm{Im}\Big(\mathrm{EXP}^*_{\mathbb{T}_\infty,E}\Big) \overset{\text{by 8.10}}{=} 0.$$

The global H^1 shares the same rank as the quotient of the local H^1 by $\mathbf{X}_{\mathbb{T}_{\infty,E}}(-)$. As a corollary, one may identify $\mathfrak{S}_{\mathbf{F}}(\mathbb{T}_{\infty,E})$ with the Λ^{wt}-torsion submodule of $H^1(\mathbf{F}_\Sigma/\mathbf{F},\ \mathbb{T}_{\infty,E})$, because the local condition $\mathbf{X}_{\mathbb{T}_{\infty,E}}(\mathbf{F}\otimes\mathbb{Z}_p)$ is saturated.

Step 2: The compact Selmer group has finite order.
Given one now knows this Selmer group to be Λ^{wt}-torsion, by the structure theory it sits inside an exact sequence

$$0 \longrightarrow \text{finite group} \longrightarrow \mathfrak{S}_{\mathbf{F}}(\mathbb{T}_{\infty,E}) \longrightarrow \bigoplus_{j=1}^{r}\frac{\Lambda^{\mathrm{wt}}}{\mathcal{P}_j^{e_j}} \longrightarrow \text{finite group} \longrightarrow 0.$$

This decomposition naturally suggests the following

Question. *Does $\mathfrak{S}_{\mathbf{F}}(\mathbb{T}_{\infty,E})$ contain pseudo-summands of the form $\Lambda^{\mathrm{wt}}/F_j^{e_j}\Lambda^{\mathrm{wt}}$ for some irreducible distinguished polynomial F_j and for $e_j\in\mathbb{N}$?*

To provide an answer, we will need to specialise at arithmetic weights inside $\mathbb{U}_E$. For any de Rham $\mathrm{Gal}(\mathbf{F}_\Sigma/\mathbf{F})$-lattice $\mathbf{T}$, the Selmer group $H^1_{g,\mathrm{Spec}\mathcal{O}_{\mathbf{F}}}(\mathbf{F}_\Sigma/\mathbf{F},\mathbf{T})$ is defined to be

$$\mathrm{Ker}\left(H^1(\mathbf{F}_\Sigma/\mathbf{F},\mathbf{T}) \overset{\oplus\mathrm{res}_\nu}{\longrightarrow} \bigoplus_{\nu\in\Sigma-S_{\mathbf{F},p}} H^1(I_{\mathbf{F}_\nu},\mathbf{T}) \oplus \bigoplus_{\nu\in S_{\mathbf{F},p}}\frac{H^1(\mathbf{F}_\nu,\mathbf{T})}{H^1_g(\mathbf{F}_\nu,\mathbf{T})}\right).$$

The following result is proved in Appendix C, and is the compact analogue of an earlier Theorem 7.15.

Theorem 9.4. *(Smith) Assume that the prime p does not ramify in the field $\mathbf{F}$. Then for all bar finitely many integral weights $k\in\mathbb{U}_E$, the induced specialisation*

$$\mathfrak{S}_{\mathbf{F}}(\mathbb{T}_{\infty,E})\otimes_{\Lambda^{\mathrm{wt}},\mathfrak{g}_k}\mathbb{Z}_p \longrightarrow H^1_{g,\mathrm{Spec}\mathcal{O}_{\mathbf{F}}}\Big(\mathbf{F}_\Sigma/\mathbf{F},\ \mathbb{T}_{\infty,E}\otimes_{\Lambda^{\mathrm{wt}},\mathfrak{g}_k}\mathbb{Z}_p\Big)$$

has finite kernel and cokernel, bounded independently of $\mathfrak{g}_k = u_0-(1+p)^{k-2}$.

N.B. This result holds for pseudo-geometric families in the sense of Definition 5.13, although it is only in the slope zero context that we need apply it here.

Corollary 9.5. *For almost all arithmetic weights $k \in \mathbb{U}_E \cap \mathbb{N}_{\geq 2}$, the order of each specialisation $\mathfrak{S}_{\mathbf{F}}(\mathbb{T}_{\infty,E}) \otimes_{\Lambda^{\mathrm{wt}},\mathfrak{g}_k} \mathbb{Z}_p$ is universally bounded.*

Proof: Let us observe that $\mathbb{T}_{\infty,E} \otimes_{\Lambda^{\mathrm{wt}},\mathfrak{g}_k} \mathbb{Z}_p$ is a lattice inside $V^*_{\mathbf{f}_k}$, the contragredient of Deligne's $G_{\mathbb{Q}}$-representation attached to the eigenform $\mathbf{f}_k \in \mathcal{S}_k^{\mathrm{ord}}(\Gamma_0(Np^r), \omega^{2-k})$. Now the Selmer group $H^1_{g,\mathrm{Spec}\mathcal{O}_{\mathbf{F}}}\left(\mathbf{F}_\Sigma/\mathbf{F}, V^*_{\mathbf{f}_k}\right)$ coincides with $H^1_{f,\mathrm{Spec}\mathcal{O}_{\mathbf{F}}}\left(\mathbf{F}_\Sigma/\mathbf{F}, V^*_{\mathbf{f}_k}\right)$ as the local condition $H^1_g(\mathbf{F}_{\mathfrak{p}}, V^*_{\mathbf{f}_k})$ is never strictly larger than $H^1_f(\mathbf{F}_{\mathfrak{p}}, V^*_{\mathbf{f}_k})$ if $\mathfrak{p}|p$.

We shall require the following deep result.

Theorem 7.16'. [Ka1, Thm 14.2] *For almost all integral weights $k \geq 3$ inside $\mathbb{U}_E$, the Bloch-Kato compact Selmer group $H^1_{f,\mathrm{Spec}\mathcal{O}_{\mathbf{F}}}\left(\mathbf{F}_\Sigma/\mathbf{F},\ \mathbb{T}_{\infty,E} \otimes_{\Lambda^{\mathrm{wt}},\mathfrak{g}_k} \mathbb{Z}_p\right)$ is finite.*

Actually this theorem was stated in terms of discrete Selmer groups instead, but they are easily seen to be equivalent. The non-vanishing of the L-value $L(\mathbf{f}_k/\mathbf{F}, 1)$ together with the residual irreducibility of $\rho_{\infty,E} \otimes_{\Lambda^{\mathrm{wt}},\mathfrak{g}_k} \mathbb{Z}_p$ at almost all $k \in \mathbb{U}_E$, forces these specialisations to be finite p-groups.

By the above $H^1_{f,\mathrm{Spec}\mathcal{O}_{\mathbf{F}}}$ is finite, so it lies in $H^1\left(\mathbf{F}_\Sigma/\mathbf{F},\ \mathbb{T}_{\infty,E} \otimes_{\Lambda^{\mathrm{wt}},\mathfrak{g}_k} \mathbb{Z}_p\right)[p^\infty]$; the latter torsion is identified with $H^0\left(\mathbf{F}_\Sigma/\mathbf{F}, (\mathbb{T}_{\infty,E} \otimes_{\Lambda^{\mathrm{wt}},\mathfrak{g}_k} \mathbb{Z}_p) \otimes \mathbb{Q}/\mathbb{Z}\right)$ via a standard technique in continuous cohomology. It follows from Theorem 9.4 that

$$\mathfrak{S}_{\mathbf{F}}(\mathbb{T}_{\infty,E}) \otimes_{\Lambda^{\mathrm{wt}},\mathfrak{g}_k} \mathbb{Z}_p \overset{\mathrm{nat}}{\to} H^1_{f,\mathrm{Spec}\mathcal{O}_{\mathbf{F}}}\left(\mathbf{F}_\Sigma/\mathbf{F},\ -\right) \hookrightarrow \left((\mathbb{T}_{\infty,E} \otimes_{\Lambda^{\mathrm{wt}},\mathfrak{g}_k} \mathbb{Z}_p) \otimes \mathbb{Q}/\mathbb{Z}\right)^{G_{\mathbf{F}}}$$

has its kernel killed by a universal power p^{ν_1} say, independent of the weight k.

Remark: This sequence respects the action of $\mathrm{Gal}(\mathbf{F}/\mathbb{Q})$ and can be decomposed. For each character $\psi \in X_{\mathbf{F}}$, the above induces a mapping

$$\mathfrak{S}_{\mathbf{F}}(\mathbb{T}_{\infty,E})^{(\psi)} \otimes_{\Lambda^{\mathrm{wt}},\mathfrak{g}_k} \mathbb{Z}_p \quad \overset{\times \mathbf{e}_\psi^{-1}}{\longrightarrow} \quad \left(\left((\psi^{-1} \otimes \mathbb{T}_{\infty,E}) \otimes_{\Lambda^{\mathrm{wt}},\mathfrak{g}_k} \mathbb{Z}_p\right) \otimes \mathbb{Q}/\mathbb{Z}\right)^{G_{\mathbb{Q}}}$$

on the ψ-components, whose kernel will again be annihilated by the integer p^{ν_1}.

Let us choose any prime number $l \nmid Np$.

By definition $1 - a_l(\mathbf{f}_k).\mathrm{Frob}_l + l < l >^{k-2} .\mathrm{Frob}_l^2$ is zero on $V^*_{\mathbf{f}_k}$, therefore $1 - \psi(l)a_l(\mathbf{f}_k) + \psi(l)^2 l < l >^{k-2}$ must kill off $\left((\psi^{-1} \otimes \mathbb{T}_{\infty,E} \otimes_{\Lambda,\mathfrak{g}_k} \mathbb{Z}_p) \otimes \mathbb{Q}/\mathbb{Z}\right)^{G_{\mathbb{Q}}}$ because Frobenius acts trivially on the $G_{\mathbb{Q}}$-invariants. We claim that there are infinitely many choices of l for which $1 - \psi(l)a_l(\mathbf{f}_k) + \psi(l)^2 l < l >^{k-2} \neq 0$. If not,

$$1 - \psi(l)a_l(\mathbf{f}_k)l^{-s} + \psi(l)^2 l < l >^{k-2} l^{-2s} = \left(1 - l^{-s}\right)\left(1 - \omega^{2-k}(l)l^{k-1-s}\right) \quad \text{for all } l \notin S$$

where S is some finite set containing Σ. Proceeding further down this cul-de-sac, we would obtain an equality of incomplete L-functions

$$L_S(\mathbf{f}_k, \psi, s) \ = \ \zeta_S(s) L_S(\omega^{2-k}, s + 1 - k)$$

which is patently ridiculous as $\mathbf{f}_k \otimes \psi$ is a cusp form, not an Eisenstein series!

Provided two weights $k, k' \in \mathbb{U}_E$ satisfy $k \equiv k' \bmod (p-1)p^{c+\mathrm{ord}_p\mathrm{radius}(\mathbb{U}_E)}$, then

$$1-\psi(l)a_l(\mathbf{f}_k)+\psi(l)^2 l < l >^{k-2} \;\equiv\; 1-\psi(l)a_l(\mathbf{f}_{k'})+\psi(l)^2 l < l >^{k'-2} \quad \text{modulo } p^{c+1}.$$

For each character $\psi \in X_{\mathbf{F}}$ and congruence class τ modulo $p-1$, we can cover weight-space by a finite collection of open disks $D_1^{\tau,\psi}, \dots, D_{n(\tau,\psi)}^{\tau,\psi}$ upon which $\mathrm{ord}_p\big(1-\psi(l)a_l(\mathbf{f}_k)+\psi(l)^2 l < l >^{k-2}\big)$ is constant for every weight $k \in D_j^{\tau,\psi}$ with $k \equiv \tau \pmod{p-1}$.

Setting $\nu_{2,\tau,\psi}$ equal to the non-negative integer

$$\max_{1\le j\le n(\tau,\psi)} \Big\{ \mathrm{ord}_p\big(1-\psi(l)a_l(\mathbf{f}_k)+\psi(l)^2 l < l >^{k-2}\big) \text{ with } k \in D_j^{\tau,\psi},\ k\equiv\tau \Big\},$$

clearly $\max_{\tau \bmod p-1}\{p^{\nu_{2,\tau,\psi}}\}$ will annihilate $\Big(\big(\psi^{-1}\otimes \mathbb{T}_{\infty,E}\otimes_{\Lambda^{\mathrm{wt}},\mathfrak{g}_k}\mathbb{Z}_p\big)\otimes\mathbb{Q}/\mathbb{Z}\Big)^{G_{\mathbb{Q}}}$. Furthermore if we put $\nu_2 := \max_{\tau,\psi}\{\nu_{2,\tau,\psi}\}$, similarly p^{ν_2} annihilates

$$\Big(\big(\mathbb{T}_{\infty,E}\otimes_{\Lambda^{\mathrm{wt}},\mathfrak{g}_k}\mathbb{Z}_p\big)\otimes\mathbb{Q}/\mathbb{Z}\Big)^{G_{\mathbf{F}}} \subset \bigoplus_{\psi\in X_{\mathbf{F}}} \Big(\big(\psi^{-1}\otimes \mathbb{T}_{\infty,E}\otimes_{\Lambda^{\mathrm{wt}},\mathfrak{g}_k}\mathbb{Z}_p\big)\otimes\mathbb{Q}/\mathbb{Z}\Big)^{G_{\mathbb{Q}}}.\mathbf{e}_{\psi}.$$

Lastly, one deduces that $p^{\nu_1+\nu_2}$ kills off $\mathfrak{S}_{\mathbf{F}}(\mathbb{T}_{\infty,E})\otimes_{\Lambda^{\mathrm{wt}},\mathfrak{g}_k}\mathbb{Z}_p$ for almost all $k\ge 3$, and the corollary is proved.

□

Remark: The answer to the question posed above is therefore negative, i.e. there can exist no pseudo-summands of the shape $\Lambda^{\mathrm{wt}}/F_j^{e_j}\Lambda^{\mathrm{wt}}$ lying inside of $\mathfrak{S}_{\mathbf{F}}(\mathbb{T}_{\infty,E})$ (otherwise the specialisations $\mathfrak{S}_{\mathbf{F}}(\mathbb{T}_{\infty,E})\otimes_{\Lambda,\mathfrak{g}_k}\mathbb{Z}_p$ would have unbounded order for varying weights k, which violates Corollary 9.5). The compact Selmer group is of finite-type over the local ring Λ^{wt}, and it follows from the structure theory that $\mathfrak{S}_{\mathbf{F}}(\mathbb{T}_{\infty,E})$ must be an abelian p-group, of order dividing $p^{\nu_1+\nu_2}$.

Step 3: The compact Selmer group is very, very finite!
We recall the definition of the degeneration maps between modular curves in §1.5. For integers $d\ge 1$ and $m,n\ge 5$ with $dm|n$, the finite map $\pi_d : X_1(n)\to X_1(m)$ operates on the affine curves $Y_1(-)$ by the rule

$$\pi_d\Big(A,\ \mu_n \overset{\theta}{\hookrightarrow} A[n]\Big) \;=\; \Big(A',\ \mu_m \overset{\theta'}{\hookrightarrow} A'[m]\Big)$$

where $A' = A/\theta(\mu_d)$, and the injection $\theta' : \mu_m \hookrightarrow \mu_{n/d} \overset{d}{\underset{\sim}{\leftarrow}} \mu_n/\mu_d \overset{\theta \bmod \mu_d}{\hookrightarrow} A/\theta(\mu_d)$.

Let's abbreviate $(\Gamma^{\mathrm{wt}})^{p^{r-1}}$ by Γ_r. Hida [Hi1] identified the Γ_r-coinvariants of $\mathbb{T}_\infty$ with the Tate module of a p-divisible subgroup of Jac $X_1(n)$ at level $n=Np^r$. The natural composition

$$H^1\Big(\mathbf{F}_\Sigma/\mathbf{F},\ \mathbb{T}_\infty\Big) \cong \varprojlim_{r\ge 1} H^1\Big(\mathbf{F}_\Sigma/\mathbf{F},\ (\mathbb{T}_\infty)_{\Gamma_r}\Big) \hookrightarrow \varprojlim_{\pi_{p\,*}} H^1\Big(\mathbf{F}_\Sigma/\mathbf{F},\ \mathrm{Ta}_p(J_r)^{\mathrm{ord}}\Big)$$

injects $\mathfrak{S}_{\mathbf{F}}(\mathbb{T}_{\infty,E})$ into the projective limit $\varprojlim_{\pi_{p\,*}}\Big(H^1\Big(\mathbf{F}_\Sigma/\mathbf{F},\ \mathrm{Ta}_p(J_r)^{\mathrm{ord}}\Big)[p^\infty]\Big)$. Again it's continuous cohomology, so the $\mathbb{Z}_p$-torsion in $H^1\Big(\mathbf{F}_\Sigma/\mathbf{F},\ \mathrm{Ta}_p(J_r)^{\mathrm{ord}}\Big)$ is isomorphic to $H^0\Big(\mathbf{F}_\Sigma/\mathbf{F},\ \mathrm{Ta}_p(J_r)^{\mathrm{ord}}\otimes\mathbb{Q}/\mathbb{Z}\Big) = J_r^{\mathrm{ord}}(\mathbf{F})[p^\infty]$ as finite groups.

The following two results of Nekovář allow us to deduce the triviality of $\mathfrak{S}_{\mathbf{F}}(\mathbb{T}_{\infty,E})$. Moreover, they are essential ingredients in our constructions (outlined in §9.3-§9.4) for the p-adic weight pairing on the Selmer group of a modular elliptic curve.

Lemma 9.6. [NP, §1.6.6] *(i)* $\pi_{1\,*}(\mathbf{e}_{\mathrm{ord}}.\mathrm{Ta}_p(J_{r+1})) \subset p\,(\mathbf{e}_{\mathrm{ord}}.\mathrm{Ta}_p(J_r))$;

Let $\frac{1}{p}\pi_{1\,*}: \mathrm{Ta}_p(J_{r+1})^{\mathrm{ord}} \to \mathrm{Ta}_p(J_r)^{\mathrm{ord}}$ *denote the map satisfying* $p\left(\frac{1}{p}\pi_{1\,*}\right) = \pi_{1\,*}$.

(ii) $\left(\frac{1}{p}\pi_{1\,*}\right)\circ\pi_1^* \ = $ *multiplication by p on* $\mathbf{e}_{\mathrm{ord}}.\mathrm{Ta}_p(J_r)$;

(iii) $\pi_1^*\circ\left(\frac{1}{p}\pi_{1\,*}\right) \ = \ \sum_{\gamma\in\Gamma_r/\Gamma_{r+1}}\langle\gamma\rangle$ *on* $\mathbf{e}_{\mathrm{ord}}.\mathrm{Ta}_p(J_{r+1})$ *where* $\Gamma_r = \left(\Gamma^{\mathrm{wt}}\right)^{p^{r-1}}$;

(iv) $\pi_{p\,*} \ = \ U_p\circ\left(\frac{1}{p}\pi_{1\,*}\right)$ *on* $\mathbf{e}_{\mathrm{ord}}.\mathrm{Ta}_p(J_{r+1})$.

Corollary 9.7. [NP, §1.6.8] *The map*

$$\varprojlim_{\pi_{p\,*}} \mathbf{e}_{\mathrm{ord}}.\mathrm{Ta}_p(J_r) \ \longrightarrow \ \varprojlim_{\frac{1}{p}\pi_{1\,*}} \mathbf{e}_{\mathrm{ord}}.\mathrm{Ta}_p(J_r), \quad (x_r)_{r\ge1} \mapsto \left(U_p^{r-1}x_r\right)_{r\ge1}$$

is an isomorphism of $\mathbf{H}^{\mathrm{ord}}_{Np^\infty}[G_{\mathbb{Q}}]$*-modules.*

We shall use these facts directly, to show the vanishing of the compact Selmer group. Because $\mathfrak{S}_{\mathbf{F}}(\mathbb{T}_{\infty,E})$ has order dividing $p^{\nu_1+\nu_2}$, for large enough $r \gg 1$ we can realise any finite subgroup $\mathbb{S} \subset \mathfrak{S}_{\mathbf{F}}(\mathbb{T}_{\infty,E})$ as a subgroup $\mathbb{S}_r$ of $\mathrm{Jac}\,X_1(Np^r)^{\mathrm{ord}}(\mathbf{F})[p^{\nu_1+\nu_2}]$.

The sequence of $\mathbb{S}_r$'s is compatible with respect to the degeneration maps $\pi_{p\,*}$ and $\pi_1^*: \mathrm{Jac}\,X_1(Np^r)(\overline{\mathbb{Q}})[p^\infty] \longrightarrow \mathrm{Jac}\,X_1(Np^{r+1})(\overline{\mathbb{Q}})[p^\infty]$, so for any $e \ge 0$

$$\mathbb{S}_r \ = \ (\pi_{p\,*})^e\left(\mathbb{S}_{r+e}\right) \ \cong \ (\pi_{p\,*})^e\circ(\pi_1^*)^e\left(\mathbb{S}_r\right).$$

By part (iv) of the lemma $(\pi_{p\,*})^e$ coincides with $\left(U_p\circ\left(\frac{1}{p}\pi_{1\,*}\right)\right)^e$, and the Albanese action of the U_p-operator is invertible on the ordinary locus. Consequently

$$\mathbb{S}_r \ \cong \ a_p(\mathbf{f})^e \times \left(\frac{1}{p}\pi_{1\,*}\right)^e \circ (\pi_1^*)^e\left(\mathbb{S}_r\right) \ \overset{\text{by 9.6(ii)}}{=} \ a_p(\mathbf{f})^e \times p^e\left(\mathbb{S}_r\right)$$

and picking $e \ge \nu_1+\nu_2$, we see that $\mathbb{S} \cong \mathbb{S}_r \subset J_r[p^{\nu_1+\nu_2}]$ must therefore be zero. But $\mathbb{S}$ was any finite subgroup of $\mathfrak{S}_{\mathbf{F}}(\mathbb{T}_{\infty,E})$, thus the Selmer group was zero too.

The proof of Theorem 9.2 is thankfully over.

9.3 The p-adic weight pairings

One of the key terms entering the Γ^{wt}-Euler characteristic is the *weight regulator.* This is the determinant of $\mathcal{C}^{\min}(\mathbf{F}) \times \mathcal{C}^{\min}(\mathbf{F})$ under a certain bilinear pairing

$$\langle -,-\rangle^{\mathrm{wt}}_{\mathbf{F},p} : H^1_f\Big(\mathbf{F}_\Sigma/\mathbf{F},\ \mathrm{Ta}_p(\mathcal{C}^{\min})\Big) \ \times \ H^1_f\Big(\mathbf{F}_\Sigma/\mathbf{F},\ \mathrm{Ta}_p(\mathcal{C}^{\min})\Big) \ \longrightarrow \ \mathbb{Q}_p.$$

To construct splittings required to define our $\langle -,-\rangle^{\mathrm{wt}}_{\mathbf{F},p}$, as in the cyclotomic case we need to view cohomology classes in terms of bi-extensions.

Interpreting one-cocycles via extension classes.
For the moment G will be an open subgroup of either $\mathrm{Gal}(\overline{\mathbf{F}}/\mathbf{F})$, or of $\mathrm{Gal}(\mathbf{F}_\nu/\mathbf{F}_\nu)$ at some place $\nu \in \Sigma$. Consider a continuous cocycle x_r lying in $H^1(G, \mathrm{Ta}_p(J_r)^{\mathrm{ord}})$. This element x_r naturally corresponds to an extension class

$$0 \longrightarrow \mathrm{Ta}_p(J_r)^{\mathrm{ord}} \longrightarrow \mathfrak{X}_{x_r} \longrightarrow \mathbb{Z}_p \longrightarrow 0$$

comprising G-modules of finite-type over $\mathbb{Z}_p$. The x_r-twisted action of G on the $\mathbb{Z}_p$-module $\mathfrak{X}_{x_r} = \mathrm{Ta}_p(J_r)^{\mathrm{ord}} \oplus \mathbb{Z}_p$ is given by $g(t,z) = \big(g(t) + z x_r(g), z\big)$.

Let us also recall from Lemma 9.6 the various transition maps $\frac{1}{p}\pi_{1\,*}$, $\pi_{p\,*}$, π_1^*. Firstly the homomorphism $\frac{1}{p}\pi_{1\,*} : \mathrm{Ta}_p(J_{r+1})^{\mathrm{ord}} \to \mathrm{Ta}_p(J_r)^{\mathrm{ord}}$ induces a transition homomorphism $H^j(G, \mathrm{Ta}_p(J_{r+1})^{\mathrm{ord}}) \longrightarrow H^j(G, \mathrm{Ta}_p(J_r)^{\mathrm{ord}})$ on the G-cohomology. Strictly speaking, one should denote the induced map by $\left(\frac{1}{p}\pi_{1\,*}\right)_*$ but it looks ugly; instead we shall abuse notation, and inaccurately refer to it as $\frac{1}{p}\pi_{1\,*}$ throughout. If there is a family of cocycles $(x_r)_{r\geq 1} \in \varprojlim_{\frac{1}{p}\pi_{1\,*}} H^1(G, \mathrm{Ta}_p(J_r)^{\mathrm{ord}})$, then in terms of extension classes we obtain

$$\begin{array}{ccccccccc}
 & & \vdots & & \vdots & & \vdots & & \\
 & & \big\downarrow \frac{1}{p}\pi_{1\,*} & & \big\downarrow & & \big\downarrow & & \\
0 & \longrightarrow & \mathrm{Ta}_p(J_{r+1})^{\mathrm{ord}} & \longrightarrow & \mathfrak{X}_{x_{r+1}} & \longrightarrow & \mathbb{Z}_p & \longrightarrow & 0 \\
 & & \big\downarrow \frac{1}{p}\pi_{1\,*} & & \big\downarrow & & \big\downarrow & & \\
0 & \longrightarrow & \mathrm{Ta}_p(J_r)^{\mathrm{ord}} & \longrightarrow & \mathfrak{X}_{x_r} & \longrightarrow & \mathbb{Z}_p & \longrightarrow & 0 \\
 & & \big\downarrow \frac{1}{p}\pi_{1\,*} & & \big\downarrow & & \big\downarrow & & \\
 & & \vdots & & \vdots & & \vdots & &
\end{array}$$

where the downward arrows represent $\cdots \mapsto x_{r+1} \overset{(\frac{1}{p}\pi_{1\,*})_*}{\mapsto} x_r \mapsto \ldots$.

Remark: If one considers the map $\pi_{p\,*} : \mathrm{Ta}_p(J_{r+1})^{\mathrm{ord}} \to \mathrm{Ta}_p(J_r)^{\mathrm{ord}}$ instead, then for every $(x_r)_{r\geq 1} \in \varprojlim_{\pi_{p\,*}} H^1(G, \mathrm{Ta}_p(J_r)^{\mathrm{ord}})$ we obtain an analogous diagram

$$\begin{array}{ccccccccc}
0 & \longrightarrow & \mathrm{Ta}_p(J_{r+1})^{\mathrm{ord}} & \longrightarrow & \mathfrak{X}_{x_{r+1}} & \longrightarrow & \mathbb{Z}_p & \longrightarrow & 0 \\
 & & \big\downarrow \pi_{p\,*} & & \big\downarrow & & \big\downarrow & & \\
0 & \longrightarrow & \mathrm{Ta}_p(J_r)^{\mathrm{ord}} & \longrightarrow & \mathfrak{X}_{x_r} & \longrightarrow & \mathbb{Z}_p & \longrightarrow & 0
\end{array}$$

on extension classes. It does not actually matter whether we choose $\frac{1}{p}\pi_{1\,*}$ or $\pi_{p\,*}$, because one already knows

$$\varprojlim_{\pi_{p\,*}} H^1(G, \mathrm{Ta}_p(J_r)^{\mathrm{ord}}) \quad \overset{\text{by } 9.7}{\cong} \quad \varprojlim_{\frac{1}{p}\pi_{1\,*}} H^1(G, \mathrm{Ta}_p(J_r)^{\mathrm{ord}}).$$

To be consistent we shall work with the former map $\frac{1}{p}\pi_{1\,*}$.

Let's focus our attention on the homomorphism $\pi_1^* : \mathrm{Ta}_p(J_r)^{\mathrm{ord}} \to \mathrm{Ta}_p(J_{r+1})^{\mathrm{ord}}$. Assume once more the cocycle $x_r \in H^1\big(G, \mathrm{Ta}_p(J_r)^{\mathrm{ord}}\big)$, which means in particular $(\pi_1^*)_*^{\,n}\, x_r \in H^1\big(G, \mathrm{Ta}_p(J_{r+n})^{\mathrm{ord}}\big)$. On extension classes $(\pi_1^*)_*^{\,n}$ induces

$$\begin{array}{ccccccccc}
 & & \vdots & & \vdots & & \vdots & & \\
 & & \uparrow{\scriptstyle (\pi_1^*)^n} & & \uparrow & & \uparrow & & \\
0 & \longrightarrow & \mathrm{Ta}_p(J_{r+n})^{\mathrm{ord}} & \longrightarrow & \mathfrak{X}_{(\pi_1^*)_*^{\,n} x_r} & \longrightarrow & \mathbb{Z}_p & \longrightarrow & 0 \\
 & & \uparrow{\scriptstyle (\pi_1^*)^n} & & \uparrow & & \uparrow & & \\
0 & \longrightarrow & \mathrm{Ta}_p(J_r)^{\mathrm{ord}} & \longrightarrow & \mathfrak{X}_{x_r} & \longrightarrow & \mathbb{Z}_p & \longrightarrow & 0 \\
 & & \uparrow{\scriptstyle (\pi_1^*)^n} & & \uparrow & & \uparrow & & \\
 & & \vdots & & \vdots & & \vdots & &
\end{array}$$

which is in the reverse direction to the transitions induced by the map $\left(\frac{1}{p}\pi_{1\,*}\right)_*^n$.
N.B. To shorten notation again, we shall often drop the subscript $_*$ from $(\pi_1^*)_*$.
Remark: These objects we have considered are all free $\mathbb{Z}_p$-modules of finite rank; the functor $\mathrm{Hom}_{\mathbb{Z}_p}\big(-,\, \mathbb{Z}_p(1)\big)$ is exact on them. Applying it to the above,

$$\begin{array}{ccccccccc}
 & & \vdots & & \vdots & & \vdots & & \\
 & & \downarrow & & \downarrow & & \downarrow{\scriptstyle \widehat{\pi_1^*}^n} & & \\
0 & \longrightarrow & \mathbb{Z}_p(1) & \longrightarrow & \mathfrak{X}^*_{(\pi_1^*)^n x_r}(1) & \longrightarrow & \mathbf{e}^*_{\mathrm{ord}}.\mathrm{Ta}_p\big({}^{\mathbf{t}}J_{r+n}\big) & \longrightarrow & 0 \\
 & & \downarrow & & \downarrow & & \downarrow{\scriptstyle \widehat{\pi_1^*}^n} & & \\
0 & \longrightarrow & \mathbb{Z}_p(1) & \longrightarrow & \mathfrak{X}^*_{x_r}(1) & \longrightarrow & \mathbf{e}^*_{\mathrm{ord}}.\mathrm{Ta}_p\big({}^{\mathbf{t}}J_r\big) & \longrightarrow & 0 \\
 & & \downarrow & & \downarrow & & \downarrow{\scriptstyle \widehat{\pi_1^*}^n} & & \\
 & & \vdots & & \vdots & & \vdots & &
\end{array}$$

where the dual mapping $\widehat{\pi_1^*}^n$ gives rise to the downward arrows in the diagram. Note ${}^{\mathbf{t}}J_r$ is the dual abelian variety to $\mathrm{Jac}X_1(Np^r)$, in fact $\mathrm{Ta}_p\big({}^{\mathbf{t}}J_r\big) \cong \mathrm{Ta}_p\big(J_r\big)^*(1)$ by the Poincaré pairing on the Jacobians (the dual projector $\mathbf{e}^*_{\mathrm{ord}}$ is then defined through the contravariant Picard action of U_p, rather than the covariant action).

Taking G-cohomology of this array, one obtains a commutative diagram

$$\begin{array}{ccccccc}
\ldots \to H^j\Big(G,\, \mathbb{Z}_p(1)\Big) & \longrightarrow & H^j\Big(G, \mathfrak{X}^*_{(\pi_1^*)^n x_r}(1)\Big) & \longrightarrow & H^j\Big(G, \mathbf{e}^*_{\mathrm{ord}}.\mathrm{Ta}_p\big({}^{\mathbf{t}}J_{r+n}\big)\Big) \to \ldots \\
\downarrow{\scriptstyle \eta_{n,r}} & & \downarrow & & \downarrow{\scriptstyle \big(\widehat{\pi_1^*}^n\big)_*} \\
\ldots \to H^j\Big(G,\, \mathbb{Z}_p(1)\Big) & \longrightarrow & H^j\Big(G,\, \mathfrak{X}^*_{x_r}(1)\Big) & \longrightarrow & H^j\Big(G, \mathbf{e}^*_{\mathrm{ord}}.\mathrm{Ta}_p\big({}^{\mathbf{t}}J_r\big)\Big) \to \ldots
\end{array}$$

which is exact along its rows.

Suppose we are given a family of elements $(x_r)_{r\geq 1} \in \varprojlim_{\frac{1}{p}\pi_{1\,*}} H^1(G, \mathrm{Ta}_p(J_r)^{\mathrm{ord}})$. The strategy now is to replace $\left\{\mathfrak{X}_{x_r}\right\}_{r\geq 1}$ with a more refined class of extensions, thereby enabling us to write down the p-adic weight pairing explicitly.

Lemma 9.8. *For each $n \geq 0$, there exists a cocycle $x'_{r+n} \in H^1(G, \mathrm{Ta}_p(J_{r+n})^{\mathrm{ord}})$ and a commutative diagram*

$$\begin{array}{ccccccc}
\ldots \to H^1\Big(G,\, \mathbb{Z}_p(1)\Big) & \longrightarrow & H^1\Big(G, \mathfrak{X}^*_{x'_{r+n}}(1)\Big) & \longrightarrow & H^1\Big(G, \mathbf{e}^*_{\mathrm{ord}}.\mathrm{Ta}_p({}^{\mathbf{t}}J_{r+n})\Big) \to \ldots \\
\Big\downarrow \eta'_n & & \Big\downarrow & & \Big\downarrow \left(\widehat{\pi_1^*}^n\right)_* \\
\ldots \to H^1\Big(G,\, \mathbb{Z}_p(1)\Big) & \longrightarrow & H^1\Big(G,\, \mathfrak{X}^*_{x_r}(1)\Big) & \longrightarrow & H^1\Big(G, \mathbf{e}^*_{\mathrm{ord}}.\mathrm{Ta}_p({}^{\mathbf{t}}J_r)\Big) \to \ldots
\end{array}$$

for which the image of η'_n is contained in $p^n.H^1(G, \mathbb{Z}_p(1))$.

Proof: By the compatibility of the x_r's,

$$(\pi_1^*)_*^n\, x_r \;=\; (\pi_1^*)_*^n \left(\frac{1}{p}\pi_{1\,*}\right)_*^n x_{r+n} \;\overset{\text{by 9.6(iii)}}{=}\; \left(\mathrm{Tr}_{\Gamma_r/\Gamma_{r+n}}\right)_* x_{r+n}$$

where the trace map is defined by $\mathrm{Tr}_{\Gamma_r/\Gamma_{r+n}} := \sum_{\gamma\in\Gamma_r/\Gamma_{r+n}} \langle\gamma\rangle$ on $\mathrm{Ta}_p(J_{r+n})^{\mathrm{ord}}$. Since this trace has degree p^n, so does the map $\left(\mathrm{Tr}_{\Gamma_r/\Gamma_{r+n}}\right)_*$ induced on cohomology. It follows that there exists a cocycle x'_{r+n} such that $p^n x'_{r+n} = \left(\mathrm{Tr}_{\Gamma_r/\Gamma_{r+n}}\right)_* x_{r+n}$. We must verify that x'_{r+n} has the required properties.

Remark: Assume M is a free $\mathbb{Z}_p$-module of finite rank, with a continuous G-action. Let $x \in H^1(G, M)$ be a one-cocycle. Then the commutative diagram

$$\begin{array}{ccccccccc}
0 & \longrightarrow & M & \longrightarrow & \mathfrak{X}_{p^n x} & \longrightarrow & \mathbb{Z}_p & \longrightarrow & 0 \\
 & & \Big\downarrow \mathrm{id} & & \Big\downarrow (\mathrm{id}, p^n) & & \Big\downarrow p^n & & \\
0 & \longrightarrow & M & \longrightarrow & \mathfrak{X}_x & \longrightarrow & \mathbb{Z}_p & \longrightarrow & 0
\end{array}$$

respects the $\mathbb{Z}_p[G]$-structure, and relates the extension class of $p^n x$ to that of x.

Substituting $M = \mathrm{Ta}_p(J_{r+n})^{\mathrm{ord}}$ and $x = x'_{r+n}$, by definition $p^n x'_{r+n} = (\pi_1^*)_*^n\, x_r$. Dualising the above diagram via $\mathrm{Hom}_{\mathbb{Z}_p}\big(-,\, \mathbb{Z}_p(1)\big)$, one obtains

$$\begin{array}{ccccccccc}
0 & \longrightarrow & \mathbb{Z}_p(1) & \longrightarrow & \mathfrak{X}^*_{x'_{r+n}}(1) & \longrightarrow & \mathbf{e}^*_{\mathrm{ord}}.\mathrm{Ta}_p({}^{\mathbf{t}}J_{r+n}) & \longrightarrow & 0 \\
 & & \Big\downarrow p^n & & \Big\downarrow (p^n,\, \mathrm{id}) & & \Big\downarrow \mathrm{id} & & \\
0 & \longrightarrow & \mathbb{Z}_p(1) & \longrightarrow & \mathfrak{X}^*_{(\pi_1^*)^n x_r}(1) & \longrightarrow & \mathbf{e}^*_{\mathrm{ord}}.\mathrm{Ta}_p({}^{\mathbf{t}}J_{r+n}) & \longrightarrow & 0 \\
 & & \Big\downarrow & & \Big\downarrow & & \Big\downarrow \widehat{\pi_1^*}^n & & \\
0 & \longrightarrow & \mathbb{Z}_p(1) & \longrightarrow & \mathfrak{X}^*_{x_r}(1) & \longrightarrow & \mathbf{e}^*_{\mathrm{ord}}.\mathrm{Ta}_p({}^{\mathbf{t}}J_r) & \longrightarrow & 0.
\end{array}$$

The right-hand column is the map $\widehat{\pi_1^*}^n$. The left-hand column is $p^n \times$ (a scalar), so the homomorphism η'_n it induces on 1-cohomology ends up in $p^n.H^1(G, \mathbb{Z}_p(1))$. □

The construction of canonical elements.
Before explaining how to associate canonical lifts, we first outline some local theory. Let ϱ denote any non-trivial character of $\mathrm{Gal}\big(\mathbf{F}(\mu_{p^\infty})/\mathbf{F}\big)$ taking values in $\mathbb{Z}_p^\times$, and set $\ell_\varrho := \log_p \circ \varrho$. If $i_{\mathbf{F}} : \mathbb{A}_{\mathbf{F}}^\times \to \mathrm{Gal}\big(\mathbf{F}(\mu_{p^\infty})/\mathbf{F}\big)$ is the reciprocity map, then

$$\widetilde{\ell_\varrho} : \mathbb{A}_{\mathbf{F}}^\times \longrightarrow \mathbb{Z}_p \qquad \text{by composing } \widetilde{\ell_\varrho} = \log_p \circ \varrho \circ i_{\mathbf{F}}.$$

One writes $\widetilde{\ell_{\varrho,\nu}}$ for the composition of $\widetilde{\ell_\varrho}$ with the inclusion $\mathbf{F}_\nu^\times \hookrightarrow \mathbb{A}_{\mathbf{F}}^\times$ at all $\nu \in \Sigma$. In particular, the kernel of $\widetilde{\ell_{\varrho,\nu}}$ is the group of universal norms relative to $\mathbf{F}(\mu_{p^\infty})/\mathbf{F}$, namely

$$\mathrm{Ker}\Big(\widetilde{\ell_{\varrho,\nu}}\Big) \;=\; \bigcap_{n\geq 0} \mathrm{Norm}_{\mathbf{F}_\nu(\mu_{p^n})/\mathbf{F}_\nu}\Big(\mathbf{F}_\nu(\mu_{p^n})^\times\Big).$$

Moreover $\widetilde{\ell_{\varrho,\nu}}$ naturally extends to $H^1\big(\mathbf{F}_\nu, \mathbb{Z}_p(1)\big) \cong \mathbf{F}_\nu^\times \,\widehat{\otimes}\, \mathbb{Z}_p$ by continuity, and its kernel is the submodule $\mathcal{N}_\infty^{\mathrm{univ}} H^1\big(\mathbf{F}_\nu, \mathbb{Z}_p(1)\big)$.

Notation: Putting $G = \mathrm{Gal}\big(\overline{\mathbf{F}_\nu}/\mathbf{F}_\nu\big)$, we write $H^1_\star(G,-) = \begin{cases} H^1_{\mathrm{nr}}(G,-) & \text{if } \nu \nmid p \\ H^1_f(G,-) & \text{if } \nu \mid p. \end{cases}$

Let us further assume our compatible family $\big(x_r\big)_{r\geq 1} \in \varprojlim_{\frac{1}{p}\pi_{1\,*}} H^1_\star\big(\mathbf{F}_\nu, \mathrm{Ta}_p(J_r)^{\mathrm{ord}}\big)$. It is an easy exercise to check both $(\pi_1^*)_*^n \left(\frac{1}{p}\pi_{1\,*}\right)_*^n x_{r+n} \in H^1_\star$ and $x'_{r+n} \in H^1_\star$, because the local condition is p-saturated.

Remark: One claims every element $y \in H^1_\star\Big(\mathbf{F}_\nu,\ \mathbf{e}^*_{\mathrm{ord}}.\mathrm{Ta}_p\big({}^{\mathbf{t}}J_{r+n}\big)\Big)$ is really the image of an element in $H^1\Big(\mathbf{F}_\nu,\ \mathfrak{X}^*_{x'_{r+n}}(1)\Big)$, under the cohomology exact sequence

$$\cdots \longrightarrow H^1\Big(\mathbf{F}_\nu, \mathbb{Z}_p(1)\Big) \longrightarrow H^1\Big(\mathbf{F}_\nu,\ \mathfrak{X}^*_{x'_{r+n}}(1)\Big)$$
$$\longrightarrow H^1\Big(\mathbf{F}_\nu,\ \mathbf{e}^*_{\mathrm{ord}}.\mathrm{Ta}_p\big({}^{\mathbf{t}}J_{r+n}\big)\Big) \stackrel{\partial}{\longrightarrow} H^2\Big(\mathbf{F}_\nu,\ \mathbb{Z}_p(1)\Big) \longrightarrow \cdots.$$

The precise reason is that the value in the H^2 of $\partial(y)$ is the cup-product $x'_{r+n} \cup y$, and this is zero because $x'_{r+n} \in H^1_\star$ is orthogonal to $H^1_\star\Big(\mathbf{F}_\nu,\ \mathbf{e}^*_{\mathrm{ord}}.\mathrm{Ta}_p\big({}^{\mathbf{t}}J_{r+n}\big)\Big)$. Consequently y came from the image of $H^1\Big(\mathbf{F}_\nu,\ \mathfrak{X}^*_{x'_{r+n}}(1)\Big)$.

We shall now consider a secondary family $\big(y_r\big)_{r\geq 1} \in \varprojlim_{\widehat{\pi_1^*}} H^1_\star\Big(\mathbf{F}_\nu,\ \mathbf{e}^*_{\mathrm{ord}}.\mathrm{Ta}_p\big({}^{\mathbf{t}}J_r\big)\Big)$. By the above remark, each element y_{r+n} possesses a lift $\widetilde{y_{r+n}} \in H^1\Big(\mathbf{F}_\nu,\ \mathfrak{X}^*_{x'_{r+n}}(1)\Big)$ such that $\widetilde{y_{r+n}} \mapsto y_{r+n}$ in our cohomology sequence.

Definition 9.9. *For each integer $r \geq 1$, the canonical lifting*

$$s_{\mathrm{can},\mathbf{F}_\nu}\big(y_r\big) \;:=\; \lim_{n\to\infty} \Big(\widehat{\pi_1^*}\Big)_*^n \Big(\widetilde{y_{r+n}}\Big)$$

converges to a a well-defined element inside the quotient $\dfrac{H^1\big(\mathbf{F}_\nu,\ \mathfrak{X}^*_{x_r}(1)\big)}{\mathcal{N}_\infty^{\mathrm{univ}} H^1\big(\mathbf{F}_\nu, \mathbb{Z}_p(1)\big)}$.

Of course, it must be checked that the above limit is truly convergent. To this end we assume that the (x'_{r+n})'s are chosen compatibly, and consider the difference

$$\left(\widehat{\pi_1^*}\right)_*^{n+m}\left(\widetilde{y_{r+n+m}}\right)-\left(\widehat{\pi_1^*}\right)_*^{n}\left(\widetilde{y_{r+n}}\right) = \left(\widehat{\pi_1^*}\right)_*^{n}\left(\left(\widehat{\pi_1^*}\right)_*^{m}\left(\widetilde{y_{r+n+m}}\right)-\widetilde{y_{r+n}}\right).$$

The interior term $\left(\widehat{\pi_1^*}\right)_*^{m}\left(\widetilde{y_{r+n+m}}\right)-\widetilde{y_{r+n}}$ clearly belongs to $H^1\left(\mathbf{F}_\nu,\ \mathfrak{X}^*_{x'_{r+n}}(1)\right)$. It follows from Lemma 9.8 that $\left(\widehat{\pi_1^*}\right)_*^{n}(\dots)$ is well-defined modulo the image of η'_n. However for some constant $c \geq 0$, one has containments

$$\begin{aligned}\eta'_n\Big(H^1(\mathbf{F}_\nu,\mathbb{Z}_p(1))\Big) &\subset p^n \times H^1(\mathbf{F}_\nu,\mathbb{Z}_p(1)) \\ &\subset \mathrm{cores}_{\mathbf{F}_\nu(\mu_{p^{n-c}})/\mathbf{F}_\nu}\big(H^1(\mathbf{F}_\nu(\mu_{p^{n-c}}),\mathbb{Z}_p(1))\big).\end{aligned}$$

Lastly as $n \to \infty$, the ambiguity lies in $\bigcap_{n\geq 0}\mathrm{cores}_{\mathbf{F}_\nu(\mu_{p^n})/\mathbf{F}_\nu}H^1(\mathbf{F}_\nu(\mu_{p^n}),\mathbb{Z}_p(1))$ i.e. inside the submodule of universal norms.

Level-compatible Selmer groups.
Unfortunately, it is not particularly apparent (to the author) how to write down a decent weight-pairing on the compact Selmer group associated to $\mathrm{Jac}X_1(Np^r)$. Out of necessity we are forced to introduce a small refinement of these objects, the advantage being that the pairing is easy to describe on the modified components.

The refined Selmer group $\widetilde{H}^1_f\left(\mathbf{F}, \mathrm{Ta}_p(J_r)^{\mathrm{ord}}\right)$ is taken to be the kernel of

$$H^1\left(\mathbf{F}_\Sigma/\mathbf{F},\ \mathrm{Ta}_p(J_r)^{\mathrm{ord}}\right) \xrightarrow{\oplus\mathrm{res}_\nu} \bigoplus_{\nu\in\Sigma}\frac{H^1(\mathbf{F}_\nu, \mathrm{Ta}_p(J_r)^{\mathrm{ord}})}{\bigcap_{n\geq 0}\left(\frac{1}{p}\pi_{1\,*}\right)^n H^1_\star(\mathbf{F}_\nu, \mathrm{Ta}_p(J_{r+n})^{\mathrm{ord}})}.$$

Similarly, on the dual locus $\overline{H}^1_f\left(\mathbf{F}, \mathbf{e}^*_{\mathrm{ord}}.\mathrm{Ta}_p({}^{\mathbf{t}}J_r)\right)$ is defined as the kernel of

$$H^1\left(\mathbf{F}_\Sigma/\mathbf{F},\ \mathbf{e}^*_{\mathrm{ord}}.\mathrm{Ta}_p({}^{\mathbf{t}}J_r)\right) \xrightarrow{\oplus\mathrm{res}_\nu} \bigoplus_{\nu\in\Sigma}\frac{H^1(\mathbf{F}_\nu,\ \mathbf{e}^*_{\mathrm{ord}}.\mathrm{Ta}_p({}^{\mathbf{t}}J_r))}{\bigcap_{n\geq 0}\left(\widehat{\pi_1^*}\right)^n H^1_\star(\mathbf{F}_\nu, \mathbf{e}^*_{\mathrm{ord}}.\mathrm{Ta}_p({}^{\mathbf{t}}J_{r+n}))}.$$

One immediately deduces that $\widetilde{H}^1_f \subset H^1_{f,\mathrm{Spec}\mathcal{O}_\mathbf{F}}$ and $\overline{H}^1_f \subset H^1_{f,\mathrm{Spec}\mathcal{O}_\mathbf{F}}$, since our local conditions are stricter than those occurring in the Bloch-Kato versions.

Remark: The optimal elliptic curve $\mathcal{C}^{\min}$ is a proper subvariety of $\mathrm{Jac}X_1(Np)$, and also appears as a subvariety of its Picard dual ${}^{\mathbf{t}}J_1$. The Tate module of $\mathcal{C}^{\min}$ must therefore occur as a direct summand in both $\mathrm{Ta}_p(J_1)^{\mathrm{ord}}$ and $\mathbf{e}^*_{\mathrm{ord}}.\mathrm{Ta}_p({}^{\mathbf{t}}J_1)$.

Consequently, in order to define a bilinear pairing

$$\langle -,-\rangle^{\mathrm{wt}}_{\mathbf{F},p} : \widetilde{H}^1_f\left(\mathbf{F},\ \mathrm{Ta}_p(\mathcal{C}^{\min})\right) \times \overline{H}^1_f\left(\mathbf{F},\ \mathrm{Ta}_p(\mathcal{C}^{\min})\right) \longrightarrow \mathbb{Z}_p$$

we can instead define more general pairings

$$\langle -,-\rangle^{(r)}_{\mathbf{F},p} : \widetilde{H}^1_f\left(\mathbf{F},\ \mathrm{Ta}_p(J_r)^{\mathrm{ord}}\right) \times \overline{H}^1_f\left(\mathbf{F},\ \mathbf{e}^*_{\mathrm{ord}}.\mathrm{Ta}_p({}^{\mathbf{t}}J_r)\right) \longrightarrow \mathbb{Z}_p$$

for all integers $r \geq 1$, then just restrict $\langle -,-\rangle^{(1)}_{\mathbf{F},p}$ to the f_E-isotypic components.

Definition 9.10. *Suppose* $x \in \widetilde{H}^1_f\Big(\mathbf{F}, \mathrm{Ta}_p(J_r)^{\mathrm{ord}}\Big)$ *and* $y \in \overline{H}^1_f\Big(\mathbf{F}, \mathbf{e}^*_{\mathrm{ord}}.\mathrm{Ta}_p({}^{\mathbf{t}}J_r)\Big)$. *If* $s_{\mathrm{glob}}(y)$ *denotes any lift of* y *to* $H^1\big(\mathbf{F}_\Sigma/\mathbf{F},\ \mathfrak{X}^*_x(1)\big)$, *then define*

$$\langle x, y\rangle^{(r)}_{\mathbf{F},p} \ := \ \sum_{\nu\in\Sigma} \widetilde{\ell_{\varrho,\nu}}\Big(s_{\mathrm{glob}}(y)_\nu - s_{\mathrm{can},\mathbf{F}_\nu}(y_\nu)\Big).$$

Does this pairing firstly make sense, and is it properly defined?

Well the difference $s_{\mathrm{glob}}(y)_\nu - s_{\mathrm{can},\mathbf{F}_\nu}(y_\nu) \in H^1\big(\mathbf{F}_\nu,\ \mathfrak{X}^*_x(1)\big)$ actually lies in the image of $H^1\big(\mathbf{F}_\nu,\ \mathbb{Z}_p(1)\big)$ via

$$H^1\Big(\mathbf{F}_\nu,\ \mathbb{Z}_p(1)\Big) \longrightarrow H^1\Big(\mathbf{F}_\nu,\ \mathfrak{X}^*_x(1)\Big) \longrightarrow H^1\Big(\mathbf{F}_\nu,\ \mathbf{e}^*_{\mathrm{ord}}.\mathrm{Ta}_p({}^{\mathbf{t}}J_r)\Big) \xrightarrow{\ \partial\ } \dots$$

because both $s_{\mathrm{glob}}(y)_\nu$ and $s_{\mathrm{can},\mathbf{F}_\nu}(y_\nu)$ are local lifts of the same element $\mathrm{res}_\nu(y)$. Whilst the element $s_{\mathrm{can},\mathbf{F}_\nu}(y_\nu)$ is only well-defined modulo $\mathcal{N}^{\mathrm{univ}}_\infty H^1\big(\mathbf{F}_\nu, \mathbb{Z}_p(1)\big)$, the universal norms comprise the kernel of $\widetilde{\ell_{\varrho,\nu}}$. Finally, changing the global lift $s_{\mathrm{glob}}(y)$ by an element $z \in H^1\big(\mathbf{F}_\Sigma/\mathbf{F},\ \mathbb{Z}_p(1)\big)$ does not affect the sum in any way, since $\sum_{\nu\in\Sigma} \widetilde{\ell_{\varrho,\nu}}(z_\nu) = 0$ by class field theory.

Question. *How do we extend the p-adic pairings on the refined Selmer groups, to encompass the full compact Selmer groups for* $\mathcal{C}^{\min}$*?*

This task will clearly be possible, if and only if the $\mathbb{Z}_p$-ranks of $\widetilde{H}^1_f\Big(\mathbf{F}, \mathrm{Ta}_p(\mathcal{C}^{\min})\Big)$ and likewise $\overline{H}^1_f\Big(\mathbf{F}, \mathrm{Ta}_p(\mathcal{C}^{\min})\Big)$ are identical to that of $H^1_{f,\mathrm{Spec}\mathcal{O}_\mathbf{F}}\Big(\mathbf{F}, \mathrm{Ta}_p(\mathcal{C}^{\min})\Big)$. Assuming these three ranks coincide, one simply defines

$$\langle -, -\rangle^{\mathrm{wt}}_{\mathbf{F},p} : H^1_{f,\mathrm{Spec}\mathcal{O}_\mathbf{F}}\Big(\mathbf{F}, \mathrm{Ta}_p(\mathcal{C}^{\min})\Big) \times H^1_{f,\mathrm{Spec}\mathcal{O}_\mathbf{F}}\Big(\mathbf{F}, \mathrm{Ta}_p(\mathcal{C}^{\min})\Big) \ \longrightarrow\ \mathbb{Q}_p$$

by the rule $\langle x, y\rangle^{\mathrm{wt}}_{\mathbf{F},p} = p^{-(n_1+n_2)} \times \langle p^{n_1}x, p^{n_2}y\rangle^{(1)}_{\mathbf{F},p}$ for integers $n_1, n_2 \gg 0$.

Lemma 9.11. *There exist integers* $c_1, c_2 \geq 0$ *such that*

(i) $p^{c_1}.\mathcal{C}^{\min}(\mathbf{F}) \,\widehat{\otimes}\, \mathbb{Z}_p \subset \widetilde{H}^1_f\Big(\mathbf{F}, \mathrm{Ta}_p(\mathcal{C}^{\min})\Big)$ *and*

(ii) $p^{c_2}.\mathcal{C}^{\min}(\mathbf{F}) \,\widehat{\otimes}\, \mathbb{Z}_p \subset \overline{H}^1_f\Big(\mathbf{F}, \mathrm{Ta}_p(\mathcal{C}^{\min})\Big)$.

Deferring the proof for a moment, the lemma implies that the ranks of $\widetilde{H}^1_f$ and $\overline{H}^1_f$ are bounded below by the Mordell-Weil rank of $\mathcal{C}^{\min}$ over $\mathbf{F}$.

Corollary 9.12. *If the p-primary part of* $\text{Ш}_\mathbf{F}(\mathcal{C}^{\min})$ *is finite, then*

$$\mathrm{rank}_{\mathbb{Z}_p} \widetilde{H}^1_f\Big(\mathbf{F}, \mathrm{Ta}_p(\mathcal{C}^{\min})\Big) \ = \ \mathrm{rank}_{\mathbb{Z}_p} \overline{H}^1_f\Big(\mathbf{F}, \mathrm{Ta}_p(\mathcal{C}^{\min})\Big) \ = \ \mathrm{rank}_{\mathbb{Z}_p} H^1_{f,\mathrm{Spec}\mathcal{O}_\mathbf{F}}.$$

Proof of 9.11: We dispose of part (i). Clearly for large enough c_1':

(I) if $\nu \nmid p$, then $\quad p^{c_1'}.H^1_{\mathrm{nr}}\Big(\mathbf{F}_\nu, \mathrm{Ta}_p(\mathcal{C}^{\min})\Big) \subset H^1_{\mathrm{nr}}\Big(\mathbf{F}_\nu, \mathbb{T}_{\infty,E}\Big)_{\Gamma^{\mathrm{wt}}}$

(II) if $\nu \mid p$, then $\quad p^{c_1'}.H^1_f\Big(\mathbf{F}_\nu, \mathrm{Ta}_p(\mathcal{C}^{\min})\Big) \subset \mathbf{X}_{\mathbb{T}_{\infty,E}}(\mathbf{F}_\nu)_{\Gamma^{\mathrm{wt}}}$

N.B. the second statement is a consequence of Corollary 5.16, the first is obvious. The left-hand sides both contain the image of $p^{c_1'}.\mathcal{C}^{\min}(\mathbf{F}) \widehat{\otimes} \mathbb{Z}_p$, hence so does

$$\mathfrak{V} = \mathrm{Ker}\left(H^1\Big(\mathbf{F}_\Sigma/\mathbf{F}, \mathrm{Ta}_p(\mathcal{C}^{\min})\Big) \overset{\oplus \mathrm{res}_\nu}{\longrightarrow} \bigoplus_{\nu \in \Sigma - S_{\mathbf{F},p}} \frac{H^1(\mathbf{F}_\nu, \mathrm{Ta}_p(\mathcal{C}^{\min}))}{H^1_{\mathrm{nr}}\Big(\mathbf{F}_\nu, \mathbb{T}_{\infty,E}\Big)_{\Gamma^{\mathrm{wt}}}} \oplus \bigoplus_{\nu \in S_{\mathbf{F},p}} \frac{H^1(\mathbf{F}_\nu, \mathrm{Ta}_p(\mathcal{C}^{\min}))}{\mathbf{X}_{\mathbb{T}_{\infty,E}}(\mathbf{F}_\nu)_{\Gamma^{\mathrm{wt}}}} \right).$$

The latter object $\mathfrak{V}$ has the $\mathbb{Z}_p$-rank of $\widetilde{H}^1_f\Big(\mathbf{F}, \mathrm{Ta}_p(\mathcal{C}^{\min})\Big)$, indeed the conditions $\mathbf{X}_{\mathbb{T}_{\infty,E}}(\mathbf{F}_\nu)_{\Gamma^{\mathrm{wt}}}$ and $\bigcap_{n \geq 0} \left(\frac{1}{p}\pi_{1\,*}\right)^n H^1_f(\mathbf{F}_\nu, \mathrm{Ta}_p(J_{1+n})^{\mathrm{ord}}) \cap H^1(\mathbf{F}_\nu, \mathrm{Ta}_p(\mathcal{C}^{\min}))$ are commensurable for all $\nu | p$. Thus there exists $c_1'' \geq 0$ satisfying $p^{c_1''}.\mathfrak{V} \subset \widetilde{H}^1_f(\ldots)$. Setting $c_1 = c_1' + c_1''$, the first bit of the lemma is proved.

To show why part (ii) is true, we just employ identical arguments to the above. The only difference is that one instead uses the local condition $H^1_{\mathcal{E}}(\mathbf{F}_\nu, \mathbb{T}_{\infty,E})_{\Gamma^{\mathrm{wt}}}$ at places $\nu | p$, which is p-commensurable with the refined submodule

$$\bigcap_{n \geq 0} \left(\widehat{\pi_1^*}\right)^n H^1_f\Big(\mathbf{F}_\nu, \mathbf{e}^*_{\mathrm{ord}}.\mathrm{Ta}_p({}^{\mathbf{t}}J_{1+n})\Big) \cap H^1\Big(\mathbf{F}_\nu, \mathrm{Ta}_p(\mathcal{C}^{\min})\Big).$$

The remaining details are the same as for 9.11(i). □

9.4 Commutativity of the bottom squares

The last remaining hurdle is to relate the p-adic weight pairing $\langle -, - \rangle^{(r)}_{\mathbf{F},p}$ at $r = 1$, to the existence of two specialisation maps

$$\begin{array}{ccc}
\displaystyle\bigoplus_{\nu \in \Sigma} \frac{H^1(\mathbf{F}_\nu, \mathcal{C}^{\min}[p^\infty])}{\mathcal{C}^{\min}(\mathbf{F}_\nu) \otimes \mathbb{Q}_p/\mathbb{Z}_p} & \overset{\oplus \delta_{\infty,E,\nu}}{\longrightarrow} & \displaystyle\bigoplus_{\nu \in \Sigma} \left(\frac{H^1(\mathbf{F}_\nu, A_{\mathbb{T}_{\infty,E}})}{H^1_\star(\mathbf{F}_\nu, A_{\mathbb{T}_{\infty,E}})} \right)^{\Gamma^{\mathrm{wt}}} \\
\downarrow & & \downarrow \\
H^1_f\Big(\mathbf{F}, \mathrm{Ta}_p(\mathcal{C}^{\min})\Big)^\vee & \overset{?}{\dashleftarrow} & \mathrm{Sel}_{\mathbf{F}}(\mathbb{T}_{\infty,E})_{\Gamma^{\mathrm{wt}}} \\
\downarrow & & \downarrow \\
H^2\Big(\mathbf{F}_\Sigma/\mathbf{F}, \mathcal{C}^{\min}[p^\infty]\Big) & \overset{??}{\dashleftarrow} & H^1\Big(\mathbf{F}_\Sigma/\mathbf{F}, A_{\mathbb{T}_{\infty,E}}\Big)_{\Gamma^{\mathrm{wt}}}
\end{array}$$

completing the bottom half of Figure 9.1.

This approach is inspired by Perrin-Riou's cyclotomic methods in [PR1, §4].

We begin with some general comments. To explicitly write down the specialisation $\mathrm{Sel}_{\mathbf{F}}\big(\mathbb{T}_{\infty,E}\big)_{\Gamma^{\mathrm{wt}}} \overset{?}{\dashrightarrow} H^1_f\Big(\mathbf{F}, \mathrm{Ta}_p(\mathcal{C}^{\min})\Big)^{\vee}$, it is first necessary to construct splittings for the *divisible versions* of the corresponding compact modules appearing in §9.3. Let us define $\overline{H}^1_f\Big(\mathbf{F},\ {}^{\mathbf{t}}J_r^{\mathrm{ord}}[p^\infty]\Big)$ to be the kernel of

$$H^1\Big(\mathbf{F}_\Sigma/\mathbf{F},\ {}^{\mathbf{t}}J_r^{\mathrm{ord}}[p^\infty]\Big) \overset{\oplus \mathrm{res}_\nu}{\longrightarrow} \bigoplus_{\nu\in\Sigma} \frac{H^1\big(\mathbf{F}_\nu,\ \mathbf{e}^*_{\mathrm{ord}}.\ {}^{\mathbf{t}}J_r^{\mathrm{ord}}[p^\infty]\big)}{\mathbb{Q}_p/\mathbb{Z}_p \otimes \bigcap_{n\geq 0}\Big(\widehat{\pi_1^*}\Big)^n H^1_\star\big(\mathbf{F}_\nu, \mathbf{e}^*_{\mathrm{ord}}.\mathrm{Ta}_p({}^{\mathbf{t}}J_{r+n})\big)}$$

at all integers $r \geq 1$.

Key Fact: For each $n \geq 0$, we claim there is a right-exact sequence

$$H^1\big(\mathbf{F}_\Sigma/\mathbf{F},\ \mu_{p^\infty}\big) \longrightarrow \mathfrak{Y}_{x'_{r+n}} \longrightarrow \overline{H}^1_f\big(\mathbf{F},\ {}^{\mathbf{t}}J_{r+n}^{\mathrm{ord}}[p^\infty]\big) \longrightarrow 0$$

where the notation $\mathfrak{Y}_{x'_{r+n}}$ denotes the inverse image of $\overline{H}^1_f\big(\mathbf{F},\ {}^{\mathbf{t}}J_{r+n}^{\mathrm{ord}}[p^\infty]\big)$ inside the lifted cohomology $H^1\Big(\mathbf{F}_\Sigma/\mathbf{F},\ \mathbb{Q}_p/\mathbb{Z}_p \otimes \mathfrak{X}^*_{x'_{r+n}}(1)\Big)$.

To justify this statement, it is enough to prove $\overline{H}^1_f$ is contained in the image of $H^1\Big(\mathbf{F}_\Sigma/\mathbf{F},\ \mathbb{Q}_p/\mathbb{Z}_p \otimes \mathfrak{X}^*_{x'_{r+n}}(1)\Big)$. If $y \in \overline{H}^1_f\Big(\mathbf{F},\ {}^{\mathbf{t}}J_{r+n}^{\mathrm{ord}}[p^\infty]\Big)$ then its image in $H^2\big(\mathbf{F}_\Sigma/\mathbf{F},\ \mu_{p^\infty}\big)$ is given by the cup-product $x'_{r+n} \cup y$. However $(x'_{r+n})_\nu \cup y_\nu = 0$ for each finite place ν, and the global Brauer group injects into $\bigoplus_\nu H^2\big(\mathbf{F}_\nu,\ \mu_{p^\infty}\big)$. It follows that $x'_{r+n} \cup y$ is itself zero, and the original claim was true.

Assume now that $y \in \overline{H}^1_f\Big(\mathbf{F},\ {}^{\mathbf{t}}J_{r+n}^{\mathrm{ord}}[p^\infty]\Big)$, and that $s_{\mathrm{glob}}(y)$ is a lift of y to $\mathfrak{Y}_{x'_{r+n}}$. Because $y \in \overline{H}^1_f$, there exists an integer $m \geq 0$ such that $\mathrm{res}_\nu(y) = p^{-m} \otimes z_\nu$ with $z_\nu \in \bigcap_{n\geq 0}\Big(\widehat{\pi_1^*}\Big)^n H^1_\star\big(\mathbf{F}_\nu, \mathbf{e}^*_{\mathrm{ord}}.\mathrm{Ta}_p({}^{\mathbf{t}}J_{r+n})\big)$. We shall set $\widehat{y}_\nu := p^{-m} \otimes s_{\mathrm{can},\mathbf{F}_\nu}(z_\nu)$ which is only properly defined modulo the group of universal norms.

Definition 9.13. *If* $x_r \in \widetilde{H}^1_f\Big(\mathbf{F}, \mathrm{Ta}_p(J_r)^{\mathrm{ord}}\Big)$ *and* $y \in \overline{H}^1_f\Big(\mathbf{F},\ {}^{\mathbf{t}}J_{r+n}^{\mathrm{ord}}[p^\infty]\Big)$ *for some integer* $n \geq 0$*, then define*

$$\theta^{(r)}_{\rho,\Sigma}(x_r)(y) \ :=\ \sum_{\nu\in\Sigma} \mathrm{id}\otimes\widetilde{\ell_{\varrho,\nu}}\Big(s_{\mathrm{glob}}(y)_\nu - \widehat{y}_\nu\Big).$$

The following are all quite straightforward.

Exercises. *(a)* $\theta^{(r)}_{\rho,\Sigma}(x_r)(y)$ *is well-defined and independent of* $s_{\mathrm{glob}}(y)$*;*

(b) The association $x_r \mapsto \theta^{(r)}_{\rho,\Sigma}(x_r)(-)$ *gives rise to a homorphism*

$$\theta^{(r)}_{\rho,\Sigma} : \widetilde{H}^1_f\Big(\mathbf{F}, \mathrm{Ta}_p(J_r)^{\mathrm{ord}}\Big) \longrightarrow \mathrm{Hom}_{\mathrm{cont}}\left(\varinjlim_n \overline{H}^1_f\Big(\mathbf{F},\ {}^{\mathbf{t}}J_{r+n}^{\mathrm{ord}}[p^\infty]\Big),\ \mathbb{Q}_p/\mathbb{Z}_p\right);$$

(c) The image of this homomorphism $\theta^{(r)}_{\rho,\Sigma}$ *is* Γ_r*-invariant.*

Before stating the main result of this section, we first introduce some notation. Assuming that $0 \to A \to B \to C \to 0$ is a short exact sequence of Λ^{wt}-modules, then multiplication by $u_0^{p^{r-1}} - 1$ and the Snake Lemma produce

$$0 \longrightarrow A^{\Gamma_r} \longrightarrow B^{\Gamma_r} \longrightarrow C^{\Gamma_r} \overset{\partial}{\longrightarrow} A_{\Gamma_r} \longrightarrow B_{\Gamma_r} \longrightarrow C_{\Gamma_r} \longrightarrow 0.$$

We write '$\mathcal{SL}_{(r)}$' for the boundary map ∂ occurring in the Γ_r-cohomology.

Proposition 9.14. *For all integers $r \geq 1$, the diagram*

$$\begin{array}{ccc}
\bigoplus_{\nu\in\Sigma} H^1\Big(\mathbf{F}_\nu, \mathrm{Ta}_p(J_r)^{\mathrm{ord}}\Big) & \overset{\mathrm{nat}}{\hookleftarrow} & \bigoplus_{\nu\in\Sigma} H^1\Big(\mathbf{F}_\nu, \mathrm{Ta}_p(J_\infty)^{\mathrm{ord}}\Big)_{\Gamma_r} \\
\Big\uparrow {\scriptstyle p^{r-1}\log_p(u_0)\times\oplus\mathrm{res}_\nu} & & \Big\uparrow {\scriptstyle \widehat{\mathcal{SL}_{(r)}}} \\
\widetilde{H}^1_f\Big(\mathbf{F}, \mathrm{Ta}_p(J_r)^{\mathrm{ord}}\Big) & \overset{\theta^{(r)}_{\rho,\Sigma}}{\longrightarrow} & \Big(\Big(\varinjlim_n \overline{H}^1_f\big(\mathbf{F}, {}^{\mathbf{t}}J^{\mathrm{ord}}_{r+n}[p^\infty]\big)\Big)^\vee\Big)^{\Gamma_r}
\end{array}$$

is a commutative square.

Let's translate this statement back in terms of the Selmer group associated to the optimal elliptic curve. The injection $\mathbb{T}_{\infty,E} \cong \varprojlim_r \Big(\mathbb{T}_{\infty,E}\Big)_{\Gamma_r} \hookrightarrow \varprojlim_r \mathrm{Ta}_p(J_r)^{\mathrm{ord}}$ induces a projection map

$$\varinjlim_r {}^{\mathbf{t}}J_r^{\mathrm{ord}}[p^\infty] \;\cong\; \varinjlim_r A_{\mathrm{Ta}_p(J_r)^{\mathrm{ord}}} \;\twoheadrightarrow\; A_{\mathbb{T}_{\infty,E}}.$$

Therefore every element of $H^1(G, A_{\mathbb{T}_{\infty,E}})$ will be represented by a cocycle from $H^1(G, {}^{\mathbf{t}}J^{\mathrm{ord}}_{r+n}[p^\infty])$ for large enough $n \geq 0$.

Corollary 9.15. *If $\theta_{\infty,E}$ is the $\mathbb{Z}_p$-linear extension of the map $\log_p(u_0)^{-1} \times \theta^{(1)}_{\rho,\Sigma}$ to the whole of $H^1_f(\mathbf{F}, -)$, then*

$$\begin{array}{ccc}
\bigoplus_{\nu\in\Sigma} H^1_\star\Big(\mathbf{F}_\nu, \mathrm{Ta}_p(\mathcal{C}^{\min})\Big) & \overset{\oplus\widehat{\delta}_{\infty,E,\nu}}{\hookleftarrow} & \bigoplus_{\nu\in\Sigma} H^1_\star\Big(\mathbf{F}_\nu, \mathbb{T}_{\infty,E}\Big)_{\Gamma^{\mathrm{wt}}} \\
\Big\uparrow {\scriptstyle \oplus\mathrm{res}_\nu} & & \Big\uparrow {\scriptstyle \widehat{\mathcal{SL}_{(1)}}} \\
H^1_f\Big(\mathbf{F}, \mathrm{Ta}_p(\mathcal{C}^{\min})\Big) & \overset{\theta_{\infty,E}}{\longrightarrow} & \Big(\mathrm{Sel}_{\mathbf{F}}(\mathbb{T}_{\infty,E})^\vee\Big)^{\Gamma^{\mathrm{wt}}}
\end{array}$$

is a commutative square.

The Pontrjagin dual of this diagram constitutes the bottom portion of Figure 9.1. Furthermore, if $\phi_{\infty,E}$ denotes the composition

$$\Big(\mathrm{Sel}_{\mathbf{F}}(\mathbb{T}_{\infty,E})^\vee\Big)^{\Gamma^{\mathrm{wt}}} \overset{\mathrm{mod}\ u_0-1}{\longrightarrow} \Big(\mathrm{Sel}_{\mathbf{F}}(\mathbb{T}_{\infty,E})^\vee\Big)_{\Gamma^{\mathrm{wt}}}$$
$$\overset{\widehat{\alpha}_{\infty,E}}{\longrightarrow} \Big(\mathrm{Sel}_{\mathbf{F}}(\mathcal{C}^{\min})[p^\infty]\Big)^\vee \longrightarrow \mathrm{Hom}_{\mathbb{Z}_p}\Big(H^1_f(\mathbf{F}, \mathrm{Ta}_p(\mathcal{C}^{\min})),\ \mathbb{Z}_p\Big)$$

then $\phi_{\infty,E} \circ \theta_{\infty,E}(x)(y) = \log_p(u_0)^{-1} \times \langle x, y\rangle^{\mathrm{wt}}_{\mathbf{F},p}$. In other words, the p-adic weight pairing factorises through the fundamental commutative diagram.

Proof of 9.14: When dealing with discrete cohomologies whose coefficients lie in $A_{\mathrm{Ta}_p(J_\infty)^{\mathrm{ord}}}$, clearly any given cocycle may be considered as taking values inside $A_{\mathrm{Ta}_p(J_{r+n})^{\mathrm{ord}}} \cong {}^{\mathbf{t}}\!J_{r+n}^{\mathrm{ord}}[p^\infty]$ with $n \gg 0$. We'll work with ${}^{\mathbf{t}}\!J_{r+n}^{\mathrm{ord}}[p^\infty]$-coefficients. Our primary task is to find a decent description of the homomorphism

$$\widehat{\mathcal{SL}_{(r)}} : \left(\overline{H}_f^1\left(\mathbf{F}, {}^{\mathbf{t}}\!J_{r+n}^{\mathrm{ord}}[p^\infty]\right)^\vee\right)^{\Gamma_r} \longrightarrow \bigoplus_{\nu\in\Sigma} H^1\left(\mathbf{F}_\nu, \mathrm{Ta}_p(J_{r+n})^{\mathrm{ord}}\right)_{\Gamma_r}.$$

Remember that its dual $\mathcal{SL}_{(r)}$ appeared as the boundary map, in the Γ_r-cohomology of the Selmer sequence defining $\overline{H}_f^1$.

Remark: Suppose that $0 \longrightarrow A \longrightarrow B \longrightarrow C \longrightarrow 0$ is a short exact sequence. Consider the diagram

$$\begin{array}{ccccccccc} 0 & \longrightarrow & A & \longrightarrow & B & \longrightarrow & C & \longrightarrow & 0 \\ & & \downarrow{\scriptstyle h_1} & & \downarrow{\scriptstyle h_2} & & \downarrow{\scriptstyle h_3} & & \\ 0 & \longrightarrow & A & \longrightarrow & B & \longrightarrow & C & \longrightarrow & 0. \end{array}$$

Then the boundary map $\mathrm{Ker}(h_3) \stackrel{\partial}{\longrightarrow} \mathrm{Coker}(h_1)$ arising from the Snake lemma is explicitly given by the formula $\partial(c) = h_2\left(c^\dagger\right) \bmod h_1(A)$, where $c^\dagger \in B$ denotes any lift of $c \in \mathrm{Ker}(h_3)$ to B.

In our situation $h_1 = h_2 = h_3 =$ multiplication by $u_0^{p^{r-1}} - 1$, moreover

$$A = \overline{H}_f^1\left(\mathbf{F}, {}^{\mathbf{t}}\!J_{r+n}^{\mathrm{ord}}[p^\infty]\right), \quad B = H^1\left(\mathbf{F}_\Sigma/\mathbf{F}, {}^{\mathbf{t}}\!J_{r+n}^{\mathrm{ord}}[p^\infty]\right), \quad C = \bigoplus_{\nu\in\Sigma} \frac{H^1(\mathbf{F}_\nu, -)}{\text{local cond's}}.$$

It follows directly from the above remark that

$$\mathcal{SL}_{(r)}(c) = \left(u_0^{p^{r-1}} - 1\right).c^\dagger \mod \left(u_0^{p^{r-1}} - 1\right) \times \text{the image of Selmer}$$

where $c^\dagger$ is any lift of c to $H^1\left(\mathbf{F}_\Sigma/\mathbf{F}, {}^{\mathbf{t}}\!J_{r+n}^{\mathrm{ord}}[p^\infty]\right)$.

Question. *What is the formula required to make the square in 9.14 commute?*

Choose elements $x_r \in \widetilde{H}_f^1\left(\mathbf{F}, \mathrm{Ta}_p(J_r)^{\mathrm{ord}}\right)$ and $(y_\nu)_\nu \in \bigoplus_{\nu\in\Sigma} H^1\left(\mathbf{F}_\nu, {}^{\mathbf{t}}\!J_r^{\mathrm{ord}}[p^\infty]\right)$. For all integers $n \geq 0$, there exists $y_n^\dagger \in H^1\left(\mathbf{F}_\Sigma/\mathbf{F}, {}^{\mathbf{t}}\!J_{r+n}^{\mathrm{ord}}[p^\infty]\right)$ such that

$$\delta_\nu^{(n)} = \mathrm{res}_\nu\left(y_n^\dagger\right) - \left(\widehat{\frac{1}{p}\pi_{1\,*}}\right)^n y_\nu \in \mathbb{Q}_p/\mathbb{Z}_p \otimes \bigcap_{m\geq 0} \left(\widehat{\pi_1^*}\right)^m H_\star^1\left(\mathbf{F}_\nu, \mathbf{e}_{\mathrm{ord}}^*.\mathrm{Ta}_p({}^{\mathbf{t}}\!J_{r+n+m})\right).$$

Observe that $\widehat{\frac{1}{p}\pi_{1\,*}}$ induces a homomorphism $\mathbf{e}_{\mathrm{ord}}^*.\mathrm{Ta}_p({}^{\mathbf{t}}\!J_r) \longrightarrow \mathbf{e}_{\mathrm{ord}}^*.\mathrm{Ta}_p({}^{\mathbf{t}}\!J_{r+1})$. Without loss of generality, one may assume the lifts $y_n^\dagger$ are chosen compatibly.

Conclusion: By the preceding discussion, the formula one requires to make 9.14 commutative is precisely

$$\theta^{(r)}_{\rho,\Sigma}(x_r)\Big(\big(u_0^{p^{r-1}}-1\big).\, y^{\dagger}_{\infty}\Big) \;=\; p^{r-1}\log_p(u_0)\times\sum_{\nu\in\Sigma}\mathrm{inv}_{\mathbf{F}_\nu}\Big(\big(x_r\big)_\nu \cup y_\nu\Big) \qquad (*)$$

where $y^{\dagger}_{\infty}=\big(y^{\dagger}_n\big)_{n\geq 0}$ belongs to $H^1\Big(\mathbf{F}_\Sigma/\mathbf{F},\, {}^{\mathbf{t}}J^{\mathrm{ord}}_{\infty}[p^\infty]\Big)$ modulo $\overline{H}^1_f$.

We start by computing the canonical lift at $\nu\in\Sigma$:

$$\Big(u_0^{p^{r-1}}-1\Big).\big(y^{\dagger}_n\big)_\nu \;=\; \Big(u_0^{p^{r-1}}-1\Big).\delta^{(n)}_\nu+\left(\widehat{\frac{1}{p}\pi_{1\,*}}\right)^n\Big(u_0^{p^{r-1}}-1\Big).y_\nu$$
$$=\; \Big(u_0^{p^{r-1}}-1\Big).\delta^{(n)}_\nu \;+\;0$$

since $u_0^{p^{r-1}}-1$ is the zero endomorphism on $\mathrm{Ta}_p(J_r)^{\mathrm{ord}}$. As a consequence

$$s_{\mathrm{can},\mathbf{F}_\nu}\Big(\big(u_0^{p^{r-1}}-1\big).\big(y^{\dagger}_n\big)_\nu\Big) \;=\; \Big(u_0^{p^{r-1}}-1\Big).\,\widehat{\delta^{(n)}_\nu}\;\in\; H^1_\star\Big(\mathbf{F}_\nu,\,\mathbb{Q}_p/\mathbb{Z}_p\otimes\mathfrak{X}^*_{x'_{r+n}}(1)\Big).$$

Let's now try and calculate the global lift of $\Big(\big(u_0^{p^{r-1}}-1\big).\, y^{\dagger}_n\Big)$ instead.

Remark: For each place $\nu\in\Sigma$, there is an array in cohomology

$$\begin{array}{ccccccc}
\ldots\to H^1\Big(\mathbf{F}_\nu,\,\mathbb{Q}_p/\mathbb{Z}_p\otimes\mathfrak{X}^*_{x'_{r+n}}(1)\Big) & \to & H^1\Big(\mathbf{F}_\nu,\,{}^{\mathbf{t}}J^{\mathrm{ord}}_{r+n}[p^\infty]\Big) & \to & H^2\Big(\mathbf{F}_\nu,\,\mu_{p^\infty}\Big)\to\ldots \\
\uparrow & & \uparrow\left(\widehat{\tfrac{1}{p}\pi_{1\,*}}\right)^n_* & & \uparrow\eta''_n \\
\ldots\to H^1\Big(\mathbf{F}_\nu,\,\mathbb{Q}_p/\mathbb{Z}_p\otimes\mathfrak{X}^*_{x_r}(1)\Big) & \longrightarrow & H^1\Big(\mathbf{F}_\nu,\,{}^{\mathbf{t}}J^{\mathrm{ord}}_{r}[p^\infty]\Big) & \longrightarrow & H^2\Big(\mathbf{F}_\nu,\,\mu_{p^\infty}\Big)\to\ldots
\end{array}$$

for which the dual homomorphism $\left(\widehat{\frac{1}{p}\pi_{1\,*}}\right)^n$ induces the upward vertical arrows. Crucially the right-hand map η''_n is of the form $p^n\times$ (a scalar), which can be shown in a similar fashion to the proof of Lemma 9.8.

Consequently if $p^n\geq\mathrm{order}(y_\nu)$, the image of $\left(\widehat{\frac{1}{p}\pi_{1\,*}}\right)^n y_\nu$ in $H^2\big(\mathbf{F}_\nu,\,\mu_{p^\infty}\big)$ is trivial. However, the restriction of $y^{\dagger}_n$ at each ν has the form

$$\mathrm{res}_\nu\left(y^{\dagger}_n\right) \;=\; \delta^{(n)}_\nu+\left(\widehat{\frac{1}{p}\pi_{1\,*}}\right)^n y_\nu$$

and we know from §9.3 how to lift the $\delta^{(n)}_\nu$'s. It follows for $p^n\geq\max_{\nu\in\Sigma}\{\mathrm{order}(y_\nu)\}$ there must exist lifts of $\mathrm{res}_\nu\left(y^{\dagger}_n\right)$ inside $H^1\Big(\mathbf{F}_\nu,\,\mathbb{Q}_p/\mathbb{Z}_p\otimes\mathfrak{X}^*_{x'_{r+n}}(1)\Big)$ at all $\nu\in\Sigma$.

Lastly, the injectivity of $\mathrm{Br}(\mathbf{F}_\Sigma/\mathbf{F}) \hookrightarrow \bigoplus_{\nu\in\Sigma} H^2(\mathbf{F}_\nu,\ \mu_{p^\infty})$ implies that there is a global lift of $y_n^\dagger$ to $H^1\Big(\mathbf{F}_\Sigma/\mathbf{F},\ \mathbb{Q}_p/\mathbb{Z}_p \otimes \mathfrak{X}^*_{x'_{r+n}}(1)\Big)$, in fact

$$s_{\mathrm{glob}}\left(\left(u_0^{p^{r-1}}-1\right).\ y_n^\dagger\right) \;=\; \left(u_0^{p^{r-1}}-1\right).s_{\mathrm{glob}}\left(y_n^\dagger\right).$$

Let's establish formula **(*)** and we will be done. The left-hand side yields

$$\theta^{(r)}_{\rho,\Sigma}(x_r)\left(\left(u_0^{p^{r-1}}-1\right).\ y_n^\dagger\right) = \sum_{\nu\in\Sigma} \widetilde{\ell_{\varrho,\nu}}\left(\left(u_0^{p^{r-1}}-1\right).s_{\mathrm{glob}}\left(y_n^\dagger\right)_\nu - \left(u_0^{p^{r-1}}-1\right).\widehat{\delta_\nu^{(n)}}\right)$$

$$= \sum_{\nu\in\Sigma} \mathrm{inv}_{\mathbf{F}_\nu}\left(\left(u_0^{p^{r-1}}-1\right).\left(s_{\mathrm{glob}}\left(y_n^\dagger\right)_\nu - \widehat{\delta_\nu^{(n)}}\right)\cup \ell_{\varrho,\nu}\right).$$

But the element $s_{\mathrm{glob}}\left(y_n^\dagger\right)_\nu - \widehat{\delta_\nu^{(n)}}$ represents a lift of $\left(\widehat{\tfrac{1}{p}\pi_{1\,*}}\right)^n y_\nu$ under the mapping $H^1\Big(\mathbf{F}_\nu,\ \mathbb{Q}_p/\mathbb{Z}_p \otimes \mathfrak{X}^*_{x'_{r+n}}(1)\Big) \longrightarrow H^1\Big(\mathbf{F}_\nu,\ {}^{\mathbf{t}}J^{\mathrm{ord}}_{r+n}[p^\infty]\Big)$.

Remark: Suppose that γ is a topological generator of $\Gamma_r \cong \mathbb{Z}_p$. Then the 2-cocycle $a_\nu \cup b_\nu \in H^2(\mathbf{F}_\nu, \mu_{p^\infty})$ is the image of $(\gamma-1).a_\nu^\dagger$ under the $H^1 \to H^2$ boundary map, where $a_\nu^\dagger$ is any lift of $a_\nu \in H^1(\mathbf{F}_\nu,\ {}^{\mathbf{t}}J^{\mathrm{ord}}_{r+n}[p^\infty])$ up to $H^1\Big(\mathbf{F}_\nu, \mathbb{Q}_p/\mathbb{Z}_p \otimes \mathfrak{X}^*_{b'_{r+n}}(1)\Big)$. This statement is proved in an identical way to [PR1, 4.5.3], albeit for the weight deformation rather than the cyclotomic one.

Substituting $\gamma = u_0^{p^{r-1}}$, $a_\nu = \left(\widehat{\tfrac{1}{p}\pi_{1\,*}}\right)^n y_\nu$ and $a_\nu^\dagger = s_{\mathrm{glob}}\left(y_n^\dagger\right)_\nu - \widehat{\delta_\nu^{(n)}}$, one deduces

$$\theta^{(r)}_{\rho,\Sigma}(x_r)\left(\left(u_0^{p^{r-1}}-1\right).\ y_n^\dagger\right) = \sum_{\nu\in\Sigma}\mathrm{inv}_{\mathbf{F}_\nu}\left(\left(\widehat{\frac{1}{p}\pi_{1\,*}}\right)^n y_\nu \cup \left(x'_{r+n}\right)_\nu\right) \times \log_p\left(u_0^{p^{r-1}}\right)$$

$$= p^{r-1}\log_p(u_0) \times \sum_{\nu\in\Sigma} p^{-n} \otimes \mathrm{inv}_{\mathbf{F}_\nu}\left(\left(\widehat{\frac{1}{p}\pi_{1\,*}}\right)^n y_\nu \cup \left(\left(\pi_1^*\right)_*^n x_r\right)_\nu\right)$$

since $p^n x'_{r+n} = \left(\pi_1^*\right)_*^n x_r$. Performing the adjoint operation:

$$\left(\widehat{\frac{1}{p}\pi_{1\,*}}\right)^n y_\nu \cup \left(\left(\pi_1^*\right)_*^n x_r\right)_\nu$$

$$= \quad y_\nu \cup \left(\left(\widehat{\frac{1}{p}\pi_{1\,*}}\right)^n \circ \left(\pi_1^*\right)_*^n x_r\right)_\nu \overset{\text{by 9.6(ii)}}{=} y_\nu \cup \left(p^n x_r\right)_\nu.$$

Therefore for all $n \gg 0$, we obtain

$$\theta^{(r)}_{\rho,\Sigma}(x_r)\left(\left(u_0^{p^{r-1}}-1\right).\ y_n^\dagger\right) \;=\; p^{r-1}\log_p(u_0) \times \sum_{\nu\in\Sigma}\mathrm{inv}_{\mathbf{F}_\nu}\left(y_\nu \cup \left(x_r\right)_\nu\right).$$

Equation **(*)** is established, and the result follows.

□

Finally, one should not forget to mention the following result, although strictly speaking it is not required in the Diamond-Euler characteristic computation.

Proposition 9.16. *The bottom segment of Figure 9.1*

$$\begin{array}{ccc} H^1_f\Big(\mathbf{F}, \mathrm{Ta}_p(\mathcal{C}^{\min})\Big)^\vee & \overset{\widehat{\theta}_{\infty,E}}{\longleftarrow} & \mathrm{Sel}_{\mathbf{F}}\big(\mathbb{T}_{\infty,E}\big)_{\Gamma^{\mathrm{wt}}} \\ \downarrow & & \downarrow \\ H^2\Big(\mathbf{F}_\Sigma/\mathbf{F},\ \mathcal{C}^{\min}[p^\infty]\Big) & \overset{\tau_{\infty,E}}{\hookleftarrow} & H^1\Big(\mathbf{F}_\Sigma/\mathbf{F},\ A_{\mathbb{T}_{\infty,E}}\Big)_{\Gamma^{\mathrm{wt}}} \end{array}$$

is a commutative square.

N.B. The proof itself is a tedious but straightforward exercise in Galois cohomology; the details are written up elsewhere in [Db3].

9.5 The leading term of $Ш_{\mathbf{F}}\big(\mathbb{T}_{\infty,E}\big)$

This lengthy calculation of the vertical Γ^{wt}-Euler characteristic for $\mathrm{Sel}_{\mathbf{F}}\big(\mathbb{T}_{\infty,E}\big)$ is almost completed. It remains for us to combine together various properties of the local and global specialisation maps, described in Chapters VIII and IX respectively. However, we first introduce a quantity which measures the (possible) discrepancy between the $\mathbb{Z}_p$-lattices $\mathrm{Ta}_p(E)$ and $\mathbb{T}_{\infty,E} \otimes_{\Lambda^{\mathrm{wt}}} \mathbb{Z}_p$.

The modular elliptic curves E and $\mathcal{C}^{\min}$ lie in the same $\mathbb{Q}$-isogeny class, hence there exists an isogeny $\vartheta : E \longrightarrow \mathcal{C}^{\min}$ defined over $\mathbb{Q}$ connecting the two objects. Without loss of generality, we may assume that ϑ is chosen so the dual isogeny $\vartheta^{\mathrm{dual}} : \mathcal{C}^{\min} \longrightarrow E$ satisfies

$$\begin{array}{ccc} X_1(Np) \twoheadrightarrow X_0(Np) & \twoheadrightarrow & X_0(N_E) \\ {\scriptstyle \text{optimal par.}}\downarrow & & \downarrow{\scriptstyle \text{strong Weil par.}} \\ \mathcal{C}^{\min} & \overset{\vartheta^{\mathrm{dual}}}{\longrightarrow} & E. \end{array}$$

Because E is the strong Weil curve, $\int_{E(\mathbb{C})^\pm} \omega_E = \Omega^\pm_{f_E}$ where ω_E denotes a Néron differential on E. Moreover, if $\omega_{\mathcal{C}^{\min}}$ is a corresponding Néron differential on $\mathcal{C}^{\min}$, then $\vartheta^*\omega_{\mathcal{C}^{\min}} = \mathfrak{c}(\vartheta)\omega_E$ for some $\mathfrak{c}(\vartheta) \in \mathbb{Q}^\times$.

Definition 9.17. *The 'defect of $\rho_{\infty,E}$' over* $\mathbf{F}$ *is the rational number*

$$\mathfrak{D}_{\mathbf{F},p}\big(\rho_{\infty,E}\big) \ := \ \left|\frac{\mathfrak{c}(\vartheta^{\mathrm{dual}})}{\mathfrak{c}(\vartheta)}\right|_p^{[\mathbf{F}:\mathbb{Q}]} \times \frac{\prod_{\nu|p} \#\mathrm{Coker}\big(\vartheta^{\mathrm{dual}}_{/\mathbf{F}_\nu}\big)}{\prod_{\nu|\infty} \#\big(\mathrm{Ker}(\vartheta)\big)^{\mathrm{Gal}(\mathbb{C}/\mathbf{F}_\nu)}}.$$

For example, if there are no rational cyclic p-isogenies between these two curves then the defect $\mathfrak{D}_{\mathbf{F},p}\big(\rho_{\infty,E}\big)$ is a p-unit, and we can discard it from our formulae. Under the assumption that E does not admit complex multiplication, when the prime $p > 37$ such cyclic p-isogenies cannot exist, therefore $\mathfrak{D}_{\mathbf{F},p}\big(\rho_{\infty,E}\big) \in \mathbb{Z}_p^\times$. Alternatively, if E has CM then beyond $p > 163$ such isogenies do not occur in nature, and again $\mathfrak{D}_{\mathbf{F},p}\big(\rho_{\infty,E}\big) \in \mathbb{Z}_p^\times$.

Let us remember that $\mathbf{F}$ was an abelian extension of $\mathbb{Q}$ in which p did not ramify. If we fix an isomorphism $\sigma : \Gamma^{\mathrm{wt}} \xrightarrow{\sim} 1 + p\mathbb{Z}_p$ sending $u_0 \mapsto 1+p$, then define

$$Ш_{\mathbf{F}}(\mathbb{T}_{\infty,E}, w) \;:=\; \mathrm{Mell}_{2,\mathbf{1}} \circ \sigma\Big(Ш_{\mathbf{F}}(\mathbb{T}_{\infty,E})\Big) \quad \text{inside of } \mathbb{Q}_{p,\mathbb{U}_E}\langle\langle w\rangle\rangle.$$

Recall also from Theorem 9.2, if $\rho_{\infty,E}$ was residually irreducible then the compact Selmer group $\mathfrak{S}_{\mathbf{F}}(\mathbb{T}_{\infty,E})$ vanished. The following is the main result of this chapter.

Theorem 9.18. *Assume that* $\mathfrak{S}_{\mathbf{F}}(\mathbb{T}_{\infty,E})$ *is zero, and the ring* $\mathcal{R}_E$ *is Gorenstein.*
(i) If $\#Ш_{\mathbf{F}}(E)[p^\infty]$ *is infinite or* $\langle -,-\rangle^{\mathrm{wt}}_{\mathbf{F},p}$ *is degenerate on* $E(\mathbf{F}) \times E(\mathbf{F})$*, then*

$$\mathrm{order}_{w=2} Ш_{\mathbf{F}}(\mathbb{T}_{\infty,E}, w) \;\geq\; \mathrm{rank}_{\mathbb{Z}} E(\mathbf{F}) + 1\,;$$

(ii) If $\#Ш_{\mathbf{F}}(E)[p^\infty]$ *is finite and* $\langle -,-\rangle^{\mathrm{wt}}_{\mathbf{F},p}$ *is non-degenerate on* $E(\mathbf{F}) \times E(\mathbf{F})$*, then* $\mathrm{order}_{w=2} Ш_{\mathbf{F}}(\mathbb{T}_{\infty,E}, w) \;=\; \mathrm{rank}_{\mathbb{Z}} E(\mathbf{F})$*, in fact*

$$\frac{Ш_{\mathbf{F}}(\mathbb{T}_{\infty,E}, w)}{(w-2)^{r_{\mathbf{F}}}}\Bigg|_{w=2} \times \prod_{\text{finite } \nu} \mathrm{Tam}^{(2)}_{\mathbf{F}_\nu}(\mathbb{T}_{\infty,E}) \times \frac{\mathrm{vol}^{(2)}_{\mathbb{T}_{\infty,E}}(H^1_{\mathcal{E}})}{\mathrm{vol}_E(H^1_e)} \times \mathfrak{D}_{\mathbf{F},p}(\rho_{\infty,E})$$
$$= \mathcal{L}^{\mathrm{wt}}_p(E) \times \mathrm{Per}^{(2)}_{\mathbf{I},\lambda_E}(\mathbf{F})\, \frac{\#Ш_{\mathbf{F}}(E)\prod_\nu [E(\mathbf{F}_\nu) : E_0(\mathbf{F}_\nu)]}{\#E(\mathbf{F})^2_{\mathrm{tors}}} \times \det\langle -,-\rangle^{\mathrm{wt}}_{\mathbf{F},p}\Big|_{E(\mathbf{F})\times E(\mathbf{F})}$$

up to an element of $\mathbb{Z}_p^\times$.

To be totally honest, we do not expect part (i) to ever occur in nature, however protocol demands that it should at least be mentioned. For those weights $k \in \mathbb{U}_E$ which are greater than two, one predicts formulae of the type

$$\frac{d^{r_?} Ш_{\mathbf{F}}(\mathbb{T}_{\infty,E})}{dX^{r_?}}\Bigg|_{X=(1+p)^{k-2}-1} \;=\; \text{arithmetic data} \;\times\; \text{a regulator term.}$$

The situation $k > 2$ is left to someone with a far steelier resolve than the author's.

Question. *How do we calculate the quantities appearing in Theorem 9.18?*

If $\mathbf{F} = \mathbb{Q}$ then the terms $r_{\mathbb{Q}}$, $\mathcal{L}^{\mathrm{wt}}_p(E)$, $\#Ш_{\mathbb{Q}}(E)$, $[E(\mathbb{Q}_l) : E_0(\mathbb{Q}_l)]$ and $\#E(\mathbb{Q})_{\mathrm{tors}}$ have been compiled in Cremona's tables, for elliptic curves of conductor $\leq 10{,}000$. More generally, the computer package MAGMA allows the user access to routines which will compute these orders numerically, even when the number field $\mathbf{F} \neq \mathbb{Q}$. This leaves us with the task of working out

(a) the value of the defect $\mathfrak{D}_{\mathbf{F},p}(-)$,

(b) the volume terms and Tamagawa factors, and

(c) the p-adic weight regulator.

Firstly, the defect is easy to compute: one simply determines the kernel/cokernel of the isogeny $\vartheta : E \longrightarrow \mathcal{C}^{\min}$, and the constants $\mathfrak{c}(\vartheta)$ and $\mathfrak{c}(\vartheta^{\mathrm{dual}}) = \frac{\deg(\vartheta)}{\mathfrak{c}(\vartheta)}$ can then be read off from the formal group of E over the p-adics.

To calculate (b) is a bit trickier, but it is possible to obtain a decent upper bound for these volumes and Tamagawa factors. By Propositions 8.9 and 8.16,

$$\#\mathrm{Ker}\Big(\oplus_{\mathfrak{p}|p}\,\delta_{\infty,E,\mathfrak{p}}\Big) \;=\; \mathcal{L}_p^{\mathrm{wt}}(E) \times \prod_{\nu\in\Sigma}\frac{\big[\mathcal{C}^{\min}(\mathbf{F}_\nu):\mathcal{C}_0^{\min}(\mathbf{F}_\nu)\big]}{\mathrm{Tam}^{(2)}_{\mathbf{F}_\nu}(\mathbb{T}_{\infty,E})}$$
$$\times\ \mathrm{Per}^{(2)}_{\mathbf{I},\lambda_E}(\mathbf{F}) \times \frac{\prod_{\mathfrak{p}|p}\mathrm{vol}_{\mathcal{C}^{\min},\mathbf{F}_\mathfrak{p}}\Big(H^1_e\big(\mathbf{F}_\mathfrak{p},\mathrm{Ta}_p(\mathcal{C}^{\min})\big)\Big)}{\mathrm{vol}^{(2)}_{\mathbb{T}_{\infty,E}}\Big(H^1_{\mathcal{E}}\big(\mathbf{F}\otimes\mathbb{Z}_p,\mathbb{T}^*_{\infty,E}(1)\big)\Big)}.$$

However, we also know that $\#\mathrm{Ker}\Big(\oplus_{\mathfrak{p}|p}\,\delta_{\infty,E,\mathfrak{p}}\Big)$ divides into

$$\Big[H^1\big(\mathbf{F}\otimes\mathbb{Z}_p,\ \mathrm{Ta}_p(\mathcal{C}^{\min})\big) : H^1\big(\mathbf{F}\otimes\mathbb{Z}_p,\ \mathbb{T}_{\infty,E}\big)\otimes_{\Lambda^{\mathrm{wt}}}\mathbb{Z}_p\Big],$$

and the above index equals $\prod_{\mathfrak{p}|p}\#\mathcal{C}^{\min}\big(\mathbf{F}_\mathfrak{p}\big)[p^\infty]$ which is certainly computable. For instance, if the field $\mathbf{F}=\mathbb{Q}$ and $\mathcal{C}^{\min}(\mathbb{Q}_p)$ contains no point of exact order p, then $\mathrm{Ker}\big(\delta_{\infty,E,p}\big)$ is trivial.

Finally, the mysterious regulator term $\mathrm{Reg}^{\mathrm{wt}}_{p,\mathbf{F}}(E) := \det\big\langle -,-\big\rangle^{\mathrm{wt}}_{\mathbf{F},p}\Big|_{E(\mathbf{F})\times E(\mathbf{F})}$ is the one quantity for which we lack a nice, general algorithm enabling its evaluation. Assuming the Selmer group of E over $\mathbf{F}$ is finite, clearly $\mathrm{Reg}^{\mathrm{wt}}_{p,\mathbf{F}}(E)$ is a p-adic unit and we are done. Alternatively if the rank $r_\mathbf{F}=1$, we are dealing with the formula

$$\frac{d\,\text{Ш}_\mathbf{F}\big(\mathbb{T}_{\infty,E},w\big)}{dw}\Bigg|_{w=2} \overset{\text{by } 9.18\text{(ii)}}{=} \text{a non-zero number} \times \big\langle P,P\big\rangle^{\mathrm{wt}}_{\mathbf{F},p}$$

where P is any point of infinite order in $E(\mathbf{F})$ which generates the free part over $\mathbb{Z}$.

Remark: In the special case where $\mathbf{F}=\mathbb{Q}\big(\sqrt{-d}\big)$ and the conductor of E splits completely in $\mathcal{O}_\mathbf{F}$, we have established a formula relating the derivative of the improved p-adic L-series over the imaginary quadratic field, with the height of a Heegner point on $\mathrm{Pic}^0 X_1(Np)$. Note the proof given in [Db3] uses analytic methods due to Gross-Zagier and B. Howard.

Example 9.19 Consider the strong Weil curve $E=X_0(11)$ over the field $\mathbf{F}=\mathbb{Q}$. The Tamagawa number of E at the bad prime 11 equals 5, elsewhere they're trivial. Let p denote a prime where E has good ordinary or bad multiplicative reduction. Then the optimal $X_1(Np)$-curve of Stevens is the modular curve $\mathcal{C}^{\min}=X_1(11)$.

(i) If $p=11$ then E has split multiplicative reduction at p, and by the theorem $\text{Ш}_\mathbb{Q}\big(\mathbb{T}_{\infty,E},2\big)$ must be a p-adic unit;

(ii) If $3\le p\le 97$ and $p\ne 11,19,29$ then E has good ordinary reduction at p, and $\text{Ш}_\mathbb{Q}\big(\mathbb{T}_{\infty,E},2\big)$ is again a p-adic unit provided $p\ne 5$.

Therefore for all ordinary primes $p<100$ other than 5, the μ-invariant over $\mathbb{Q}$ associated to the vertical deformation of $\mathrm{Ta}_p\big(X_0(11)\big)$ is identically zero.

Proof of Theorem 9.18: The Herbrand quotient of a homomorphism $g : A \to B$ is the ratio of integers

$$\mathcal{H}(g) \;=\; \frac{\#\mathrm{Coker}(g)}{\#\mathrm{Ker}(g)}$$

whenever it is defined, of course.

Let us now take $A = H^0\Big(\Gamma^{\mathrm{wt}},\, \mathrm{Sel}_{\mathbf{F}}(\mathbb{T}_{\infty,E})^{\vee}\Big)$ and $B = H^1\Big(\Gamma^{\mathrm{wt}},\, \mathrm{Sel}_{\mathbf{F}}(\mathbb{T}_{\infty,E})^{\vee}\Big)$. Assuming $\xi : A \longrightarrow B$, $a \mapsto a \bmod u_0 - 1$ is a quasi-isomorphism, then

$$\left.\frac{d^r\text{Ш}_{\mathbf{F}}(\mathbb{T}_{\infty,E})}{dX^r}\right|_{X=0} \;=\; \mathcal{H}(\xi) \;=\; \mathcal{H}\Big(\mathrm{Sel}_{\mathbf{F}}(\mathbb{T}_{\infty,E})^{\Gamma^{\mathrm{wt}}} \longrightarrow \mathrm{Sel}_{\mathbf{F}}(\mathbb{T}_{\infty,E})_{\Gamma^{\mathrm{wt}}}\Big)^{-1}$$

up to a p-adic unit, where the order of vanishing $r = \dim\big(\mathbb{Q}_p \otimes A\big) = \dim\big(\mathbb{Q}_p \otimes B\big)$. Note that the mapping ξ has finite kernel and cokernel, if and only if $\mathrm{Sel}_{\mathbf{F}}(\mathbb{T}_{\infty,E})^{\vee}$ possesses a semi-simple Λ^{wt}-structure at the prime ideal (X).

Evaluating Herbrand quotients around the commutative diagram

$$\begin{array}{ccc}
0 & & 0 \\
\downarrow & & \downarrow \\
\mathrm{Sel}_{\mathbf{F}}(\mathcal{C}^{\min})[p^\infty] & \xrightarrow{\alpha_{\infty,E}} & \mathrm{Sel}_{\mathbf{F}}(\mathbb{T}_{\infty,E})^{\Gamma^{\mathrm{wt}}} \\
\downarrow & & \downarrow \\
H^1\Big(\mathbf{F}_\Sigma/\mathbf{F},\, \mathcal{C}^{\min}[p^\infty]\Big) & \overset{\beta_{\infty,E}}{\twoheadrightarrow} & H^1\Big(\mathbf{F}_\Sigma/\mathbf{F},\, A_{\mathbb{T}_{\infty,E}}\Big)^{\Gamma^{\mathrm{wt}}} \\
\downarrow \lambda_{\mathcal{C}^{\min}} & & \downarrow \lambda_{\infty,E} \\
\displaystyle\bigoplus_{\nu\in\Sigma} \frac{H^1\big(\mathbf{F}_\nu, \mathcal{C}^{\min}[p^\infty]\big)}{\mathcal{C}^{\min}(\mathbf{F}_\nu)\otimes\mathbb{Q}_p/\mathbb{Z}_p} & \overset{\oplus\delta_{\infty,E,\nu}}{\twoheadrightarrow} & \displaystyle\bigoplus_{\nu\in\Sigma}\left(\frac{H^1\big(\mathbf{F}_\nu, A_{\mathbb{T}_{\infty,E}}\big)}{H^1_\star\big(\mathbf{F}_\nu, A_{\mathbb{T}_{\infty,E}}\big)}\right)^{\Gamma^{\mathrm{wt}}} \\
\downarrow & & \downarrow \\
H^1_f\Big(\mathbf{F}, \mathrm{Ta}_p(\mathcal{C}^{\min})\Big)^{\vee} & \overset{\widehat{\theta}_{\infty,E}}{\longleftarrow} & \mathrm{Sel}_{\mathbf{F}}(\mathbb{T}_{\infty,E})_{\Gamma^{\mathrm{wt}}} \\
\downarrow & & \downarrow \\
H^2\Big(\mathbf{F}_\Sigma/\mathbf{F},\, \mathcal{C}^{\min}[p^\infty]\Big) & \hookleftarrow & H^1\Big(\mathbf{F}_\Sigma/\mathbf{F},\, A_{\mathbb{T}_{\infty,E}}\Big)_{\Gamma^{\mathrm{wt}}}
\end{array}$$

a straightforward arrow-chasing argument reveals

$$\mathcal{H}(\xi)^{-1} \;=\; \mathcal{H}\Big(\widehat{\phi}_{\infty,E}\Big)\,\mathcal{H}(\beta_{\infty,E})^{-1}\mathcal{H}\Big(\oplus\,\delta_{\infty,E,\nu}\Big)\mathcal{H}\Big(\widehat{\theta}_{\infty,E}\Big).$$

However $\mathcal{H}\Big(\widehat{\phi}_{\infty,E}\Big)\,\mathcal{H}\Big(\widehat{\theta}_{\infty,E}\Big) = \mathcal{H}\Big(\phi_{\infty,E}\circ\theta_{\infty,E}\Big)^{-1}$, and one already knows that the composition $\phi_{\infty,E}\circ\theta_{\infty,E}(x)(y) = \log_p(u_0)^{-1} \times \langle x, y\rangle^{\mathrm{wt}}_{\mathbf{F},p}$. It therefore follows $\mathcal{H}\Big(\phi_{\infty,E}\circ\theta_{\infty,E}\Big)$ equals $\log_p(u_0)^{-r} \times$ the discriminant of the p-adic weight pairing.

Remark: For simplicity we now suppose that the p-Sylow subgroup of $\text{Ш}_{\mathbf{F}}(\mathcal{C}^{\min})$ is finite, in which case by [BK, Prop 5.4]

$$\mathcal{C}^{\min}(\mathbf{F}) \,\widehat{\otimes}\, \mathbb{Z}_p \;\cong\; H^1_f\big(\mathbf{F}, \mathrm{Ta}_p(\mathcal{C}^{\min})\big).$$

Furthermore, the order of vanishing of $\text{Ш}_{\mathbf{F}}(\mathbb{T}_{\infty,E})$ at $X=0$ will coincide exactly with the Mordell-Weil rank $r_{\mathbf{F}}$ of the elliptic curves E and $\mathcal{C}^{\min}$.

Applying Propositions 8.7(i), 8.9 and 8.16, one deduces that

$$\mathcal{H}\Big(\oplus\,\delta_{\infty,E,\nu}\Big)^{-1} \;=\; \mathcal{L}_p^{\mathrm{wt}}(E) \times \prod_{\nu\in\Sigma} \frac{\big[\mathcal{C}^{\min}(\mathbf{F}_\nu) : \mathcal{C}_0^{\min}(\mathbf{F}_\nu)\big]}{\mathrm{Tam}^{(2)}_{\mathbf{F}_\nu}(\mathbb{T}_{\infty,E})}$$
$$\times\; \mathrm{Per}^{(2)}_{\mathbf{I},\lambda_E}(\mathbf{F}) \times \frac{\prod_{\mathfrak{p}|p} \mathrm{vol}_{\mathcal{C}^{\min},\mathbf{F}_\mathfrak{p}}\Big(H^1_e\big(\mathbf{F}_\mathfrak{p}, \mathrm{Ta}_p(\mathcal{C}^{\min})\big)\Big)}{\mathrm{vol}^{(2)}_{\mathbb{T}_{\infty,E}}\Big(H^1_{\mathcal{E}}\big(\mathbf{F}\otimes\mathbb{Z}_p, \mathbb{T}^*_{\infty,E}(1)\big)\Big)}.$$

On the other hand,

$$\mathcal{H}(\beta_{\infty,E})^{-1} \;=\; \#\mathcal{C}^{\min}(\mathbf{F})[p^\infty] \qquad \text{by Lemma 7.7.}$$

N.B. Throughout the remainder of this proof, the equalities which are mentioned are only assumed to hold true modulo $\mathbb{Z}_p^\times$.

Tying these loose ends together, we obtain

$$\frac{d^{r_{\mathbf{F}}}\text{Ш}_{\mathbf{F}}(\mathbb{T}_{\infty,E})}{dX^{r_{\mathbf{F}}}}\bigg|_{X=0} \;=\; \mathcal{L}_p^{\mathrm{wt}}(E) \times \mathrm{Per}^{(2)}_{\mathbf{I},\lambda_E}(\mathbf{F}) \times \prod_{\nu\in\Sigma} \mathrm{Tam}^{(2)}_{\mathbf{F}_\nu}(\mathbb{T}_{\infty,E})^{-1} \times \log_p(u_0)^{-r_{\mathbf{F}}}$$
$$\times\; \frac{\mathrm{vol}_{\mathcal{C}^{\min},\mathbf{F}}\Big(H^1_e\big(\mathbf{F}\otimes\mathbb{Z}_p, \mathrm{Ta}_p(\mathcal{C}^{\min})\big)\Big)}{\mathrm{vol}^{(2)}_{\mathbb{T}_{\infty,E}}\Big(H^1_{\mathcal{E}}\big(\mathbf{F}\otimes\mathbb{Z}_p, \mathbb{T}^*_{\infty,E}(1)\big)\Big)} \times \mathbf{BSD}^{\mathrm{wt}}_{p,\mathbf{F}}(\mathcal{C}^{\min})$$

where the notation $\mathbf{BSD}^{\mathrm{wt}}_{p,\mathbf{F}}(\mathcal{C}^{\min})$ indicates the p-adic number

$$\frac{\#\text{Ш}_{\mathbf{F}}(\mathcal{C}^{\min})[p^\infty] \times \prod_\nu \big[\mathcal{C}^{\min}(\mathbf{F}_\nu) : \mathcal{C}_0^{\min}(\mathbf{F}_\nu)\big]}{\#\mathcal{C}^{\min}(\mathbf{F})[p^\infty]^2} \times \det\big\langle -,-\big\rangle^{\mathrm{wt}}_{\mathbf{F},p}\Big|_{\mathcal{C}^{\min}(\mathbf{F})\times\mathcal{C}^{\min}(\mathbf{F})}.$$

Proposition 9.20. *Under our hypothesis that* $\text{Ш}_{\mathbf{F}}(\mathcal{C}^{\min})[p^\infty]$ *is finite,*

$$\mathrm{vol}_{\mathcal{C}^{\min},\mathbf{F}}(H^1_e) \times \mathbf{BSD}^{\mathrm{wt}}_{p,\mathbf{F}}(\mathcal{C}^{\min}) \;=\; \mathfrak{D}_{\mathbf{F},p}(\rho_{\infty,E})^{-1} \times \mathrm{vol}_{E,\mathbf{F}}(H^1_e) \times \mathbf{BSD}^{\mathrm{wt}}_{p,\mathbf{F}}(E).$$

This statement is basically Cassels' isogeny theorem in disguise – the modified proof is supplied in the next section.

The above proposition allows us to exchange the optimal $X_1(Np)$-curve $\mathcal{C}^{\min}$ with the strong Weil curve E, although this scales the formula by the defect $\mathfrak{D}_{\mathbf{F},p}(\rho_{\infty,E})$. Finally,

$$\log_p(u_0)^{r_{\mathbf{F}}} \times \frac{d^{r_{\mathbf{F}}}\text{Ш}_{\mathbf{F}}(\mathbb{T}_{\infty,E})}{dX^{r_{\mathbf{F}}}}\bigg|_{X=0} \;=\; \frac{d^{r_{\mathbf{F}}}\Big(\sigma^{w-2}\big(\text{Ш}_{\mathbf{F}}(\mathbb{T}_{\infty,E})\big)\Big)}{dw^{r_{\mathbf{F}}}}\bigg|_{w=2}$$

and the theorem is proved. □

9.6 Variation under the isogeny $\vartheta : E \to \mathcal{C}^{\min}$

In this final section we shall supply the missing demonstration of Proposition 9.20. Under the condition that $\mathrm{III}_{\mathbf{F}}(\mathcal{C}^{\min})$ is finite, we are required to show

$$\frac{\mathrm{vol}_{E,\mathbf{F}}\Big(H^1_e\big(\mathbf{F}\otimes\mathbb{Z}_p, \mathrm{Ta}_p(E)\big)\Big)}{\mathrm{vol}_{\mathcal{C}^{\min},\mathbf{F}}\Big(H^1_e\big(\mathbf{F}\otimes\mathbb{Z}_p, \mathrm{Ta}_p(\mathcal{C}^{\min})\big)\Big)} \times \frac{\mathbf{BSD}^{\mathrm{wt}}_{p,\mathbf{F}}(E)}{\mathbf{BSD}^{\mathrm{wt}}_{p,\mathbf{F}}(\mathcal{C}^{\min})} \stackrel{?}{=} \mathfrak{D}_{\mathbf{F},p}\big(\rho_{\infty,E}\big).$$

Note that all equalities here are assumed to be true *only up to p-adic units.*

Remark: The rational isogeny $\vartheta : E \longrightarrow \mathcal{C}^{\min}$ admits a factorisation

$$E \xrightarrow{\vartheta_p} \frac{E}{\mathrm{Ker}(\theta)[p^\infty]} \xrightarrow{\vartheta'_p} \frac{E}{\mathrm{Ker}(\theta)} \cong \mathcal{C}^{\min}$$

where $\deg(\vartheta_p)$ is a power of p, and the degree of the complement ϑ'_p is coprime to p. Changing both the volume term and $\mathbf{BSD}^{\mathrm{wt}}_{p,\mathbf{F}}$ by the isogeny ϑ'_p contributes factors which are p-adic units; since we are only interested in the p-part of the formula, without loss of generality we may assume that $\deg(\vartheta)$ is a power of p.

For an elliptic curve $\mathcal{A}_{/\mathbb{Q}}$, we defined $\mathbf{BSD}_{\infty,\mathbf{F}}(\mathcal{A})$ to be the complex number

$$\frac{\#\mathrm{III}_{\mathbf{F}}(\mathcal{A}) \times \prod_\nu \big[\mathcal{A}(\mathbf{F}_\nu) : \mathcal{A}_0(\mathbf{F}_\nu)\big]}{\sqrt{\mathrm{disc}_{\mathbf{F}}} \times \#\mathcal{A}(\mathbf{F})^2_{\mathrm{tors}}} \times \det\big\langle -,-\big\rangle^{\mathrm{NT}}_{\mathbf{F},\infty}\Big|_{\mathcal{A}(\mathbf{F})\times\mathcal{A}(\mathbf{F})}.$$

Then up to p-adic units, there is an equality

$$\frac{\mathbf{BSD}^{\mathrm{wt}}_{p,\mathbf{F}}(E)}{\mathbf{BSD}^{\mathrm{wt}}_{p,\mathbf{F}}(\mathcal{C}^{\min})} = \frac{\mathbf{BSD}_{\infty,\mathbf{F}}(E)}{\mathbf{BSD}_{\infty,\mathbf{F}}(\mathcal{C}^{\min})} \times \left\{\frac{\mathrm{Reg}_{\infty,\mathbf{F}}(\mathcal{C}^{\min})}{\mathrm{Reg}_{\infty,\mathbf{F}}(E)} \times \frac{\mathrm{Reg}^{\mathrm{wt}}_{p,\mathbf{F}}(E)}{\mathrm{Reg}^{\mathrm{wt}}_{p,\mathbf{F}}(\mathcal{C}^{\min})}\right\}.$$

However the bracketted term $\{\dots\}$ above lies inside $\mathbb{Z}_p^\times$, which follows because the Néron-Tate pairing and the p-adic weight pairing are non-degenerate bilinear forms on $E(\mathbf{F})\otimes\mathbb{Q}$.

We next apply an old result of Cassels [Ca2, Thms 1.1-1.5].

Theorem 1.19. *The quantity* $\Omega_{\mathcal{A}}(\mathbf{F}) \times \mathbf{BSD}_{\infty,\mathbf{F}}(\mathcal{A})$ *is* $\mathbf{F}$*-isogeny invariant.*

As a corollary, one deduces that

$$\frac{\mathbf{BSD}^{\mathrm{wt}}_{p,\mathbf{F}}(E)}{\mathbf{BSD}^{\mathrm{wt}}_{p,\mathbf{F}}(\mathcal{C}^{\min})} = \frac{\Omega_{\mathcal{C}^{\min}}(\mathbf{F})}{\Omega_E(\mathbf{F})} \times \text{ a } p\text{-adic unit} \qquad (**)$$

N.B. the ratio $\frac{\Omega_{\mathcal{C}^{\min}}(\mathbf{F})}{\Omega_E(\mathbf{F})}$ is viewed as an $\mathbf{F}$-rational number, not a complex number! One should also recall from Definition 9.17, the defect is

$$\mathfrak{D}_{\mathbf{F},p}\big(\rho_{\infty,E}\big) = \left|\frac{\mathfrak{c}(\vartheta^{\mathrm{dual}})}{\mathfrak{c}(\vartheta)}\right|_p^{[\mathbf{F}:\mathbb{Q}]} \times \frac{\prod_{\nu|p}\#\mathrm{Coker}\big(\vartheta^{\mathrm{dual}}_{/\mathbf{F}_\nu}\big)}{\prod_{\nu|\infty}\#\big(\mathrm{Ker}(\vartheta)\big)^{\mathrm{Gal}(\mathbb{C}/\mathbf{F}_\nu)}}$$

where $\vartheta^{\mathrm{dual}}_{/\mathbf{F}_\nu} : \mathcal{C}^{\min}(\mathbf{F}_\nu) \longrightarrow E(\mathbf{F}_\nu)$ is the homomorphism induced on local points.

Lemma 9.21. *With the same assumptions as before, up to p-adic units*

$$(a)\quad \frac{\Omega_E(\mathbf{F})}{\Omega_{\mathcal{C}^{\min}}(\mathbf{F})} = \prod_{\nu\ \text{real}} \#\Big(\mathrm{Ker}(\vartheta)^+\Big) \times \prod_{\nu\ \text{complex}} \#\Big(\mathrm{Ker}(\vartheta)\Big) \times \prod_{\nu|p} \Big|\mathfrak{c}(\vartheta)\Big|_p^{[\mathbf{F}_\nu:\mathbb{Q}_p]};$$

$$(b)\quad \frac{\mathrm{vol}_{E,\mathbf{F}}\Big(H^1_e\big(\mathbf{F}\otimes\mathbb{Z}_p, \mathrm{Ta}_p(E)\big)\Big)}{\mathrm{vol}_{\mathcal{C}^{\min},\mathbf{F}}\Big(H^1_e\big(\mathbf{F}\otimes\mathbb{Z}_p, \mathrm{Ta}_p(\mathcal{C}^{\min})\big)\Big)} = \prod_{\nu|p} \Big|\mathfrak{c}(\vartheta^{\mathrm{dual}})\Big|_p^{[\mathbf{F}_\nu:\mathbb{Q}_p]} \times \#\mathrm{Coker}\big(\vartheta^{\mathrm{dual}}_{/\mathbf{F}_\nu}\big).$$

In fact this lemma implies Proposition 9.20 – simply substitute 9.21(a) and 9.21(b) into Equation **(**)**, then directly apply the definition of the defect.

Proof of 9.21: We start with part (a). The isogeny ϑ induces a homomorphism $\vartheta^* : H^1_{\mathrm{dR}}(\mathcal{C}^{\min}/\mathbb{Q}) \longrightarrow H^1_{\mathrm{dR}}(E/\mathbb{Q})$ of $\mathbb{Q}$-vector spaces. This map respects the Hodge filtration and as $\mathrm{Fil}^0 H^1_{\mathrm{dR}}$ is one-dimensional, there exists $\mathfrak{c}(\vartheta) \neq 0$ such that $\vartheta^*\omega_{\mathcal{C}^{\min}} = \mathfrak{c}(\vartheta)\omega_E$. We may assume $\deg(\vartheta)$ is a power of p, in which case $\mathfrak{c}(\vartheta) \in p^{\mathbb{Z}}$. For all places ν of $\mathbf{F}$, by [Ca2, 3.4]

$$\frac{\int_{E(\mathbf{F}_\nu)} \vartheta^*\big(\text{diff. form on } \mathcal{C}^{\min}\big)}{\int_{\mathcal{C}^{\min}(\mathbf{F}_\nu)} \text{diff. form on } \mathcal{C}^{\min}} = \frac{\#\mathrm{Ker}\Big(E(\mathbf{F}_\nu) \xrightarrow{\vartheta} \mathcal{C}^{\min}(\mathbf{F}_\nu)\Big)}{\#\mathrm{Coker}\Big(E(\mathbf{F}_\nu) \xrightarrow{\vartheta} \mathcal{C}^{\min}(\mathbf{F}_\nu)\Big)}.$$

If the place $\nu|\infty$ and $\mathbf{F}_\nu = \mathbb{C}$, then

$$\frac{\int_{E(\mathbb{C})} \vartheta^*\omega_{\mathcal{C}^{\min}} \wedge \overline{\vartheta^*\omega_{\mathcal{C}^{\min}}}}{\int_{\mathcal{C}^{\min}(\mathbb{C})} \omega_{\mathcal{C}^{\min}} \wedge \overline{\omega_{\mathcal{C}^{\min}}}} = \frac{\#\mathrm{Ker}\Big(E(\mathbb{C}) \xrightarrow{\vartheta} \mathcal{C}^{\min}(\mathbb{C})\Big)}{\#\mathrm{Coker}\Big(E(\mathbb{C}) \xrightarrow{\vartheta} \mathcal{C}^{\min}(\mathbb{C})\Big)} = \#\mathrm{Ker}(\vartheta);$$

as an immediate consequence

$$\Big|\mathfrak{c}(\vartheta)\Big|_\nu \times \frac{\int_{E(\mathbb{C})} \omega_E \wedge \overline{\omega_E}}{\int_{\mathcal{C}^{\min}(\mathbb{C})} \omega_{\mathcal{C}^{\min}} \wedge \overline{\omega_{\mathcal{C}^{\min}}}} = \#\mathrm{Ker}(\vartheta).$$

If the place $\nu|\infty$ and $\mathbf{F}_\nu = \mathbb{R}$, then

$$\frac{\int_{E(\mathbb{R})} \vartheta^*\omega_{\mathcal{C}^{\min}}}{\int_{\mathcal{C}^{\min}(\mathbb{R})} \omega_{\mathcal{C}^{\min}}} = \frac{\#\mathrm{Ker}\Big(E(\mathbb{R}) \xrightarrow{\vartheta} \mathcal{C}^{\min}(\mathbb{R})\Big)}{\#\mathrm{Coker}\Big(E(\mathbb{R}) \xrightarrow{\vartheta} \mathcal{C}^{\min}(\mathbb{R})\Big)} = \frac{\#\Big(\mathrm{Ker}(\vartheta) \cap E(\mathbb{R})\Big)}{1};$$

note the sequence $0 \to \mathrm{Ker}(\vartheta) \cap E(\mathbb{R}) \to E(\mathbb{R}) \xrightarrow{\vartheta} \mathcal{C}^{\min}(\mathbb{R}) \to H^1\big(\mathbb{C}/\mathbb{R}, \mathrm{Ker}(\vartheta)\big)$ is exact, and the right-hand cohomology is trivial as $\#\mathrm{Gal}(\mathbb{C}/\mathbb{R})$ is prime to $\#\mathrm{Ker}(\vartheta)$. It follows that

$$\Big|\mathfrak{c}(\vartheta)\Big|_\nu \times \frac{\int_{E(\mathbb{R})} \omega_E}{\int_{\mathcal{C}^{\min}(\mathbb{R})} \omega_{\mathcal{C}^{\min}}} = \#\mathrm{Ker}(\vartheta)^+.$$

The archimedean periods $\Omega_{\mathcal{A}}(\mathbf{F})$ decompose into a product over the infinite places. To summarise we have just shown

$$\frac{\Omega_E(\mathbf{F})}{\Omega_{\mathcal{C}^{\min}}(\mathbf{F})} = \prod_{\nu|\infty}\left|\frac{1}{\mathfrak{c}(\vartheta)}\right|_{\nu} \times \prod_{\nu \text{ real}} \#\Big(\mathrm{Ker}(\vartheta)^{+}\Big) \times \prod_{\nu \text{ complex}} \#\Big(\mathrm{Ker}(\vartheta)\Big).$$

Finally, using the product formula over the number field $\mathbf{F}$:

$$\prod_{\nu|\infty}\left|\frac{1}{\mathfrak{c}(\vartheta)}\right|_{\nu} = \prod_{\nu|p}\Big|\mathfrak{c}(\vartheta)\Big|_{\nu} = \prod_{\nu|p}\Big|\mathfrak{c}(\vartheta)\Big|_{p}^{[\mathbf{F}_\nu:\mathbb{Q}_p]} \quad \text{modulo } \mathbb{Z}_p^{\times}$$

and part (a) follows easily.

Remark: To prove that (b) is true, for all $\nu|p$ there is a commutative diagram

$$\begin{array}{ccc}
\Big(\mathcal{C}^{\min}(\mathbf{F}_\nu)\,\widehat{\otimes}\,\mathbb{Z}_p\Big)_{/\mathbb{Z}_p-\mathrm{tors}} & \xrightarrow{\vartheta^{\mathrm{dual}}} & \Big(E(\mathbf{F}_\nu)\,\widehat{\otimes}\,\mathbb{Z}_p\Big)_{/\mathbb{Z}_p-\mathrm{tors}} \\
\downarrow \cong & & \downarrow \cong \\
\text{image of } H^1_e\big(\mathbf{F}_\nu, \mathrm{Ta}_p(\mathcal{C}^{\min})\big) & \xrightarrow{\vartheta^{\mathrm{dual}}_{/\mathbf{F}_\nu}} & \text{image of } H^1_e\big(\mathbf{F}_\nu, \mathrm{Ta}_p(E)\big)
\end{array}$$

where the image is taken inside of $H^1\big(\mathbf{F}_\nu, V_p(E)\big)$, and the downward arrows are induced by the Kummer map.

This time $\left(\vartheta^{\mathrm{dual}}\right)^{*}\omega_E = \mathfrak{c}\big(\vartheta^{\mathrm{dual}}\big)\omega_{\mathcal{C}^{\min}}$, so in terms of Haar measures

$$\mathrm{vol}_{(\vartheta^{\mathrm{dual}})^{*}\omega_E,\mathbf{F}_\nu}\big(-\big) = \Big|\mathfrak{c}\big(\vartheta^{\mathrm{dual}}\big)\Big|_p \times \mathrm{vol}_{\omega_{\mathcal{C}^{\min}},\mathbf{F}_\nu}\big(-\big).$$

In addition, we also know (up to elements of $\mathbb{Z}_p^{\times}$) that

$$\frac{\mathrm{vol}_{(\vartheta^{\mathrm{dual}})^{*}\omega_E,\mathbf{F}_\nu}\Big(H^1_e\big(\mathbf{F}_\nu, \mathrm{Ta}_p(\mathcal{C}^{\min})\big)\Big)}{\mathrm{vol}_{\omega_E,\mathbf{F}_\nu}\Big(H^1_e\big(\mathbf{F}_\nu, \mathrm{Ta}_p(E)\big)\Big)} = \frac{\#\mathrm{Ker}\big(\vartheta^{\mathrm{dual}}_{/\mathbf{F}_\nu}\big)}{\#\mathrm{Coker}\big(\vartheta^{\mathrm{dual}}_{/\mathbf{F}_\nu}\big)}.$$

But the mapping $\vartheta^{\mathrm{dual}}_{/\mathbf{F}_\nu}$ is injective, since $\mathrm{Ker}\big(\vartheta^{\mathrm{dual}}\big) \subset \mathcal{C}^{\min}(\mathbb{Q})[p^\infty]$ lies in the $\mathbb{Z}_p$-torsion subgroup of $H^1_e\big(\mathbf{F}_\nu, \mathrm{Ta}_p(\mathcal{C}^{\min})\big)$. Therefore one obtains

$$\Big|\mathfrak{c}\big(\vartheta^{\mathrm{dual}}\big)\Big|_p^{\dim_{\mathbb{Q}_p}(H^1_e)} \times \frac{\mathrm{vol}_{\omega_{\mathcal{C}^{\min}},\mathbf{F}_\nu}\Big(H^1_e\big(\mathbf{F}_\nu, \mathrm{Ta}_p(\mathcal{C}^{\min})\big)\Big)}{\mathrm{vol}_{\omega_E,\mathbf{F}_\nu}\Big(H^1_e\big(\mathbf{F}_\nu, \mathrm{Ta}_p(E)\big)\Big)} = \frac{1}{\#\mathrm{Coker}\big(\vartheta^{\mathrm{dual}}_{/\mathbf{F}_\nu}\big)}.$$

Moreover $\dim_{\mathbb{Q}_p}\Big(H^1_e\big(\mathbf{F}_\nu, V_p(E)\big)\Big) = [\mathbf{F}_\nu : \mathbb{Q}_p]$, whence

$$\frac{\mathrm{vol}_{\omega_{\mathcal{C}^{\min}},\mathbf{F}_\nu}\Big(H^1_e\big(\mathbf{F}_\nu, \mathrm{Ta}_p(\mathcal{C}^{\min})\big)\Big)}{\mathrm{vol}_{\omega_E,\mathbf{F}_\nu}\Big(H^1_e\big(\mathbf{F}_\nu, \mathrm{Ta}_p(E)\big)\Big)} = \frac{1}{\Big|\mathfrak{c}\big(\vartheta^{\mathrm{dual}}\big)\Big|_p^{[\mathbf{F}_\nu:\mathbb{Q}_p]} \times \#\mathrm{Coker}\big(\vartheta^{\mathrm{dual}}_{/\mathbf{F}_\nu}\big)}.$$

Taking the product over all prime ideals ν above p, the lemma is proved.

□

CHAPTER X

Two-Variable Iwasawa Theory of Elliptic Curves

Our study of the p-ordinary part of the Selmer group for E is almost at an end. However there is one last twist in the tale. Up to this point, the ground field $\mathbf{F}$ we have been working over has remained fixed, but this assumption can now be relaxed. In particular, allowing the number field $\mathbf{F}_n$ to vary in a cyclotomic $\mathbb{Z}_p$-extension adjoins an extra variable, Y say, to the ambient deformation ring.

Why is this cyclotomic variable Y important, and what net benefits are gained? The principal motivation has its origins in the Iwasawa theory of elliptic curves admitting complex multiplication. Specifically, if E has CM by an order in $\mathcal{O}_{\mathbb{Q}(\sqrt{-D})}$ then $\mathrm{Gal}\Big(\mathbb{Q}(E[p^\infty])/\mathbb{Q}(E[p])\Big) \cong \mathbb{Z}_p \times \mathbb{Z}_p$. In terms of completed group algebras

$$\mathbb{Z}_p[[G]] \;\cong\; \mathbb{Z}_p[[\Gamma^{\mathrm{wt}} \times \Gamma^{\mathrm{cy}}]][\Delta] \;\cong\; \bigoplus_{\delta \in \Delta} \mathbb{Z}_p[[X,Y]].\delta$$

where $G = \mathrm{Gal}\Big(\mathbb{Q}(E[p^\infty])/\mathbb{Q}(\sqrt{-D})\Big)$, and Δ is finite with order coprime to p. Therefore in the CM case, combining the cyclotomic deformation with the weight deformation is *equivalent* to studying the arithmetic of E over the field of definition of its p-power division points.

If E does not admit complex multiplication, then $\mathrm{Gal}\big(\mathbb{Q}(E[p^\infty])/\mathbb{Q}\big)$ is an open subgroup of $\mathrm{GL}_2(\mathbb{Z}_p)$, and its underlying Iwasawa algebra is a non-commutative power series ring in four variables. In this scenario $\mathbb{Z}_p\big[\big[\Gamma^{\mathrm{wt}} \times \Gamma^{\mathrm{cy}}\big]\big]$ represents the completed group ring of its maximal torus, i.e. the cyclotomic-weight deformation encodes the abelianisation of GL_2-Iwasawa theory.

A major goal of this chapter is to formulate three new **Main Conjectures**: the first corresponding to the vertical line $s = 1$, another for the line of symmetry $s = k/2$, and the last for the whole (s,k)-plane. These are fundamental statements reconciling the 'algebraic world' with the 'analytic world' (as a useful mnemonic). More precisely, there should be a one-to-one correspondence

$$\Big\{\text{characteristic ideals of Selmer}\Big\} \quad \overset{\text{Main Conj.}}{\longleftrightarrow} \quad \Big\{p\text{-adic } L\text{-functions}\Big\}.$$

Warning: Before we forget it entirely, let's first remember to compute the leading term of the algebraic L-series $\mathbf{III_F}\Big(\mathbb{T}_{\infty,E}^{-1/2}\Big)$. Indeed this p-adic L-value is a vital constituent of the Euler characteristic over the two-variable deformation of $E[p^\infty]$. As many of the details are akin to the vertical calculation (in the previous chapter), we shall attempt to keep the exposition relatively brief.

10.1 The half-twisted Euler characteristic formula

Throughout one takes $\mathbf{F}$ to be an abelian number field, and the prime number $p \geq 3$. For simplicity assume $\rho_{\infty,E}$ is residually irreducible, and $\mathcal{R}_E$ is a Gorenstein ring. The fundamental work of Flach and Nekovář [Fl,Ne3] guarantees the existence of a (nearly) self-dual, symplectic form

$$\left\langle -,-\right\rangle^{\mathrm{Nek}}_{\mathbf{F},p,\mathbb{T}^{-1/2}_{\infty,E}} : \widetilde{H}^1_f\left(\mathbf{F}_\Sigma/\mathbf{F},\, \mathbb{T}^{-1/2}_{\infty,E}\right) \times \widetilde{H}^1_f\left(\mathbf{F}_\Sigma/\mathbf{F},\, \mathbb{T}^{-1/2}_{\infty,E}\right) \longrightarrow \mathcal{R};$$

its manifestation at weight two, yields a $\mathbb{Z}_p$-bilinear pairing

$$\left\langle -,-\right\rangle^{-1/2}_{\mathbf{F},p} : H^1_f\left(\mathbf{F}_\Sigma/\mathbf{F},\, V_pE\right)^{\mathrm{gen}} \times H^1_f\left(\mathbf{F}_\Sigma/\mathbf{F},\, V_pE\right)^{\mathrm{gen}} \longrightarrow \mathbb{Q}_p$$

on the 'generic part' of the vector-space H^1_f.

Notation: Let us denote by $\widetilde{\lambda}^{-1/2}_{\infty,E}$ the direct sum of the restriction maps

$$H^1\left(\mathbf{F}_\Sigma/\mathbf{F},\, A^{-1/2}_{\mathbb{T}_{\infty,E}}\right) \longrightarrow \bigoplus_{\nu\in\Sigma-S_{\mathbf{F},p}}\left(\frac{H^1\left(\mathbf{F}_\nu, A^{-1/2}_{\mathbb{T}_{\infty,E}}\right)}{H^1_{\mathrm{nr}}\left(\mathbf{F}_\nu, A^{-1/2}_{\mathbb{T}_{\infty,E}}\right)}\right) \oplus \bigoplus_{\nu\in S_{\mathbf{F},p}}\left(\frac{H^1\left(\mathbf{F}_\nu, A^{-1/2}_{\mathbb{T}_{\infty,E}}\right)}{H^1_+(\mathbf{F}_\nu)^\perp}\right).$$

The next result is effectively the half-twisted analogue of Theorem 9.18.

Theorem 10.1. *Suppose the following four conditions hold:*

(i) $\#\text{Ш}_{\mathbf{F}}(E)[p^\infty]$ *is finite;*

(ii) $\left\langle -,-\right\rangle^{-1/2}_{\mathbf{F},p}$ *is non-degenerate;*

(iii) the homomorphism $\widetilde{\lambda}^{-1/2}_{\infty,E}$ *is surjective;*

(iv) E does **not** *attain split multiplicative reduction at any place above p.*

Then $\mathrm{order}_{w=2}\text{Ш}_{\mathbf{F}}\left(\mathbb{T}^{-1/2}_{\infty,E}, w\right) = \mathrm{rank}_{\mathbb{Z}}E(\mathbf{F}) - \mathrm{corank}_{\Lambda^{\mathrm{wt}}}\mathrm{Sel}_{\mathbf{F}}\left(\mathbb{T}^{-1/2}_{\infty,E}\right) = r^\dagger_{\mathbf{F}}$ *say, and its leading term is given by*

$$\left.\frac{\text{Ш}_{\mathbf{F}}\left(\mathbb{T}^{-1/2}_{\infty,E}, w\right)}{(w-2)^{r^\dagger_{\mathbf{F}}}}\right|_{w=2} \times \prod_\nu \mathrm{Tam}^{(2)}_{\mathbf{F}_\nu}\left(\mathbb{T}^{-1/2}_{\infty,E}\right) \times \mathrm{covol}^{(2)}_{\mathbb{T}^{-1/2}_{\infty,E}}\left(\mathbf{F}\otimes\mathbb{Z}_p\right) \times \mathfrak{D}_{\mathbf{F},p}\left(\rho_{\infty,E}\right)$$

$$= \prod_{\mathfrak{p}\in S^{\mathrm{good}}_{\mathbf{F},p}} \#\widetilde{E}\left(\mathbb{F}_{q_\mathfrak{p}}\right)^2 \times \frac{\#\text{Ш}_{\mathbf{F}}(E)\prod_\nu\left[E(\mathbf{F}_\nu):E_0(\mathbf{F}_\nu)\right]}{\#E(\mathbf{F})^2_{\mathrm{tors}}} \times \det\left.\left\langle -,-\right\rangle^{-1/2}_{\mathbf{F},p}\right|_{E(\mathbf{F})\times E(\mathbf{F})}$$

modulo $\mathbb{Z}_p^\times$, *where* $S^{\mathrm{good}}_{\mathbf{F},p} = S_{\mathbf{F},p} \cap \left\{\mathfrak{p} \nmid N_E\right\}$.

N.B. The quantity $\prod_{\mathfrak{p}\in S^{\mathrm{good}}_{\mathbf{F},p}}\left(\#\widetilde{E}\left(\mathbb{F}_{q_\mathfrak{p}}\right)\right)^2$ is trivial when $p\big|N_E$ as the set $S^{\mathrm{good}}_{\mathbf{F},p} = \emptyset$. Alternatively, if $p \nmid N_E$ then the subset $S^{\mathrm{good}}_{\mathbf{F},p}$ is the whole of $S_{\mathbf{F},p}$, and the product is therefore non-empty.

Proof: The main ingredients are identical to the demonstration of Theorem 9.18. Firstly condition (iv) implies the set of split multiplicative primes $S^{\mathrm{mult}}_{\mathbf{F},p}$ is void, in which case Conjecture 7.11 holds. As a nice consequence, the natural mapping $\mathrm{Sel}_{\mathbf{F}}\left(\mathcal{C}^{\min}\right)[p^\infty] \overset{\alpha^{-1/2}_{\infty,E}}{\longrightarrow} \mathrm{Sel}_{\mathbf{F}}\left(\mathbb{T}_{\infty,E}\otimes\Psi^{-1/2}\right)^{\Gamma^{\mathrm{wt}}}$ will then be a quasi-isomorphism of cofinitely-generated groups over $\mathbb{Z}_p$.

Fortunately, condition (iii) ensures we have the short exact sequence

$$0 \longrightarrow \mathrm{Sel}_{\mathbf{F}}\left(\mathbb{T}^{-1/2}_{\infty,E}\right) \longrightarrow H^1\left(\mathbf{F}_\Sigma/\mathbf{F},\ A^{-1/2}_{\mathbb{T}_{\infty,E}}\right) \overset{\widetilde{\lambda}^{-1/2}_{\infty,E}}{\longrightarrow} \bigoplus_{\nu\in\Sigma} \frac{H^1\left(\mathbf{F}_\nu, A^{-1/2}_{\mathbb{T}_{\infty,E}}\right)}{H^1_{\star,+}\left(\mathbf{F}_\nu,\mathbb{T}^{-1/2}_{\infty,E}\right)^{\perp}} \longrightarrow 0.$$

Taking its Γ^{wt}-cohomology, produces a commutative diagram with exact columns:

$$\begin{array}{ccc}
0 & & 0 \\
\downarrow & & \downarrow \\
\mathrm{Sel}_{\mathbf{F}}\left(\mathcal{C}^{\min}\right)[p^\infty] & \overset{\alpha^{-1/2}_{\infty,E}}{\longrightarrow} & \mathrm{Sel}_{\mathbf{F}}\left(\mathbb{T}^{-1/2}_{\infty,E}\right)^{\Gamma^{\mathrm{wt}}} \\
\downarrow & & \downarrow \\
H^1\left(\mathbf{F}_\Sigma/\mathbf{F},\ \mathcal{C}^{\min}[p^\infty]\right) & \overset{\beta^{-1/2}_{\infty,E}}{\longrightarrow} & H^1\left(\mathbf{F}_\Sigma/\mathbf{F},\ A^{-1/2}_{\mathbb{T}_{\infty,E}}\right)^{\Gamma^{\mathrm{wt}}} \\
\downarrow \lambda^{-1/2}_{\mathcal{C}^{\min}} & & \downarrow \lambda^{-1/2}_{\infty,E} \\
\displaystyle\bigoplus_{\nu\in\Sigma-S_{\mathbf{F},p}} \frac{H^1\left(\mathbf{F}_\nu,\mathcal{C}^{\min}[p^\infty]\right)}{H^1_{\mathrm{nr}}\left(\mathbf{F}_\nu,\mathcal{C}^{\min}[p^\infty]\right)} & \overset{\oplus\delta^{-1/2}_{\infty,E,\nu}}{\longrightarrow} & \displaystyle\bigoplus_{\nu\in\Sigma-S_{\mathbf{F},p}} \left(\frac{H^1\left(\mathbf{F}_\nu, A^{-1/2}_{\mathbb{T}_{\infty,E}}\right)}{H^1_{\mathrm{nr}}\left(\mathbf{F}_\nu, A^{-1/2}_{\mathbb{T}_{\infty,E}}\right)}\right)^{\Gamma^{\mathrm{wt}}} \\
\displaystyle\oplus\bigoplus_{\nu\in S_{\mathbf{F},p}} \frac{H^1\left(\mathbf{F}_\nu,\mathcal{C}^{\min}[p^\infty]\right)}{\mathcal{W}_{\mathcal{C}^{\min},+}\left(\mathbf{F}_\nu\right)^{\perp}} & & \displaystyle\oplus\bigoplus_{\nu\in S_{\mathbf{F},p}} \left(\frac{H^1\left(\mathbf{F}_\nu, A^{-1/2}_{\mathbb{T}_{\infty,E}}\right)}{H^1_{+}\left(\mathbf{F}_\nu\right)^{\perp}}\right)^{\Gamma^{\mathrm{wt}}} \\
\downarrow & & \downarrow \\
H^1_f\left(\mathbf{F},\mathrm{Ta}_p\left(\mathcal{C}^{\min}\right)\right)^{\vee} & \overset{\widehat{\theta}^{-1/2}_{\infty,E}}{\dashleftarrow} & \mathrm{Sel}_{\mathbf{F}}\left(\mathbb{T}^{-1/2}_{\infty,E}\right)_{\Gamma^{\mathrm{wt}}}
\end{array}$$

Figure 10.1

Under the conditions (i) and (ii), the order of vanishing of $\text{Ш}_{\mathbf{F}}\left(\mathbb{T}^{-1/2}_{\infty,E},\ w\right)$ at $w=2$ is precisely equal to the dimension of $\mathbb{Q}_p\otimes H^0\left(\Gamma^{\mathrm{wt}},\ \left(\mathrm{Sel}_{\mathbf{F}}\left(\mathbb{T}^{-1/2}_{\infty,E}\right)_{/\Lambda^{\mathrm{wt}}\text{-div}}\right)^{\vee}\right)$.

In fact, the half-twisted pairing $\langle -,-\rangle^{-1/2}_{\mathbf{F},p}$ factorises through the dual map $\theta^{-1/2}_{\infty,E}$. Its discriminant is therefore intertwined with that of

$$H^0\left(\Gamma^{\mathrm{wt}},\mathrm{Tors}_{\Lambda^{\mathrm{wt}}}\left(\mathrm{Sel}_{\mathbf{F}}\left(\mathbb{T}^{-1/2}_{\infty,E}\right)^{\vee}\right)\right) \overset{\mathrm{nat}}{\longrightarrow} H^1\left(\Gamma^{\mathrm{wt}},\mathrm{Tors}_{\Lambda^{\mathrm{wt}}}\left(\mathrm{Sel}_{\mathbf{F}}\left(\mathbb{T}^{-1/2}_{\infty,E}\right)^{\vee}\right)\right)$$

which can now be computed via Figure 10.1.

Remarks: (a) Every horizontal arrow in the above diagram is a quasi-isomorphism.

(b) If the map $\xi_{\rm div}^{-1/2} : \mathcal{S}^{\Gamma^{\rm wt}} \longrightarrow \mathcal{S}_{\Gamma^{\rm wt}}$ where $\mathcal{S} = {\rm Sel}_{\mathbf{F}}\left(\mathbb{T}_{\infty,E}^{-1/2}\right)_{/\Lambda^{\rm wt}\text{-div}}$, then

$$\log_p(u_0)^{-r_{\mathbf{F}}^{\dagger}} \times {\rm disc}\left(\langle -,-\rangle_{\mathbf{F},p}^{-1/2}\right) \;=\; \mathcal{H}\left(\xi_{\rm div}^{-1/2}\right)^{-1} \times \mathcal{H}\left(\oplus\, \delta_{\infty,E,\nu}^{-1/2}\right) \times \mathcal{H}\left(\beta_{\infty,E}^{-1/2}\right)^{-1}.$$

(c) The Herbrand quotient of $\beta_{\infty,E}^{-1/2}$ was calculated in Lemma 7.12(a).

(d) The Herbrand quotient of $\delta_{\infty,E,\nu}^{-1/2}$ was calculated in Proposition 8.7(ii) if $\nu \nmid p$, and in Proposition 8.13 when $\nu | p$.

Combining (a)-(b)-(c)-(d) together, one obtains a formula relating the half-twisted arithmetic invariants of $\mathcal{C}^{\min}$ to the reciprocal Herbrand quotient $\mathcal{H}\left(\xi_{\rm div}^{-1/2}\right)^{-1}$. However the latter quantity is the leading term of $\text{Ш}_{\mathbf{F}}\left(\mathbb{T}_{\infty,E}^{-1/2}\right)$ at $X = 0$.

Finally, one may exchange the optimal curve $\mathcal{C}^{\min}$ with the strong Weil curve E using Cassels' Isogeny Theorem 1.19, which explains why the defect term $\mathfrak{D}_{\mathbf{F},p}$ rears its ugly head once more.

□

Example 10.2. Let's consider the modular curve $E = X_0(11)$ over the field $\mathbf{F} = \mathbb{Q}$. Since the analytic rank is zero, the half-twisted regulators must therefore be trivial. If $7 \leq p \leq 97$ and $p \neq 11, 19, 29$ then the elliptic curve E has good ordinary reduction at p, and our theorem implies that $\text{Ш}_{\mathbb{Q}}\left(\mathbb{T}_{\infty,E}^{-1/2}, 2\right)$ will be a p-adic unit. It follows at all good ordinary primes $p < 100$ excluding 5, the μ-invariant over $\mathbb{Q}$ associated to the half-twisted deformation of ${\rm Ta}_p\big(X_0(11)\big)$ vanishes (c.f. Ex. 9.19).

The reader will notice that we have studiously avoided the prime $p = 11$ in the above example, because Theorem 10.1 does not cover split multiplicative reduction. Nevertheless in the semistable case, one has a shrewd idea what to expect.

Conjecture 10.3. *Assume that the prime number p divides N_E. Then*

$${\rm order}_{w=2}\text{Ш}_{\mathbf{F}}\left(\mathbb{T}_{\infty,E}^{-1/2}, w\right) \;=\; {\rm rank}_{\mathbb{Z}}E(\mathbf{F}) \;+\; \#S_{\mathbf{F},p}^{\rm split} \;-\; {\rm corank}_{\Lambda^{\rm wt}}{\rm Sel}_{\mathbf{F}}\big(\mathbb{T}_{\infty,E}^{-1/2}\big);$$

moreover, the leading term of $\text{Ш}_{\mathbf{F}}^{-1/2}$ satisfies

$$\left.\frac{\text{Ш}_{\mathbf{F}}\big(\mathbb{T}_{\infty,E}^{-1/2}, w\big)}{(w-2)^{r_{\mathbf{F}}^{\dagger}+\#S_{\mathbf{F},p}^{\rm split}}}\right|_{w=2} \times \prod_{\nu} {\rm Tam}_{\mathbf{F}_\nu}^{(2)}\left(\mathbb{T}_{\infty,E}^{-1/2}\right) \times {\rm covol}_{\mathbb{T}_{\infty,E}^{-1/2}}^{(2)}(\mathbf{F}\otimes\mathbb{Z}_p) \times \mathfrak{D}_{\mathbf{F},p}$$

$$= \prod_{\nu \in S_{\mathbf{F},p}^{\rm split}} \frac{\log_p\left(\mathbf{q}_{E/\mathbf{F}_\nu}\right)}{{\rm ord}_\nu\left(\mathbf{q}_{E/\mathbf{F}_\nu}\right)} \times \frac{\#\text{Ш}_{\mathbf{F}}(E)\prod_\nu\left[E(\mathbf{F}_\nu):E_0(\mathbf{F}_\nu)\right]}{\#E(\mathbf{F})_{\rm tors}^{2}} \times \det\langle -,-\rangle_{\mathbf{F},p}^{-1/2}$$

again up to p-adic units.

N.B. Over the rationals, the $\mathcal{L}$-invariant $\frac{\log_p(\mathbf{q}_E)}{{\rm ord}_p(\mathbf{q}_E)} \neq 0$ courtesy of the result in [StÉ]. For a general number field $\mathbf{K} \neq \mathbb{Q}$ the non-vanishing is still an open problem.

Question. *When is the global-to-local map* $\widetilde{\lambda}^{-1/2}_{\infty,E}$ *surjective?*

In general, the answer to this question lies fairly deep.

If the representation $\rho_{\infty,E}: G_{\mathbb{Q}} \longrightarrow \mathrm{GL}_2\big(\mathbb{Z}_{p,\mathbb{U}_E}\langle\langle w\rangle\rangle\big)$ is of CM-type by $\mathcal{O}_{\mathbb{Q}(\sqrt{-D})}$, this reduces to the existence or otherwise of non-trivial pseudo-null submodules in the (Pontrjagin dual of) the two-variable Selmer group over $\mathbf{F}\Big(\sqrt{-D},\ E[p^\infty]\Big)$. Alternatively, if we demand that the p-adic L-function is suitably well-behaved along the line $s = k/2$, one can still make some progress:

Proposition 10.4. *Assume Greenberg's conjecture holds for the family* $\{\mathbf{f}_k\}_{k\in\mathbb{U}_E\cap\mathbb{N}}$ *with its generic rank* $r_{\text{gen}}(\mathbf{F}) = 0$, *and in addition that* $\mathcal{R}_E$ *is a regular local ring. Then the homomorphism* $\widetilde{\lambda}^{-1/2}_{\infty,E}$ *is always surjective.*

Proof: Let $\mathfrak{S}_{\mathbf{F}}\Big(\mathbb{T}^{-1/2}_{\infty,E}\Big)$ denote the compact Selmer group associated to the half-twisted deformation of $\rho_{E,p}$. Its Pontrjagin dual sits inside the Poitou-Tate sequence

$$0 \longrightarrow \mathrm{Sel}_{\mathbf{F}}\Big(\mathbb{T}^{-1/2}_{\infty,E}\Big) \longrightarrow H^1\Big(\mathbf{F}_\Sigma/\mathbf{F},\ A^{-1/2}_{\mathbb{T}_{\infty,E}}\Big)$$
$$\xrightarrow{\widetilde{\lambda}^{-1/2}_{\infty,E}} \bigoplus_{\nu\in\Sigma} \frac{H^1\big(\mathbf{F}_\nu, A^{-1/2}_{\mathbb{T}_{\infty,E}}\big)}{H^1_{\star,+}\big(\mathbf{F}_\nu, \mathbb{T}^{-1/2}_{\infty,E}\big)^{\perp}} \longrightarrow \mathfrak{S}_{\mathbf{F}}\Big(\mathbb{T}^{-1/2}_{\infty,E}\Big)^{\vee} \longrightarrow \ \cdots.$$

To establish the surjectivity of $\widetilde{\lambda}^{-1/2}_{\infty,E}$, clearly it is enough to show $\mathfrak{S}_{\mathbf{F}}\Big(\mathbb{T}^{-1/2}_{\infty,E}\Big) = 0$. There are two stages:

(A) Prove that $\mathfrak{S}_{\mathbf{F}}\Big(\mathbb{T}^{-1/2}_{\infty,E}\Big)$ is Λ^{wt}-torsion, and has universally p-bounded order;

(B) Prove that all proper subgroups of $H^1\Big(\mathbf{F}_\Sigma/\mathbf{F},\ \mathbb{T}^{-1/2}_{\infty,E}\Big)[p^\infty]$ are zero.

Since (A) implies the compact Selmer group lies in the $\mathbb{Z}_p$-torsion submodule of $H^1\Big(\mathbf{F}_\Sigma/\mathbf{F},\ \mathbb{T}^{-1/2}_{\infty,E}\Big)$, any finite subgroup of $\mathfrak{S}_{\mathbf{F}}\Big(\mathbb{T}^{-1/2}_{\infty,E}\Big)$ must then be zero by (B). As a corollary, the whole Selmer group was trivial to begin with.

Remark: One can identify the p^∞-torsion submodule of $H^1\Big(\mathbf{F}_\Sigma/\mathbf{F},\ \mathbb{T}^{-1/2}_{\infty,E}\Big)$ with the 0th-cohomology group

$$H^0\Big(\mathbf{F}_\Sigma/\mathbf{F},\ \mathbb{T}^{-1/2}_{\infty,E}\otimes\mathbb{Q}_p/\mathbb{Z}_p\Big) \subset \left(\Psi^{-1/2}_{\mathcal{R}}\otimes\mathbf{e}_{\mathcal{R}}.\varinjlim_{\pi_1^*} {}^{\mathbf{t}}J_r^{\mathrm{ord}}[p^\infty]^{\vee}\otimes_{\mathbb{Z}_p}\mu_{p^\infty}\right)(\mathbf{F}).$$

The right-hand side has no finite p-subgroups, which can be seen by using the same argument to Step 3 of Theorem 9.2 (N.B. as an $\mathcal{R}_E$-module $\mathbf{e}_{\mathcal{R}}.\varinjlim_{\pi_1^*} {}^{\mathbf{t}}J_r^{\mathrm{ord}}[p^\infty]^{\vee}$ is identical to $\Psi^{-1/2}_{\mathcal{R}}\otimes\mathbf{e}_{\mathcal{R}}.\varinjlim_{\pi_1^*} {}^{\mathbf{t}}J_r^{\mathrm{ord}}[p^\infty]^{\vee}$, therefore the transition maps π_1^*, $\pi_{p\,*}$ must again act nilpotently on the p-torsion). It immediately follows (B) is true.

This leaves us with the task of establishing (A).

Key Claim: For almost all arithmetic weights $k \in \mathbb{U}_E \cap \mathbb{N}_{\geq 2}$, the natural map

$$\mathfrak{S}_{\mathbf{F}}\left(\mathbb{T}^{-1/2}_{\infty,E}\right) \otimes_{\Lambda^{\mathrm{wt}},\mathfrak{g}_k} \mathbb{Z}_p \quad \stackrel{\mathrm{nat}}{\longrightarrow} \quad H^1_{f,\mathrm{Spec}\mathcal{O}_{\mathbf{F}}}\left(\mathbf{F}_\Sigma/\mathbf{F},\ \mathbb{T}^{-1/2}_{\infty,E} \otimes_{\Lambda^{\mathrm{wt}},\mathfrak{g}_k} \mathbb{Z}_p\right)$$

has finite kernel and cokernel, bounded independently of k.

When $\mathbf{F} = \mathbb{Q}$ and $\mathcal{R}_E = \Lambda^{\mathrm{wt}}$ this statement is shown in the article [NP, Sect.4]. However the proof generalises seamlessly to Noetherian regular local rings, as all their height one prime ideals are principal. Similarly, the argument for abelian number fields $\mathbf{F}$ reduces to the analogous result over $\mathbb{Q}$, since one may decompose the Selmer group into its $X_{\mathbf{F}}$-eigencomponents.

Remark: If $k \equiv 2 \mod 2(p-1)$ then $\mathbb{T}^{-1/2}_{\infty,E} \otimes_{\Lambda^{\mathrm{wt}},\mathfrak{g}_k} \mathbb{Z}_p$ is a lattice inside $V_{\mathbf{f}_k}(k/2)$. Provided that the L-value $L(\mathbf{f}_k/\mathbf{F}, k/2)$ is non-zero, Kato has shown [Ka1, 14.2(2)] the vector-space Selmer group

$$H^1_{f,\mathrm{Spec}\mathcal{O}_{\mathbf{F}}}\left(\mathbf{F}_\Sigma/\mathbf{F},\ V_{\mathbf{f}_k}(k/2)\right) \;=\; 0.$$

In particular, the above specialisation $H^1_{f,\mathrm{Spec}\mathcal{O}_{\mathbf{F}}}\left(\mathbf{F}_\Sigma/\mathbf{F},\ \mathbb{T}^{-1/2}_{\infty,E} \otimes_{\Lambda^{\mathrm{wt}},\mathfrak{g}_k} \mathbb{Z}_p\right)$ must therefore be a finite abelian p-group.

Because we assumed that the generic rank $r_{\mathrm{gen}}(\mathbf{F}) = 0$, the L-value at $s = k/2$ is non-zero at all bar finitely many exceptional weights k. Moreover the 'Key Claim' together with Kato's result implies $\#\left(\mathfrak{S}_{\mathbf{F}}\left(\mathbb{T}^{-1/2}_{\infty,E}\right) \otimes_{\Lambda^{\mathrm{wt}},\mathfrak{g}_k} \mathbb{Z}_p\right) < \infty$ for such k, in which case the compact Selmer group over $\mathbf{F}$ is Λ^{wt}-torsion.

Lastly, an identical argument to Theorem 9.2 (Step 2) establishes that its order is bounded above by $p^{\nu_1^\dagger + \nu_2^\dagger}$; here $p^{\nu_1^\dagger}$ is chosen large enough to kill off the kernel of

$$\mathfrak{S}_{\mathbf{F}}\left(\mathbb{T}^{-1/2}_{\infty,E}\right) \otimes_{\Lambda^{\mathrm{wt}},\mathfrak{g}_k} \mathbb{Z}_p \stackrel{\mathrm{nat}}{\to} H^1_{f,\mathrm{Spec}\mathcal{O}_{\mathbf{F}}} \hookrightarrow \left(\left(\mathbb{T}^{-1/2}_{\infty,E} \otimes_{\Lambda^{\mathrm{wt}},\mathfrak{g}_k} \mathbb{Z}_p\right) \otimes \mathbb{Q}/\mathbb{Z}\right)^{G_{\mathbf{F}}}$$

and likewise $\nu_2^\dagger = \max_{\tau,\psi}\left\{\nu^\dagger_{2,\tau,\psi}\right\}$, where each integer $\nu^\dagger_{2,\tau,\psi}$ is set equal to

$$\max_{1\leq j\leq n(\tau,\psi)} \left\{\mathrm{ord}_p\left(1 - \psi(l)a_l(\mathbf{f}_k)l^{1-k/2} + \psi(l)^2 l^{2-k/2} < l >^{k-2}\right) \,\middle|\, k \in D_j^{\tau,\psi},\ k \equiv \tau\right\}$$

N.B. the above notation is the same as was employed in the proof of Theorem 9.2.

The surjectivity of the homomorphism $\widetilde{\lambda}^{-1/2}_{\infty,E}$ follows immediately.

□

If $r_{\mathrm{gen}}(\mathbf{F}) > 0$ then it is unclear whether or not the global-to-local map surjects. One possibility is to mimic the approach of §6.5, and construct an embedding

$$\mathfrak{S}_{\mathbf{F}}\left(\mathbb{T}^{-1/2}_{\infty,E}\right) \quad \stackrel{?}{\hookrightarrow} \quad \mathrm{Hom}_{\Lambda^{\mathrm{wt}}}\left(\mathrm{Sel}_{\mathbf{F}}\left(\mathbb{T}^{-1/2}_{\infty,E}\right)^\vee,\ \mathbb{Z}_p\langle\langle w\rangle\rangle\right)^{\square}.$$

The right-hand side is a free $\mathbb{Z}_p[\![\Gamma^{\mathrm{wt}}]\!]$-module of finite-type; one could then compute Λ^{wt}-ranks along the Poitou-Tate sequence, to deduce triviality of $\mathrm{Coker}\left(\widetilde{\lambda}^{-1/2}_{\infty,E}\right)$.

10.2 The p-adic height over a double deformation

It is fair to say we now possess a pretty good understanding of the arithmetic of E along the vertical line $s = 1$, and also along the line of symmetry $s = k/2$ (subject to Greenberg's Conjecture 7.5, of course). The very last piece of the jigsaw is to allow the number field $\mathbf{F}$ to vary in a cyclotomic $\mathbb{Z}_p$-extension of the ground field. The finite sub-layers in this $\mathbb{Z}_p$-extension '$\mathbf{F}^{\text{cy}}$' are still abelian extensions of $\mathbb{Q}$, however they are no longer unramified at the prime p.

We'll begin by recalling the structure theory for modules over a two-variable deformation ring. Let R be a normal integral domain, finite and flat over $\mathbb{Z}_p[\![X]\!]$. An $R[\![Y]\!]$-module is called *pseudo-null* if its localisations at all height one primes in Spec $R[\![Y]\!]$ are trivial. Further, a homomorphism $\phi : M \longrightarrow N$ of $R[\![Y]\!]$-modules is said to be a *pseudo-isomorphism* if Kerϕ and Cokerϕ are both pseudo-null.

Notation: $M \overset{\text{ps}\sim}{\longrightarrow} N.$

Assume that M denotes some compact, finitely-generated, torsion $R[\![Y]\!]$-module. The structure theory informs us there exists a pseudo-isomorphism

$$M \overset{\text{ps}\sim}{\longrightarrow} \bigoplus_{i=1}^{t} \frac{R[\![Y]\!]}{\mathcal{P}_i^{e_i}}$$

where the $\mathcal{P}_i$'s are height one prime ideals of $R[\![Y]\!]$, and the e_i's are positive integers. The product $\prod_{i=1}^t \mathcal{P}_i^{e_i}$ is the characteristic ideal of M (it need not be principal).

Hypothesis. *The integral domain R is regular.*

As an immediate consequence, the two-variable ring $R[\![Y]\!]$ must also be regular. This condition guarantees that the ideal $\text{char}_{R[\![Y]\!]}(M)$ has a principal generator. We also observe that all regular CNL rings are automatically Gorenstein.

Lemma 10.5. *Let $\mathfrak{C}_M(Y)$ denote the characteristic power series of M. Then*

$$r \;=\; \text{order}_{Y=0}\Big(\mathfrak{C}_M(Y)\Big) \;\geq\; \text{rank}_R\Big(M/Y.M\Big)$$

with equality iff $\alpha : M[\times Y] \longrightarrow M/Y.M$ has an R-torsion kernel and cokernel; assuming this does occur, one has the leading-term formula

$$\left.\frac{\mathfrak{C}_M(Y)}{Y^r}\right|_{Y=0} \;=\; \frac{\text{char}_R(\text{Coker}(\alpha))}{\text{char}_R(\text{Ker}(\alpha))} \quad \text{modulo } R^\times.$$

The proof is fairly standard, and can be found in any decent algebra textbook.

Remark: In the next section we shall take $R = \mathcal{R}_E$, the universal p-ordinary deformation ring for $\rho_{E,p} : G_{\mathbb{Q}} \to \text{GL}_2(\mathbb{Z}_p)$. The $R[\![Y]\!]$-module M will then either be the dual of $\text{Sel}_{\mathbf{F}^{\text{cy}}}\Big(\mathbb{T}_{\infty,E}\Big)$, or in the half-twisted case the dual of $\text{Sel}_{\mathbf{F}^{\text{cy}}}\Big(\mathbb{T}_{\infty,E}^{-1/2}\Big)$.

A formal definition of the p-adic height pairing.
Let $\mathbf{F}^{\text{cy}}$ denote the cyclotomic $\mathbb{Z}_p$-extension of $\mathbf{F}$, with sub-layers $\mathbf{F}_n$ of degree p^n. Recall (from Chapter IX) at every non-trivial character $\varrho : \text{Gal}(\mathbf{F}^{\text{cy}}/\mathbf{F}) \xrightarrow{\sim} 1+p\mathbb{Z}_p$ and for each place $\nu \in \Sigma$, we defined

$$\ell_{\varrho,\nu} : G_{\mathbf{F}_\nu} \longrightarrow \mathbb{Z}_p \qquad \text{by restricting } \log_p \circ \varrho \circ i_{\mathbf{F}} \text{ to the group } G_{\mathbf{F}_\nu}.$$

In particular, the homomorphism $\ell_{\varrho,\nu}$ belongs to $\text{Hom}_{\text{cont}}\big(G_{\mathbf{F}_\nu}, \mathbb{Z}_p\big) = H^1\big(\mathbf{F}_\nu, \mathbb{Z}_p\big)$. Extending scalars the element $\ell_{R,\nu} = \ell_{\varrho,\nu} \otimes 1$ lies in $H^1\big(\mathbf{F}_\nu, \mathbb{Z}_p\big) \otimes_{\mathbb{Z}_p} R \cong H^1\big(\mathbf{F}_\nu, R\big)$, thence under the cup-product pairing

$$H^1\big(\mathbf{F}_\nu,\ R(1)\big) \times H^1\big(\mathbf{F}_\nu,\ R\big) \overset{\text{inv}_{\mathbf{F}_\nu} \circ \cup}{\longrightarrow} R$$

it gives rise to another homomorphism $\widetilde{\ell_{R,\nu}} : H^1\big(\mathbf{F}_\nu, R(1)\big) \longrightarrow R$. The kernel of $\widetilde{\ell_{R,\nu}}$ will be precisely the R-submodule $\mathcal{N}_\infty^{\text{univ}} H^1\big(\mathbf{F}_\nu, R(1)\big)$ of universal norms.

Let $\mathbf{T}$ be a free R-module of finite type, equipped with a continuous $G_{\mathbf{F}}$-action.

Notation: For places ν of $\mathbf{F}_n$ with $n \geq 0$, fix local conditions $X_{\nu,n} \subset H^1\big(\mathbf{F}_{\nu,n}, \mathbf{T}\big)$ which are compatible with respect to corestriction in $\mathbf{F}_\nu^{\text{cy}}$. Then one defines

$$H^1_{X,\Sigma}\big(\mathbf{F},\mathbf{T}\big)^0 \;=\; \text{Ker}\left(H^1\big(\mathbf{F}_\Sigma/\mathbf{F},\mathbf{T}\big) \overset{\oplus \text{res}_\nu}{\longrightarrow} \bigoplus_{\nu \in \Sigma} \frac{H^1\big(\mathbf{F}_\nu,\mathbf{T}\big)}{\mathcal{N}_\infty^{\text{univ}}\big(X_\nu\big)} \right)$$

where $\mathcal{N}_\infty^{\text{univ}}\big(X_\nu\big) = \bigcap_{n \geq 0} \text{cores}_{\mathbf{F}_n/\mathbf{F}}\big(X_{\nu,n}\big)$. The corresponding dual object is

$$H^1_{X^\perp,\Sigma}\big(\mathbf{F},\mathbf{T}^*(1)\big)^0 \;=\; \text{Ker}\left(H^1\big(\mathbf{F}_\Sigma/\mathbf{F},\mathbf{T}^*(1)\big) \overset{\oplus \text{res}_\nu}{\longrightarrow} \bigoplus_{\nu \in \Sigma} \frac{H^1\big(\mathbf{F}_\nu,\mathbf{T}^*(1)\big)}{\mathcal{N}_\infty^{\text{univ}}\big(X_\nu^\perp\big)} \right)$$

N.B. here $\mathcal{N}_\infty^{\text{univ}}\big(X_\nu^\perp\big) = \bigcap_{n \geq 0} \text{cores}_{\mathbf{F}_n/\mathbf{F}}\big(X_{\nu,n}^\perp\big)$ under the cup-product pairing.

For example, if $\nu \nmid p$ then a natural choice of $X_{\nu,n}$'s are the unramified cocycles. The *formal height pairing* is a bilinear form on $H^1_{X,\Sigma}\big(\mathbf{F},\mathbf{T}\big)^0 \times H^1_{X^\perp,\Sigma}\big(\mathbf{F},\mathbf{T}^*(1)\big)^0$ which will take values in the coefficient ring R.

Fix a local field $K = \mathbf{F}_\nu$, and let x be an element of the R-module $\mathcal{N}_\infty^{\text{univ}}\big(X_\nu\big)$. Since $\mathcal{N}_\infty^{\text{univ}}\big(X_\nu\big) \subset H^1\big(K,\mathbf{T}\big)$, it corresponds to an extension class

$$0 \longrightarrow \mathbf{T} \longrightarrow \mathfrak{X}_x \longrightarrow R \longrightarrow 0$$

of $R[G_K]$-modules. Applying $\text{Hom}_R\big(-, R(1)\big)$ and taking G_K-invariants, then

$$\begin{aligned} \ldots \longrightarrow \mathbf{T}^*(1)^{G_K} \longrightarrow H^1\Big(G_K,\ R(1)\Big) &\longrightarrow H^1\Big(G_K,\ \mathfrak{X}_x^*(1)\Big) \\ &\longrightarrow H^1\Big(G_K,\ \mathbf{T}^*(1)\Big) \longrightarrow H^2\Big(G_K,\ R(1)\Big) \longrightarrow \ldots \end{aligned}$$

arises as a long exact sequence in Galois cohomology.

Assume further that $y \in \mathcal{N}_\infty^{\text{univ}}(X_\nu^\perp)$. As the one-cocycle y is orthogonal to x, clearly $y \mapsto 0$ inside $H^2(G_K, R(1))$ and so must lie in the image of $H^1(G_K, \mathfrak{X}_x^*(1))$. Therefore there exists some lift $\widetilde{y}_0$ of the original element $y = y_0$.

Remark: We can apply the same argument to any norm compatible sequence $(y_n)_n \in \varprojlim_n X_{\nu,n}^\perp$, to obtain classes $\widetilde{y}_n \in H^1(K_n, \mathfrak{X}_x^*(1))$ such that $\widetilde{y}_n \mapsto y_n$. For each $n \geq 0$, one sets $\widehat{y}_n = \text{cores}_{K_n/K}(\widetilde{y}_n) \in H^1(K, \mathfrak{X}_x^*(1))$ which is a lift of y. The sequence $\{\widehat{y}_n\}$ converges to an element

$$s_{\text{can},K}(y) = \lim_{n\to\infty}(\widehat{y}_n) \in H^1(K, \mathfrak{X}_x^*(1)) \big/ \mathcal{N}_\infty^{\text{univ}} H^1(K, R(1))$$

which is well-defined, and independent of all the choices made above.

To define the p-adic height pairing, choose a point $P \in H^1_{X,\Sigma}(\mathbf{F}, \mathbf{T})^0$ corresponding to an extension class $0 \to \mathbf{T} \to \mathfrak{X}_P \to R \to 0$ of $R[G_\mathbf{F}]$-modules, so that

$$H^1(\mathbf{F}_\Sigma/\mathbf{F}, R(1)) \to H^1(\mathbf{F}_\Sigma/\mathbf{F}, \mathfrak{X}_P^*(1)) \to H^1(\mathbf{F}_\Sigma/\mathbf{F}, \mathbf{T}^*(1)) \to H^2(\mathbf{F}_\Sigma/\mathbf{F}, R(1))$$

is the associated sequence in the $\text{Gal}(\mathbf{F}_\Sigma/\mathbf{F})$-cohomology of the Kummer dual.

Now take $Q \in H^1_{X^\perp,\Sigma}(\mathbf{F}, \mathbf{T}^*(1))^0$, and let $s_{\text{glob}}(Q)$ be any lift to the cohomology group $H^1(\mathbf{F}_\Sigma/\mathbf{F}, \mathfrak{X}_P^*(1))$. It follows that $s_{\text{glob}}(Q)_\nu - s_{\text{can},\mathbf{F}_\nu}(Q_\nu)$ belongs to the quotient $H^1(\mathbf{F}_\nu, R(1)) \big/ \mathcal{N}_\infty^{\text{univ}} H^1(\mathbf{F}_\nu, R(1))$ at every place $\nu \in \Sigma$.

Definition 10.6. *One defines the formal p-adic height pairing*

$$\langle -, - \rangle_{R,\mathbf{T},\{X_\nu\}_\nu} : H^1_{X,\Sigma}(\mathbf{F}, \mathbf{T})^0 \times H^1_{X^\perp,\Sigma}(\mathbf{F}, \mathbf{T}^*(1))^0 \longrightarrow R$$

by the summation formula

$$\langle P, Q \rangle_{R,\mathbf{T},\{X_\nu\}_\nu} := \sum_{\nu\in\Sigma} \widetilde{\ell_{R,\nu}}\Big(s_{\text{glob}}(Q)_\nu - s_{\text{can},\mathbf{F}_\nu}(Q_\nu)\Big).$$

Crucially this pairing is independent of the global lift $s_{\text{glob}}(Q)$, via the product rule. If $R = \mathbb{Z}_p$ and $X_\nu = H^1_f$, one obtains the classical p-adic height pairing [Sn1,PR1].

10.3 Behaviour of the characteristic ideals

The main objective of this chapter is to evaluate the leading term of the algebraic p-adic L-function, associated to a 2-variable deformation (cyclotomic and weight). An intrinsic component of the leading term is the p-adic regulator, which is built out of the formal height pairing we have just introduced. Depending on whether one is considering the vertical or the half-twisted deformation, the local conditions $\{X_\nu\}_{\nu\in\Sigma}$ must be chosen carefully at each finite place.

It makes good sense to give a short review of our hypotheses to date.

Assume that the Hypothesis ($\mathbf{F}$-ab) holds, and the prime $p \geq 3$ is unramified in $\mathbf{F}$. Recall that $E_{/\mathbb{Q}}$ denoted a strong Weil curve of conductor N_E, possessing either good ordinary or bad multiplicative reduction over the field of p-adic numbers $\mathbb{Q}_p$. To simplify the exposition considerably, we now further suppose that the universal deformation ring $\mathcal{R}_E$ is *regular.*

Variation of the height pairing in a half-twisted lifting.
The primary object of study will be the two-variable Selmer group

$$\mathrm{Sel}_{\mathbf{F}^{\mathrm{cy}}}\big(\mathbb{T}_{\infty,E}^{-1/2}\big) \quad := \quad \varinjlim_{n} \mathrm{Sel}_{\mathbf{F}_n}\big(\mathbb{T}_{\infty,E}^{-1/2}\big).$$

This is naturally a discrete module under the action of $\Gamma^{\mathrm{cy}} = \mathrm{Gal}\big(\mathbf{F}^{\mathrm{cy}}/\mathbf{F}\big) \cong \mathbb{Z}_p$ and the group of diamond operators $\Gamma^{\mathrm{wt}} \cong 1 + p\mathbb{Z}_p$.

Proposition 10.7. *The Pontrjagin dual of* $\mathrm{Sel}_{\mathbf{F}^{\mathrm{cy}}}\big(\mathbb{T}_{\infty,E}^{-1/2}\big)$ *is* $\mathbb{Z}_p\big[\big[\Gamma^{\mathrm{cy}}, \Gamma^{\mathrm{wt}}\big]\big]$*-torsion.*

In fact, this result is a surprisingly easy consequence of its difficult p-adic version. One knows that the $\mathbb{Z}_p\big[\big[\Gamma^{\mathrm{cy}}, \Gamma^{\mathrm{wt}}\big]\big]$-corank of Selmer must be bounded above by the $\mathbb{Z}_p\big[\big[\Gamma^{\mathrm{cy}}\big]\big]$-corank of its specialisations over weight-space. At weight two for example, we are led to consider the Λ^{cy}-corank of $\mathrm{Sel}_{\mathbf{F}^{\mathrm{cy}}}\big(\mathcal{C}^{\min}\big)[p^\infty]$. In the CM case Rubin showed the specialised coranks are zero infinitely often, whilst in the non-CM situation there is a similar result due to Kato [Ru1,Ka1].

(In the latter scenario, it may become necessary to demand residual irreducibility of the big Galois representation $\rho_\infty : \mathrm{Gal}\big(\overline{\mathbb{Q}}/\mathbb{Q}\big) \to \mathrm{GL}_2\big(\mathcal{R}_E\big)$, since this ensures the absolute irreducibility of the mod p Galois representation associated to the eigenforms $\mathbf{f}_k$, at infinitely many weights $k \in \mathbb{U}_E \cap \mathbb{N}_{\geq 2}$.)

For each place $\nu \in \Sigma$, one employs the local conditions

$$X_\nu^\dagger \quad = \quad \begin{cases} \mathrm{Ker}\Big(H^1\big(\mathbf{F}_\nu, \mathbb{T}_{\infty,E}^{-1/2}\big) \longrightarrow H^1\big(I_{\mathbf{F}_\nu}, \mathbb{T}_{\infty,E}^{-1/2}\big)\Big) & \text{if } \nu \nmid p \\ \mathrm{Ker}\Big(H^1\big(\mathbf{F}_\nu, \mathbb{T}_{\infty,E}^{-1/2}\big) \longrightarrow H^1\big(I_{\mathbf{F}_\nu}, \mathbb{T}_{\infty,E}^{-1/2}/\mathrm{F}^+\big)\Big) & \text{if } \nu \mid p. \end{cases}$$

Centering the formal height pairing $\big\langle -,-\big\rangle_{\mathcal{R}_E, \mathbb{T}_{\infty,E}^{-1/2}, \{X_\nu^\dagger\}_\nu}$ at the base weight $k_0 = 2$, the pairing extends naturally to a Λ^{wt}-bilinear form

$$\big\langle -,-\big\rangle_{\mathbb{T}_{\infty,E}^{-1/2}}^{\mathrm{cy}} : \mathfrak{S}_{\mathbf{F}}\big(\mathbb{T}_{\infty,E}^{-1/2}\big) \times \mathfrak{S}_{\mathbf{F}}\big(\mathbb{T}_{\infty,E}^{-1/2}\big) \quad \longrightarrow \quad \mathrm{Frac}\Big(\mathbb{Z}_{p,\mathbb{U}_E}\langle\langle w\rangle\rangle\Big).$$

Finally, following Plater [Pl, Sect.5] one defines the **Tamagawa ideal** at $\nu \in \Sigma$ by

$$\ell_\nu\big(\mathbb{T}_{\infty,E}^{-1/2}\big) \quad := \quad \mathrm{char}_{\Lambda^{\mathrm{wt}}}\left(\frac{X_{\nu,\mathrm{sat}}^\dagger}{\mathcal{N}_\infty^{\mathrm{univ}}\big(X_{\nu,\mathrm{sat}}^\dagger\big)}\right)$$

where $X_{\nu,\mathrm{sat}}^\dagger$ denotes the Λ^{wt}-saturation of $X_\nu^\dagger$ inside the group $H^1\Big(\mathbf{F}_\nu,\ \mathbb{T}_{\infty,E}^{-1/2}\Big)$.

The proof of the following result is postponed until the next section.

Theorem 10.8. *If we set* $\mathfrak{C}^{\mathrm{alg}}_{\mathbf{F},p}\left(\mathbb{T}^{-1/2}_{\infty,E}, Y\right) := \mathrm{char}_{\Lambda^{\mathrm{wt}}[\![Y]\!]}\left(\mathrm{Sel}_{\mathbf{F}^{\mathrm{cy}}}\left(\mathbb{T}^{-1/2}_{\infty,E}\right)^{\vee}\right)$, *then the characteristic power series has a zero of order* $\mathfrak{r} = \mathrm{rank}_{\Lambda^{\mathrm{wt}}}\mathfrak{S}_{\mathbf{F}}\left(\mathbb{T}^{-1/2}_{\infty,E}\right)$ *at* $Y = 0$; *furthermore, there is an equality*

$$\left.\frac{\mathfrak{C}^{\mathrm{alg}}_{\mathbf{F},p}\left(\mathbb{T}^{-1/2}_{\infty,E}, Y\right)}{Y^{\mathfrak{r}}}\right|_{Y=0} = \log_p \varrho(\gamma_0)^{-\mathfrak{r}} \prod_{\nu\in\Sigma} \ell_\nu\left(\mathbb{T}^{-1/2}_{\infty,E}\right) \mathrm{disc}\langle -,-\rangle^{\mathrm{cy}}_{\mathbb{T}^{-1/2}_{\infty,E}} \text{Ш}_{\mathbf{F}}\left(\mathbb{T}^{-1/2}_{\infty,E}\right)$$

which is taken modulo $\left(\Lambda^{\mathrm{wt}}\right)^{\times}$.

Of course if Conjecture 7.18 is true, the order of vanishing should be $\mathfrak{r} \overset{?}{=} r_{\mathrm{gen}}(\mathbf{F})$. Let us now consider the two-variable algebraic p-adic L-function

$$\mathbf{L}^{\mathrm{alg}}_{p,\mathbf{F}}\left(\mathbb{T}^{-1/2}_{\infty,E}, s, w\right) := \left(\varrho^{s-1} \times \mathrm{Mell}_{2,1} \circ \sigma\right)\left(\mathrm{char}_{\mathbb{Z}_p[\![\Gamma^{\mathrm{cy}},\Gamma^{\mathrm{wt}}]\!]}\left(\mathrm{Sel}_{\mathbf{F}^{\mathrm{cy}}}\left(\mathbb{T}^{-1/2}_{\infty,E}\right)^{\vee}\right)\right)$$

which is an analytic function of $(s,w) \in \mathbb{Z}_p \times \mathbb{U}_E$, and is well-defined up to units.

Corollary 10.9. *Under the same assumptions as Theorem 10.1,*

$$\left.\frac{\mathbf{L}^{\mathrm{alg}}_{p,\mathbf{F}}\left(\mathbb{T}^{-1/2}_{\infty,E}, s, w\right)}{(s-1)^{\mathfrak{r}}(w-2)^{r^{\dagger}_{\mathbf{F}}}}\right|_{(s,w)=(1,2)} = \prod_{\nu|p} \frac{\ell_\nu\left(\mathbb{T}^{-1/2}_{\infty,E}\right)\Big|_{w=2}}{\mathrm{Tam}^{(2)}_{\mathbf{F}_\nu}\left(\mathbb{T}^{-1/2}_{\infty,E}\right)} \times \left(\mathrm{covol}^{(2)}_{\mathbb{T}^{-1/2}_{\infty,E}} \times \mathfrak{D}_{\mathbf{F},p}\right)^{-1}$$

$$\times \prod_{\mathfrak{p}\in S^{\mathrm{good}}_{\mathbf{F},p}} \#\widetilde{E}\left(\mathbb{F}_{q_{\mathfrak{p}}}\right)^2 \times \frac{\#\text{Ш}_{\mathbf{F}}(E)\prod_\nu\left[E(\mathbf{F}_\nu) : E_0(\mathbf{F}_\nu)\right]}{\#E(\mathbf{F})^2_{\mathrm{tors}}}$$

$$\times\, \mathrm{disc}\langle -,-\rangle^{\mathrm{cy}}_{\mathbb{T}^{-1/2}_{\infty,E}}\Big|_{w=2} \times \det\langle -,-\rangle^{-1/2}_{\mathbf{F},p}\Big|_{E(\mathbf{F})\times E(\mathbf{F})}.$$

Again this formula is correct only up to p-adic units.

The first line consists of fudge factors arising from specialisation to weight two. The second line is essentially the Birch and Swinnerton-Dyer formula for E over $\mathbf{F}$. Curiously, the last line is the value of a regulator map at the point $(s,k) = (1,2)$ and splits into two pieces: the value of the formal p-adic regulator at weight two, multiplied by the half-twisted regulator arising from Nekovář's theory.

N.B. We already have numerical algorithms to compute the first two lines, but regrettably the last line is still a mystery (unless both ranks are zero).

Proof of 10.9: An elementary calculation reveals

$$\log_p \varrho(\gamma_0)^{\mathfrak{r}} \times \left.\frac{\mathfrak{C}^{\mathrm{alg}}_{\mathbf{F},p}\left(\mathbb{T}^{-1/2}_{\infty,E}, Y\right)}{Y^{\mathfrak{r}}}\right|_{Y=0} = \left.\frac{\mathfrak{C}^{\mathrm{alg}}_{\mathbf{F},p}\left(\mathbb{T}^{-1/2}_{\infty,E}, \varrho(\gamma_0)^{s-1}-1\right)}{(s-1)^{\mathfrak{r}}}\right|_{s=1}$$

which means that the $\log_p \varrho(\gamma_0)^{-\mathfrak{r}}$ term drops out of the formulae completely.

As a direct consequence of Theorem 10.8, one has the equality

$$\left.\frac{\mathbf{L}_{p,\mathbf{F}}^{\mathrm{alg}}\left(\mathbb{T}_{\infty,E}^{-1/2},s,w\right)}{(s-1)^{\mathfrak{r}}(w-2)^{r_{\mathbf{F}}^{\dagger}}}\right|_{(s,w)=(1,2)} = \prod_{\nu\in\Sigma}\mathrm{Mell}_{2,\mathbf{1}}\circ\sigma\left(\ell_\nu\left(\mathbb{T}_{\infty,E}^{-1/2}\right)\right)\Big|_{w=2}$$

$$\times\ \mathrm{disc}\langle-,-\rangle^{\mathrm{cy}}_{\mathbb{T}_{\infty,E}^{-1/2}}\Big|_{w=2}\times\left.\frac{\mathrm{Mell}_{2,\mathbf{1}}\circ\sigma\left(\text{Ш}_{\mathbf{F}}\left(\mathbb{T}_{\infty,E}^{-1/2}\right)\right)}{(w-2)^{r_{\mathbf{F}}^{\dagger}}}\right|_{w=2}.$$

At those places $\nu\in\Sigma$ which do not lie above p, it is an easy exercise to show that the Γ^{wt}-Euler characteristic of $X_{\nu,\mathrm{sat}}^{\dagger}\big/\mathcal{N}_{\infty}^{\mathrm{univ}}\left(X_{\nu,\mathrm{sat}}^{\dagger}\right)$ equals $\mathrm{Tam}_{\mathbf{F}_\nu}^{(2)}\left(\mathbb{T}_{\infty,E}^{-1/2}\right)$. In other words

$$\prod_{\nu\in\Sigma-S_{\mathbf{F},p}}\ell_\nu\left(\mathbb{T}_{\infty,E}^{-1/2}\right)\Big|_{w=2} = \prod_{\nu\in\Sigma-S_{\mathbf{F},p}}\mathrm{Tam}_{\mathbf{F}_\nu}^{(2)}\left(\mathbb{T}_{\infty,E}^{-1/2}\right)\quad\text{up to a } p\text{-unit.}$$

Fortunately the leading term of $\text{Ш}_{\mathbf{F}}\left(\mathbb{T}_{\infty,E}^{-1/2}\right)$ was computed by us in Theorem 10.1; substituting in this value, the corollary is proved.

□

Variation of the height pairing in a vertical deformation.
Let's now switch our attention to the vertical lifting $\mathbb{T}_{\infty,E}$, corresponding to $s=1$. The local conditions to pick at $\nu\in\Sigma$ are

$$X_\nu^{\mathrm{wt}} = \begin{cases}\mathrm{Ker}\Big(H^1\left(\mathbf{F}_\nu,\mathbb{T}_{\infty,E}\right)\longrightarrow H^1\left(I_{\mathbf{F}_\nu},\mathbb{T}_{\infty,E}\right)\Big) & \text{if } \nu\nmid p\\ \mathrm{Ker}\Big(H^1\left(\mathbf{F}_\nu,\mathbb{T}_{\infty,E}\right)\longrightarrow H^1\left(\mathbf{F}_\nu,\mathbb{T}_{\infty,E}\otimes_{\Lambda^{\mathrm{wt}}}\mathbb{B}_{\mathrm{dR}}^{\mathrm{wt}}\right)\Big) & \text{if } \nu\mid p.\end{cases}$$

To cut a long story short, here are the analogies with the half-twisted scenario:

Proposition 10.10. *The dual of* $\mathrm{Sel}_{\mathbf{F}^{\mathrm{cy}}}\left(\mathbb{T}_{\infty,E}\right)=\varinjlim_n\mathrm{Sel}_{\mathbf{F}_n}\left(\mathbb{T}_{\infty,E}\right)$ *has trivial* $\mathbb{Z}_p\left[\left[\Gamma^{\mathrm{cy}},\Gamma^{\mathrm{wt}}\right]\right]$*-rank, i.e.* $\mathrm{Sel}_{\mathbf{F}^{\mathrm{cy}}}\left(\mathbb{T}_{\infty,E}\right)^{\vee}$ *is torsion over the two-variable ring.*

Theorem 10.11. *The power series* $\mathfrak{C}_{\mathbf{F},p}^{\mathrm{alg}}\left(\mathbb{T}_{\infty,E},Y\right):=\mathrm{char}_{\Lambda^{\mathrm{wt}}[[Y]]}\left(\mathrm{Sel}_{\mathbf{F}^{\mathrm{cy}}}\left(\mathbb{T}_{\infty,E}\right)^{\vee}\right)$ *does not vanish at the point $Y=0$, and has constant term*

$$\mathfrak{C}_{\mathbf{F},p}^{\mathrm{alg}}\left(\mathbb{T}_{\infty,E}^{-1/2},\,0\right) = \prod_{\nu\in\Sigma}\ell_\nu\left(\mathbb{T}_{\infty,E}\right)\times\text{Ш}_{\mathbf{F}}\left(\mathbb{T}_{\infty,E}\right)\qquad\mathrm{mod}\ \left(\Lambda^{\mathrm{wt}}\right)^{\times}$$

where $\ell_\nu\left(\mathbb{T}_{\infty,E}\right)$ *is the characteristic power series of* $\dfrac{X_{\nu,\mathrm{sat}}^{\mathrm{wt}}}{\mathcal{N}_{\infty}^{\mathrm{univ}}\left(X_{\nu,\mathrm{sat}}^{\mathrm{wt}}\right)}$ *at each* $\nu\in\Sigma$.

Corollary 10.12. *Under identical conditions to Theorem 9.18(ii),*

$$\left.\frac{\mathbf{L}_{p,\mathbf{F}}^{\mathrm{alg}}\left(\mathbb{T}_{\infty,E},s,w\right)}{(w-2)^{r_{\mathbf{F}}}}\right|_{(s,w)=(1,2)} = \prod_{\nu\mid p}\frac{\ell_\nu\left(\mathbb{T}_{\infty,E}\right)\Big|_{w=2}}{\mathrm{Tam}_{\mathbf{F}_\nu}^{(2)}\left(\mathbb{T}_{\infty,E}\right)}\times\left(\frac{\mathrm{vol}_{\mathbb{T}_{\infty,E}}^{(2)}\left(H_{\mathcal{E}}^1\right)}{\mathrm{vol}_E\left(H_e^1\right)}\times\mathfrak{D}_{\mathbf{F},p}\right)^{-1}$$

$$\times\,\mathcal{L}_p^{\mathrm{wt}}(E)\,\mathrm{Per}_{\mathbf{I},\lambda_E}^{(2)}(\mathbf{F})\times\frac{\#\text{Ш}_{\mathbf{F}}(E)\prod_\nu\left[E(\mathbf{F}_\nu):E_0(\mathbf{F}_\nu)\right]}{\#E(\mathbf{F})_{\mathrm{tors}}^{2}}\times\det\langle-,-\rangle_{\mathbf{F},p}^{\mathrm{wt}}\Big|_{E(\mathbf{F})\times E(\mathbf{F})}.$$

Remarks: (a) The proof of Proposition 10.10 uses exactly the same specialisation argument to that of 10.7. However this time one exploits the Λ^{cy}-cotorsion of the Selmer groups $H^1_{f,\mathrm{Spec}\mathcal{O}_{\mathbf{F}}}\Big(\mathbf{F}_\Sigma/\mathbf{F}^{\mathrm{cy}},\ A_{\mathbb{T}_{\infty,E}}\big[\times\mathfrak{g}_k\big]\Big)$ at infinitely many $k\in\mathbb{U}_E$.

(b) The reader may well be wondering why there is no discriminant term in 10.11, yet there is in Theorem 10.8? The reason is simple – it's there but you can't see it! The formal height pairing is between

$$\big\langle -,-\big\rangle_{\mathcal{R},\mathbb{T}_\infty,\{X_\nu^{\mathrm{wt}}\}_\nu}\ :\ H^1_{X^{\mathrm{wt}},\Sigma}\big(\mathbf{F},\mathbb{T}_\infty\big)^0\ \times\ H^1_{X^{\mathrm{wt},\perp},\Sigma}\big(\mathbf{F},\mathbb{T}^*_\infty(1)\big)^0\ \longrightarrow\ \mathcal{R}.$$

If one specialises $H^1_{X^{\mathrm{wt}},\Sigma}\big(\mathbf{F},\mathbb{T}_\infty\big)^0$ to the λ_E-isotypic component, then its image is contained inside the compact Selmer group $\mathfrak{S}_{\mathbf{F}}\big(\mathbb{T}_{\infty,E}\big)$. The latter object vanishes courtesy of Theorem 9.2, in which case $\mathrm{disc}\big\langle -,-\big\rangle^{\mathrm{cy}}_{\mathbb{T}_{\infty,E}}$ must be a $\mathbb{Z}_p\langle\langle w\rangle\rangle$-unit.

(c) Corollary 10.12 follows directly from 10.11, and the formula given in 9.18(ii). Once again the $\mathrm{Tam}^{(2)}_{\mathbf{F}_\nu}\big(\mathbb{T}_{\infty,E}\big)$'s cancel with $\ell_\nu\big(\mathbb{T}_{\infty,E}\big)\Big|_{w=2}$ at those places $\nu\nmid p$.

10.4 The proof of Theorems 10.8 and 10.11

To avoid bamboozling the reader, we have elected to supply the proof *only* for the half-twisted representation rather than the vertical one. The vertical situation is actually a lot easier: there is no p-adic height pairing to contend with here, since the Selmer group for $A_{\mathbb{T}_{\infty,E}}$ over the ground field $\mathbf{F}$ contains no Λ^{wt}-divisible piece. The arguments we will exploit are entirely adapted from A. Plater's work in [Pl], i.e. this forces us to work over the universal deformation rings themselves.

For simplicity, assume that the local factor $\mathcal{R}$ through which $\mathbf{H}^{\mathrm{ord}}_{Np^\infty}$ acts on $\mathbb{T}_\infty$ is a regular local ring – at the end, we will explain how to remove this hypothesis. To be consistent with Plater's article [Pl], we shall employ his notation throughout; for the lattice $\mathbf{T}:=\mathbb{T}_\infty\otimes\Psi_{\mathcal{R}}^{-1/2}$ one defines

$$\mathfrak{X}_{\infty,f}\big(A_{\mathbf{T}}\big)\ =\ H^1_{f,\Sigma}\Big(\mathbf{F}^{\mathrm{cy}},A_{\mathbf{T}}\Big)^\vee,\quad \mathfrak{X}^1_{\infty,\Sigma}\big(A_{\mathbf{T}}\big)\ =\ H^1\Big(\mathbf{F}_\Sigma/\mathbf{F}^{\mathrm{cy}},A_{\mathbf{T}}\Big)^\vee$$

$$\mathcal{Z}_{\infty,f,\nu}\big(\mathbf{T}\big)\ =\ \varprojlim_n H^1_f\Big(\mathbf{F}_{\nu,n},\mathbf{T}\Big),\quad \mathcal{Z}_{\infty,f,\Sigma}\big(\mathbf{T}\big)=\bigoplus_{\nu\in\Sigma}\mathcal{Z}_{\infty,f,\nu}\big(\mathbf{T}\big).$$

Strategy: Providing that the height pairing $\big\langle -,-\big\rangle_{\mathcal{R},\mathbf{T},\{X_\nu^\dagger\}_\nu}$ is non-degenerate, our aim is to establish the formula

$$\frac{\mathrm{char}_{\mathcal{R}[\![Y]\!]}\Big(\mathfrak{X}_{\infty,f}\big(A_{\mathbf{T}}\big)\Big)}{Y^r}\Bigg|_{Y=0} = \log_p\varrho(\gamma_0)^{-r}\times\prod_{\nu\in\Sigma}\mathrm{char}_{\mathcal{R}}\Big(X^\dagger_{\nu,\mathrm{sat}}\Big/\mathcal{N}^{\mathrm{univ}}_\infty X^\dagger_{\nu,\mathrm{sat}}\Big)\quad (*)$$
$$\times\ \mathrm{disc}\big\langle -,-\big\rangle_{\mathcal{R},\mathbf{T},\{X_\nu^\dagger\}_\nu}\ \times\ \mathrm{char}_{\mathcal{R}}\left(\mathrm{Tors}_{\mathcal{R}}\Big(H^1_{f,\Sigma}\big(\mathbf{F},A_{\mathbf{T}}\big)^\vee\Big)\right)$$

with $r=\mathrm{rank}_{\mathcal{R}}H^1_{f,\Sigma}\big(\mathbf{F},\mathbf{T}\big)$. The statement of Theorem 10.8 will then follow by restricting this formula to its λ_E-isotype, and centering it at weight two.

Equation **(*)** is an indirect consequence of the following three results.

Lemma 10.13. *Both* $\mathrm{Tors}_{\mathcal{R}} H^1\Big(\mathbf{F}_\Sigma/\mathbf{F}, \mathbf{T}^*(1)\Big)$ *and* $H^1\Big(\Gamma^{\mathrm{cy}}, A_{\mathbf{T}}(\mathbf{F}^{\mathrm{cy}})\Big)^\vee$ *share the same characteristic ideal as* $\mathcal{R}$*-modules.*

Let $\phi_{\mathcal{R}}$ denote the sequence of compositions

$$\mathfrak{X}_{\infty,f}(A_{\mathbf{T}})^{\Gamma^{\mathrm{cy}}} \xrightarrow{\mathrm{nat}} \mathfrak{X}_{\infty,f}(A_{\mathbf{T}})_{\Gamma^{\mathrm{cy}}} \longrightarrow \mathrm{Hom}_{\mathrm{cont}}\Big(H^1_{f,\Sigma}(\mathbf{F}, A_{\mathbf{T}})^\sim, \mathbb{Q}_p/\mathbb{Z}_p\Big) \longrightarrow \mathrm{Hom}_{\mathcal{R}}\Big(H^1_{f,\Sigma}(\mathbf{F}, \mathbf{T}^*(1)), \mathcal{R}\Big)$$

where $H^1_{f,\Sigma}(\mathbf{F}, A_{\mathbf{T}})^\sim := \mathrm{Ker}\left(H^1(\mathbf{F}_\Sigma/\mathbf{F}, A_{\mathbf{T}}) \longrightarrow \bigoplus_{\nu\in\Sigma} \dfrac{H^1(\mathbf{F}_\nu, A_{\mathbf{T}})}{\mathcal{N}^{\mathrm{univ}}_\infty(X^\dagger_{\nu,\mathrm{sat}})^\perp}\right)$.

Lemma 10.14. *There exists a map* $\theta_{\gamma_0} : H^1_{f,\Sigma}(\mathbf{F}, \mathbf{T})^0 \longrightarrow \mathfrak{X}_{\infty,f}(A_{\mathbf{T}})^{\Gamma^{\mathrm{cy}}}$ *through which the formal height pairing factorises:*

$$\phi_{\mathcal{R}}\Big(\theta_{\gamma_0}(x)\Big)(y) \quad = \quad \log_p \varrho(\gamma_0)^{-1} \times \langle x, y\rangle_{\mathcal{R},\mathbf{T},\{X^\dagger_\nu\}_\nu}$$

at all points $x \in H^1_{f,\Sigma}(\mathbf{F}, \mathbf{T})^0$ *and* $y \in H^1_{f,\Sigma}(\mathbf{F}, \mathbf{T}^*(1))$.

Lemma 10.15. *For the homomorphism* θ_{γ_0} *as above, the following diagram:*

$$\begin{array}{ccccccc} 0 \longrightarrow & \mathfrak{X}^1_{\infty,\Sigma}(A_{\mathbf{T}})^{\Gamma^{\mathrm{cy}}} & \longrightarrow & \mathfrak{X}_{\infty,f}(A_{\mathbf{T}})^{\Gamma^{\mathrm{cy}}} & \longrightarrow & \mathcal{Z}_{\infty,f,\Sigma}(\mathbf{T})_{\Gamma^{\mathrm{cy}}} \\ & \uparrow{\scriptstyle \alpha_{\Sigma,\gamma_0}} & & \uparrow{\scriptstyle \theta_{\gamma_0}} & & \downarrow{\scriptstyle \pi} \\ 0 \longrightarrow & H^2(\mathbf{F}_\Sigma/\mathbf{F}, A_{\mathbf{T}})^\vee & \longrightarrow & H^1_{f,\Sigma}(\mathbf{F}, \mathbf{T})^0 & \longrightarrow & \bigoplus_{\nu\in\Sigma} \mathcal{N}^{\mathrm{univ}}_\infty\Big(X^\dagger_{\nu,\mathrm{sat}}\Big) \end{array}$$

is commutative with exact rows.

N.B. Here the dual homomorphism to α_{Σ,γ_0} is given by the sequence

$$H^1(\mathbf{F}_\Sigma/\mathbf{F}^{\mathrm{cy}}, A_{\mathbf{T}})_{\Gamma^{\mathrm{cy}}} \xrightarrow{\sim} H^1\Big(\Gamma^{\mathrm{cy}}, H^1(\mathbf{F}_\Sigma/\mathbf{F}^{\mathrm{cy}}, A_{\mathbf{T}})\Big) \longrightarrow H^2(\mathbf{F}_\Sigma/\mathbf{F}, A_{\mathbf{T}})$$

and π denotes the natural projection.

We now explain how these three technical lemmas imply the desired formula. The plan is to compare the Selmer group over $\mathbf{F}$ with that of its cousin over the cyclotomic $\mathbb{Z}_p$-extension of $\mathbf{F}$, but first we need to introduce some extra notation. By Pontrjagin duality, the restriction map $H^1(\mathbf{F}_\Sigma/\mathbf{F}, A_{\mathbf{T}}) \longrightarrow H^1(\mathbf{F}_\Sigma/\mathbf{F}^{\mathrm{cy}}, A_{\mathbf{T}})^{\Gamma^{\mathrm{cy}}}$ yields a dual homomorphism

$$\pi_\Sigma : \mathfrak{X}^1_{\infty,\Sigma}(A_{\mathbf{T}})_{\Gamma^{\mathrm{cy}}} \cong \left(\varinjlim_n H^1(\mathbf{F}_\Sigma/\mathbf{F}_n, A_{\mathbf{T}})^\vee\right)_{\Gamma^{\mathrm{cy}}} \longrightarrow H^1(\mathbf{F}_\Sigma/\mathbf{F}, A_{\mathbf{T}})^\vee.$$

Similarly for the discrete Selmer groups themselves, the natural restriction map $H^1_{f,\Sigma}(\mathbf{F}, A_{\mathbf{T}})^\sim \longrightarrow H^1_{f,\Sigma}(\mathbf{F}^{\mathrm{cy}}, A_{\mathbf{T}})^{\Gamma^{\mathrm{cy}}}$ dualises to give

$$\pi_0 : \mathfrak{X}_{\infty,f}(A_{\mathbf{T}})_{\Gamma^{\mathrm{cy}}} \longrightarrow \Big(H^1_{f,\Sigma}(\mathbf{F}, A_{\mathbf{T}})^\sim\Big)^\vee.$$

Linking these various strands together, one obtains a large commutative diagram with exact rows and columns:

$$
\begin{array}{ccccccc}
 & & & & 0 & & 0 \\
 & & & & \downarrow & & \downarrow \\
 & & 0 & \longrightarrow & H^2\big(\mathbf{F}_\Sigma/\mathbf{F}, A_{\mathbf{T}}\big)^\vee & \stackrel{\alpha_{\Sigma,\gamma_0}}{\longrightarrow} & \mathfrak{X}^1_{\infty,\Sigma}\big(A_{\mathbf{T}}\big)^{\Gamma^{\mathrm{cy}}} \\
 & & & & \downarrow & & \downarrow \\
 & & & & H^1_{f,\Sigma}\big(\mathbf{F}, \mathbf{T}\big)^0 & \stackrel{\theta_{\gamma_0}}{\longrightarrow} & \mathfrak{X}_{\infty,f}\big(A_{\mathbf{T}}\big)^{\Gamma^{\mathrm{cy}}} \\
 & & & & \downarrow & & \downarrow \\
 & & 0 & \longleftarrow & \bigoplus_{\nu\in\Sigma} \mathcal{N}^{\mathrm{univ}}_\infty\Big(X^\dagger_{\nu,\mathrm{sat}}\Big) & \stackrel{\pi}{\longleftarrow} & \mathcal{Z}_{\infty,f,\Sigma}\big(\mathbf{T}\big)_{\Gamma^{\mathrm{cy}}} \\
 & & & & \downarrow & & \downarrow \\
0 & \longleftarrow & H^1\Big(\Gamma^{\mathrm{cy}}, A_{\mathbf{T}}\big(\mathbf{F}^{\mathrm{cy}}\big)\Big)^\vee & \longleftarrow & H^1\big(\mathbf{F}_\Sigma/\mathbf{F}, A_{\mathbf{T}}\big)^\vee & \stackrel{\pi_\Sigma}{\longleftarrow} & \mathfrak{X}^1_{\infty,\Sigma}\big(A_{\mathbf{T}}\big)_{\Gamma^{\mathrm{cy}}} \\
 & & & & \downarrow & & \downarrow \\
 & & & & \Big(H^1_{f,\Sigma}\big(\mathbf{F}, A_{\mathbf{T}}\big)^\sim\Big)^\vee & \stackrel{\pi_0}{\longleftarrow} & \mathfrak{X}_{\infty,f}\big(A_{\mathbf{T}}\big)_{\Gamma^{\mathrm{cy}}} \\
 & & & & \downarrow & & \downarrow \\
 & & & & 0 & & 0
\end{array}
$$

Figure 10.2

The left vertical arrows arise from the dual Poitou-Tate sequence for $H^1_{f,\Sigma}\big(\mathbf{F}, A_{\mathbf{T}}\big)^\sim$, whilst the right vertical arrows above come from taking the Γ^{cy}-cohomology of the two-variable Selmer group. Lastly the top two squares commute by Lemma 10.15, and the bottom two commute because it's true at each finite layer $\mathbf{F}_n$.

Remark: Let $\xi_{\mathcal{R}}$ denote the natural homomorphism between the Γ^{cy}-invariants of $\mathfrak{X}_{\infty,f}\big(A_{\mathbf{T}}\big)$ and its Γ^{cy}-coinvariants. Taking determinants of the various maps occurring throughout Figure 10.2, then standard nonsense informs us

$$
\frac{\mathrm{char}_{\mathcal{R}}\big(\mathrm{Coker}(\xi_{\mathcal{R}})\big)}{\mathrm{char}_{\mathcal{R}}\big(\mathrm{Ker}(\xi_{\mathcal{R}})\big)} = \prod_{\nu\in\Sigma} \mathrm{char}_{\mathcal{R}}\Big(X^\dagger_{\nu,\mathrm{sat}}\Big/\mathcal{N}^{\mathrm{univ}}_\infty X^\dagger_{\nu,\mathrm{sat}}\Big) \times \mathrm{disc}\Big(\phi_{\mathcal{R}}\big(\theta_{\gamma_0}(-)\big)(-)\Big)
$$
$$
\times\, \mathrm{char}_{\mathcal{R}}\Big(\mathrm{Tors}_{\mathcal{R}}\Big(H^1_{f,\Sigma}\big(\mathbf{F}, A_{\mathbf{T}}\big)^\vee\Big)\Big) \times \frac{\mathrm{char}_{\mathcal{R}}(M_1)}{\mathrm{char}_{\mathcal{R}}(M_2)} \quad \mathrm{mod}\ \mathcal{R}^\times
$$

where we have set $M_1 := H^1\big(\mathbf{F}_\Sigma/\mathbf{F}, A_{\mathbf{T}}\big)^\vee_{/\mathcal{R}\text{-div}}$ and $M_2 := H^1\big(\Gamma^{\mathrm{cy}}, A_{\mathbf{T}}(\mathbf{F}^{\mathrm{cy}})\big)^\vee$. Since $H^1\big(\mathbf{F}_\Sigma/\mathbf{F}, \mathbf{T}^*(1)\big) \otimes_{\mathcal{R}} \mathcal{R}^\vee$ can be identified with the maximal $\mathcal{R}$-divisible submodule of $H^1\big(\mathbf{F}_\Sigma/\mathbf{F}, A_{\mathbf{T}}\big)$, in fact $M_1 \cong \mathrm{Tors}_{\mathcal{R}} H^1\Big(\mathbf{F}_\Sigma/\mathbf{F}, \mathbf{T}^*(1)\Big)$.

The left-hand side is the leading term of the algebraic L-function, that is

$$\frac{\mathrm{char}_{\mathcal{R}}\big(\mathrm{Coker}(\xi_{\mathcal{R}})\big)}{\mathrm{char}_{\mathcal{R}}\big(\mathrm{Ker}(\xi_{\mathcal{R}})\big)} \quad = \quad \left.\frac{\mathrm{char}_{\mathcal{R}[\![Y]\!]}\Big(\mathfrak{X}_{\infty,f}\big(A_{\mathbf{T}}\big)\Big)}{Y^r}\right|_{Y=0} \qquad \text{modulo } \mathcal{R}^{\times}.$$

Applying Lemma 10.13, the $\mathrm{char}_{\mathcal{R}}(M_1)/\mathrm{char}_{\mathcal{R}}(M_2)$-term represents an $\mathcal{R}$-unit. Moreover, inside the group $\mathrm{Frac}(\mathcal{R})/\mathcal{R}^{\times}$ there is an equality

$$\mathrm{disc}\Big(\phi_{\mathcal{R}}\big(\theta_{\gamma_0}(-)\big)(-)\Big) \;=\; \log_p \varrho(\gamma_0)^{-r} \times \mathrm{disc}\big\langle -,-\big\rangle_{\mathcal{R},\mathbf{T},\{X_\nu^{\dagger}\}_\nu} \quad \text{by Lemma 10.14.}$$

Thankfully this is strong enough to establish Formula **(*)**.

Remark: The same argument works fine for coefficient rings $\mathcal{R}$ that are Gorenstein but not necessarily regular. However, one must rewrite the whole argument in terms of $\mathcal{R}$-divisors rather than in terms of characteristic power series (in all other ways, the details are identical).

The demonstration of Lemma 10.13.
If λ is an irreducible element of $\mathcal{R}$, then write $\mathcal{O}_\lambda$ for the quotient ring $\mathcal{R}/(\lambda)$. Because $\mathcal{R}$ is Gorenstein so is $\mathcal{O}_\lambda$, in which case

$$\mathbf{T}_\lambda^*(1) = \mathrm{Hom}_{\mathcal{O}_\lambda}\big(\mathbf{T}_\lambda, \mathcal{O}_\lambda(1)\big) \cong \frac{\mathbf{T}^*(1)}{\lambda \mathbf{T}^*(1)} \quad \text{and} \quad A_{\mathbf{T}_\lambda} = A_{\mathbf{T}}[\times\lambda] \cong \mathbf{T}_\lambda^*(1) \otimes \mathbb{Q}_p/\mathbb{Z}_p.$$

For a fixed finite set $\mathfrak{B}$ of height one primes and two sheaves of abelian groups Y_1, Y_2 over $\mathrm{Spec}(\mathcal{R})$, we'll write $Y_1(\lambda) \overset{\mathfrak{B}}{\sim} Y_2(\lambda)$ if the ratio $\#Y_1(\lambda)/\#Y_2(\lambda)$ is finite and bounded independently of all height one primes $(\lambda) \notin \mathfrak{B}$.

We claim that the statement of Lemma 10.13 follows from the following facts:

(A) $\#\mathrm{Tors}_{\mathbb{Z}_p} H^1\Big(\mathbf{F}_\Sigma/\mathbf{F}, \mathbf{T}_\lambda^*(1)\Big) \;=\; \#H^1\Big(\Gamma^{\mathrm{cy}}, A_{\mathbf{T}_\lambda}\big(\mathbf{F}^{\mathrm{cy}}\big)\Big)$ at almost all (λ);

(B) $M_1 \otimes_{\mathcal{R}} \mathcal{O}_\lambda \overset{\mathfrak{B}_1}{\sim} \mathrm{Tors}_{\mathbb{Z}_p} H^1\Big(\mathbf{F}_\Sigma/\mathbf{F},\ \mathbf{T}^*(1) \otimes_{\mathcal{R}} \mathcal{O}_\lambda\Big)$ for some finite set $\mathfrak{B}_1$;

(C) $M_2 \otimes_{\mathcal{R}} \mathcal{O}_\lambda \overset{\mathfrak{B}_2}{\sim} H^1\Big(\Gamma^{\mathrm{cy}},\ A_{\mathbf{T}}[\times\lambda]\big(\mathbf{F}^{\mathrm{cy}}\big)\Big)$ for another finite set $\mathfrak{B}_2$.

Assuming that (A),(B),(C) hold true, one easily deduces $M_1 \otimes_{\mathcal{R}} \mathcal{O}_\lambda \overset{\mathfrak{B}'}{\sim} M_2 \otimes_{\mathcal{R}} \mathcal{O}_\lambda$ where $\mathfrak{B}'$ is the union of the sets $\mathfrak{B}_1$, $\mathfrak{B}_2$ and the primes excluded from part (A). However, it is well-known that if the orders of the specialisations of two $\mathcal{R}$-modules coincide (up to a bounded factor) outside a finite bad set of primes, then the two $\mathcal{R}$-modules share the same characteristic ideal. In our situation we conclude that $\mathrm{char}_{\mathcal{R}}(M_1) = \mathrm{char}_{\mathcal{R}}(M_2)$, which proves Lemma 10.13.

It therefore remains to justify the three claims above. We start with part (A). As a general comment, let us first observe that $A_{\mathbf{T}}\big(\mathbf{F}^{\mathrm{cy}}\big)$ is an $\mathcal{R}$-cotorsion module since at the weight two, its specialisation $\mathcal{C}^{\min}\big(\mathbf{F}^{\mathrm{cy}}\big)[p^\infty]$ is a finite abelian p-group thanks to a very useful result in [Im].

Provided λ is chosen outside the support of the $\mathcal{R}$-modules M_1 and M_2, then

$$\#\mathrm{Tors}_{\mathbb{Z}_p} H^1\Big(\mathbf{F}_\Sigma/\mathbf{F}, \mathbf{T}^*_\lambda(1)\Big) \;=\; \#H^0\Big(\mathbf{F}_\Sigma/\mathbf{F},\, \mathbf{T}^*_\lambda(1)\otimes\mathbb{Q}_p/\mathbb{Z}_p\Big) \;=\; \#A_{\mathbf{T}_\lambda}(\mathbf{F})$$

and as Γ^{cy} is pro-cyclic,

$$\#H^1(\Gamma^{\mathrm{cy}}, A_{\mathbf{T}_\lambda}(\mathbf{F}^{\mathrm{cy}})) \;=\; \#H^0(\Gamma^{\mathrm{cy}}, A_{\mathbf{T}_\lambda}(\mathbf{F}^{\mathrm{cy}})) \;=\; \#A_{\mathbf{T}_\lambda}(\mathbf{F}).$$

Combining these two equalities together, part (A) follows at once.

To prove (B) one examines the short exact sequence

$$0\rightarrow \frac{H^1(\mathbf{F}_\Sigma/\mathbf{F}, \mathbf{T}^*(1))}{\lambda.H^1(\mathbf{F}_\Sigma/\mathbf{F}, \mathbf{T}^*(1))} \longrightarrow H^1\Big(\mathbf{F}_\Sigma/\mathbf{F}, \mathbf{T}^*_\lambda(1)\Big) \longrightarrow H^2\Big(\mathbf{F}_\Sigma/\mathbf{F}, \mathbf{T}^*(1)\Big)[\times\lambda]\rightarrow 0.$$

Isolating the exceptional set $\mathfrak{B}_1^{(0)} = \mathrm{Ass}_{\mathcal{R}}\Big(H^2(\mathbf{F}_\Sigma/\mathbf{F}, \mathbf{T}^*(1))\Big)$, clearly

$$\mathrm{Tors}_{\mathbb{Z}_p}\Big(H^1\Big(\mathbf{F}_\Sigma/\mathbf{F}, \mathbf{T}^*(1)\otimes_{\mathcal{R}}\mathcal{O}_\lambda\Big)\Big) \;\overset{\mathfrak{B}_1^{(0)}}{\sim}\; \mathrm{Tors}_{\mathbb{Z}_p}\left(\frac{H^1(\mathbf{F}_\Sigma/\mathbf{F}, \mathbf{T}^*(1))}{\lambda.H^1(\mathbf{F}_\Sigma/\mathbf{F}, \mathbf{T}^*(1))}\right).$$

On the other hand, if $\mathfrak{B}_1^{(1)}$ denotes the support of M_1 then

$$\frac{\mathrm{Tors}_{\mathcal{R}} H^1(\mathbf{F}_\Sigma/\mathbf{F}, \mathbf{T}^*(1))}{\lambda.\mathrm{Tors}_{\mathcal{R}} H^1(\mathbf{F}_\Sigma/\mathbf{F}, \mathbf{T}^*(1))} \;\overset{\mathfrak{B}_1^{(1)}}{\sim}\; \mathrm{Tors}_{\mathbb{Z}_p}\left(\frac{H^1(\mathbf{F}_\Sigma/\mathbf{F}, \mathbf{T}^*(1))}{\lambda.H^1(\mathbf{F}_\Sigma/\mathbf{F}, \mathbf{T}^*(1))}\right).$$

Setting $\mathfrak{B}_1 := \mathfrak{B}_1^{(0)}\cup\mathfrak{B}_1^{(1)}$, statement (B) is now proved. This leaves us with (C).

Remarks: (i) Applying the Snake Lemma to the diagram

$$\begin{array}{ccccccccc}
0 & \longrightarrow & H^1(\Gamma^{\mathrm{cy}}, A_{\mathbf{T}}(\mathbf{F}^{\mathrm{cy}})) & \longrightarrow & H^1(\mathbf{F}_\Sigma/\mathbf{F}, A_{\mathbf{T}}) & \longrightarrow & H^1(\mathbf{F}_\Sigma/\mathbf{F}^{\mathrm{cy}}, A_{\mathbf{T}})^{\Gamma^{\mathrm{cy}}} & \longrightarrow & 0\\
 & & \downarrow{\times\lambda} & & \downarrow{\times\lambda} & & \downarrow{\times\lambda} & & \\
0 & \longrightarrow & H^1(\Gamma^{\mathrm{cy}}, A_{\mathbf{T}}(\mathbf{F}^{\mathrm{cy}})) & \longrightarrow & H^1(\mathbf{F}_\Sigma/\mathbf{F}, A_{\mathbf{T}}) & \longrightarrow & H^1(\mathbf{F}_\Sigma/\mathbf{F}^{\mathrm{cy}}, A_{\mathbf{T}})^{\Gamma^{\mathrm{cy}}} & \longrightarrow & 0
\end{array}$$

and carefully avoiding $\mathfrak{B}_2^{(0)} = \mathrm{Ass}_{\mathcal{R}}\Big(H^1(\Gamma^{\mathrm{cy}}, A_{\mathbf{T}}(\mathbf{F}^{\mathrm{cy}}))\Big)$, one deduces that

$$H^1(\Gamma^{\mathrm{cy}}, A_{\mathbf{T}}(\mathbf{F}^{\mathrm{cy}}))[\times\lambda].\; H^1(\mathbf{F}_\Sigma/\mathbf{F}^{\mathrm{cy}}, A_{\mathbf{T}})^{\Gamma^{\mathrm{cy}}}[\times\lambda] \;\overset{\mathfrak{B}_2^{(0)}}{\sim}\; H^1(\mathbf{F}_\Sigma/\mathbf{F}, A_{\mathbf{T}})[\times\lambda].$$

(ii) From $0\rightarrow H^1(\Gamma^{\mathrm{cy}}, A_{\mathbf{T}_\lambda}(\mathbf{F}^{\mathrm{cy}}))\rightarrow H^1(\mathbf{F}_\Sigma/\mathbf{F}, A_{\mathbf{T}_\lambda})\rightarrow H^1(\mathbf{F}_\Sigma/\mathbf{F}^{\mathrm{cy}}, A_{\mathbf{T}_\lambda})^{\Gamma^{\mathrm{cy}}}\rightarrow 0$ we readily see $\#H^1(\Gamma^{\mathrm{cy}}, A_{\mathbf{T}_\lambda}(\mathbf{F}^{\mathrm{cy}})).\#H^1(\mathbf{F}_\Sigma/\mathbf{F}^{\mathrm{cy}}, A_{\mathbf{T}_\lambda})^{\Gamma^{\mathrm{cy}}} = \#H^1(\mathbf{F}_\Sigma/\mathbf{F}, A_{\mathbf{T}_\lambda})$.

(iii) Lastly $H^1(\mathbf{F}_\Sigma/\mathbf{F}, A_{\mathbf{T}})[\times\lambda] \overset{\mathfrak{B}_2^{(1)}}{\sim} H^1(\mathbf{F}_\Sigma/\mathbf{F}, A_{\mathbf{T}_\lambda})$ with $\mathfrak{B}_2^{(1)} = \mathrm{Ass}_{\mathcal{R}} A_{\mathbf{T}}(\mathbf{F})^\vee$ and $H^1(\mathbf{F}_\Sigma/\mathbf{F}^{\mathrm{cy}}, A_{\mathbf{T}})^{\Gamma^{\mathrm{cy}}}[\times\lambda] \overset{\mathfrak{B}_2^{(2)}}{\sim} H^1(\mathbf{F}_\Sigma/\mathbf{F}^{\mathrm{cy}}, A_{\mathbf{T}_\lambda})^{\Gamma^{\mathrm{cy}}}$ with $\mathfrak{B}_2^{(2)} = \mathrm{Ass}_{\mathcal{R}} A_{\mathbf{T}}(\mathbf{F}^{\mathrm{cy}})^\vee$, thence part (C) follows by choosing $\mathfrak{B}_2 = \mathfrak{B}_2^{(0)}\cup\mathfrak{B}_2^{(1)}\cup\mathfrak{B}_2^{(2)}$.

The demonstration of Lemma 10.14.
Fix a global point $x \in H^1_{f,\Sigma}(\mathbf{F},\mathbf{T})^0$, and let $0 \to \mathbf{T} \to \mathfrak{X}_x \to \mathcal{R} \to 0$ be the corresponding extension class sequence of $\mathcal{R}[\mathrm{Gal}(\mathbf{F}_\Sigma/\mathbf{F})]$-modules. Our aim is to split terms in the 1-cohomology of its dual sequence $0 \to A_\mathcal{R} \to A_{\mathfrak{X}_x} \to A_\mathbf{T} \to 0$. Firstly at each finite level $\mathbf{F}_n$, there is a right-exact sequence

$$H^1(\mathbf{F}_\Sigma/\mathbf{F}_n, A_\mathcal{R}) \longrightarrow \mathfrak{Y}_x(\mathbf{F}_n) \longrightarrow H^1_{f,\Sigma}(\mathbf{F}_n, A_\mathbf{T}) \longrightarrow 0$$

where $\mathfrak{Y}_x(\mathbf{F}_n)$ denotes the pre-image of $H^1_{f,\Sigma}(\mathbf{F}_n, A_\mathbf{T})$ inside $H^1(\mathbf{F}_\Sigma/\mathbf{F}_n, A_{\mathfrak{X}_x})$ N.B. the surjectivity on the right follows by the same reasoning as [PR1, §4.3.2].

Let $y \in H^1_{f,\Sigma}(\mathbf{F}^{\mathrm{cy}}, A_\mathbf{T})$, so in particular $y \in H^1_{f,\Sigma}(\mathbf{F}_m, A_\mathbf{T})$ for some $m \gg 0$. One now chooses a global lift $s_{\mathrm{glob}}(y)$ of y to the group $\mathfrak{Y}_x(\mathbf{F}_m)$.

Question. *Can we find canonical lifts of the point y over all completions $\mathbf{F}_{\nu,m}$?*

For each place $\nu \in \Sigma$, the natural map

$$\mathcal{N}^{\mathrm{univ}}_\infty\left(X^\dagger_{\nu,m,\mathrm{sat}}\right) \otimes_\mathcal{R} \mathcal{R}^\vee \quad\longrightarrow\quad X^\dagger_{\nu,m,\mathrm{sat}} \otimes_\mathcal{R} \mathcal{R}^\vee$$

is surjective as $\mathcal{R}^\vee = \mathrm{Hom}_{\mathrm{cont}}(\mathcal{R}, \mathbb{Q}_p/\mathbb{Z}_p)$ is $\mathcal{R}$-divisible. It follows that the element $y_\nu = \mathrm{res}_\nu(y) \in H^1_\star(\mathbf{F}_{\nu,m}, A_\mathbf{T}) = X^\dagger_{\nu,m,\mathrm{sat}} \otimes_\mathcal{R} \mathcal{R}^\vee$ can be written in the form

$$y_\nu \;=\; z_\nu \otimes \eta_\nu \qquad \text{where } z_\nu \in \mathcal{N}^{\mathrm{univ}}_\infty\left(X^\dagger_{\nu,m,\mathrm{sat}}\right) \text{ and } \eta_\nu : \mathcal{R} \to \mathbb{Q}_p/\mathbb{Z}_p.$$

One may therefore define $\widehat{y}_\nu \in H^1(\mathbf{F}_{\nu,m}, A_{\mathfrak{X}_x})$ by the rule $\widehat{y}_\nu = s_{\mathrm{can},\mathbf{F}_{\nu,m}}(z_\nu) \otimes \eta_\nu$. In addition, we also know $\widehat{y}_\nu$ projects down to $\mathrm{res}_\nu(y)$, and is well-defined modulo $\mathcal{N}^{\mathrm{univ}}_\infty H^1(\mathbf{F}_{\nu,m}, \mathcal{R}(1)) \otimes_\mathcal{R} \mathcal{R}^\vee$.

Definition 10.16. *The map $\theta_{\gamma_0} : H^1_{f,\Sigma}(\mathbf{F},\mathbf{T})^0 \to \mathrm{Hom}_{\mathrm{cont}}\left(H^1_{f,\Sigma}(\mathbf{F}^{\mathrm{cy}}, A_\mathbf{T}), \mathbb{Q}_p/\mathbb{Z}_p\right)$ is given by the summation*

$$\theta_{\gamma_0}(x)(y) \quad=\quad \log_p \varrho(\gamma_0)^{-1} \times \sum_{\nu\in\Sigma} \widetilde{\ell_{\mathcal{R},\nu,\infty}} \otimes \mathrm{id}\Big(s_{\mathrm{glob}}(y)_\nu - \widehat{y}_\nu\Big)$$

where $\widetilde{\ell_{\mathcal{R},\nu,\infty}} : H^1(\mathbf{F}^{\mathrm{cy}}_\nu, R(1)) \longrightarrow R$ through $\widetilde{\ell_{\mathcal{R},\nu,\infty}} := \lim_{m\to\infty} p^{-m}.\, \widetilde{\ell_{\mathcal{R},\nu,\mathbf{F}_{\nu,m}}}$.

It is a fairly basic exercise to check the $\mathcal{R}$-homomorphism θ_{γ_0} is Γ^{cy}-equivariant. As a corollary, we deduce that $\mathrm{Im}(\theta_{\gamma_0})$ is contained in $H^0\Big(\Gamma^{\mathrm{cy}}, \mathfrak{X}_{\infty,f}(A_\mathbf{T})\Big)$.

Finally, to show the height pairing admits a factorisation

$$\langle x, y\rangle_{\mathcal{R},\mathbf{T},\{X^\dagger_\nu\}_\nu} \;=\; \log_p \varrho(\gamma_0) \times \phi_\mathcal{R}\Big(\theta_{\gamma_0}(x)\Big)(y) \quad \text{with } (x,y) \in H^1_{f,\Sigma}(-)^0 \times H^1_{f,\Sigma}$$

one simply compares Definitions 10.6 and 10.16 (that they coincide is immediate).

The demonstration of Lemma 10.15.

We begin by pointing out that under our hypothesis $\mathcal{R}$ is regular, all its height one prime ideals will be principal. To prove the commutativity of the diagram in 10.15, it is equivalent to establish that

$$\begin{array}{ccccccc}
0 \to \Big(H^1\big(\mathbf{F}_\Sigma/\mathbf{F}^{\mathrm{cy}}, A_{\mathbf{T}_\lambda}\big)^\vee\Big)^{\Gamma^{\mathrm{cy}}} & \longrightarrow & \Big(H^1_{f,\Sigma}\big(\mathbf{F}^{\mathrm{cy}}, A_{\mathbf{T}_\lambda}\big)^\vee\Big)^{\Gamma^{\mathrm{cy}}} & \longrightarrow & \bigoplus_{\nu\in\Sigma}\Big(\varprojlim_n H^1_f(\mathbf{F}_{\nu,n}, \mathbf{T}_\lambda)\Big)_{\Gamma^{\mathrm{cy}}} \\
\uparrow \alpha_{\Sigma,\gamma_0} \bmod \lambda & & \uparrow \theta_{\gamma_0} \bmod \lambda & & \downarrow \pi \bmod \lambda \\
0 \longrightarrow \quad H^2\Big(\mathbf{F}_\Sigma/\mathbf{F}, A_{\mathbf{T}_\lambda}\Big)^\vee & \longrightarrow & H^1_{f,\Sigma}(\mathbf{F}, \mathbf{T}_\lambda)^0 & \longrightarrow & \bigoplus_{\nu\in\Sigma}\mathcal{N}^{\mathrm{univ}}_\infty\Big(H^1_f(\mathbf{F}_\nu, \mathbf{T}_\lambda)\Big)
\end{array}$$

is commutative with exact rows, at infinitely many height one primes $(\lambda) \in \operatorname{Spec}\mathcal{R}$. For instance, let's prove the same statement modulo $\mathcal{P} = (\lambda)$ for all bar finitely many bad arithmetic weights. The problem naturally subdivides into two parts:

(I) Prove the commutativity of the left-hand square modulo λ;

(II) Prove the commutativity of the right-hand square modulo λ.

Its resolution boils down to a couple of rather delicate cohomological calculations, which were undertaken by Perrin-Riou in [PR1, §4.4-§4.5]. Nonetheless we shall include them here, to ensure the demonstration of Theorem 10.8 is self-contained.

Proof of (I): Given a profinite group G acting on some Galois representation M, one writes $C^j(G, M)$ for the set of continuous j-cochains, and likewise $Z^j(G, M)$ for the set of continuous j-cocycles.

We start with an explicit description of $H^2\big(\mathbf{F}_\Sigma/\mathbf{F}, A_{\mathbf{T}_\lambda}\big)^\vee \longrightarrow H^1_{f,\Sigma}(\mathbf{F}, \mathbf{T}_\lambda)^0$. For each $\varphi \in H^2\big(\mathbf{F}_\Sigma/\mathbf{F}, A_{\mathbf{T}_\lambda}\big)^\vee$ there exists $a = a(\varphi) \in H^1_{f,\Sigma}(\mathbf{F}, \mathbf{T}_\lambda)^0$ such that $a_\nu \in H^1(\mathbf{F}_\nu, \mathbf{T}_\lambda)$ is zero at all $\nu \in \Sigma$; moreover, if $b \in H^2\big(\mathbf{F}_\Sigma/\mathbf{F}, A_{\mathbf{T}_\lambda}\big)$ restricts to zero at all $\nu \in \Sigma$, then $\varphi(b)$ is given by the formula

$$\varphi(b) \quad = \quad -\sum_{\nu\in\Sigma} \operatorname{inv}_{\mathbf{F}_\nu}\left(\varepsilon_\nu - \alpha'_\nu \cup \beta_\nu\right) \qquad \text{c.f. [PR1, §4.4.4].}$$

Note that $\beta \in Z^2\big(\mathrm{Gal}(\mathbf{F}_\Sigma/\mathbf{F}), A_{\mathbf{T}_\lambda}\big)$ represents b, and $\alpha \in Z^1\big(\mathrm{Gal}(\mathbf{F}_\Sigma/\mathbf{F}), \mathbf{T}_\lambda\big)$ is a representative of a for which $\alpha_\nu = d\alpha'_\nu$ with $\alpha'_\nu \in C^0\big(G_{\mathbf{F}_\nu}, \mathbf{T}_\lambda\big)$. Lastly, the cochain $\varepsilon \in C^2\big(\mathrm{Gal}(\mathbf{F}_\Sigma/\mathbf{F}), \mu_{p^\infty}\big)$ satisfies $d\varepsilon = \alpha \cup \beta$.

Let's now describe explicitly the image of the element $a \in H^1_{f,\Sigma}(\mathbf{F}, \mathbf{T}_\lambda)^0$ above under the homomorphism $\theta_{\gamma_0} \bmod \lambda : H^1_{f,\Sigma}(\mathbf{F}, \mathbf{T}_\lambda)^0 \longrightarrow \Big(H^1_{f,\Sigma}(\mathbf{F}^{\mathrm{cy}}, A_{\mathbf{T}_\lambda})_{\Gamma^{\mathrm{cy}}}\Big)^\vee$. Because a trivialises at all $\nu \in \Sigma$, there exist splittings in cohomology

$$H^1_f\big(K, \mathfrak{X}^*_{a_\nu}(1)\big) \quad = \quad H^1_f\big(K, \mathbb{Z}_p(1)\big) \,\oplus\, H^1_f\big(K, \mathbf{T}^*_\lambda(1)\big)$$

for every extension $\mathbf{F}_\nu \subset K \subset \mathbf{F}^{\mathrm{cy}}_\nu$. It follows $s_{\mathrm{can},K}$ splits into a direct sum, whence $s_{\mathrm{can},K}(y)$ is the image of $(0 \oplus \gamma) - \alpha'_\nu \cup \gamma_\nu \in Z^1\big(G_{\mathbf{F}^{\mathrm{cy}}_\nu}, \mu_{p^\infty}\big)$ inside $H^1\big(\mathbf{F}^{\mathrm{cy}}_\nu, A_{\mathfrak{X}_{a_\nu}}\big)$ N.B. here the continuous cocycle $\gamma \in Z^1\big(G_{\mathbf{F}^{\mathrm{cy}}}, A_{\mathbf{T}_\lambda}\big)$ denotes a representative of y.

One can choose for $s_{\mathrm{glob}}(y)$ the class $0 \oplus \gamma - \delta$, provided $\delta \in C^1\big(\mathrm{Gal}(\mathbf{F}_\Sigma/\mathbf{F}^{\mathrm{cy}}), \mu_{p^\infty}\big)$ satisfies $d\delta = \alpha \cup \gamma$ under the boundary map. As a corollary,

$$(\theta_{\gamma_0} \bmod \lambda)(a)(y) \;=\; -\log_p \varrho(\gamma_0)^{-1} \sum_{\nu \in \Sigma(\mathbf{F}^{\mathrm{cy}})} \widetilde{\ell_{\mathcal{R},\nu,\infty} \bmod \lambda}\Big(\delta_\nu - \alpha'_\nu \cup \gamma_\nu\Big).$$

Remark: To visualise the map $\alpha_{\Sigma,\gamma_0} \bmod \lambda$, fix an element $y \in H^1_{f,\Sigma}(\mathbf{F}^{\mathrm{cy}}, A_{\mathbf{T}_\lambda})$. For integers $m \geq 0$ we consider $l_m \in \mathrm{Hom}\Big(\mathrm{Gal}(\mathbf{F}^{\mathrm{cy}}/\mathbf{F}_m), \mathbb{Z}_p\Big)$ where $l_m : \gamma_0^{p^m} \mapsto 1$, e.g. $l_m = \log_p \varrho(\gamma_0)^{-1} \times p^{-m} \ell_{\varrho,\mathbf{F}_{\nu,m}}$ in our previous notation. If $m \gg 0$ then

$$(\alpha_{\Sigma,\gamma_0} \bmod \lambda)(y) = \mathrm{cores}_{\mathbf{F}_m/\mathbf{F}}\big(y \cup l_m\big) \qquad \text{by [PR1, §4.4.7],}$$

since the cup-product $y \cup l_m$ belongs to $H^2\big(\mathbf{F}_\Sigma/\mathbf{F}_m, A_{\mathbf{T}_\lambda}\big)$ for suitably large m.

The commutativity of the left-hand square in Lemma 10.15 will follow, assuming one can prove the formula

$$\sum_{\nu \in \Sigma(\mathbf{F}_m)} \widetilde{\ell_{\mathcal{R},\nu,\infty} \bmod \lambda}\Big(\delta_\nu - \alpha'_\nu \cup \gamma_\nu\Big) \;\stackrel{?}{=}\; \sum_{\nu \in \Sigma} \mathrm{inv}_{\mathbf{F}_\nu}\big(\varepsilon_\nu - \alpha'_\nu \cup \beta_\nu\big)$$

with $b = \mathrm{cores}_{\mathbf{F}_m/\mathbf{F}}\big(y \cup p^{-m}\ell_{\varrho,\mathbf{F}_{\nu,m}}\big)$. We'll take $\varepsilon = \mathrm{cores}_{\mathbf{F}_m/\mathbf{F}}\big(\delta \cup p^{-m}\ell_{\varrho,\mathbf{F}_{\nu,m}}\big)$. Crucially the problem reduces to verifying for any $x \in H^1\big(\mathbf{F}_{\nu,m}, \mu_{p^\infty}\big)$, that

$$\widetilde{\ell_{\mathcal{R},\nu,\infty} \bmod \lambda}(x) \;\stackrel{?}{=}\; \mathrm{inv}_{\mathbf{F}_\nu}\Big(\mathrm{cores}_{\mathbf{F}_{\nu,m}/\mathbf{F}_\nu}\big(x \cup p^{-m}\ell_{\varrho,\mathbf{F}_{\nu,m}}\big)\Big).$$

By construction $\widetilde{\ell_{\mathcal{R},\nu,\infty} \bmod \lambda}(x)$ agrees with $x \cup p^{-m}\ell_{\varrho,\mathbf{F}_{\nu,m}}$ on $H^1\big(\mathbf{F}_{\nu,m}, \mathcal{O}_\lambda(1)\big)$; further, it is a basic fact in Galois cohomology that $\mathrm{inv}_{\mathbf{F}_\nu} \circ \mathrm{cores}_{\mathbf{F}_{\nu,m}/\mathbf{F}_\nu} = \mathrm{inv}_{\mathbf{F}_{\nu,m}}$. The proof of (I) then follows from the commutativity of

$$\begin{array}{ccc} H^1\big(\mathbf{F}_{\nu,m}, \mathbb{Q}_{p,\lambda}(1)\big) & \xrightarrow{-\cup \log_p(\varrho_{\nu,m})} & H^2\big(\mathbf{F}_{\nu,m}, \mathbb{Q}_{p,\lambda}(1)\big) \\ \big\downarrow {\scriptstyle \log_p(\varrho_{\nu,m}) \circ \mathrm{Art}_{\mathbf{F}_{\nu,m}}} & & \big\downarrow {\scriptstyle \mathrm{inv}_{\mathbf{F}_{\nu,m}}} \\ \mathbb{Q}_{p,\lambda} & \xrightarrow{=} & \mathbb{Q}_{p,\lambda} \end{array}$$

which was shown by Serre in [Se; Chap XI, Prop. 2].

Proof of (II): Firstly, observe $\mathcal{N}_\infty^{\mathrm{univ}} H^1_{f,\Sigma}(\mathbf{F}, \mathbf{T}_\lambda)$ injects into $\bigoplus_{\nu \in \Sigma} H^1(\mathbf{F}_\nu, \mathbf{T}_\lambda)$. We'll abbreviate the homomorphism θ_{γ_0} modulo λ by the shorter version $\theta_{\gamma_0,[\lambda]}$; the crux of the argument is to show commutativity of the square

$$\begin{array}{ccc} H^1_{f,\Sigma}(\mathbf{F}, \mathbf{T}_\lambda)^0 & \xrightarrow{\oplus \mathrm{res}_\nu} & \displaystyle\bigoplus_{\nu \in \Sigma} \frac{\mathcal{N}_\infty^{\mathrm{univ}}\Big(H^1_f(\mathbf{F}_\nu, \mathbf{T}_\lambda)\Big)}{\mathcal{N}_\infty^{\mathrm{univ}}\Big(H^1_{f,\Sigma}(\mathbf{F}, \mathbf{T}_\lambda)\Big)} \\ \big\downarrow {\scriptstyle \theta_{\gamma_0,[\lambda]}} & & \big\uparrow {\scriptstyle \mathrm{proj}} \\ \Big(H^1_{f,\Sigma}(\mathbf{F}^{\mathrm{cy}}, A_{\mathbf{T}_\lambda})_{\Gamma^{\mathrm{cy}}}\Big)^\vee & \longrightarrow & \displaystyle\bigoplus_{\nu \in \Sigma} \left(\frac{\varprojlim_n H^1_f(\mathbf{F}_{\nu,n}, \mathbf{T}_\lambda)}{\varprojlim_n H^1_{f,\Sigma_n}(\mathbf{F}_n, \mathbf{T}_\lambda)}\right)_{\Gamma^{\mathrm{cy}}} \end{array}$$

at infinitely many distinct points $\mathcal{P} = (\lambda)$, which densely populate $\mathrm{Spec}(\mathcal{R})$.

This can certainly be achieved whenever $\mathbf{T}_\lambda$ is a pseudo-geometric representation i.e. for the subset of all arithmetic primes. The modulo $\mathcal{P}$ argument itself has been cannibalised verbatim from [PR1, Prop. 4.5.2].

Let $a_\nu \in H^1(\mathbf{F}_\nu, \mathbf{T}_\lambda)$ and $b_\nu \in H^1(\mathbf{F}_\nu, A_{\mathbf{T}_\lambda})$ – we need a description of $a_\nu \cup b_\nu$ under the inverse map to the isomorphism $\mu_\nu : H^1(\mathbf{F}_\nu^{\mathrm{cy}}, \mu_{p^\infty})_{\Gamma^{\mathrm{cy}}} \xrightarrow{\sim} H^2(\mathbf{F}_\nu, \mu_{p^\infty})$. Because $H^2(\mathbf{F}_\nu^{\mathrm{cy}}, \mu_{p^\infty})$ is trivial, there exists $\widehat{b}_\nu \in H^1(\mathbf{F}_\nu^{\mathrm{cy}}, A_{\mathfrak{X}_{a_\nu}})$ whose image in $H^1(\mathbf{F}_\nu^{\mathrm{cy}}, A_{\mathbf{T}_\lambda})$ is precisely $\mathrm{res}_{\mathbf{F}_\nu^{\mathrm{cy}}/\mathbf{F}_\nu}(b_\nu)$. Clearly $(\gamma_0-1).\widehat{b}_\nu$ is zero in $H^1(\mathbf{F}_\nu^{\mathrm{cy}}, A_{\mathbf{T}_\lambda})$, so it must have originated from $H^1(\mathbf{F}_\nu^{\mathrm{cy}}, \mu_{p^\infty})$. As a direct consequence

$$a_\nu \cup b_\nu \;=\; \mu_\nu\Big((\gamma_0 - 1).\widehat{b}_\nu\Big), \qquad \text{i.e. } (\gamma_0 - 1).\widehat{b}_\nu \;=\; \mu_\nu^{-1}(a_\nu \cup b_\nu).$$

Remark: Fix an element $a \in H^1_{f,\Sigma}(\mathbf{F}, \mathbf{T}_\lambda)^0$. Consider any $y \in H^1(\mathbf{F}_\Sigma/\mathbf{F}^{\mathrm{cy}}, A_{\mathbf{T}_\lambda})$ and $b = \{b_\nu\} \in \bigoplus_{\nu\in\Sigma} H^1(\mathbf{F}_\nu, A_{\mathbf{T}_\lambda})$ satisfying

$$\delta_\omega \;=\; y_\omega - \mathrm{res}_{\mathbf{F}_\nu^{\mathrm{cy}}/\mathbf{F}_\nu}(b_\nu) \;\in\; \mathbb{Q}_p/\mathbb{Z}_p \otimes_{\mathbb{Z}_p} \mathcal{N}_\infty^{\mathrm{univ}}\Big(H^1_f(\mathbf{F}_\nu, \mathbf{T}_\lambda)\Big)$$

at primes ω lying over ν. In particular, $(\gamma_0 - 1).y \in H^1_{f,\Sigma}(\mathbf{F}^{\mathrm{cy}}, A_{\mathbf{T}_\lambda})$.

For (a, y, b) as above, to establish the square commutes it is equivalent to prove

$$\theta_{\gamma_0,[\lambda]}(a)\Big((\gamma_0 - 1).y\Big) \;\stackrel{??}{=}\; \sum_{\nu\in\Sigma} \mathrm{inv}_{\mathbf{F}_\nu}(a_\nu \cup b_\nu).$$

Well it is immediate that $(\gamma_0 - 1).y_\omega = (\gamma_0 - 1).\delta_\omega$; if $\widehat{\delta}_\omega$ is a canonical lift of δ_ω, then $(\gamma_0 - 1).\widehat{\delta}_\omega$ must be a canonical lift of $(\gamma_0 - 1).y_\omega$. Similarly, if $\widetilde{y}$ denotes a global lift of y to $H^1(\mathbf{F}_\Sigma/\mathbf{F}^{\mathrm{cy}}, A_{\mathfrak{X}_a})$, then $(\gamma_0 - 1).\widetilde{y}$ is a global lift of $(\gamma_0 - 1).y$. Thus picking $m \gg 0$, the quantity $\theta_{\gamma_0,[\lambda]}(a)\Big((\gamma_0 - 1).y\Big)$ coincides by 10.16 with

$$\log_p \varrho(\gamma_0)^{-1} \sum_{\nu\in\Sigma} \widetilde{\ell_{\mathcal{R},\nu,\infty}} \bmod \lambda \otimes \mathrm{id}\Big((\gamma_0 - 1).\big(s_{\mathrm{glob}}(y)_\nu - \widehat{y}_\nu\big)\Big)$$

$$\stackrel{\text{by (I)}}{=} \log_p \varrho(\gamma_0)^{-1}\sum_{\nu\in\Sigma}\sum_{\omega|\nu} \mathrm{inv}_{\mathbf{F}_\nu}\Big(\mathrm{cores}_{\mathbf{F}_{m,\omega}/\mathbf{F}_\nu}\Big((\gamma_0 - 1).(\widetilde{y}_\omega - \widehat{\delta}_\omega) \cup p^{-m}.\ell_{\varrho,\mathbf{F}_{m,\omega}}\Big)\Big).$$

The projection of $\widetilde{y}_\omega - \widehat{\delta}_\omega$ onto $H^1(\mathbf{F}_{m,\omega}, A_{\mathbf{T}_\lambda})$ will be equal to $(b_\nu + \delta_\omega) - \delta_\omega = b_\nu$. In other words $\widehat{b}_\nu := \widetilde{y}_\omega - \widehat{\delta}_\omega$ is a canonical lift of b_ν, whence

$$(\gamma_0 - 1).\Big(\widetilde{y}_\omega - \widehat{\delta}_\omega\Big) \;=\; (\gamma_0 - 1).\widehat{b}_\nu \;=\; \mu_\nu^{-1}(a_\nu \cup b_\nu).$$

We are inexorably led to the conclusion

$$\theta_{\gamma_0,[\lambda]}(a)\Big((\gamma_0 - 1).y\Big) \;=\; \sum_{\nu\in\Sigma}\sum_{\omega|\nu} \mathrm{inv}_{\mathbf{F}_{m,\omega}}\left(\mu_\nu^{-1}(a_\nu \cup b_\nu) \;\cup\; \frac{p^{-m}.\ell_{\varrho,\mathbf{F}_{m,\omega}}}{\log_p \varrho(\gamma_0)}\right)$$

which is none other than $\sum_{\nu\in\Sigma} \mathrm{inv}_{\mathbf{F}_\nu}(a_\nu \cup b_\nu)$. The statement (II) now follows.

10.5 The main conjectures over weight-space

To complement our intended predictions, let us first review the classical Iwasawa main conjecture for elliptic curves. Whilst one can formulate the statement purely in terms of $\Lambda^{\mathrm{cy}} = \mathbb{Z}_p[\![\Gamma^{\mathrm{cy}}]\!]$, we prefer to work over the affinoid algebra of the disk. Recall that $\mathbf{L}_p(E,s) \in \mathbb{Q}_p\langle\!\langle s \rangle\!\rangle$ denoted the Mazur-Tate-Teitelbaum p-adic L-series interpolating Dirichlet twists of $E_{/\mathbb{Q}}$ by characters of p-power conductor.

Main Conjecture 10.17. *(The Cyclotomic Version)*

$$\varrho^{s-1} \circ \mathrm{char}_{\mathbb{Z}_p[\![\Gamma^{\mathrm{cy}}]\!]}\left(\mathrm{Sel}_{\mathbb{Q}^{\mathrm{cy}}}(E)^{\vee}\right) \quad = \quad \mathbf{L}_p(E,s) \qquad \mathrm{mod}\ \mathbb{Z}_p\langle\!\langle s \rangle\!\rangle^{\times}.$$

Due to fundamental work of Rubin, Kato, Greenberg-Vatsal and Skinner-Urban, many instances of this conjecture have actually been proven [Ru1,Ka1,GV,SU]. Moreover one should notice that the left-hand side is p-integral, which would then imply integrality for the right-hand side of the formula too (see [St, Thm 4.6]).

Let's find the analogues of 10.17 over weight-space. Throughout, all conjectured equalities will be stated **only up to units in the underlying affinoid algebra**. It makes good sense to compile a list of hypotheses:

- The curve E has either good ordinary or bad multiplicative reduction at $p \geq 5$.
- The number field $\mathbf{F}$ is an abelian extension of $\mathbb{Q}$ in which p is unramified.
- The modular deformation ring $\mathcal{R}_E$ lifting $\rho_{E,p} : G_{\mathbb{Q}} \to \mathrm{GL}_2(\mathbb{Z}_p)$ is Gorenstein.
- The two-variable Selmer group is $\mathbb{Z}_p\left[\!\left[\Gamma^{\mathrm{cy}} \times \Gamma^{\mathrm{wt}}\right]\!\right]$-cotorsion.

For example, in the non-CM case this last condition certainly holds when the big Galois representation $\rho_{\infty,E}$ is residually absolutely irreducible.

N.B. Let $\mathbf{err}_{\mathbf{F}}(\mathbb{T}_{\infty,E})$ denote the reciprocal of $\prod_{\nu\in\Sigma} \mathrm{Tam}^{(2)}_{\mathbf{F}_\nu}(\mathbb{T}_{\infty,E}) \times \dfrac{\mathrm{vol}^{(2)}_{\mathbb{T}_{\infty,E}}\left(H^1_{\mathcal{E}}\right)}{\mathrm{vol}_E\left(H^1_e\right)}$, and $\mathbf{err}_{\mathbf{F}}\left(\mathbb{T}^{-1/2}_{\infty,E}\right)$ the inverse of $\prod_\nu \mathrm{Tam}^{(2)}_{\mathbf{F}_\nu}\left(\mathbb{T}^{-1/2}_{\infty,E}\right) \times \mathrm{covol}^{(2)}_{\mathbb{T}^{-1/2}_{\infty,E}}(\mathbf{F}) \times \mathrm{Per}^{(2)}_{\mathbf{I},\lambda_E}(\mathbf{F})$.

Main Conjecture 10.18. *(The Vertical Deformation)*

$$\text{Ш}_{\mathbf{F}}(\mathbb{T}_{\infty,E}, w) \quad = \quad \frac{\mathbf{err}_{\mathbf{F}}(\mathbb{T}_{\infty,E})}{\mathfrak{D}_{\mathbf{F},p}(\rho_{\infty,E})} \times \mathbf{L}^{\mathrm{imp}}_p(\mathbf{f}/\mathbf{F}, w, 1).$$

Main Conjecture 10.19. *(The Half-Twisted Deformation)*

$$\text{Ш}_{\mathbf{F}}\left(\mathbb{T}^{-1/2}_{\infty,E}, w\right) \quad = \quad \frac{\mathbf{err}_{\mathbf{F}}\left(\mathbb{T}^{-1/2}_{\infty,E}\right)}{\mathfrak{D}_{\mathbf{F},p}(\rho_{\infty,E})} \times \left.\frac{\mathbf{L}^{\mathrm{GS}}_{p,\mathbf{F}}(\mathbf{f}, w, s)}{(s - w/2)^{r_{\mathrm{gen}}(\mathbf{F})}}\right|_{s=w/2}.$$

One strongly suspects both $\mathbf{err}_{\mathbb{Q}}(\mathbb{T}_{\infty,E})$ and $\mathbf{err}_{\mathbb{Q}}\left(\mathbb{T}^{-1/2}_{\infty,E}\right)$ are always p-adic units. We have checked this extensively for elliptic curves of conductor $< 10{,}000$ and ordinary primes in the range $5 \leq p \leq 97$, and have yet to find a counterexample. Of course, this does not mean that there isn't one in general.

Remarks: (a) With both these conjectures, the presence of the defect term $\mathfrak{D}_{\mathbf{F},p}$ (defined in 9.17) is essential. It can be non-trivial whenever $\mathrm{Ta}_p(E) \not\cong \mathbb{T}_{\infty,E} \otimes_{\Lambda^{\mathrm{wt}}} \mathbb{Z}_p$ on an integral level, and reflects the fact that there is no intrinsic choice of lattice. By stark contrast, there are no such complications for the cyclotomic deformation, as the $G_{\mathbb{Q}}$-module $\mathrm{Ta}_p(E)[\![\Gamma^{\mathrm{cy}}]\!]$ represents the only canonical lattice to pick.

(b) In the vertical main conjecture, the fudge-factor $\mathbf{err}_{\mathbf{F}}(\mathbb{T}_{\infty,E})$ has been carefully chosen so as to be compatible with the BSD formula. If the former is true, then

$$\left.\frac{\mathbf{L}_p^{\mathrm{imp}}(\mathbf{f}/\mathbf{F}, w, 1)}{(w-2)^{r_{\mathbf{F}}}}\right|_{w=2} = \left(\text{multiplier in } \mathbf{L}_p^{\mathrm{imp}}\right) \times \left(p\text{-part of BSD}\right) \times \det\langle -,-\rangle_{\mathbf{F},p}^{\mathrm{wt}}$$

up to a p-adic unit. From an arithmetic standpoint, one could very well view the above equation as a **vertical BSD-conjecture** along the line $s = 1$.

(c) Analogously, the half-twisted main conjecture implies (at weight two) that the dominant $(w-2)^{??}$-term in the Taylor series for $\left.\frac{\mathbf{L}_{p,\mathbf{F}}^{\mathrm{GS}}(\mathbf{f},w,s)}{(s-w/2)^{r_{\mathrm{gen}}(\mathbf{F})}}\right|_{s=w/2}$ equals

$$\left(\text{the multiplier in } \mathbf{L}_{p,\mathbf{F}}^{\mathrm{GS}}\right) \times \left(\text{the } p\text{-primary part of BSD}\right) \times \det\langle -,-\rangle_{\mathbf{F},p}^{-1/2}.$$

Thus one may reinterpret this statement as a **half-twisted BSD-conjecture**, along the central line $s = w/2$.

Warning: It is really commonsense to take these conjectures with a pinch of salt. Our sole method of calculating the constant terms in 10.18 and 10.19 was through the technique of Γ^{wt}-Euler characteristics. In particular, the special values of these two L-functions were computed only at the base weight $w = 2$, not at higher weight. One must therefore believe that at weight $w = k > 2$, the Bloch-Kato conjectures continue to hold for all classical eigenforms $\mathbf{f}_k$ in the Hida family.

The main conjecture for the whole (s, k)-plane.

The next logical step is to formulate a 'universal Main Conjecture', which describes the behaviour of the L-functions $L(\mathbf{f}_k, s)$ over the critical region in the (s, k)-plane. Recall in §10.3 we associated an algebraic p-adic L-function by

$$\mathbf{L}_{p,\mathbf{F}}^{\mathrm{alg}}\left(\mathbb{T}_{\infty,E}^{-1/2}, s, w\right) = \left(\varrho^{s-1} \times \mathrm{Mell}_{2,\mathbf{1}} \circ \sigma\right)\left(\mathrm{char}_{\mathbb{Z}_p[\![\Gamma^{\mathrm{cy}},\Gamma^{\mathrm{wt}}]\!]}\left(\mathrm{Sel}_{\mathbf{F}^{\mathrm{cy}}}\left(\mathbb{T}_{\infty,E}^{-1/2}\right)^{\vee}\right)\right)$$

which was analytic on the affinoid domain $(s, w) \in \mathbb{Z}_p \times \mathbb{U}_E$.

À priori, the two-variable main conjecture should be of the form

$$\mathbf{L}_{p,\mathbf{F}}^{\mathrm{alg}}\left(\mathbb{T}_{\infty,E}^{-1/2}, s, w\right) \stackrel{?}{=} \mathbf{ERR}_{???}\left(\mathbb{T}_{\infty,E}^{-1/2}\right) \times \mathbf{L}_{p,\mathbf{F}}^{\mathrm{GS}}\left(\mathbf{f}, w, s + \frac{w}{2} - 1\right)$$

where the error function $\mathbf{ERR}_{???}\left(\mathbb{T}_{\infty,E}^{-1/2}\right)$ will be some element of $\mathrm{Frac}\left(\mathbb{Z}_{p,\mathbb{U}_E}\langle\!\langle w\rangle\!\rangle\right)$. The reason why the cyclotomic variable has been shifted via $s \mapsto s + w/2 - 1$, is because $\mathbb{T}_{\infty,E}^{-1/2}$ carries a half-twist through the character $\langle \chi_{\mathrm{cy}}^{-1/2} \circ \mathrm{Res}_{\mathbb{Q}^{\mathrm{cyc}}}(-)\rangle$.

To get a clearer idea of the shape of the factor $\mathbf{ERR}_{???}$, it is enough to understand its behaviour at arithmetic weights. In order to achieve this, we first mention an important result of Emerton, Pollack and Weston.

Theorem 10.20. [EPW] *If the cyclotomic μ-invariant vanishes for one modular form $\mathbf{f}_{k_0}$ in the Hida family, then it vanishes for all classical forms $\{\mathbf{f}_k\}_{k\in\mathbb{U}_E\cap\mathbb{N}_{\geq 2}}$.*

A conjecture of Greenberg predicts the triviality of the μ-invariant for $\mathbf{f}_2$ if $\rho_{E,p}$ is a residually irreducible Galois representation. In general, one would expect the difference of the algebraic and analytic μ-invariants to be consistent, over those branches of the Hida family $\mathcal{H}(\overline{\rho}_\infty)$ exhibiting common ramification behaviour. Here the notation $\mathcal{H}(\overline{\rho}_\infty)$ indicates the set of all p-stabilised ordinary newforms with prescribed mod p Galois representation isomorphic to $\overline{\rho}_\infty$.

It follows that the valuation of $\mathbf{ERR}_{???}$ should be constant over weight-space, which means it lies inside $p^{\mathbb{Z}}$. To determine it exactly, there are several approaches. One could attempt to work out both the algebraic and the analytic μ-invariants for some eigenform $\mathbf{f}_{k_0}$ in the family, but in complete generality this is too difficult. It is far better to compare valuations of the leading terms of both analytic and algebraic L-functions, i.e. the discrepancy should equal $\mu^{\mathrm{cy,alg}}(\mathbf{f}_k)-\mu^{\mathrm{cy,an}}(\mathbf{f}_k)$ for all forms $\mathbf{f}_k\in\mathcal{H}(\overline{\rho}_\infty)$ lying on branches sharing the same ramification type as $\mathbf{f}_2$.

Let's define the constant term $\mathbf{ERR}_{\mathbf{F}^{\mathrm{cy}}}\left(\mathbb{T}_{\infty,E}^{-1/2}\right)$ to be the *reciprocal* of

$$\mathrm{covol}^{(2)}_{\mathbb{T}_{\infty,E}^{-1/2}}(\mathbf{F}\otimes\mathbb{Z}_p)\times\mathrm{Per}^{(2)}_{\mathbf{I},\lambda_E}(\mathbf{F})\times\mathfrak{D}_{\mathbf{F},p}(\rho_{\infty,E})\times\prod_{\nu|p}\frac{\mathrm{Tam}^{(2)}_{\mathbf{F}_\nu}\left(\mathbb{T}_{\infty,E}^{-1/2}\right)}{\ell_\nu\left(\mathbb{T}_{\infty,E}^{-1/2}\right)\Big|_{w=2}}$$

N.B. Assuming the various incarnations of the Birch and Swinnerton-Dyer formulae, then the Corollary 10.9 suggests this must be the correct fudge-factor.

Main Conjecture 10.21. *(The Two-Variable Version)*

$$\mathbf{L}^{\mathrm{alg}}_{p,\mathbf{F}}\left(\mathbb{T}_{\infty,E}^{-1/2},s,w\right)\quad=\quad\mathbf{ERR}_{\mathbf{F}^{\mathrm{cy}}}\left(\mathbb{T}_{\infty,E}^{-1/2}\right)\times\mathbf{L}^{\mathrm{GS}}_{p,\mathbf{F}}\left(\mathbf{f},w,s+\frac{w}{2}-1\right).$$

In support of the two-variable conjecture, consider the algebraic and analytic orders along the line of symmetry in the functional equation. Then

$$\begin{aligned}\mathrm{order}_{w=2}\left(\frac{\mathbf{L}^{\mathrm{alg}}_{p,\mathbf{F}}\left(\mathbb{T}_{\infty,E}^{-1/2},s,w\right)}{(s-1)^{\mathfrak{r}}}\Bigg|_{s=1}\right)&\overset{\text{by } 10.8}{=}\quad\mathrm{order}_{w=2}\,\text{Ш}_{\mathbf{F}}\left(\mathbb{T}_{\infty,E}^{-1/2},w\right)\\&\overset{\text{by } 10.19}{=}\quad\mathrm{order}_{w=2}\left(\frac{\mathbf{L}^{\mathrm{GS}}_{p,\mathbf{F}}(\mathbf{f},w,s)}{(s-w/2)^{r_{\mathrm{gen}}(\mathbf{F})}}\Bigg|_{s=w/2}\right)\\&=\quad\mathrm{order}_{w=2}\left(\frac{\mathbf{L}^{\mathrm{GS}}_{p,\mathbf{F}}(\mathbf{f},w,s+w/2-1)}{(s-1)^{r_{\mathrm{gen}}(\mathbf{F})}}\Bigg|_{s=1}\right).\end{aligned}$$

However Conjecture 7.18 predicted that $\mathfrak{r}=r_{\mathrm{gen}}(\mathbf{F})$, which ties up the orders of vanishing for $\mathbf{L}^{\mathrm{alg}}_{p,\mathbf{F}}$ and $\mathbf{L}^{\mathrm{GS}}_{p,\mathbf{F}}$ quite nicely.

Exercise. *Let's assume that the two-variable Main Conjecture 10.21 holds true. By applying Corollary 10.9, show that the* $\left(s-\frac{w}{2}\right)^{r_{\mathrm{gen}}(\mathbf{F})} \times (w-2)^{r^{\dagger}_{\mathbf{F}}}$*-coefficient in the expansion of the Greenberg-Stevens L-function, equals*

$$\mathrm{Per}^{(2)}_{\mathbf{I},\lambda_E}(\mathbf{F}) \prod_{\mathfrak{p}\in S^{\mathrm{good}}_{\mathbf{F},p}} \#\widetilde{E}\left(\mathbb{F}_{q_{\mathfrak{p}}}\right)^2 \times (p\text{-part of BSD}) \times \mathrm{disc}\langle -,-\rangle^{\mathrm{cy}}_{\mathbb{T}^{-1/2}_{\infty,E}}\Big|_{w=2} \det\langle -,-\rangle^{-1/2}_{\mathbf{F},p}$$

in the case where E has good ordinary reduction at p.

10.6 Numerical examples, open problems

We now seek a deformation-theoretic explanation for the result of Emerton et al. Recall that the μ-invariant associated to an element $G \in \mathbb{Z}_p[\![Y]\!]$ is the exponent for the power of p occurring in the decomposition

$$G(Y) \quad = \quad p^{\mu} \times \Big(\text{distinguished polynomial in } \mathbb{Z}_p[Y]\Big) \times \Big(\text{element of } \mathbb{Z}_p[\![Y]\!]^{\times}\Big).$$

Equivalently, if $G(Y) = \sum_{j=0}^{\infty} g_j Y^j$ then $\mu = \min_{j\geq 0}\left\{\mathrm{ord}_p(g_j)\right\}$.

Over a 2-variable deformation ring, this notion generalises in an obvious manner. If $F \in \mathbb{Z}_p[\![X,Y]\!]$ has the power series expansion $F(X,Y) = \sum_{i=0}^{\infty}\sum_{j=0}^{\infty} f_{i,j}X^iY^j$, then one defines $\mu = \mu(F)$ to be the non-negative number $\min_{i,j\geq 0}\left\{\mathrm{ord}_p(f_{i,j})\right\}$. Under the double p-adic Mellin transform $\widehat{F}(s,w) = F\Big(\sigma^{w-2}(u_0)-1,\ \varrho^{s-1}(\gamma_0)-1\Big)$, it is straightforward to check

$$\mu(F) \;=\; \min_{i,j\geq 0}\Big\{\mathrm{ord}_p(\mathcal{F}_{i,j}) - \log_p \sigma^i(u_0) - \log_p \varrho^j(\gamma_0)\Big\} \;=\; \min_{i,j\geq 0}\Big\{\mathrm{ord}_p(\mathcal{F}_{i,j}) - (i+j)\Big\}$$

where $\mathcal{F}_{i,j}$ denotes the coefficient of $(w-2)^i(s-1)^j$ in the Taylor series for $\widehat{F}(s,w)$.

Definition 10.22. *The 'generic μ-invariant' over the double deformation of $\rho_{E,p}$ is defined to be the non-negative integer*

$$\mu^{\mathrm{gen}}_{p,\mathbf{F}}(\rho_{\infty,E}) \quad := \quad \min_{i,j\geq 0}\Big\{\mathrm{ord}_p(\mathcal{L}_{i,j}) - (i+j)\Big\}$$

with the $\mathcal{L}_{i,j}$'s appearing as $\mathbf{L}^{\mathrm{alg}}_{p,\mathbf{F}}\left(\mathbb{T}^{-1/2}_{\infty,E}, s, w\right) \;=\; \sum_{i=0}^{\infty}\sum_{j=0}^{\infty} \mathcal{L}_{i,j}(w-2)^i(s-1)^j$.

What is the connection between this number, and the statement of Theorem 10.20? Morally the equality $\mu^{\mathrm{gen}}_{p,\mathbf{F}}(\rho_{\infty,E}) = 0$ is determined by the simultaneous vanishing of all algebraic invariants $\mu^{\mathrm{cy,alg}}(\mathbf{f}_k)$ for the whole p-ordinary family $\left\{\mathbf{f}_k\right\}_{k\in\mathbb{U}_E}$. Therefore the immediate task facing us is to find some means of calculating this generic μ-invariant, via an explicit formula.

The following cheat works well over $\mathbf{F} = \mathbb{Q}$, for elliptic curves of small conductor.

Proposition 10.23. *If $L(E,1)\neq 0$ and E has good ordinary reduction at $p\geq 5$, then the generic μ-invariant is bounded above by*

$$\mu_{p,\mathbb{Q}}^{\mathrm{gen}}(\rho_{\infty,E}) \quad\leq\quad \delta_p(E)+\mathrm{ord}_p\Big(\#\text{Ш}_{\mathbb{Q}}(E)\times \mathrm{Tam}_{\mathbb{Q}}(E)\Big)-\mathrm{ord}_p\Big(\mathfrak{D}_{\mathbb{Q},p}(\rho_{\infty,E})\Big)$$

where the first term $\delta_p(E) \;=\; \begin{cases} 0 & \text{if } a_p(E)\neq +1 \\ 0 & \text{if } a_p(E)=+1 \text{ and } E(\mathbb{Q})[p]\neq\{O_E\} \\ 2 & \text{otherwise.}\end{cases}$

Proof: Under the stated conditions, each of $E(\mathbb{Q})$ and $\text{Ш}_{\mathbb{Q}}(E)$ are finite groups. It follows directly that $\mathfrak{r}=0$ and $r_{\mathbb{Q}}^{\dagger}=0$, in which case

$$\frac{\mathbf{L}_{p,\mathbb{Q}}^{\mathrm{alg}}(\mathbb{T}_{\infty,E}^{-1/2},s,w)}{(s-1)^{\mathfrak{r}}(w-2)^{r_{\mathbb{Q}}^{\dagger}}}\Bigg|_{(s,w)=(1,2)} \;=\; \frac{\ell_p\left(\mathbb{T}_{\infty,E}^{-1/2}\right)\Big|_{w=2}}{\mathrm{Tam}_{\mathbb{Q}_p}^{(2)}\left(\mathbb{T}_{\infty,E}^{-1/2}\right)}\times\left(\mathrm{covol}_{\mathbb{T}_{\infty,E}^{-1/2}}^{(2)}\times\mathfrak{D}_{\mathbb{Q},p}\right)^{-1}$$
$$\times\,\#\widetilde{E}(\mathbb{F}_p)^2\times\frac{\#\text{Ш}_{\mathbb{Q}}(E)\prod_l\left[E(\mathbb{Q}_l):E_0(\mathbb{Q}_l)\right]}{\#E(\mathbb{Q})^2}$$

from Corollary 10.9. Thus the constant term $\mathbf{L}_{p,\mathbb{Q}}^{\mathrm{alg}}(\mathbb{T}_{\infty,E}^{-1/2},1,2)$ has valuation

$$\delta_p(E)\;+\;\xi_p(E)\;+\;\mathrm{ord}_p\Big(\#\text{Ш}_{\mathbb{Q}}(E)\times\mathrm{Tam}_{\mathbb{Q}}(E)\Big)\;-\;\mathrm{ord}_p\Big(\mathfrak{D}_{\mathbb{Q},p}(\rho_{\infty,E})\Big)$$

where $\delta_p(E)=2\,\mathrm{ord}_p(\#\widetilde{E}(\mathbb{F}_p))-2\,\mathrm{ord}_p(\#E(\mathbb{Q}))$, and $\xi_p(E)$ is equal to the p-adic order of $\mathrm{covol}_{\mathbb{T}_{\infty,E}^{-1/2}}^{(2)}(\mathbb{Q}_p)^{-1}\times\ell_p\left(\mathbb{T}_{\infty,E}^{-1/2}\right)\Big|_{w=2}\times\mathrm{Tam}_{\mathbb{Q}_p}^{(2)}\left(\mathbb{T}_{\infty,E}^{-1/2}\right)^{-1}$.

It is an easy exercise to verify this definition of $\delta_p(E)$ coincides with that in the statement of the proposition. On the other hand, by Proposition 8.13(a) we find

$$\frac{\left(\#\widetilde{E}(\mathbb{F}_p)[p^\infty]\right)^2\times\left|\left[\mathcal{C}^{\min}(\mathbb{Q}_p):\mathcal{C}_0^{\min}(\mathbb{Q}_p)\right]\right|_p^{-1}}{\mathrm{covol}_{\mathbb{T}_{\infty,E}^{-1/2}}^{(2)}(\mathbb{Q}_p)\times\mathrm{Tam}_{\mathbb{Q}_p}^{(2)}\left(\mathbb{T}_{\infty,E}^{-1/2}\right)} \;=\; \#\mathrm{Ker}\left(\delta_{\infty,E,p}^{-1/2}\right)\;\geq\;0.$$

Further, the valuation of $\mathrm{Tam}_{\mathbb{Q}_p}^{(2)}\left(\mathbb{T}_{\infty,E}^{-1/2}\right)$ is bounded above by the valuation of $\left[\mathcal{C}^{\min}(\mathbb{Q}_p):\mathcal{C}_0^{\min}(\mathbb{Q}_p)\right]$; the latter quantity must be zero since E (and therefore $\mathcal{C}^{\min}$) has good reduction at p. As a direct consequence $-2\mathrm{ord}_p\left(\widetilde{E}(\mathbb{F}_p)\right)\leq\xi_p(E)\leq 0$, thus one may deduce $\mathrm{ord}_p\left(\mathbf{L}_{p,\mathbb{Q}}^{\mathrm{alg}}(\mathbb{T}_{\infty,E}^{-1/2},1,2)\right)$ at least satisfies the bound predicted in 10.23.

By its very definition, the generic μ-invariant is bounded by $\mathrm{ord}_p\Big(\mathcal{L}_{i,j}-(i+j)\Big)$ for any term $\mathcal{L}_{i,j}$ in the Taylor series expansion of $\mathbf{L}_{p,\mathbb{Q}}^{\mathrm{alg}}(\mathbb{T}_{\infty,E}^{-1/2},s,w)$; in particular it is bounded above by the valuation of the constant term.

The result follows.

□

If the p-part of Ш is large, then this upper bound on $\mu_{p,\mathbb{Q}}^{\text{gen}}$ becomes very poor. However for elliptic curves of conductor < 1000 the support of Ш is usually tiny, in which case the proposition yields a majoration that can be used by a computer. We now include a couple of worked examples, where one can show triviality of the generic μ-invariant over a double deformation (for the field $\mathbf{F} = \mathbb{Q}$).

Example 10.24 Let's return to our favourite modular elliptic curve $E = X_0(11)$. As a consequence of 10.23, one discovers that

$$\mu_{p,\mathbb{Q}}^{\text{gen}}(\rho_{\infty,E}) = 0 \qquad \text{for all primes } p \text{ in the range } 3 \le p \le 97,\ \ p \ne 5, 11, 19, 29.$$

Recall that 19 and 29 were two primes of good supersingular reduction for E.

If $p = 5$ then $E[5] \cong \mu_5 \times \mathbb{Z}/5\mathbb{Z}$ as a $G_{\mathbb{Q}}$-module, the group $\text{Ш}_{\mathbb{Q}}(E)[5]$ is trivial and $\left[E(\mathbb{Q}_{11}) : E_0(\mathbb{Q}_{11})\right] = 5$. In this case Proposition 10.23 tells us

$$\mu_{5,\mathbb{Q}}^{\text{gen}}(\rho_{\infty,E}) \ \le\ \delta_5(E) + \text{ord}_5(1 \times 5) - \text{ord}_5(\mathfrak{D}_{\mathbb{Q},5}) \ =\ 0 + 1 - 0$$

i.e. the generic μ-invariant at $p = 5$ will be either zero or one. Were it to be zero(!) it would then follow $\mu_5^{\text{cy,alg}}(\mathbf{f}_k) = 0$ for infinitely many weights $k \in \mathbb{Z}_p \cap \mathbb{N}$, in fact **for all such** k courtesy of Theorem 10.20; however Greenberg [Gr2, p71] has shown that $\mu_5^{\text{cy,alg}}(f_{X_0(11)}) = 1$ (N.B. for $E = X_0(11)$ and $p = 5$, then 10.17 holds true). One therefore concludes $\mu_{5,\mathbb{Q}}^{\text{gen}}(\rho_{\infty,E}) = 1$.

Lastly, if $p = 11$ then the reduction type for E is split multiplicative, and 10.23 tells us nothing whatsoever. Fortunately, the situation is not totally irretrievable. An explicit computation reveals $\text{Sel}_{\mathbb{Q}^{\text{cy}}}(E)[11^\infty] \cong \mathbb{Q}_{11}/\mathbb{Z}_{11}$ equipped with trivial Galois action, in which case $\mu_{11}^{\text{cy,alg}}(f_{X_0(11)})$ must vanish. The earlier theorem of Emerton-Pollack-Weston implies that $\mu_{11}^{\text{cy,alg}}(\mathbf{f}_k) = 0$ for every $k \in \mathbb{Z}_p$ with $k \ge 2$. Since these numbers together constitute an upper bound on the generic μ-invariant, clearly we must have $\mu_{11,\mathbb{Q}}^{\text{gen}}(\rho_{\infty,E}) = 0$.

Remarks: (a) The prime $p = 5$ is in some sense an exceptional one for $X_0(11)$; the curve possesses a rational 5-torsion point, whence $\rho_{E,5}$ is residually reducible.

(b) For $p = 11$, the Γ^{cy}-invariants of $\text{Sel}_{\mathbb{Q}^{\text{cy}}}(E)[11^\infty]$ have $\mathbb{Z}_{11}$-corank equal to one. Greenberg proved that the algebraic λ^{cy}-invariant is one; this property is reflected by a simple zero along the line $s = k/2$ in the two-variable p-adic L-function.

Example 10.25 Consider the strong Weil elliptic curve

$$E_1 : y^2 = x^3 - 412x + 3316$$

which has conductor 280. It is well known that its Mordell-Weil rank over $\mathbb{Q}$ is one, e.g. $(x, y) = (-18, 70)$ is a point of infinite order. Unfortunately since $L(E_1, 1) = 0$, we cannot directly apply Proposition 10.23 in order to evaluate generic μ-invariants for those Hida families lifting $f_{E_1} \in \mathcal{S}_2^{\text{new}}(\Gamma_0(280))$.

However all is not lost ... Let us consider instead the non-isogenous elliptic curve

$$E_2 : y^2 \ = \ x^3 \ - \ x^2 \ - \ 4$$

which is usually labelled $56B(A)$ in the tables. This time the Mordell-Weil rank is trivial, and by 10.23 we deduce $\mu^{\rm gen}_{p,\mathbb{Q}}(\rho_{\infty,E_2}) = 0$ at good ordinary primes $p < 100$. In particular $p = 5$ is a good ordinary prime for E_2, hence $\mu^{\rm gen}_{5,\mathbb{Q}}(\rho_{\infty,E_2})$ vanishes for the family $\{\mathbf{f}^{(2)}_k\}_{k\in\mathbb{U}_{E_2}}$ deforming $\rho_{E_2,5} : G_{\mathbb{Q}} \longrightarrow \mathrm{Aut}(V^*_{f_{E_2}})$.

By Ribet's level-lowering results, there exists a congruence $f_{E_1} \equiv f_{E_2} \mod 5$ between the two newforms. In this situation, the $\mathrm{rank}_{\mathbb{Z}_5[\![X]\!]}\mathcal{R}_{E_1} = 2$ as there are exactly two congruent 5-stabilised newforms of tame level 56, namely f_{E_1} and $\mathbf{f}^{(2)}_2$. The triviality of $\mu^{\rm gen}_{5,\mathbb{Q}}$ for ρ_{∞,E_2} – in tandem with Theorem 10.20 – implies that

$$0 \ \overset{\text{by } 10.23}{=} \ \mu^{\rm gen}_{5,\mathbb{Q}}(\rho_{\infty,E_2}) \ = \ \mu^{\rm cy,alg}_{5}(\mathbf{f}^{(2)}_k) \ \overset{\text{by } 10.20}{=} \ \mu^{\rm cy,alg}_{5}(\mathbf{f}^{(1)}_k)$$

because both the families $\{\mathbf{f}^{(1)}_k\}_{k\in\mathbb{U}_{E_1}}$ and $\{\mathbf{f}^{(2)}_k\}_{k\in\mathbb{U}_{E_2}}$ have the same modulo 5 Galois representation. Meanwhile the right-most numbers form an upper bound on the generic μ-invariant of ρ_{∞,E_1}. Via the scenic route, one concludes that $\mu^{\rm gen}_{5,\mathbb{Q}}(\rho_{\infty,E_1}) = 0$, despite the complex L-series $L(E_1, s)$ possessing a zero at $s = 1$.

Example 10.26 It is worthwhile speculating how often (on average) the generic μ-invariant vanishes at a given prime. To this end, John Cremona performed a computer search in MAGMA over the first few strong Weil curves E of rank zero, and good ordinary primes p for E in the range $3 < p < 100$.

Up to conductor $N_E \leq 500$, there were exactly 11306 candidate pairs $\{E, p\}$. Out of this ensemble, the upper bound in Proposition 10.23 automatically forced the vanishing of the generic μ-invariant for 11072 pairs, i.e. the silent majority. Unsurprisingly for the remaining 234 exceptional pairs $\{E, p\}$, the upper bound $\widetilde{\mu}_p \ := \ \delta_p + \mathrm{ord}_p(\#\text{Ш} \times \mathrm{Tam}) - \mathrm{ord}_p(\mathfrak{D}_{\mathbb{Q},p})$ was too weak to imply $\mu^{\rm gen}_{p,\mathbb{Q}} = 0$. Tabulated below are the 21 specimens for which $\widetilde{\mu}_p = 1$:

$\{11A1, 5\}, \{26B1, 7\}, \{38B1, 5\}, \{57C1, 5\}, \{58B1, 5\}, \{114C1, 5\}, \{118B1, 5\}, \{158C1, 5\}, \{170C1, 7\}, \{174A1, 7\}, \{182A1, 5\}, \{203A1, 5\}, \{264D1, 7\}, \{330D1, 7\}, \{354E1, 11\}, \{366D1, 7\}, \{378G1, 5\}, \{406D1, 5\}, \{426C1, 5\}, \{442E1, 11\}, \{483A1, 5\}$.

The leftover pairs all shared the value $\widetilde{\mu}_p = 2$, save for the isolated example $246B1$ which yielded a disappointing $\widetilde{\mu}_p = 3$ at the prime $p = 5$.

Remark: For any pair $\{E, p\}$ satisfying $\widetilde{\mu}_p \leq 1$, to deduce that $\mu^{\rm gen}_{p,\mathbb{Q}}(\rho_{\infty,E})$ equals 1 it is enough to exhibit a single eigenform $\mathbf{f}_k$ whose cyclotomic μ-invariant is positive (this follows by exactly the same reasoning we deployed for $\{E, p\} = \{X_0(11), 5\}$). Unfortunately, if it is merely known that $\widetilde{\mu}_p \leq 2$ then it is impossible to deduce $\mu^{\rm gen}_{p,\mathbb{Q}}(\rho_{\infty,E}) = 2$ simply by determining that one of the $\mu^{\rm cy,alg}_{p}(\mathbf{f}_k)$'s is 2.

This book has really only scratched the surface of what is a fairly involved topic. There is much current research into the behaviour of L-functions over weight-space, and the author apologises profusely for rather neglecting the higher weights $k > 2$. By way of a conclusion, we shall now discuss five problems which are fundamentally important to further progress in the subject.

To make the exposition shorter, let us restrict ourselves to the field $\mathbf{F} = \mathbb{Q}$.

Problem 10.1. *Formulate a generic version of the λ-invariant, and relate it to each of the individual $\lambda_p^{\mathrm{cy,alg}}(\mathbf{f}_k)$'s for all arithmetic weights $k \in \mathbb{U}_E$.*

The zeros of $\mathbf{L}_{p,\mathbb{Q}}^{\mathrm{alg}}(\mathbb{T}_{\infty,E}^{-1/2}, s, k)$ form a rigid subvariety, $\mathcal{W}_0^{\mathrm{zer}}$ say, of the (s,k)-plane. Presumably, the global invariant $\lambda_{p,\mathbb{Q}}^{\mathrm{gen}}(\rho_{\infty,E})$ should be the number of irreducible components constituting $\mathcal{W}_0^{\mathrm{zer}}$. Emerton et al [EPW] proved that $\lambda_p^{\mathrm{cy,alg}}(\mathbf{f}_k)$ is constant along branches of $\mathcal{H}(\overline{\rho}_\infty)$ (provided that the $\mu^{\mathrm{cy,alg}}$-invariant vanishes). However, one cannot totally discount the possibility that certain of these zeros arise as families of isolated points (s_k, k), where $k \geq 2$ ranges over the disk $\mathbb{U}_E$.

T. Ochiai [Oc2] has recently shown that the characteristic ideal of Selmer divides into an analytic p-adic L-function $\mathbf{L}_{p\text{-adic}}^{\mathrm{Och}}(\mathbf{f}, w, s)$, which is built out of non-primitive Kato-Beilinson elements. Precisely how this function is related to $\mathbf{L}_{p,\mathbb{Q}}^{\mathrm{GS}}(\mathbf{f}, w, s)$ is a moot point, but it should produce inequalities

$$\lambda_p^{\mathrm{cy,alg}}(\mathbf{f}_k) \leq \lambda_p^{\mathrm{cy,anal}}(\mathbf{f}_k) \qquad \text{at almost all arithmetic weights } k \in \mathbb{U}_E.$$

Another major stumbling block is that one does not know whether Greenberg's conjecture for $\mathbf{L}_{p,\mathbb{Q}}^{\mathrm{GS}}(\mathbf{f}, w, s)$ holds true in general, along the central line $s = w/2$. If E has complex multiplication and its analytic rank is zero, then the conjecture is actually a theorem for the Katz-Yager p-adic L (N.B. in the non-CM case we cannot even prove Conjecture 7.11, let alone 7.5).

Problem 10.2. *Prove that* $\dfrac{\mathbf{L}_{p,\mathbb{Q}}^{\mathrm{GS}}(\mathbf{f}, w, s)}{(s - w/2)^{r_{\mathrm{gen}}(\mathbb{Q})}}$ *is not identically zero.*

Where exceptional zeros do occur, the natural approach to adopt when studying arithmetic is through the application of what are now called 'Selmer complexes'. These complexes were introduced by Nekovář in [Ne3]. Amongst other things, they reflect stability for the $\mathbb{Z}_p$-corank of Selmer at split multiplicative primes.

There is a very general notion of p-adic height pairing in this context, and both $\langle -, - \rangle_{\mathbb{Q},p}^{\mathrm{wt}}$ and $\langle -, - \rangle_{\mathbb{Q},p}^{-1/2}$ are just earthly manifestations of his formal dualities. For example if $\mathcal{R}_E = \Lambda^{\mathrm{wt}}$, the existence of a non-degenerate alternating pairing on $\mathrm{Sel}_{\mathbb{Q}}(\mathbb{T}_{\infty,E}^{-1/2})$ guarantees the algebraic L-function $\text{Ш}_{\mathbb{Q}}(\mathbb{T}_{\infty,E}^{-1/2})$ is actually a square. Moreover, to eventually prove Conjecture 10.3 it is probably necessary to dispense with Selmer groups entirely, and work instead with complexes (to read about this viewpoint further we recommend the monograph [MR]).

Heuristically one expects the vast majority of elliptic curves over $\mathbb{Q}$, to either have rank zero or to have rank one (although ranks > 1 also exhibit positive density). We already discussed in this section how to bound $\mu_{p,\mathbb{Q}}^{\text{gen}}$ in the rank zero scenario.

Now let E be a rank one elliptic curve defined over the rationals.

Problem 10.3. *Find a formula (upper bound) relating $\mu_{p,\mathbb{Q}}^{\text{gen}}(\rho_{\infty,E})$ to the vertical height $\langle P, P\rangle_{\mathbb{Q},p}^{\text{wt}}$, for some point P of infinite order in $E(\mathbb{Q})$.*

Over an imaginary quadratic extension $\mathbb{Q}(\sqrt{-D})$ in which the conductor N_E splits completely, we have a Gross-Zagier formula relating the derivative of the improved p-adic L-function at $s = 1$ with the vertical height of a Heegner point (see [Db3]). The techniques underpinning the proof are based upon work of Bertolini-Darmon and Howard [BD,Ho] in the anti-cyclotomic setting. Of course, developing such a numerical algorithm to compute the vertical height is quite another matter.

The solution to the following is probably still some way off.

Problem 10.4. *Prove the two-variable Main Conjecture in the non-CM case.*

Perhaps more realistically, one could first attempt to show either the vertical or half-twisted main conjectures. The latter are offspring of the two-variable version and should be a lot more tractable. For the vertical main conjecture, we have already established (in certain special cases) the divisibility of $\text{Ш}_{\mathbb{Q}}(\mathbb{T}_{\infty,E})$ into the improved p-adic L-function.

Finally, what happens if $p \geq 3$ is a prime of good supersingular reduction for E? Stevens (and also Panchiskin) has found analogues of the two-variable L-function. Together with Iovita he has contructed integral lattices, which play an analogous rôle to the Galois representations $(\mathbb{T}_{\infty,E})^*$. In addition, they have developed a Coleman power series interpolation for their local cohomologies.

Problem 10.5. *Extend the various arithmetic results in this book, to encompass overconvergent families of modular forms of slope $< k - 1$.*

Included as Appendix C is a one-variable control theorem governing the p^∞-Selmer groups associated to these lattices. However, it no longer makes sense to work over the local ring $\mathbb{Z}_p[\![X]\!]$, instead one works rigid analytically with Tate algebras. By associating a p-adic height pairing, in some sense grafted onto branches of the eigencurve, one would obtain regulator formulae in the spirit of 9.18, 10.1 and 10.9. This would be a significant achievement in itself.

APPENDIX A

The Primitivity of Zeta Elements

Recall from Chapter III, we omitted the proof of Theorem 3.8. In this appendix the missing argument is supplied. We assume that the reader is reasonably familiar with Kato's paper [Ka1], and follow his notations wherever possible.

As in §3.2, f shall denote a Hecke eigenform of weight $k \geq 2$ and character ϵ. Throughout we write $\mathbf{T}$ for the $G_{\mathbb{Q}}$-stable $\mathcal{O}_f$-submodule generated by the image of $H^1\Big(Y_1(N)(\mathbb{C}),\ \mathfrak{F}_p^{(k)}\Big)$ inside the p-adic realisation. If $\gamma \in V_{\mathrm{Betti}}(\mathbb{C})$, then $\gamma^{\pm}$ denotes the projection to the eigenspace on which complex conjugation acts through ± 1. Let's fix an element $\underline{\delta} = \underline{\delta}^+ + \underline{\delta}^-$ inside $\mathbf{T}$.

Theorem 3.8'. *For all integers $n \geq 0$ and every integer $M \geq 1$ coprime to p, there exist modified zeta-elements $\mathbf{z}^{(p)}_{\underline{\delta},Mp^n} \in H^1_{\text{ét}}\Big(\mathbb{Z}\big[\zeta_{Mp^n}, 1/p\big],\ V_f\Big)$ satisfying:*

(i) If $m = Mp^n$ and l is any prime number, then

$$\mathrm{cores}_{(\mu_{ml})/(\mu_m)} \mathbf{z}^{(p)}_{\underline{\delta},ml} = \begin{cases} \Big(1 - \overline{a_l(f)} l^{-k} \sigma_l^{-1} + \overline{\epsilon}(l) l^{-k-1} \sigma_l^{-2}\Big) \cdot \mathbf{z}^{(p)}_{\underline{\delta},m} & \text{if } l \nmid pm \\ \mathbf{z}^{(p)}_{\underline{\delta},m} & \text{if } l | pm; \end{cases}$$

(ii) For each index $m = Mp^n$, there exists a constant $\nu_M \geq 0$ depending on the support of M, such that $\mathbf{z}^{(p)}_{\underline{\delta},Mp^n} \in H^1_{\text{ét}}\Big(\mathbb{Z}\big[\zeta_{Mp^n}, 1/p\big],\ p^{-\nu_M}\mathbf{T}\Big)$;

(iii) For any primitive character ψ modulo Mp^n and integer $r \in \{1, \ldots, k-1\}$, choosing $\pm$ to be the sign of $\psi(-1)(-1)^{k-r-1}$ we have

$$\mathrm{per}_{\infty}\left(\sum_{\sigma \in \mathrm{Gal}\left(\mathbb{Q}(\mu_{Mp^n})/\mathbb{Q}\right)} \psi(\sigma) \exp^*\Big(\mathbf{z}^{(p)}_{\underline{\delta},Mp^n} \otimes \zeta_{p^n}^{\otimes k-r}\Big)^{\sigma}\right)^{\pm} = (2\pi i)^{k-r-1} L_{\{p\}}\big(f^*, \psi, r\big)\underline{\delta}^{\pm}.$$

Regrettably the normalisations in Theorems 3.8 and 3.8' are somewhat different. Firstly, we are considering special values of the dual cusp form f^* rather than f. Moreover, the elements $\mathbf{z}^{(p)}_{\underline{\delta},Mp^n}$ above are related to $\mathbf{z}_{Mp^n} \in H^1_{\text{ét}}\Big(\mathbb{Z}\big[\zeta_{Mp^n}, 1/p\big], V_f^*\Big)$ in §3.2 via the Tate twist $\Big(\mathbf{z}_{Mp^n}\Big)_n = \Big(\mathbf{z}^{(p)}_{\gamma,Mp^n} \otimes \zeta_{p^n}^{\otimes k-1}\Big)_n$. It is straightforward but tedious to check that the two definitions agree.

To simplify the exposition, we now assume the following two conditions hold:

(I) The cusp form f does not have complex multiplication, i.e. $L(f,s) \neq L(\theta,s)$ for any Hecke character $\theta : \mathrm{Pic}\big(\mathcal{O}_{\mathbb{Q}(\sqrt{d})}\big) \to \mathbb{C}^{\times}$ with $d < 0$;

(II) The residual representation $\overline{\rho_f} : G_{\mathbb{Q}} \longrightarrow \mathrm{Aut}\big(\mathbf{T}/\mathfrak{m}_{\mathcal{O}_f}\big)$ is irreducible.

At the end, we explain how to remove these hypotheses.

Fix matrices α_1, α_2 in $\mathrm{SL}_2(\mathbb{Z})$ and also positive integers $1 \le j_1, j_2 \le k-1$ such that $\delta(f, j_1, \alpha_1)^+ \neq 0$ and $\delta(f, j_2, \alpha_2)^- \neq 0$. Clearly, we may write $\underline{\delta}$ above in terms of

$$\underline{\delta} \;=\; b_1\delta(f, j_1, \alpha_1)^+ \;+\; b_2\delta(f, j_2, \alpha_2)^- \quad \text{for certain scalars } b_1, b_2 \in \mathcal{K}_f.$$

As before Λ_f^{cy} is the Iwasawa algebra $\mathcal{O}_f[\![G_\infty]\!]$, and Δ_M denotes the Galois group of the extension $\mathbb{Q}(\mu_M)$ over $\mathbb{Q}$. Consider the non-zero divisor of Λ_f^{cy} given by

$$\nu(c,d,j) \;=\; \left(c^2 - c^{k+1-j}.\sigma_c\right) \times \left(d^2 - d^{j+1}.\sigma_d\right) \times \prod_{l|N,\, l\neq p} \left(1 - \overline{a_l(f)}l^{-k}.\sigma_l^{-1}\right).$$

Let us choose $c, d \neq \pm 1$ satisfying $\gcd(cd, 6pM) = 1$ and $c \equiv d \equiv 1 (\mathrm{mod}\ N)$.

Definition A.1. *We define primitive zeta-elements* $\left(\mathbf{z}^{(p)}_{\underline{\delta}, Mp^n}\right)_n$ *lying inside the projective limit* $\mathrm{Frac}\left(\Lambda_f^{\mathrm{cy}}\right) \otimes_{\Lambda_f^{\mathrm{cy}}} \varprojlim_n H^1_{\text{ét}}\left(\mathbb{Z}\left[\zeta_{Mp^n}, 1/p\right], \mathbf{T}\right)$ *via*

$$\begin{aligned}\left(\mathbf{z}^{(p)}_{\underline{\delta}, Mp^n}\right)_n \;&=\; \nu(c,d,j_1)^{-1} b_1 \otimes \left({}_{c,d}\mathbf{z}^{(p)}_{Mp^n}(f, k, j_1, \alpha_1, \mathrm{supp}(pNM))\right)_n^- \\ &\quad + \nu(c,d,j_2)^{-1} b_2 \otimes \left({}_{c,d}\mathbf{z}^{(p)}_{Mp^n}(f, k, j_2, \alpha_2, \mathrm{supp}(pNM))\right)_n^+ .\end{aligned}$$

Crudely speaking, our aim is to replace the '$\mathrm{Frac}(\Lambda^{\mathrm{cy}}) \otimes_{\Lambda^{\mathrm{cy}}} -$' with a '$\mathbb{Q}_p \otimes_{\mathbb{Z}_p} -$'. We must therefore ensure that the primitive zeta-elements have no poles in the cyclotomic direction.

Theorem A.2. *Under Hypotheses (I),(II) the compact finitely-generated module* $\varprojlim_n H^1_{\text{ét}}\left(\mathbb{Z}\left[\zeta_{Mp^n}, 1/p\right], \mathbf{T}\right)$ *is free of rank one over* $\mathcal{O}_f\left[\left[\mathrm{Gal}\left(\mathbb{Q}(\mu_{Mp^\infty})/\mathbb{Q}\right)\right]\right]$.

Proof: Let ψ be any character of Δ_M. Consider the ψ-eigenspace

$$\varprojlim_n H^1_{\text{ét}}\left(\mathbb{Z}\left[\zeta_{Mp^n}, 1/p\right], \mathbf{T}\right)^{(\psi)} \;\cong\; \varprojlim_n H^1_{\text{ét}}\left(\mathbb{Z}\left[\zeta_{p^n}, 1/p\right], \mathbf{T} \otimes \psi^{-1}\right)$$

where the identification is as a pseudo-isomorphism, in the sense of Λ_f^{cy}-modules. Now $\rho_f : G_\mathbb{Q} \to \mathrm{Aut}(\mathbf{T})$ is assumed to be residually irreducible, hence so are its twists $\rho_f \otimes \psi^{-1} = \rho_{f\otimes\psi}$. By [Ka1, Thm 12.4(3)] applied to the cusp form $f \otimes \psi$, the right-hand side above is free of rank one over $\Lambda^{\mathrm{cy}}_{f,\psi}$.

□

Thus to determine $\left(\mathbf{z}^{(p)}_{\underline{\delta}, Mp^n}\right)_n$ uniquely, via Theorem A.2 it is enough to compute special values at all finite characters ψ of $\mathrm{Gal}\left(\mathbb{Q}(\mu_{Mp^\infty})/\mathbb{Q}\right)$.

Let us suppose that the conditions

$$c^2 - c^{k-j}\overline{\psi}(c) \neq 0, \quad d^2 - d^j\overline{\psi}(d) \neq 0 \quad \text{and} \quad 1 - \overline{a_l(f)}l^{1-k+j}\psi(l) \neq 0 \ \text{ if } l|N$$

are met for all $j \in \{0, \ldots, k-2\}$, which may exclude finitely many choices of ψ.

If $\psi|_{G_\infty}$ has conductor p^m, then we have a sequence of evaluation homomorphisms

$$\Lambda^{\mathrm{cy}}\left\{\frac{1}{\nu(c,d,j_1)\nu(c,d,j_2)}\right\}\otimes_{\Lambda^{\mathrm{cy}}}\varprojlim_n H^1_{\text{ét}}\Big(\mathbb{Z}[\zeta_{Mp^n},1/p],\ \mathbf{T}\Big)$$

$$\Big\downarrow \ \mathrm{proj}_m(-\otimes\zeta_{p^n}^{\otimes k-r})$$

$$\mathcal{K}_f\otimes_{\mathcal{O}_f} H^1_{\text{ét}}\Big(\mathbb{Z}[\zeta_{Mp^m},1/p],\ \mathbf{T}(k-r)\Big)\left\{\frac{1}{\mathrm{proj}_m(\underline{\nu})}\right\}$$

$$\Big\downarrow \ \textstyle\sum_{\sigma\in\Delta_{Mp^m}}\psi(\sigma)\exp^*(-)^\sigma$$

$$\overline{\mathcal{K}_f}\otimes_{\mathcal{O}_f}\Big(M_k(X_1(N))\otimes_{\lambda_{f^*}}\mathcal{O}_f\Big)$$

whose composition with the period mapping per_∞ will be abbreviated by $\mathcal{L}_{\psi,M,r}$. Note the action of complex conjugation on $\varprojlim_n H^1_{\text{ét}}\Big(\mathbb{Z}[\zeta_{Mp^n},1/p],\ \mathbf{T}\Big)$ becomes the action of $-\psi(-1)$ on the target eigenspace $\overline{\mathcal{K}_f}\otimes_{\mathcal{O}_f}\Big(M_k(X_1(N))\otimes_{\lambda_{f^*}}\mathcal{O}_f\Big)$.

Remark: By [Ka1,Thm 6.6] the image of each $\Big({}_{c,d}\mathbf{z}^{(p)}_{Mp^n}(f,k,j,\alpha,\mathrm{supp}(pNM))\Big)^{\mp}_n$ under $\mathcal{L}_{\psi,M,r}$ is precisely
$(c^2-c^{1+r-j}\overline{\psi}(c))\times(d^2-d^{j+1+r-k}\overline{\psi}(d))\times(2\pi i)^{k-r-1}L_{\{MNp\}}(f^*,\psi,r).\delta(f,j,\alpha)^{\pm}$
where $\pm$ is the sign of $\psi(-1)(-1)^{k-r-1}$.

If $\psi(-1)(-1)^{k-r-1}$ has even parity, it follows that $\mathcal{L}_{\psi,M,r}\Big(\mathbf{z}^{(p)}_{\underline{\delta},Mp^n}\Big)_n$ equals the special value $\psi\chi_{\mathrm{cy}}^{k-r}\Big(\nu(c,d,j_1)^{-1}b_1\Big)\times\mathcal{L}_{\psi,M,r}\Big({}_{c,d}\mathbf{z}^{(p)}_{Mp^n}(f,k,j_1,\alpha_1,\mathrm{supp}(pNM))\Big)^{-}_n$, i.e.

$$\psi\chi_{\mathrm{cy}}^{k-r}\Big(\nu(c,d,j_1)^{-1}\Big)\times(c^2-c^{1+r-j_1}\overline{\psi}(c))\times(d^2-d^{j_1+1+r-k}\overline{\psi}(d))$$
$$\times(2\pi i)^{k-r-1}\times\prod_{l|N,\ l\neq p}(1-\overline{a_l(f)}\psi(l)l^{-r})\times L_{\{Mp\}}(f^*,\psi,r).\ b_1\delta(f,j_1,\alpha_1)^+.$$

Cancelling out this collection of extraneous Euler factors, yields the simplification

$$\mathcal{L}_{\psi,M,r}\Big(\mathbf{z}^{(p)}_{\underline{\delta},Mp^n}\Big)_n\ =\ (2\pi i)^{k-r-1}\times L_{\{Mp\}}(f^*,\psi,r).\ \underline{\delta}^+\ .$$

The above coincides exactly with the formula given in 3.8'(iii), certainly when ψ is a primitive character modulo Mp^m, because $\psi(l)=0$ for all primes l dividing M. In similar fashion, when $\psi(-1)(-1)^{k-r-1}$ has odd parity the formulae also agree. Thus the primitive zeta-elements are independent of the choice of $(c,d)\in\mathbb{Z}\times\mathbb{Z}$, and Theorem 3.8'(iii) is established.

Fortunately, there is nothing to prove for 3.8'(i) because from [Ka1, Prop 8.12], the elements ${}_{c,d}\mathbf{z}^{(p)}_{ml}(f,k,j_1,\alpha_1,\mathrm{supp}(pNml))$ satisfy identical vertical and sideways Euler system relations. The $\mathbf{z}^{(p)}_{\underline{\delta},ml}$'s are obtained from the imprimitive ${}_{c,d}\mathbf{z}^{(p)}_{ml}$'s via scalars lying in $\mathrm{Frac}(\Lambda_f^{\mathrm{cy}})$, and trivially the Iwasawa algebra Λ_f^{cy} is commutative! This leaves us with the remaining task of justifying Theorem 3.8'(ii).

Let's start with two technicalities, proved in the same way as [Ka1, Lemma 13.10].

Fact #1. *Assume* $a, A \geq 1$ *are integers, and* $\gcd(c, 6pA) = \gcd(d, 6pN) = 1$. *Then*

$$\left({}_{c,d}\mathbf{z}^{(p)}_{Mp^n}(f,k,j,a(A),\mathrm{supp}(pAM))\right)_n = \prod_{l|A,\, l\neq p}\left(1-\overline{a_l(f)}l^{-k}\sigma_l^{-1}+\overline{\epsilon}(l)l^{-k-1}\sigma_l^{-2}\right)$$
$$\times\left\{c^2d^2\left(\mathbf{z}^{(p)}_{\gamma_1,Mp^n}\right)_n - c^{k+1-j}d^2\sigma_c\left(\mathbf{z}^{(p)}_{\gamma_2,Mp^n}\right)_n - c^2d^{j+1}\epsilon(d)\sigma_d\left(\mathbf{z}^{(p)}_{\gamma_3,Mp^n}\right)_n \right.$$
$$\left. + c^{k+1-j}d^{j+1}\epsilon(d)\sigma_{cd}\left(\mathbf{z}^{(p)}_{\gamma_4,Mp^n}\right)_n\right\}$$

where the subscripts are $\gamma_1 = \delta(f,j,a(A))$, $\gamma_2 = \delta(f,j,ac(A))$, $\gamma_3 = \delta(f,j,ad^{-1}(A))$ *and lastly* $\gamma_4 = \delta(f,j,acd^{-1}(A))$.

Fact #2. *If* $\alpha \in SL_2(\mathbb{Z})$, $\gcd(cd, 6pM) = 1$ *and* $c \equiv d \equiv 1 (\mathrm{mod}\ N)$, *then*

$$\left({}_{c,d}\mathbf{z}^{(p)}_{Mp^n}(f,k,j,\alpha,\mathrm{supp}(pNM))\right)_n = \nu(c,d,j)\times\left(\mathbf{z}^{(p)}_{\delta(f,j,\alpha),Mp^n}\right)_n.$$

We write $\mathcal{Z}_M^{\mathrm{imp}}$ for the Λ_f^{cy}-submodule of $\varprojlim_n H^1_{\text{ét}}\left(\mathbb{Z}[\zeta_{Mp^n}, 1/p], \mathbf{T}\right)$ generated by:

(1) $\left({}_{c,d}\mathbf{z}^{(p)}_{Mp^n}(f,k,j,a(A),\mathrm{supp}(pAM))\right)_n$ with $a, A \geq 1$ and c, d as in Fact #1;

(2) $\left({}_{c,d}\mathbf{z}^{(p)}_{Mp^n}(f,k,j,\alpha,\mathrm{supp}(pNM))\right)_n$ with $\alpha \in \mathrm{SL}_2(\mathbb{Z})$ and c, d as in Fact #2.

Similarly, let us write $\mathcal{Z}_M^{\mathrm{prim}} \subset \mathrm{Frac}(\Lambda^{\mathrm{cy}}) \otimes_{\Lambda^{\mathrm{cy}}} \varprojlim_n H^1_{\text{ét}}\left(\mathbb{Z}[\zeta_{Mp^n}, 1/p], \mathbf{T}\right)$ for the Λ_f^{cy}-submodule generated by $\left(\mathbf{z}^{(p)}_{\underline{\delta},Mp^n}\right)_n$ for all $\underline{\delta} \in \mathbf{T}$. We deduce immediately from #1 and #2, that $\mathcal{Z}_M^{\mathrm{imp}}$ is contained inside $\mathcal{Z}_M^{\mathrm{prim}}$.

Key Claim: The quotient $\mathcal{Z}_M^{\mathrm{prim}}/\mathcal{Z}_M^{\mathrm{imp}}$ is a finite p-group.

Assuming the truth of this assertion, we may set $\nu_M := \mathrm{ord}_p\left(\#\frac{\mathcal{Z}_M^{\mathrm{prim}}}{\mathcal{Z}_M^{\mathrm{imp}}}\right)$. Then

$$p^{\nu_M}.\left(\mathbf{z}^{(p)}_{\underline{\delta},Mp^n}\right)_n \in \mathcal{Z}_M^{\mathrm{imp}} \subset \varprojlim_n H^1_{\text{ét}}\left(\mathbb{Z}[\zeta_{Mp^n}, 1/p], \mathbf{T}\right)$$

and Theorem 3.8'(ii) is proved at all layers n, simultaneously.

In order to justify our Key Claim, we mimic Kato's argument in the case $M = 1$. By localisation considerations, it is enough to establish that $\mathcal{Z}_M^{\mathrm{imp}} \otimes_{\Lambda_f^{\mathrm{cy}}} \left(\Lambda_f^{\mathrm{cy}}\right)_{\mathfrak{p}}$ equals $\mathcal{Z}_M^{\mathrm{prim}} \otimes_{\Lambda_f^{\mathrm{cy}}} \left(\Lambda_f^{\mathrm{cy}}\right)_{\mathfrak{p}}$ for any height one prime ideal $\mathfrak{p} \in \mathrm{Spec}(\Lambda_f^{\mathrm{cy}})$, $p \notin \mathfrak{p}$.

Suppose that we have made such a choice of $\mathfrak{p}$, so we obtain a homomorphism $h_{\mathfrak{p}} : \Lambda_f^{\mathrm{cy}} \to \Lambda_f^{\mathrm{cy}}/\mathfrak{p} \hookrightarrow \overline{K_f}$. Pick a pair of integers $(c, d) \in \mathbb{Z} \times \mathbb{Z}$ satisfying the conditions $\gcd(c, 6pM) = \gcd(d, 6pMN) = 1$ and $c^2 \neq 1 \neq d^2$.

Fact #3. *For a sufficiently large power* A *of* p, *there exists a Dirichlet character* $\theta : (\mathbb{Z}/A\mathbb{Z})^\times \to \overline{\mathbb{Q}}^\times$ *such that:*

- $L_{\{Mp\}}(f^*, \theta^{-1}, k-1) \neq 0$;
- $c^2 - c^2\theta^{-1}(c)h_{\mathfrak{p}}(\sigma_c) \neq 0$ *and* $d^2 - d^k\epsilon(d)\theta(d)h_{\mathfrak{p}}(\sigma_d) \neq 0$.

In fact if $k \geq 3$, any choice of character ensures the twisted L-value is non-zero. Moreover, we can always choose θ so that $\theta(-1) = \pm 1$, where $\mp$ denotes the image of complex conjugation inside $\Lambda_f^{\mathrm{cy}}/\mathfrak{p}$.

Fix an $\mathcal{O}_f$-integral $\underline{\delta} \in V_{\mathrm{Betti}}$. By [Ka1, 13.11(2)] since $L_{\{Mp\}}(f^*, \theta^{-1}, k-1)$ is non-zero, there exists $b \in \overline{\mathbb{Q}}$ such that $\underline{\delta}^{\pm} = b \times \sum_{a \in (\mathbb{Z}/A\mathbb{Z})^\times} \theta(a)\delta(f, k-1, a(A))$.

Let L indicate the field extension generated over $\mathrm{Frac}\left(\Lambda_f^{\mathrm{cy}}/\mathfrak{p}\right)$ by the values of θ. We shall define $\mathfrak{p}' := \mathrm{Ker}\left(\mathcal{O}_L[\![G_\infty]\!] \xrightarrow{h_\mathfrak{p}} L\right)$ to be the thickening of the ideal $\mathfrak{p}$. Working inside $\mathrm{Frac}\left(\left(\Lambda_f^{\mathrm{cy}}\right)_\mathfrak{p}\right) \otimes_{\Lambda_f^{\mathrm{cy}}} \varprojlim_n H^1_{\text{ét}}\left(\mathbb{Z}[\zeta_{Mp^n}, 1/p], \mathbf{T}\right) \otimes_{\mathcal{O}_f} L$, one finds

$$\left(\mathbf{z}^{(p)}_{\underline{\delta}, Mp^n}\right)_n = \left(\mathbf{z}^{(p)}_{\underline{\delta}^{\pm}, Mp^n}\right)_n = b \times \sum_{a \in (\mathbb{Z}/A\mathbb{Z})^\times} \theta(a) \left(\mathbf{z}^{(p)}_{\delta(f,k-1,a(A)), Mp^n}\right)_n.$$

On the other hand, a simple generalisation of [Ka1, Lemma 13.11(1)] shows that $\sum_{a \in (\mathbb{Z}/A\mathbb{Z})^\times} \theta(a) \left(\mathbf{z}^{(p)}_{\delta(f,k-1,a(A)), Mp^n}\right)_n$ coincides with the summation

$$h_\mathfrak{p}(\mathfrak{u})^{-1} \times \sum_{a \in (\mathbb{Z}/A\mathbb{Z})^\times} \theta(a) \left({}_{c,d}\mathbf{z}^{(p)}_{Mp^n}(f, k, j, a(A), \mathrm{supp}(pAM))\right)_n,$$

where the non-zero divisor

$$\mathfrak{u} = \left(c^2 - c^2\theta^{-1}(c)\sigma_c\right)\left(d^2 - d^k\epsilon(d)\theta(d)\sigma_d\right) \prod_{l|A,\, l \neq p} \left(1 - \overline{a_l(f)} l^{-k}\sigma_l^{-1} + \overline{\epsilon}(l) l^{-k-1}\sigma_l^{-2}\right).$$

By Fact #3 above, our choice of θ ensured that $h_\mathfrak{p}(\mathfrak{u})$ is a unit of $\mathcal{O}_L[\![G_\infty]\!]_{\mathfrak{p}'}$. It follows that $\left(\mathbf{z}^{(p)}_{\underline{\delta}, Mp^n}\right)_n$ belongs to the localised module $\mathcal{Z}_M^{\mathrm{imp}} \otimes_{\Lambda_f^{\mathrm{cy}}} \mathcal{O}_L[\![G_\infty]\!]_{\mathfrak{p}'}$, whence

$$\mathcal{Z}_M^{\mathrm{prim}} \otimes_{\Lambda_f^{\mathrm{cy}}} \left(\Lambda_f^{\mathrm{cy}}\right)_\mathfrak{p} \subset \mathcal{Z}_M^{\mathrm{imp}} \otimes_{\Lambda_f^{\mathrm{cy}}} \left(\Lambda_f^{\mathrm{cy}}\right)_\mathfrak{p} \quad \text{as required.}$$

The proof of Theorem 3.8'(ii) is finished.

Remarks: (a) Hypothesis (I) requiring f to be without complex multiplication, is not actually necessary. However in the proof of Theorem A.2, one needs to utilise the results in [Ka1, Section 15] instead.

(b) Likewise Hypothesis (II) may be removed, provided one replaces Theorem A.2 with the weaker statement that $\mathbb{Q} \otimes \varprojlim_n H^1_{\text{ét}}\left(\mathbb{Z}[\zeta_{Mp^n}, 1/p], \mathbf{T}\right)$ is free of rank one over $\mathbb{Q} \otimes \mathcal{O}_f\left[\left[\mathrm{Gal}(\mathbb{Q}(\mu_{Mp^\infty})/\mathbb{Q})\right]\right]$. This causes the constants of integrality p^{ν_M} to increase slightly, sufficient to kill off the excess p^∞-torsion in the projective limit.

(c) Even though the special values of the $\mathbf{z}^{(p)}_{\underline{\delta}, Mp^n}$'s have the missing Euler factors at $l|N$ restored, they are totally useless for bounding the ranks of Selmer groups. The reason is that the lattices $p^{-\nu_M}\mathbf{T}$ are not fixed as the support of M grows, so one cannot construct global derivative classes in the sense of Kolyvagin [Kv].

APPENDIX B

Specialising the Universal Path Vector

In Chapter VI we had stated (without proof) two technical lemmas on zeta-elements. The first of these was a renormalisation result, necessary to define a canonical Kato-Beilinson element with $\mathcal{R}$-adic coefficients. The second result was used when showing that the whole $\mathcal{R}$-adic Euler system, itself had no cyclotomic poles.

Without further ado, let us fill in the missing demonstrations.

Lemma B.1. *There exists a vector* $\underline{\delta}_{1,Np^\infty}^{\mathrm{univ}\ *}$ *in the space* $\overline{\mathcal{MS}}^{\mathrm{ord}}(Np^\infty)$, *such that*

$$\left(\underline{\delta}_{1,Np^\infty}^{\mathrm{univ}\ *} \otimes {}_{c,d}\mathfrak{Z}_{Mp^n}^{\dagger}\right)[\lambda] \mod P_{k,\epsilon} \quad = \quad \nabla_{k-2}\left(\gamma^*_{\mathbf{f}^*_{P_{k,\epsilon}}}\right) \otimes {}_{c,d}\mathbf{z}_{Mp^n}$$

for all primitive $\lambda : \mathbf{h}^{\mathrm{ord}}(Np^\infty; \mathcal{O}_{\mathcal{K}}) \otimes_{\Lambda^{\mathrm{wt}}_{\mathcal{K}}} \mathbf{I} \to \mathbf{I}$, *and points* $P_{k,\epsilon} \in \mathrm{Spec}\,\mathbf{I}(\mathcal{O}_{\mathcal{K}})^{\mathrm{alg}}$.

Proof: Again we are forced to assume that the reader is familiar with the notations and conventions of the mammoth article [Ka1]. Recall in §2.5, we introduced certain path elements $\delta_{1,Np^r}(k,j',a(\mathbf{a}))$ lying inside of $H^1\Big(Y_1(Np^r)(\mathbb{C}),\ \mathfrak{F}^{(k)}\Big)$ as follows. Firstly, let β_1, β_2 be sections of $\phi^*_{a(\mathbf{a})}\mathrm{Sym}^{k-2}_{\mathbb{Z}}(\underline{H}^1)$ where $\phi_{a(\mathbf{a})} : (0,\infty) \to Y_1(Np^r)$ sent $y \mapsto \mathrm{pr}\big((yi+a)/\mathbf{a}\big)$ – the stalk of β_1 at y was yi, and the stalk of β_2 was 1. Then $\delta_{1,Np^r}(k,j',a(\mathbf{a}))$ was the image of the class of $\Big(\phi_{a(\mathbf{a})},\ \beta_1^{j'-1}\beta_2^{k-j'-1}\Big)$ under

$$\begin{array}{ccc} H_1\Big(X_1(Np^r)(\mathbb{C}),\ \{\mathrm{cusps}\},\ \mathrm{Sym}^{k-2}_{\mathbb{Z}}(\underline{H}^1)\Big) & \cong & H^1\Big(Y_1(Np^r)(\mathbb{C}),\ \mathrm{Sym}^{k-2}_{\mathbb{Z}}(\underline{H}^1)\Big) \\ & & \Big\downarrow \text{trace} \\ & & H^1\Big(Y_1(Np^r)(\mathbb{C}),\ \mathfrak{F}^{(k)}\Big). \end{array}$$

In the situation of Chapter VI, we had subsequently taken $(\mathbf{a}, a) = (1, 0)$.

An advantage of the p-ordinary control theory is that to obtain elements at different weights k, it is enough to deform elements at weight $k = 2$ and then vary the levels Np^r uniformly. We thus work at weight two throughout this proof. Consider the compatible sequence

$$\Big(\phi_{a(\mathbf{a})},\ \beta_1^{j'-1}\beta_2^{k-j'-1}\Big)_{r\ge 1} \ \in\ \varprojlim_r H_1\Big(X_1(Np^r)(\mathbb{C}),\ \{\mathrm{cusps}\},\ \mathrm{Sym}^{k-2}_{\mathbb{Z}}(\underline{H}^1)\Big);$$

whenever $(k,j') = (2,1)$ and $(\mathbf{a}, a) = (1,0)$, its image $\underline{\delta}_{1,Np^\infty}^{\mathrm{univ}}$ lies inside $\mathcal{UM}(\mathbb{Z}_p)$. Since $\overline{\mathcal{MS}}^{\mathrm{ord}}$ is $\mathbb{Z}_p$-dual to $\mathcal{UM}^{\mathrm{ord}}$, denote by $\underline{\delta}_{1,Np^\infty}^{\mathrm{univ}\ *}$ the dual base to $\mathbf{e}.\ \underline{\delta}_{1,Np^\infty}^{\mathrm{univ}}$. Fortunately after this type of normalisation, the periods behave well modulo $P_{k,\epsilon}$.

Remark: We shall now try to show for primitive $\lambda : \mathbf{h}^{\mathrm{ord}}(Np^{\infty};\mathcal{O}_{\mathcal{K}}) \otimes_{\Lambda^{\mathrm{wt}}_{\mathcal{K}}} \mathbf{I} \to \mathbf{I}$,

$$\left(\underline{\delta}^{\mathrm{univ}}_{1,Np^{\infty}}{}^{*} \otimes {}_{c,d}\mathfrak{Z}^{\dagger}_{Mp^n}\right)[\lambda] \mod P_{k,\epsilon} \;=\; \nabla_{k-2}(\gamma^*) \otimes {}_{c,d}\mathbf{z}_{Mp^n}$$

where ${}_{c,d}\mathbf{z}_{Mp^n} \in H^1\left(\mathbb{Q}(\mu_{Mp^n}), V^*_{\mathbf{f}_{P_{k,\epsilon}}}\right)$ represent the integral but (c,d)-dependent p-adic zeta-elements. In fact, it is sufficient to verify

$$\gamma_{\mathbf{f}^*_{P_{k,\epsilon}}} \otimes \nabla^{-1}_{k-2}\left(\left(\underline{\delta}^{\mathrm{univ}}_{1,Np^{\infty}}{}^{*} \otimes {}_{c,d}\mathfrak{Z}^{\dagger}_{Mp^n}\right)[\lambda] \mod P_{k,\epsilon}\right)$$

shares the same special values as ${}_{c,d}\mathbf{z}_{Mp^n}$ under the dual exponential maps.

Examining its construction closely, the Tate twist ${}_{c,d}\mathfrak{Z}^{\dagger}_{Mp^n} \otimes \zeta^{\otimes -j}_{p^n}[\lambda] \mod P_{k,\epsilon}$ must coincide with the image of the K_2-system

$$\left({}_{c,d}z_{Mp^{n+m},MNp^{n+r+m}}\right)_{r,m} \quad \text{under the map} \quad \varepsilon^{\otimes -j} \otimes \mathfrak{c}^{\mathrm{univ}}_{\infty,p}[\lambda] \mod P_{k,\epsilon},$$

i.e. under the realisation map $\mathfrak{c}_{k,j,k-1,0(1),p}$ which was described in Definition 2.14. Furthermore, the norm-compatible elements $\left({}_{c,d}z_{-,-}\right)_{r,m}$ are sent via $\mathfrak{c}_{k,j,k-1,0(1),p}$ to their cohomological versions

$${}_{c,d}\mathbf{z}^{(p)}_{\underline{\delta},Mp^n} \otimes \zeta^{\otimes -j}_{p^n} \;\in\; H^1\left(\mathbb{Q}(\mu_{Mp^n}),\, \mathbf{T}(-j)\right) \quad \text{with } \underline{\delta} = \delta_{1,Np^r}(k,k-1,0(1)).$$

Let ψ be a Dirichlet character modulo Mp^n, and choose a sign $\pm = \psi(-1)(-1)^j$. Then

$$\mathrm{per}_{\infty}\left(\sum_{b\in(\mathbb{Z}/Mp^n\mathbb{Z})^{\times}} \psi(b)\exp^*\left({}_{c,d}\mathbf{z}^{(p)}_{\underline{\delta},Mp^n} \otimes \zeta^{\otimes -j}_{p^n}\right)^{\sigma_b}\right)^{\pm}$$

takes the precise value

$$(c^2-c^2\psi^{-1}(c))\,(d^2-d^k\psi^{-1}(d))\,(2\pi i)^{k-2-j}L_{\{Np\}}(\mathbf{f}_{P_{k,\epsilon}},\psi,1+j).\delta_{1,Np^r}(k,k-1,0(1))^{\pm}$$

inside the Betti realisation. We also know that

$$\begin{aligned}(2\pi i)^{2-k}\mathrm{per}_{\infty}(\mathbf{f}^*_{P_{k,\epsilon}}) &= \Omega^+_{\mathbf{f}_{P_{k,\epsilon}}}\gamma^+ + \Omega^-_{\mathbf{f}_{P_{k,\epsilon}}}\gamma^- \\ &= \Omega^+_{0(1)}\delta_{1,Np^r}(k,k-1,0(1))^+ + \Omega^-_{0(1)}\delta_{1,Np^r}(k,k-1,0(1))^-\end{aligned}$$

hence one may replace $\delta_{1,Np^r}(k,k-1,0(1))^{\pm}$ above with $\dfrac{\Omega^{\pm}_{\mathbf{f}_{P_{k,\epsilon}}}}{\Omega^{\pm}_{0(1)}} \times \gamma^{\pm}$. Conversely,

$$\begin{aligned}\left\langle \gamma^{\pm}_{\mathbf{f}^*_{P_{k,\epsilon}}},\, \nabla^{-1}_{k-2}\left(\underline{\delta}^{\mathrm{univ}}_{1,Np^{\infty}}{}^{*}\right)\right\rangle &= \left\langle \nabla_{k-2}\left(\frac{\Omega^{\pm}_{0(1)}}{\Omega^{\pm}_{\mathbf{f}_{P_{k,\epsilon}}}}.\delta_{1,Np^r}(k,k-1,0(1))^{\pm}\right),\, \underline{\delta}^{\mathrm{univ}}_{1,Np^{\infty}}{}^{*}\right\rangle \\ &= \frac{\Omega^{\pm}_{0(1)}}{\Omega^{\pm}_{\mathbf{f}_{P_{k,\epsilon}}}} \times \left\langle \delta_{1,Np^r}(2,1,0(1))^{\pm},\, \underline{\delta}^{\mathrm{univ}}_{1,Np^{\infty}}{}^{*}[\lambda] \mod P_{k,\epsilon}\right\rangle = \frac{\Omega^{\pm}_{0(1)}}{\Omega^{\pm}_{\mathbf{f}_{P_{k,\epsilon}}}}.\end{aligned}$$

These two period ratios cancel each other out, so we conclude that the special values of $\left(\underline{\delta}^{\mathrm{univ}}_{1,Np^{\infty}}{}^{*} \otimes {}_{c,d}\mathfrak{Z}^{\dagger}_{Mp^n}\right)[\lambda] \mod P_{k,\epsilon}$ and of $\nabla_{k-2}(\gamma^*) \otimes {}_{c,d}\mathbf{z}_{Mp^n}$ are identical.

□

Lemma B.2. *The* $\nu_M\left(\mathbf{f}_{P_{k,\epsilon}}, \mathbf{T}\right)$*'s are locally constant over weight-space.*

Proof: Fix the prime number $p \geq 5$, the tame level N, and a positive integer M. Let's further assume the p-stabilised eigenform $\mathbf{f}_{P_{k,\epsilon}}$ has level Np^r for some $r \geq 1$. By its very definition, the quantity $\nu_M\left(\mathbf{f}_{P_{k,\epsilon}}, \mathbf{T}\right)$ was the index

$$\left[\mathcal{Z}_M^{\mathrm{prim}}\left(\mathbf{f}_{P_{k,\epsilon}}, \mathbf{T}\right) : \mathcal{Z}_M^{\mathrm{imp}}\left(\mathbf{f}_{P_{k,\epsilon}}, \mathbf{T}\right)\right] \quad \text{for each lattice } \mathbf{T} = \mathbb{T}_\infty[\lambda] \otimes_{\mathbf{I}} \mathbf{I}/P_{k,\epsilon}.$$

Recall that $\mathcal{Z}_M^{\mathrm{prim}}$ denoted the Λ_f^{cy}-submodule of $\mathbb{Q} \otimes \varprojlim_n H^1_{\mathrm{\acute{e}t}}\Big(\mathbb{Z}[\zeta_{Mp^n}, 1/p], \mathbf{T}\Big)$ generated by the $\left(\mathbf{z}^{(p)}_{\underline{\delta}, Mp^n}\right)_n$'s with $\underline{\delta} \in \mathbf{T}$. Likewise $\mathcal{Z}_M^{\mathrm{imp}}$ was generated by both

(1) $\left({}_{c,d}\mathbf{z}^{(p)}_{Mp^n}(f, k, j, a(A), \mathrm{supp}(pAM))\right)_n$ with $a, A \geq 1$;

(2) $\left({}_{c,d}\mathbf{z}^{(p)}_{Mp^n}(f, k, j, \alpha, \mathrm{supp}(pNM))\right)_n$ with $\alpha \in \mathrm{SL}_2(\mathbb{Z})$.

The precise background information on these is given in Appendix A.

One now examines in detail, integrality properties of the étale realisation maps $\mathfrak{c}^{\mathrm{univ}}_{\infty,p}$ and $\mathfrak{c}_{k,1,k-1,0(1),p}$ (rather painstakingly defined in §6.1 and §2.5, respectively). After analysis, the above indices $\nu_M(\dots)$ depend on two fundamental variations:

- the relative index of $\mathbb{T}_\infty[\lambda] \otimes_{\mathbf{I}} \mathbf{I}/P_{k,\epsilon}$ in $H^1_{\mathrm{\acute{e}t}}\left(Y_1(Np^r)_{\overline{\mathbb{Q}}},\ \mathrm{Sym}^{k-2}_{\mathbb{Z}_p}\underline{H}^1_p\right)[\lambda_{P_{k,\epsilon}}]$;
- the Manin-Drinfeld constant of integrality, required to split the inclusion

$$\mathbb{Q}_p \otimes_{\mathbb{Z}_p} H^1_{\mathrm{\acute{e}t}}\left(X_1(Np^r)_{\overline{\mathbb{Q}}},\ \mathrm{Sym}^{k-2}_{\mathbb{Z}_p}\underline{H}^1_p\right) \hookrightarrow \mathbb{Q}_p \otimes_{\mathbb{Z}_p} H^1_{\mathrm{\acute{e}t}}\left(Y_1(Np^r)_{\overline{\mathbb{Q}}},\ \mathrm{Sym}^{k-2}_{\mathbb{Z}_p}\underline{H}^1_p\right).$$

The latter constant is an invariant of the Hecke algebra at weight k and level Np^r acting on the ordinary component. The former index depends on the structure of the parabolic cohomology, with coefficients in the p-adic sheaf $\mathfrak{F}_p^{(k)}$.

Therefore the result will follow, provided we can show both the Hecke algebra $h_k^{\mathrm{ord}}(\Gamma_1(Np^r), \epsilon; \mathcal{O}_\mathcal{K})$ and the cohomologies $\mathbf{e}.H^1_{\mathrm{par}}(\Gamma_1(Np^r), \mathrm{Sym}^{k-2}_{\mathbb{Z}_p})$, are locally controlled over the weight-space. We now quote two key control theorems.

- [Hi1, Corr 4.5] – There are isomorphisms for each $r > s > 0$:

$$\begin{aligned}
\mathrm{res} : \mathbf{e}.H^1_{\mathrm{par}}\left(\Phi_r^s, \mathrm{Sym}^{k-2}(\mathbb{Z}/p^r\mathbb{Z})\right) &\xrightarrow{\sim} \mathbf{e}.H^1_{\mathrm{par}}\left(\Gamma_1(Np^s), \mathrm{Sym}^{k-2}(\mathbb{Z}/p^r\mathbb{Z})\right)\\
\mathrm{res} : \mathbf{e}.H^1_{\mathrm{par}}\left(\Phi_1, \mathrm{Sym}^{k-2}(\mathbb{Z}/p^r\mathbb{Z})\right) &\xrightarrow{\sim} \mathbf{e}.H^1_{\mathrm{par}}\left(\Phi_r, \mathrm{Sym}^{k-2}(\mathbb{Z}/p^r\mathbb{Z})\right)\\
\iota_r \circ \mathrm{res} : \mathbf{e}.H^1_{\mathrm{par}}\left(\Phi_1, \mathrm{Sym}^{k-2}(\mathbb{Z}/p^r\mathbb{Z})\right) &\xrightarrow{\sim} \mathbf{e}.H^1_{\mathrm{par}}\Big(\Phi_r,\ \mathbb{Z}/p^r\mathbb{Z} \otimes \mathcal{X}_{k-2}\Big).
\end{aligned}$$

- [Oh, Thm 3.1.3] – For some neighborhood $\mathbb{D} \subset \mathbb{Z}_p$ of k, there are isomorphisms

$$\mathbf{e}.\mathbf{h}(Np, \epsilon; \mathcal{O}_\mathcal{K}) \xrightarrow{\sim} h^{\mathrm{ord}}_{k'}(Np^\infty, \epsilon\omega^{k-k'}; \mathcal{O}_\mathcal{K}) \quad \text{for all } k' \in \mathbb{D} \text{ with } k' \geq 2.$$

It follows there exists a sufficiently small neighborhood $\mathbb{D}' = \mathbb{D} \cap \mathbb{U}_{k_0}$ containing the weight k, so that both the parabolic cohomology and Hecke algebras are locally isomorphic for all arithmetic weights $k' \in \mathbb{D}'$. Since weight-space $\mathbb{U}_{k_0}$ is completely covered by such disks $\mathbb{D}'$, the lemma is true.

□

APPENDIX C

The Weight-Variable Control Theorem

by Paul A. Smith

The purpose of this appendix is to prove control theorems for Selmer groups over affinoid algebras. The obvious application is to representations arising from analytic families of automorphic forms. There already exist numerous control theorems for representations over standard deformation rings due to Mazur, Hida, and others. Note that the switch to affinoids forces us to exploit the primary decomposition of modules over a Tate algebra, rather than Iwasawa's finer structure theory.

The payback is that one can now study overconvergent families of positive slope. Suppose we are given a big Selmer group 'Sel_∞' associated to an analytic family. One needs to confirm the specialisation of Sel_∞ at an arithmetic weight $k \geq 2$ coincides with the usual Selmer group 'Sel_k' associated to each p-stabilisation at k, at least *to within some finite bound.*

Using the local conditions described in [DS], one may define both compact and discrete Selmer groups quite generally for representations over affinoid algebras. We then prove (Theorem C.15) that for all bar a finite set of bad weights X'_{bad}, the specialisation of this big Selmer group at a de Rham weight k will equal the Bloch-Kato Selmer group, up to a controlled factor of course.

We begin with some general comments regarding the terminology used here, which differs somewhat from the main text.

C.1 Notation and assumptions

Let F be a finite extension of $\mathbb{Q}$, $p \geq 3$ a prime, and K a finite extension of $\mathbb{Q}_p$. We will write $\mathcal{O}_K$ to represent the ring of integers of K. For a fixed weight $k_0 \geq 2$, let $\mathbb{U}_{k_0}$ be a neighbourhood of k_0 in $\mathbb{Z}_p$.

Definition C.1. *We define the affinoid algebra of power series defined over K and converging on $\mathbb{U}_{k_0}$ as*

$$K\langle\langle s\rangle\rangle_{\mathbb{U}_{k_0}} \ := \ \left\{ f(s) = \sum_{n\geq 0} a_n s^n : a_n \in K, f(s) \textit{ converges for all } s \in \mathbb{U}_{k_0} \right\}.$$

Via an affine linear transformation, one may identify this with the *Tate algebra*

$$K\langle\langle s\rangle\rangle \ := \ \left\{ \sum_{n\geq 0} a_n s^n : a_n \in K, |a_n|_p \to 0 \text{ as } n \to \infty \right\}.$$

In an analogous fashion, for any order $\mathcal{O} = \mathbb{Z} + p^m\mathcal{O}_K$ then $\mathcal{O}\langle\langle s\rangle\rangle$ denotes the subring of $K\langle\langle s\rangle\rangle$ for which all coefficients $a_n \in \mathcal{O}$.

The primary object of study will be a free G_F-representation $\mathbb{W}$, unramified outside a finite set of places, and of positive rank over $K\langle\langle s\rangle\rangle$. In particular, let's take our bad set of primes Σ to include all primes of F which are either divisible by p, are archimedean, or at which $\mathbb{W}$ is ramified. We shall need to consider completions of F at each $v \in \Sigma$, and we denote these fields by F_v.

Definition C.2. *For any irreducible distinguished polynomial $\lambda \in K\langle\langle s\rangle\rangle$, define the specialisation of $\mathbb{W}$ at λ to be*

$$\mathbb{W}_\lambda \ := \ \mathbb{W} \otimes \frac{K\langle\langle s\rangle\rangle}{\lambda K\langle\langle s\rangle\rangle}.$$

In particular, we often take $\lambda = \lambda_k := s - (k - k_0)$ for any $k \in \mathbb{U}_{k_0}$.

We will assume that $\mathbb{W}$ satisfies the following additional properties:

(1) $\mathbb{W}$ is *pseudo-geometric* at every prime $v|p$, i.e.

(a) For each $v|p$, there exists a finite extension L of F_v with $L \subset F_{v,\infty}$, such that as a G_L-module $\mathbb{W}$ satisfies

$$\mathbb{W} \otimes_{K\langle\langle s\rangle\rangle} \mathbb{C}_p\langle\langle s\rangle\rangle[s^{-1}] \ \cong \ \bigoplus_{i,j \in \mathbb{Z}} \mathbb{C}_p\langle\langle s\rangle\rangle[s^{-1}](\chi^i \Psi^j)^{\oplus e_{i,j}^{(v)}}.$$

Here $e_{i,j}^{(v)} \in \mathbb{Z}_{\geq 0}$ for all pairs $(i,j) \in \mathbb{Z}^2$, with only finitely many non-zero.

(b) For every place $v|p$ and all but finitely many $k \in \mathbb{U}_{k_0} \cap \mathbb{N}$, the representation $\mathbb{W}_{\lambda_k}$ is de Rham.

(2) There exists a lattice $\mathbb{L}$ of $\mathbb{W}$ defined over $\mathcal{O}\langle\langle s\rangle\rangle$ for some $\mathcal{O} = \mathbb{Z} + p^m\mathcal{O}_K$; in other words, $\mathbb{L}$ is a free $\mathcal{O}\langle\langle s\rangle\rangle$-module with an action of G_F such that as $K\langle\langle s\rangle\rangle[G_F]$-modules,

$$\mathbb{L} \otimes_{\mathcal{O}\langle\langle s\rangle\rangle} K\langle\langle s\rangle\rangle \ = \ \mathbb{W}.$$

(3) Finally, for primes $v \in \Sigma$ which do not lie over p, we assume that both $\mathbb{W}^{G_{F_v}}$ and $\mathbb{W}^*(1)^{G_{F_v}}$ are zero, where $\mathbb{W}^* := \mathrm{Hom}_{K\langle\langle s\rangle\rangle}\big(\mathbb{W}, K\langle\langle s\rangle\rangle\big)$.

Examples: (i) The $G_\mathbb{Q}$-representations $\mathbb{T}_{\infty,\underline{v}}$ over $K\langle\langle s\rangle\rangle_{\mathbb{U}_{k_0}}$, which were the image of the Hecke eigenspaces $\mathbb{T}_\infty[\lambda]$ under the analytic transform $\widetilde{\mathrm{Mell}}_{k_0,\epsilon}$ for a primitive triple $\underline{v} = (\lambda, k_0, \epsilon)$ (see Definition 7.1).

(ii) The vertical representations $\mathbb{T}_{\infty,E}$ deforming $\rho_{E,p}$, are just special cases of (i).

(iii) The overconvergent Galois representations attached to certain Coleman families of modular forms of fixed positive slope; the integral lattices $\mathbb{L}$ themselves are constructed by Iovita and Stevens in [IS].

(iv) Tensor products, direct sums/summands of representations satisfying (1)–(3).

N.B. The half-twisted families $\mathbb{T}_{\infty,\underline{v}} \otimes \Psi^{-1/2}$ are **not** covered by the above situation.

We shall work mostly with the lattice $\mathbb{L}$, rather than $\mathbb{W}$ itself. Let λ be an irreducible distinguished polynomial whose coefficients lie in $\mathcal{O}$. In particular, we will often take $\lambda = \lambda_k := (s - (k - k_0))$, for some integral weight $k \in \mathbb{U}_{k_0}$. Let us consider the effect of multiplication by λ on various $\mathcal{O}\langle\langle s\rangle\rangle$-modules. For a given module M, we define $M_\lambda := M/\lambda M$ which is a module over the ring $\mathcal{O}_\lambda := \frac{\mathcal{O}\langle\langle s\rangle\rangle}{\lambda\mathcal{O}\langle\langle s\rangle\rangle}$.

Our methods are closely related to those of Nekovář and Plater in their article [NP, Sect.4], where they prove a similar result for the half-twisted form $\mathbb{T}_{\infty,E}^{-1/2}$. Moreover, our main theorem transplants the one-variable weight control theorems due to Hida et al, to Galois representations defined over affinoid algebras.

C.2 Properties of affinoids

We start by reviewing some properties of affinoid algebras. Fully detailed proofs of all results stated here can be found in [Sm]. The following lemma will play a crucial role in our proof of the control theorem.

Lemma C.3. *Let $f, g \in \mathcal{O}\langle\langle s\rangle\rangle$ be coprime elements. Then the ideal they generate (f, g) is of finite index in $\mathcal{O}\langle\langle s\rangle\rangle$.*

Proof: The argument relies on the fact that we have both a division theorem and a Weierstrass preparation theorem for $\mathcal{O}\langle\langle s\rangle\rangle$-modules. Using these, one may assume without loss of generality that f and g are polynomials. We say that a polynomial $h = \sum_{i=0}^{n} a_i s^i \in \mathcal{O}[s]$ has Weierstass degree t if $|a_t|_p = \max_{i\geq 0} |a_i|_p$ and $|a_t|_p > \max_{i>t} |a_i|_p$. Let $h \in (f, g)$ be of minimal Weierstrass degree. Again one may assume by the preparation theorem that h is of the form $h = \pi^\nu H$, with $H = 1$ or H a monic polynomial.

Suppose that $H \neq 1$. Since f and g are relatively prime, we can surmise that H does not divide f. Then by the division theorem there exist q, r such that $f = Hq + r$, with $\deg(r) < \deg(H)$. Multiplying by π^ν and rearranging gives us $\pi^\nu r = \pi^\nu f - hq$, which implies that $\pi^\nu r \in (f, g)$. However, $\deg(\pi^\nu r) < \deg(h)$ which contradicts the choice of h. Therefore $h = \pi^\nu$.

Since f and g are coprime, one may assume that f is a monic polynomial, and we consider the subgroup $(\pi^\nu, f) \subseteq (f, g)$. Let $a(s) \in \mathcal{O}\langle\langle s\rangle\rangle$. Using the division theorem again we have $a = fq + r$ for some q, r with $\deg(r) < \deg(f)$, so that $a \equiv r$ modulo f. Since $\mathcal{O}/\pi^\nu\mathcal{O}$ is finite, there are only finitely many possibilities for r modulo π^ν, thus we see that (π^ν, f) – and also (f, g) – has finite index in $\mathcal{O}\langle\langle s\rangle\rangle$. □

Of course in the control theorem, we will need to consider cohomological properties of these $\mathcal{O}\langle\langle s\rangle\rangle$-modules. The first issue to address is to check for a finitely-generated $\mathcal{O}\langle\langle s\rangle\rangle$-module M, its interesting cohomologies are themselves finitely-generated. This is relatively simple to verify by extending p-adic results of Tate [Ta2, 2.1]. The following is a basic, yet highly useful result.

Proposition C.4. *Let M be a finitely generated free $A\langle\langle s\rangle\rangle$-module with a G-action. Then there exists a finite set of $\lambda \in X$, such that for all λ not in this set:*

(1) if $A = K$, then $H^i(G, M_\lambda) \cong H^i(G, M)_\lambda$;

(2) if $A = \mathcal{O}$ then $H^i(G, M)_\lambda \hookrightarrow H^i(G, M_\lambda)$, and we have a finite index $[H^i(G, M_\lambda) : H^i(G, M)_\lambda]$ which is bounded independently of λ.

Proof: Since M is free over $A\langle\langle s\rangle\rangle$, multiplication by λ gives a short exact sequence

$$0 \to M \stackrel{\times\lambda}{\to} M \to M_\lambda \to 0.$$

Taking G-invariants and truncating, one cuts out short exact sequences

$$0 \to H^i(G, M)_\lambda \to H^i(G, M_\lambda) \stackrel{\delta}{\to} H^{i+1}(G, M)[\lambda] \to 0.$$

Thus the index $[H^i(G, M_\lambda) : H^i(G, M)_\lambda]$ is determined by the size of the group $H^{i+1}(G, M)[\lambda]$. The following lemma then completes the proof.

Lemma C.5. *Let M be an $A\langle\langle s\rangle\rangle$-module.*

(1) If $A = K$, then for all but finitely many λ, we have $M[\lambda] = 0$.

(2) If $A = \mathcal{O}$, then for all but finitely many λ, the groups $M[\lambda]$ are finite and bounded independently of λ.

Proof: If A is an order $\mathcal{O}$, then let us write $M_K := M \otimes_{\mathcal{O}\langle\langle s\rangle\rangle} K\langle\langle s\rangle\rangle$ which is a $K\langle\langle s\rangle\rangle$-module. If $A = K$ we set $M_K := M$. We may view each irreducible distinguished polynomial λ as generating an ideal in $K\langle\langle s\rangle\rangle$, where we have the additional property that (λ) is a maximal ideal (see [BGR, 7.1.2]). We will first show that $M_K[\lambda] = 0$ for almost all λ, thus proving the first part of the lemma. The second part will follow readily from this.

Since M_K is a module of finite type over a noetherian ring, we have [Mt,Th.6.4] an ascending chain

$$0 = M_0 \subset M_1 \subset \ldots \subset M_n = M_K$$

of $K\langle\langle s\rangle\rangle$-submodules, such that $\frac{M_{i+1}}{M_i} \cong \frac{K\langle\langle s\rangle\rangle}{\mathcal{P}_i}$ for some prime ideal $\mathcal{P}_i$ of $K\langle\langle s\rangle\rangle$. Consequently, one obtains short exact sequences

$$0 \to M_i \to M_{i+1} \to \left(\frac{K\langle\langle s\rangle\rangle}{\mathcal{P}_i}\right) \to 0$$

for each i. Multiplying by λ and applying the Snake lemma, yields exact sequences

$$0 \to M_i[\lambda] \to M_{i+1}[\lambda] \to \left(\frac{K\langle\langle s\rangle\rangle}{\mathcal{P}_i}\right)[\lambda] \to \cdots$$

Since the M_i's form a finite chain, we see that $M_K[\lambda] = 0$ as long as each map $M_i[\lambda] \to M_{i+1}[\lambda]$ is an isomorphism. This inevitably leads us to consider the (λ)-torsion quotients $\left(\frac{K\langle\langle s\rangle\rangle}{\mathcal{P}_i}\right)[\lambda]$.

Suppose one has a power series $f \in K\langle\langle s\rangle\rangle$ with $f \notin \mathcal{P}_i$. Then $\lambda.f \in \mathcal{P}_i$ if and only if $\lambda \in \mathcal{P}_i$ (since $\mathcal{P}_i$ is prime), that is if and only if $(\lambda) = \mathcal{P}_i$ (since (λ) is maximal). Hence for $(\lambda) \neq \mathcal{P}_i$ we have $\left(\frac{K\langle\langle s\rangle\rangle}{\mathcal{P}_i}\right)[\lambda] = 0$. We may apply this to each term in our decreasing sequence, so that unless $(\lambda) = \mathcal{P}_i$ for some $0 \leq i \leq n-1$, we have $M_K[\lambda] = 0$ as required.

To finish the proof of the second part, we return to our original $\mathcal{O}\langle\langle s\rangle\rangle$-module which was M. There is a commutative diagram

$$\begin{array}{ccccccccc} 0 & \rightarrow & M[\lambda] & \rightarrow & M & \stackrel{\lambda}{\rightarrow} & M & \rightarrow & 0 \\ & & \tau\downarrow & & -\otimes 1\downarrow & & -\otimes 1\downarrow & & \\ 0 & \rightarrow & M_K[\lambda] & \rightarrow & M_K & \stackrel{\lambda}{\rightarrow} & M_K & \rightarrow & 0 \end{array}$$

which induces a map $\tau : M[\lambda] \to M_K[\lambda]$, say. For every $\lambda \neq \mathcal{P}_i$ $(i = 0, \ldots, n-1)$ one knows that $M_K[\lambda] = 0$, whence

$$M[\lambda] \hookrightarrow \operatorname{Ker}\left(M \to M \otimes_{\mathcal{O}\langle\langle s\rangle\rangle} K\langle\langle s\rangle\rangle\right) = M[\pi^\infty].$$

Here $M[\pi^\infty]$ consists of the π^n-torsion elements of M for all $n > 0$. Thus any element which is λ-torsion is also π^m-torsion for some m, and since M is finitely generated, there must be a finite maximum N such that multiplication by π^N kills all π^∞-torsion elements. By Lemma C.3 the ideal (λ, π^N) has finite index in $\mathcal{O}\langle\langle s\rangle\rangle$, and this index is bounded by p^N and the Weierstrass degree of λ.

Since these are both independent of λ, the result follows. □

C.3 The cohomology of a lattice $\mathbb{L}$

Suppose that $\mathbb{W}$ is a G_F-representation of pseudo-geometric type, with lattice $\mathbb{L}$. Consider the short exact sequence

$$0 \to \mathbb{L} \stackrel{\lambda}{\to} \mathbb{L} \to \mathbb{L}_\lambda \to 0. \qquad (1)$$

Taking Galois cohomology, we arrive at a second sequence

$$0 \to \frac{H^n(G, \mathbb{L})}{\lambda H^n(G, \mathbb{L})} \to H^n(G, \mathbb{L}_\lambda) \to H^{n+1}(G, \mathbb{L})[\lambda] \to 0. \qquad (2)$$

Here G denotes either $G_{F_v} := G(\bar{F}_v/F_v)$ for some prime v, or the group $G(F_\Sigma/F)$ where F_Σ is the maximal algebraic extension of F unramified outside of Σ.

Our aim is to use this sequence to compare the cohomology groups $H^n(G, \mathbb{L}_\lambda)$ and $H^n(G, \mathbb{L})_\lambda = \frac{H^n(G,\mathbb{L})}{\lambda H^n(G,\mathbb{L})}$. In order to do this, we must first try to understand the groups $H^{n+1}(G, \mathbb{L})[\lambda]$. For this we use Lemma C.5 which tells us that for almost all λ, these groups are finite and bounded independently of λ.

Let X be the set of all polynomials $\lambda_k = \left(s - (k - k_0)\right)$ for weights $k \in \mathbb{U}_{k_0}$.

Definition C.6. *We define the subset X^n_{bad} of X by the condition that $\lambda_k \in X^n_{bad}$ whenever there exists some $v \in \Sigma$ such that $H^n(F_v, \mathbb{L})[\lambda_k]$ has positive $\mathcal{O}\langle\langle s\rangle\rangle$-rank. Further, let's put $X_{bad} = \bigcup_{n=0}^{2} X^n_{bad}$ and likewise $X_{good} := X - X_{bad}$.*

By Lemma C.5 and the fact that Σ is a finite set, clearly X^n_{bad} is also a finite set. It then follows that X_{good} consists of all but a finite number of λ_k in X.

Upon applying Proposition C.4, one immediately deduces

Lemma C.7. *For all λ in X_{good}, there is an injective homomorphism*

$$\frac{H^1(F_v, \mathbb{L})}{\lambda H^1(F_v, \mathbb{L})} \hookrightarrow H^1(F_v, \mathbb{L}_\lambda)$$

which has finite cokernel, bounded independently of λ.

C.4 Local conditions

To be able to define compact Selmer groups, one must write down local conditions in the 1-cohomology of $\mathbb{L}$. In accordance with the main text, we shall employ

Definition C.8. *(i) When $v \mid p$, define $H^1_G(F_v, \mathbb{L})$ to be the $\mathcal{O}\langle\langle s\rangle\rangle$-saturation inside $H^1(F_v, \mathbb{L})$ of*

$$H^1_G(F_v, \mathbb{L})^0 := \mathrm{Ker}\left(H^1(F_v, \mathbb{L}) \to H^1(F_v, \mathbb{L} \otimes_{\mathcal{O}\langle\langle s\rangle\rangle} \mathbb{B}_{dR})\right).$$

For the specialisations one has

$$H^1_g(F_v, \mathbb{L}_\lambda)^0 := \mathrm{Ker}\left(H^1(F_v, \mathbb{L}_\lambda) \to H^1(F_v, \mathbb{L}_\lambda \otimes_{\mathcal{O}_\lambda} B_{dR})\right).$$

N.B. this group is automatically $\mathcal{O}_\lambda$-saturated since B_{dR} is a field.

(ii) For $v \nmid p$ we define $H^1_{nr}(F_v, \mathbb{L})$ (resp. $H^1_{nr}(F_v, \mathbb{L}_\lambda)$) to be the $\mathcal{O}\langle\langle s\rangle\rangle$-saturation (resp. $\mathcal{O}_\lambda$-saturation) of $H^1_{nr}(F_v, \mathbb{L})^0$ (resp. $H^1_{nr}(F_v, \mathbb{L}_\lambda)^0$), where as usual for a G_{F_v}-module M

$$H^1_{nr}(F_v, M)^0 := \mathrm{Ker}\left(H^1(F_v, M) \to H^1(I_v, M)^{\mathrm{Frob}_v}\right).$$

Our next task is to show that there exist homomorphisms for these subgroups

$$H^1_*(F_v, \mathbb{L})_\lambda \to H^1_\#(F_v, \mathbb{L}_\lambda)$$

that are injective, with finite cokernel bounded independently of the choice of λ. The notation H^1_* (resp. $H^1_\#$) represents either H^1_G (resp. H^1_g) or H^1_{nr} (resp. H^1_{nr}). We also require the same results for quotient groups, i.e. that the induced maps

$$\frac{H^1(F_v, \mathbb{L})_\lambda}{H^1_*(F_v, \mathbb{L})_\lambda} \to \frac{H^1(F_v, \mathbb{L}_\lambda)}{H^1_\#(F_v, \mathbb{L}_\lambda)}$$

are injective, with finite and bounded cokernels.

Local conditions away from p.
First let us consider those $v \in \Sigma$ which do not lie over p. Since $p \neq 2$, the local conditions at the infinite primes are trivial, so we have our required homomorphisms at these primes automatically.

By assumption (3) one knows that $H^0(F_v, \mathbb{W})$ is zero, and it follows from this that $H^0(F_v, \mathbb{L})$ must be a torsion $\mathcal{O}\langle\langle s\rangle\rangle$-module. Furthermore, we may use assumption (3) again, along with the pairing

$$H^0(F_v, \mathbb{W}^*(1)) \times H^2(F_v, \mathbb{W}) \stackrel{\cup}{\to} H^2(F_v, K\langle\langle s\rangle\rangle(1)) \stackrel{\mathrm{inv}_{F_v}}{\to} K\langle\langle s\rangle\rangle$$

to deduce that $H^2(F_v, \mathbb{W})$ has zero $K\langle\langle s\rangle\rangle$-rank. Therefore each $H^2(F_v, \mathbb{L})$ must have zero $\mathcal{O}\langle\langle s\rangle\rangle$-rank (since they are finitely-generated).

Now for each λ in X_{good}, there are local Euler characteristic formulae

$$\sum_{i=0}^{2} \mathrm{rank}_{\mathcal{O}\langle\langle s\rangle\rangle} H^i(F_v, \mathbb{L}) \;=\; \sum_{i=0}^{2} \mathrm{rank}_{\mathcal{O}_\lambda} H^i(F_v, \mathbb{L}_\lambda)$$

(this follows directly from equation (2)). We also have the short exact sequence

$$0 \to \mathbb{L}_\lambda \stackrel{\times p}{\to} \mathbb{L}_\lambda \to \mathbb{L}_\lambda / p\mathbb{L}_\lambda \to 0$$

induced by multiplication by p (the kernel of this map is trivial, since λ and p are coprime in $\mathcal{O}\langle\langle s\rangle\rangle$ always). Combining these facts, one quickly discovers

$$\sum_{i=0}^{2} \mathrm{rank}_{\mathcal{O}_\lambda} H^i(F_v, \mathbb{L}_\lambda) \;=\; \left(\mathrm{rank}_{\mathbb{Z}_p} \mathcal{O}_\lambda\right)^{-1} \times \sum_{i=0}^{2} \dim_{\mathbb{F}_p} H^i(F_v, \mathbb{L}_\lambda / p\mathbb{L}_\lambda). \qquad (3)$$

The groups $\mathbb{L}_\lambda / p\mathbb{L}_\lambda$ are finite G_{F_v}-modules, and it is a very well known result (see for example [Mi, I.2.9]) that the sum on the right-hand side is zero when $v \nmid p$. In particular, since $H^0(F_v, \mathbb{L})$ and $H^2(F_v, \mathbb{L})$ both have zero rank, it follows immediately that $\mathrm{rank}_{\mathcal{O}\langle\langle s\rangle\rangle} H^1(F_v, \mathbb{L}) = 0$ too.

This means that our subgroup $H^1_{nr}(F_v, \mathbb{L})$ has the same rank as $H^1(F_v, \mathbb{L})$, and by definition it is $\mathcal{O}\langle\langle s\rangle\rangle$-saturated in $H^1(F_v, \mathbb{L})$. We therefore have

$$H^1_{nr}(F_v, \mathbb{L}) \;=\; H^1(F_v, \mathbb{L}).$$

In fact equation (2) tells us that for all λ in X_{good}, the $\mathcal{O}_\lambda$-ranks of the specialised cohomology groups $H^i(F_v, \mathbb{L}_\lambda)$ (for $i = 0, 1, 2$) will be equal to the $\mathcal{O}\langle\langle s\rangle\rangle$-ranks of the generic groups $H^i(F_v, \mathbb{L})$. Since we have just seen that these latter groups all have rank zero, the same is true of the specialised groups, and in particular $\mathrm{rank}_{\mathcal{O}_\lambda} H^1(F_v, \mathbb{L}_\lambda) = 0$. It is then easy to see that $H^1_{nr}(F_v, \mathbb{L}_\lambda) = H^1(F_v, \mathbb{L}_\lambda)$.

We have established the following result.

Lemma C.9. *Let $v \in \Sigma$ with $v \nmid p$. Then for all $\lambda \in X_{good}$, the homomorphisms*

$$\frac{H^1_{nr}(F_v,\mathbb{L})}{\lambda H^1_{nr}(F_v,\mathbb{L})} \rightarrow H^1_{nr}(F_v,\mathbb{L}_\lambda)$$

are injective with finite cokernel, bounded independently of λ.

On the quotient groups, we have $H^1(F_v,\mathbb{L})/H^1_{nr}(F_v,\mathbb{L}) = 0$ and likewise for the specialised quotients, $H^1(F_v,\mathbb{L}_\lambda)/H^1_{nr}(F_v,\mathbb{L}_\lambda) = 0$ also.

Local conditions above p.

Let us now turn our attention to those $v \in \Sigma$ which lie above p. We know from [DS] that at all $v \mid p$ and almost all $\lambda \in \mathbb{U}_{k_0} \cap \mathbb{N}$,

$$\mathrm{rank}_{\mathcal{O}\langle\langle s\rangle\rangle} H^1_G(F_v,\mathbb{L}) \;=\; \mathrm{rank}_{\mathcal{O}_\lambda} H^1_g(F_v,\mathbb{L}_\lambda).$$

For all $\lambda \in X_{good}$, there exists a commutative diagram with left-exact rows

$$\begin{array}{ccccccc} 0 & \rightarrow & H^1_G(F_v,\mathbb{L})^0 & \rightarrow & H^1(F_v,\mathbb{L}) & \rightarrow & H^1(F_v,\mathbb{L}\otimes_{\mathcal{O}\langle\langle s\rangle\rangle}\mathbb{B}_{dR}) \\ & & \downarrow & & \downarrow & & \downarrow \\ 0 & \rightarrow & H^1_g(F_v,\mathbb{L}_\lambda) & \rightarrow & H^1(F_v,\mathbb{L}_\lambda) & \rightarrow & H^1(F_v,\mathbb{L}_\lambda\otimes_{\mathcal{O}} B_{dR}). \end{array}$$

We see immediately that the image of $H^1_G(F_v,\mathbb{L})^0$ in $H^1(F_v,\mathbb{L}_\lambda)$ is contained in $H^1_g(F_v,\mathbb{L}_\lambda)$. Since $H^1_g(F_v,\mathbb{L}_\lambda)$ is $\mathcal{O}_\lambda$-saturated, we may replace $H^1_G(F_v,\mathbb{L})^0$ by its $\mathcal{O}\langle\langle s\rangle\rangle$-saturation $H^1_G(F_v,\mathbb{L})$, and one obtains a map

$$\tilde{\alpha} : H^1_G(F_v,\mathbb{L}) \longrightarrow H^1_g(F_v,\mathbb{L}_\lambda).$$

The kernel of $\tilde{\alpha}$ is contained within $\lambda H^1(F_v,\mathbb{L})$, because $H^1(F_v,\mathbb{L})_\lambda$ injects into $H^1(F_v,\mathbb{L}_\lambda)$. By definition this kernel is also contained within $H^1_G(F_v,\mathbb{L})$, and since $H^1_G(F_v,\mathbb{L}) \cap \lambda H^1(F_v,\mathbb{L}) = \lambda H^1_G(F_v,\mathbb{L})$, we deduce that the homomorphism

$$\alpha : H^1_G(F_v,\mathbb{L})_\lambda \;\rightarrow\; H^1_g(F_v,\mathbb{L}_\lambda)$$

must be injective.

Now $H^1_G(F_v,\mathbb{L})$ is $\mathcal{O}\langle\langle s\rangle\rangle$-saturated, hence $\frac{H^1(F_v,\mathbb{L})}{H^1_G(F_v,\mathbb{L})}$ must be $\mathcal{O}\langle\langle s\rangle\rangle$-torsion free; in particular $\left(\frac{H^1(F_v,\mathbb{L})}{H^1_G(F_v,\mathbb{L})}\right)[\lambda] = 0$. Applying the Snake lemma to the diagram

$$\begin{array}{ccccccccc} 0 & \rightarrow & H^1_G(F_v,\mathbb{L}) & \rightarrow & H^1(F_v,\mathbb{L}) & \rightarrow & \left(\dfrac{H^1(F_v,\mathbb{L})}{H^1_G(F_v,\mathbb{L})}\right) & \rightarrow & 0 \\ & & \times\lambda\downarrow & & \times\lambda\downarrow & & \times\lambda\downarrow & & \\ 0 & \rightarrow & H^1_G(F_v,\mathbb{L}) & \rightarrow & H^1(F_v,\mathbb{L}) & \rightarrow & \left(\dfrac{H^1(F_v,\mathbb{L})}{H^1_G(F_v,\mathbb{L})}\right) & \rightarrow & 0, \end{array}$$

we get an exact sequence

$$0 = \left(\frac{H^1(F_v,\mathbb{L})}{H^1_G(F_v,\mathbb{L})}\right)[\lambda] \rightarrow H^1_G(F_v,\mathbb{L})_\lambda \rightarrow H^1(F_v,\mathbb{L})_\lambda \rightarrow \left(\frac{H^1(F_v,\mathbb{L})}{H^1_G(F_v,\mathbb{L})}\right)_\lambda \rightarrow 0$$

from which it may be deduced that $\left(\frac{H^1(F_v,\mathbb{L})}{H^1_G(F_v,\mathbb{L})}\right)_\lambda \cong \frac{H^1(F_v,\mathbb{L})_\lambda}{H^1_G(F_v,\mathbb{L})_\lambda}$.

One can therefore use these two groups interchangeably hereafter.

Lemma C.10. *For all $\lambda \in X_{good}$ and all $v \mid p$, the homomorphism*

$$\alpha : H^1_G(F_v, \mathbb{L})_\lambda \to H^1_g(F_v, \mathbb{L}_\lambda)$$

is injective, with cokernel that is finite and bounded independently of λ.

The induced map on the quotient groups

$$\bar{\beta} : \left(\frac{H^1(F_v, \mathbb{L})}{H^1_G(F_v, \mathbb{L})}\right)_\lambda \to \left(\frac{H^1(F_v, \mathbb{L}_\lambda)}{H^1_g(F_v, \mathbb{L}_\lambda)}\right)$$

has finite kernel and cokernel, each bounded independently of λ.

Proof: The following diagram nicely describes the link between the three maps of interest to us:

$$\begin{array}{ccccccccc} 0 & \to & H^1_G(F_v, \mathbb{L})_\lambda & \to & H^1(F_v, \mathbb{L})_\lambda & \to & \left(\frac{H^1(F_v, \mathbb{L})}{H^1_G(F_v, \mathbb{L})}\right)_\lambda & \to & 0 \\ & & {\scriptstyle\alpha}\downarrow & & {\scriptstyle\beta}\downarrow & & {\scriptstyle\bar{\beta}}\downarrow & & \\ 0 & \to & H^1_g(F_v, \mathbb{L}_\lambda) & \to & H^1(F_v, \mathbb{L}_\lambda) & \to & \left(\frac{H^1(F_v, \mathbb{L}_\lambda)}{H^1_g(F_v, \mathbb{L}_\lambda)}\right) & \to & 0. \end{array} \tag{4}$$

We have already seen that β is injective with finite bounded cokernel, and that α is injective. It remains to prove that $\bar{\beta}$ has finite kernel and cokernel, bounded independently of λ, and that α has finite and bounded cokernel. To prove this, we define a subgroup Z of $H^1(F_v, \mathbb{L})_\lambda$ as the pre-image of $H^1_g(F_v, \mathbb{L}_\lambda)$ under β.

The subgroup $H^1_g(F_v, \mathbb{L}_\lambda)$ is $\mathcal{O}_\lambda$-saturated, and since $\lambda \in X_{good}$, we know that $\mathrm{rank}_{\mathcal{O}\langle\langle s\rangle\rangle} H^1_G(F_v, \mathbb{L}) = \mathrm{rank}_{\mathcal{O}_\lambda} H^1_g(F_v, \mathbb{L}_\lambda)$. As the diagram (4) is commutative, we see that Z must equivalently be the $\mathcal{O}_\lambda$-saturation of $H^1_G(F_v, \mathbb{L})_\lambda$ inside $H^1(F_v, \mathbb{L})_\lambda$. Let us consider the kernel of $\bar{\beta}$; then one has

$$\begin{aligned} \mathrm{Ker}(\bar{\beta}) &\cong \frac{\{x \in H^1(F_v, \mathbb{L})_\lambda : \beta(x) \in H^1_g(F_v, \mathbb{L}_\lambda)\}}{H^1_G(F_v, \mathbb{L})_\lambda} \\ &= \frac{\beta^{-1}(H^1_g(F_v, \mathbb{L}_\lambda))}{H^1_G(F_v, \mathbb{L})_\lambda} \\ &= \frac{Z}{H^1_G(F_v, \mathbb{L})_\lambda}. \end{aligned}$$

Thus to show that $\mathrm{Ker}(\bar{\beta})$ is finite and bounded independently of λ, we must show that the same is true of the index of $H^1_G(F_v, \mathbb{L})_\lambda$ inside its saturation in $H^1(F_v, \mathbb{L})_\lambda$. The size of this index is bounded by the size of the maximal λ-torsion subgroup of $H^1(F_v, \mathbb{L})$, because $H^1_G(F_v, \mathbb{L})$ is $\mathcal{O}\langle\langle s\rangle\rangle$-saturated. By Lemma C.5 we see that $H^1(F_v, \mathbb{L})[\lambda]$ is indeed finite and bounded independently of λ, and the same is true of the kernel of $\bar{\beta}$.

It only remains to consider the cokernels of α and of $\bar{\beta}$. The property that these are finite and bounded independently of λ, follows immediately upon applying the Snake lemma to diagram (4) above.

□

C.5 Dualities via the Ext-pairings

In this section we consider instead a discrete dual of the lattice $\mathbb{L}$, namely

$$\mathbb{A} \ := \ \mathrm{Hom}_{\mathrm{cont}}(\mathbb{L}, \mu_{p^\infty})$$

which is the Pontryagin dual of $\mathbb{L}$ with its Galois action twisted by the cyclotomic character. Clearly $\mathbb{A}$ is a discrete $\mathcal{O}\langle\langle s\rangle\rangle$-module. We define the specialisation of $\mathbb{A}$ at λ as the discrete dual of the specialisation of $\mathbb{L}$ at λ, i.e. $\mathbb{A}_\lambda := \mathrm{Hom}_{\mathrm{cont}}(\mathbb{L}_\lambda, \mu_{p^\infty})$. Note that this is not the same as $\mathbb{A}/\lambda\mathbb{A}$, in fact we have $\mathbb{A}_\lambda = \mathbb{A}[\lambda]$.

The first task is to establish homomorphisms $H^1(G, \mathbb{A}_\lambda) \to H^1(G, \mathbb{A})[\lambda]$, and to show that for almost all λ in X_{good}, these homomorphisms have finite kernel and cokernel, bounded independently of λ. Our starting point is the sequence (1), to which we apply the left-exact functor $\mathrm{Hom}(-, \mu_{p^\infty})$. This gives us another exact sequence

$$0 \to \mathbb{A}_\lambda \to \mathbb{A} \xrightarrow{\lambda} \mathbb{A} \to \mathrm{Ext}^1(\mathbb{L}_\lambda, \mu_{p^\infty}) \to \mathrm{Ext}^1(\mathbb{L}, \mu_{p^\infty}) \to \ \dots \qquad (5)$$

We also have an analogous sequence naturally extending G-invariants of $\mathbb{A}$ and $\mathbb{A}_\lambda$ (where G here is either G_{F_v} or $G(F_\Sigma/F)$), defined as

$$0 \to H^0(G, \mathbb{A}_\lambda) \to H^0(G, \mathbb{A}) \xrightarrow{\lambda} H^0(G, \mathbb{A}) \to$$
$$\mathrm{Ext}^1_G(\mathbb{L}_\lambda, \mu_{p^\infty}) \to \mathrm{Ext}^1_G(\mathbb{L}, \mu_{p^\infty}) \xrightarrow{\lambda} \ \dots \qquad (6)$$

There is a spectral sequence for Ext groups [Mi, II.0] which for G-modules M, N is given by

$$H^r(G, \mathrm{Ext}^s(M, N)) \Rightarrow \mathrm{Ext}^{r+s}_G(M, N).$$

This gives rise to (yet another) exact sequence

$$0 \to H^1(G, \mathrm{Hom}(M, N)) \to \mathrm{Ext}^1_G(M, N) \to H^0(G, \mathrm{Ext}^1(M, N)) \to \ \dots \qquad (7)$$

In particular, we note $H^1(G, \mathbb{A})$ injects into $\mathrm{Ext}^1_G(\mathbb{L}, \mu_{p^\infty})$, and $H^1(G, \mathbb{A}_\lambda)$ injects into $\mathrm{Ext}^1_G(\mathbb{L}_\lambda, \mu_{p^\infty})$. Furthermore, from (6) there are natural homomorphisms

$$\mathrm{Ext}^1_G(\mathbb{L}_\lambda, \mu_{p^\infty}) \to \mathrm{Ext}^1_G(\mathbb{L}, \mu_{p^\infty})[\lambda] \qquad (8)$$

which are surjective.

Local duality.
In order to examine the kernel of these maps, we shall assume for the moment that $G = G_{F_v}$ for some $v \in \Sigma$. We then have Yoneda (or Ext) perfect pairings

$$\mathrm{Ext}^1_{G_{F_v}}(\mathbb{L}, \mu_{p^\infty}) \times H^1(F_v, \mathbb{L}) \xrightarrow{\cup} H^2(F_v, \mu_{p^\infty}) \xrightarrow{\mathrm{inv}_{F_v}} \mathbb{Q}_p/\mathbb{Z}_p,$$

and similarly

$$\mathrm{Ext}^1_{G_{F_v}}(\mathbb{L}_\lambda, \mu_{p^\infty}) \times H^1(F_v, \mathbb{L}_\lambda) \xrightarrow{\mathrm{inv}_{F_v} \circ \cup} \mathbb{Q}_p/\mathbb{Z}_p$$

at all specialisations $\lambda \in X$.

Combining these with the exact sequences (6) and (2), we arrive at the following diagram with exact columns:

$$\begin{array}{ccccc}
H^0(F_v, \mathbb{A}) & & 0 & & \\
\lambda \big\downarrow & & \big\uparrow & & \\
H^0(F_v, \mathbb{A}) & & H^2(F_v, \mathbb{L})[\lambda] & & \\
\delta \big\downarrow & & \big\uparrow & & \\
\mathrm{Ext}^1_{G_{F_v}}(\mathbb{L}_\lambda, \mu_{p^\infty}) & \times & H^1(F_v, \mathbb{L}_\lambda) & \to & \mathbb{Q}_p/\mathbb{Z}_p \\
\big\downarrow & & \big\uparrow & & \\
\mathrm{Ext}^1_{G_{F_v}}(\mathbb{L}, \mu_{p^\infty}) & \times & H^1(F_v, \mathbb{L}) & \to & \mathbb{Q}_p/\mathbb{Z}_p.
\end{array}$$

Let us consider the image of the map δ. As both pairings in the diagram are perfect, this subgroup must have an orthogonal complement $Y \subset H^1(F_v, \mathbb{L}_\lambda)$, and $\#H^1(F_v, \mathbb{L}_\lambda)/Y = \#\mathrm{Im}(\delta)$. Since the columns are exact, it follows that $H^1(F_v, \mathbb{L}_\lambda)/Y$ maps bijectively onto $H^2(F_v, \mathbb{L})[\lambda]$, and this group is finite and bounded independently of λ for all $\lambda \in X_{good}$. Thus the same is true of $\mathrm{Im}(\delta)$. Because the kernel of our homomorphism (8) must be equal to $\mathrm{Im}(\delta)$, we have proved the following:

Lemma C.11. *For all $\lambda \in X_{good}$ the homomorphism*

$$\mathrm{Ext}^1_{G_{F_v}}(\mathbb{L}_\lambda, \mu_{p^\infty}) \ \to\ \mathrm{Ext}^1_{G_{F_v}}(\mathbb{L}, \mu_{p^\infty})[\lambda]$$

is surjective, with kernel that is finite and bounded independently of λ.

We now wish to study the restriction of this homomorphism, as a map from the subgroup $H^1(F_v, \mathbb{A}_\lambda)$ to the subgroup $H^1(F_v, \mathbb{A})[\lambda]$. First we must check that the image of the former group is contained in the latter. This follows from a result of Milne [Mi, II.0], which states that if $M \times N \to P$ is a pairing of G-modules, then the diagram

$$\begin{array}{ccccll}
H^1(G, M) & \times & H^1(G, N) & \to & H^2(G, P) & \text{(cup-product pairing)} \\
\big\downarrow & & \big\downarrow & & \big\downarrow & \\
\mathrm{Ext}^1_G(N, P) & \times & H^1(G, N) & \to & H^2(G, P) & \text{(Ext pairing)}
\end{array}$$

commutes. Using this result alongside Lemma C.11, one obtains a homomorphism

$$H^1(F_v, \mathbb{A}_\lambda) \ \to\ H^1(F_v, \mathbb{A})[\lambda] \qquad (9)$$

as was specified above.

To calculate the kernel and cokernel of (9), one should apply the Snake lemma to the diagram

$$\begin{array}{ccccc} 0 \to H^1(F_v, \mathbb{A}_\lambda) & \to & \mathrm{Ext}^1_{G_{F_v}}(\mathbb{L}_\lambda, \mu_{p^\infty}) & \to & \mathrm{Ext}^1(\mathbb{L}_\lambda, \mu_{p^\infty})^{G_{F_v}} \\ \downarrow & & \downarrow & & \downarrow \\ 0 \to H^1(F_v, \mathbb{A})[\lambda] & \to & \mathrm{Ext}^1_{G_{F_v}}(\mathbb{L}, \mu_{p^\infty})[\lambda] & \to & \mathrm{Ext}^1(\mathbb{L}, \mu_{p^\infty})[\lambda]^{G_{F_v}} \end{array}$$

where the rows are derived from the exact sequence (7). Thus the kernel of (9) is contained in the kernel of (8), which we know by Lemma C.11 to be finite and bounded independently of λ. Its cokernel is then bounded by the cokernel of (8), together with the kernel of

$$\mathrm{Ext}^1(\mathbb{L}_\lambda, \mu_{p^\infty})^{G_{F_v}} \to \mathrm{Ext}^1(\mathbb{L}, \mu_{p^\infty})[\lambda]^{G_{F_v}}. \tag{10}$$

Moreover by Lemma C.11, the cokernel of (8) is finite and bounded independently of λ. To determine the kernel of (10), truncate the exact sequence (5) to obtain

$$0 \to \mathbb{A}/\lambda\mathbb{A} \to \mathrm{Ext}^1(\mathbb{L}_\lambda, \mu_{p^\infty}) \to \mathrm{Ext}^1(\mathbb{L}, \mu_{p^\infty})[\lambda] \to 0.$$

Taking G_{F_v}-invariants, we see that the kernel of (10) is equal to $(\mathbb{A}/\lambda\mathbb{A})^{G_{F_v}}$.

To make any further progress, unfortunately we are forced to assume:

Finiteness Hypothesis. *For each $v \in \Sigma$, the group $(\mathbb{A}/\lambda\mathbb{A})^{G_{F_v}}$ is finite and bounded independently of λ, at almost all $\lambda \in X$.*

This condition should not be too restrictive. Note for example that if $\mathbb{A}$ is λ-divisible for all bar finitely many λ, then the condition is satisfied with $(\mathbb{A}/\lambda\mathbb{A})^{G_{F_v}} = 0$. In fact, in this situation we will have equality between the Ext groups and the cohomology groups. This is certainly the case for Hida families, when the relevant component of the Hecke algebra is Gorenstein **and** the number field F is an abelian extension of $\mathbb{Q}$, in which the prime p does not ramify.

Let's now define a subset X'_{bad} of X by adjoining to X_{bad} all those λ for which $(\mathbb{A}/\lambda\mathbb{A})^{G_{F_v}}$ is infinite, and set $X'_{good} := X - X'_{bad}$. We have therefore shown

Lemma C.12. *For all $\lambda \in X'_{good}$ and $v \in \Sigma$, the homomorphism*

$$H^1(F_v, \mathbb{A}_\lambda) \to H^1(F_v, \mathbb{A})[\lambda]$$

has finite kernel and cokernel, bounded independently of λ.

Global duality.
We now consider the global Galois group $G_\Sigma := G(F_\Sigma/F)$. For each place $v \in \Sigma$ and for each G_Σ-module M, there exist restriction maps $H^i(G_\Sigma, M) \to H^i(G_{F_v}, M)$. One defines $\text{Ш}^i(G_\Sigma, M)$ to be the kernel of the map

$$H^i(G_\Sigma, M) \to \bigoplus_{v \in \Sigma} H^i(G_{F_v}, M).$$

Choosing $i = 2$ and $M = \mathbb{L}$, there is trivially a diagram

$$\begin{array}{ccccc} 0 \to \text{Ш}^2(G_\Sigma, \mathbb{L}) & \to & H^2(G_\Sigma, \mathbb{L}) & \overset{\oplus \mathrm{res}_v}{\to} & \bigoplus_{v\in\Sigma} H^2(G_{F_v}, \mathbb{L}) \\ \downarrow{\times\lambda} & & \downarrow{\times\lambda} & & \downarrow{\times\lambda} \\ 0 \to \text{Ш}^2(G_\Sigma, \mathbb{L}) & \to & H^2(G_\Sigma, \mathbb{L}) & \overset{\oplus \mathrm{res}_v}{\to} & \bigoplus_{v\in\Sigma} H^2(G_{F_v}, \mathbb{L}), \end{array}$$

from which one obtains (via the Snake lemma) the exact sequence

$$0 \to Z \to \text{Ш}^2(G_\Sigma, \mathbb{L})_\lambda \to H^2(G_\Sigma, \mathbb{L})_\lambda \to \bigoplus_{v\in\Sigma} \frac{\mathrm{res}_v\left(H^2(G_{F_v}, \mathbb{L})\right)}{\lambda \mathrm{res}_v\left(H^2(G_{F_v}, \mathbb{L})\right)}.$$

The module Z is a subquotient of $\bigoplus_{v\in\Sigma} \mathrm{res}_v\left(H^2(G_\Sigma, \mathbb{L})\right)[\lambda]$, which must itself be contained within $\bigoplus_{v\in\Sigma} H^2(G_{F_v}, \mathbb{L})[\lambda]$. By Lemma C.5 this latter object is finite and bounded independently of λ, so the same is true of Z.

Now from the exact sequence (2), there exist injections

$$H^2(G, \mathbb{L})_\lambda \hookrightarrow H^2(G, \mathbb{L}_\lambda)$$

with G either G_{F_v} or G_Σ. The cokernels of these maps are contained in $H^3(G, \mathbb{L})$, which is trivial as G has cohomological dimension 2. Thus $H^2(G, \mathbb{L})_\lambda \cong H^2(G, \mathbb{L}_\lambda)$ and one obtains a commutative diagram

$$\begin{array}{ccccc} 0 \to Z \to \text{Ш}^2(G_\Sigma, \mathbb{L})_\lambda & \to & H^2(G_\Sigma, \mathbb{L})_\lambda & \to & \bigoplus_{v\in\Sigma} H^2(G_{F_v}, \mathbb{L})_\lambda \\ \downarrow{\tau} & & \| & & \| \\ 0 \to \text{Ш}^2(G_\Sigma, \mathbb{L}_\lambda) & \to & H^2(G_\Sigma, \mathbb{L}_\lambda) & \to & \bigoplus_{v\in\Sigma} H^2(G_{F_v}, \mathbb{L}_\lambda). \end{array}$$

We can easily determine from this the kernel and cokernel of τ. Because both other vertical maps in the diagram are equalities, the cokernel must therefore be trivial. The kernel of τ must be equal to Z, and we have already seen that this is finite and bounded independently of λ, for all λ in X'_{good}.

To translate these dualities in terms discrete cohomology, we use the fact that for a G_Σ-module M there is a perfect pairing [Mi, II.4.14]

$$\text{Ш}^2(G_\Sigma, M) \times \text{Ш}^1(G_\Sigma, \mathrm{Hom}(M, \mu_{p^\infty})) \to \mathbb{Q}_p/\mathbb{Z}_p.$$

Examining the diagram

$$\begin{array}{ccccc} \text{Ш}^2(G_\Sigma, \mathbb{L}) & \times & \text{Ш}^1(G_\Sigma, \mathbb{A}) & \to & \mathbb{Q}_p/\mathbb{Z}_p \\ \downarrow{\times\lambda} & & \uparrow{\times\lambda} & & \\ \text{Ш}^2(G_\Sigma, \mathbb{L}) & \times & \text{Ш}^1(G_\Sigma, \mathbb{A}) & \to & \mathbb{Q}_p/\mathbb{Z}_p \\ \downarrow & & \uparrow & & \\ \text{Ш}^2(G_\Sigma, \mathbb{L})_\lambda & \times & \text{Ш}^1(G_\Sigma, \mathbb{A})[\lambda] & \to & \mathbb{Q}_p/\mathbb{Z}_p \end{array}$$

we deduce that the pairing $\text{Ш}^2(G_\Sigma, \mathbb{L})_\lambda \times \text{Ш}^1(G_\Sigma, \mathbb{A})[\lambda] \to \mathbb{Q}_p/\mathbb{Z}_p$ is also perfect.

Taken in conjunction with the analogous pairings for all the λ-specialised groups, one easily obtains the commutativity of

$$\begin{array}{ccccc} Ш^2(G_\Sigma,\mathbb{L})_\lambda & \times & Ш^1(G_\Sigma,\mathbb{A})[\lambda] & \to & \mathbb{Q}_p/\mathbb{Z}_p \\ \tau\downarrow & & \tau^\vee\uparrow & & \\ Ш^2(G_\Sigma,\mathbb{L}_\lambda) & \times & Ш^1(G_\Sigma,\mathbb{A}_\lambda) & \to & \mathbb{Q}_p/\mathbb{Z}_p. \end{array}$$

Since τ is surjective with finite and bounded kernel, the dual map $\tau^\vee$ must be injective with finite cokernel, bounded independently of λ.

Lemma C.13. *For all $\lambda \in X'_{good}$, the homomorphism*

$$b : H^1(G_\Sigma, \mathbb{A}_\lambda) \;\to\; H^1(G_\Sigma, \mathbb{A})[\lambda]$$

has finite kernel and cokernel, each bounded independently of λ.

Proof: We can relate $\tau^\vee$ to the map b via the diagram

$$\begin{array}{ccccc} 0 \to Ш^1(G_\Sigma,\mathbb{A}_\lambda) & \to & H^1(G_\Sigma,\mathbb{A}_\lambda) & \to & \bigoplus_{v\in\Sigma} H^1(G_{F_v},\mathbb{A}_\lambda) \\ \tau^\vee\downarrow & & b\downarrow & & \downarrow \\ 0 \to Ш^1(G_\Sigma,\mathbb{A})[\lambda] & \to & H^1(G_\Sigma,\mathbb{A})[\lambda] & \to & \bigoplus_{v\in\Sigma} H^1(G_{F_v},\mathbb{A})[\lambda]. \end{array}$$

Now $\tau^\vee$ is injective with a finite cokernel, and furthermore each of the local maps $H^1(G_{F_v},\mathbb{A}_\lambda) \to H^1(G_{F_v},\mathbb{A})[\lambda]$ has finite kernel and cokernel.

The result follows upon applying the Snake lemma.

□

C.6 Controlling the Selmer groups

To define discrete Selmer groups one needs to dualise these local conditions, which can be accomplished courtesy of the cup-product pairing

$$H^1(F_v,\mathbb{A}) \times H^1(F_v,\mathbb{L}) \;\to\; \mathbb{Q}_p/\mathbb{Z}_p.$$

For $v \mid p$ we define $H^1_E(F_v,\mathbb{A})$ as the orthogonal complement of $H^1_G(F_v,\mathbb{L})$ under this pairing. For $v \nmid p$ we define $H^1_{nr}(F_v,\mathbb{A})$ to be the orthogonal complement of $H^1_{nr}(F_v,\mathbb{L})$, which means that these groups are actually trivial!

Similarly, we define $H^1_e(F_v,\mathbb{A}_\lambda)$ and $H^1_{nr}(F_v,\mathbb{A}_\lambda)$ using the p-adic pairings

$$H^1(F_v,\mathbb{A}_\lambda) \times H^1(F_v,\mathbb{L}_\lambda) \to \mathbb{Q}_p/\mathbb{Z}_p.$$

By dualising our previous results, it is immediate that the maps

$$H^1_e(F_v,\mathbb{A}_\lambda) \;\to\; H^1_E(F_v,\mathbb{A})[\lambda]$$

and likewise

$$\frac{H^1(F_v,\mathbb{A}_\lambda)}{H^1_e(F_v,\mathbb{A}_\lambda)} \;\to\; \frac{H^1(F_v,\mathbb{A})}{H^1_E(F_v,\mathbb{A})}[\lambda]$$

have finite kernel and cokernel, bounded independently of λ.

Definition C.14. *For a general discrete module M, the Selmer group of M is defined to be*

$$\mathrm{Sel}(F_\Sigma/F, M) \;:=\; \mathrm{Ker}\left(H^1(G_\Sigma, M) \to \bigoplus_{v\in\Sigma} \frac{H^1(F_v, M)}{H^1_*(F_v, M)} \right)$$

where the $H^1_*(F_v, M)$ denote the local conditions for M.

This means (in our case) we form $\mathrm{Sel}(F_\Sigma/F, \mathbb{A})$ by taking $H^1_*(F_v, \mathbb{A}) = H^1_E(F_v, \mathbb{A})$ if $v|p$, and $H^1_{nr}(F_v, \mathbb{A})$ otherwise. For the specialisations, we define $\mathrm{Sel}(F_\Sigma/F, \mathbb{A}_\lambda)$ using $H^1_e(F_v, \mathbb{A}_\lambda)$ if $v|p$, and $H^1_{nr}(F_v, \mathbb{A}_\lambda)$ for all other $v \in \Sigma$.

The main result of this appendix is the following.

Theorem C.15. *Under the finiteness hypothesis, for all $\lambda \in X'_{good}$ the maps*

$$a : \mathrm{Sel}\big(F_\Sigma/F, \mathbb{A}_\lambda\big) \longrightarrow \mathrm{Sel}\big(F_\Sigma/F, \mathbb{A}\big)[\lambda]$$

have finite kernel and cokernel, each bounded independently of λ.

Proof: At every specialisation $\lambda \in X$, consider the commutative diagram

$$\begin{array}{ccccc}
0 \to \mathrm{Sel}(F_\Sigma/F, \mathbb{A}_\lambda) & \to H^1(G_\Sigma, \mathbb{A}_\lambda) & \to & \displaystyle\bigoplus_{v\in\Sigma} \frac{H^1(F_v, \mathbb{A}_\lambda)}{H^1_*(F_v, \mathbb{A}_\lambda)} \\
\Big\downarrow a & \Big\downarrow b & & \Big\downarrow c \\
0 \to \mathrm{Sel}(F_\Sigma/F, \mathbb{A})[\lambda] & \to H^1(G_\Sigma, \mathbb{A})[\lambda] & \to & \displaystyle\bigoplus_{v\in\Sigma} \left(\frac{H^1(F_v, \mathbb{A})}{H^1_*(F_v, \mathbb{A})}\right)[\lambda].
\end{array}$$

We have already established that each of the maps b and c, has finite and bounded kernel/cokernel at every $\lambda \in X'_{good}$.

The theorem immediately follows from an application of the Snake lemma. □

Bibliography

[Be] A. Beilinson, *Higher regulators and values of L-functions*, J. Soviet Math. **30** (1985), 2036-2070.

[BD] M. Bertolini and H. Darmon, *A rigid analytic Gross-Zagier formula and arithmetic applications*, Ann. Math. (2) **146** (1997), 111-147.

[BK] S. Bloch and K. Kato, *L-functions and Tamagawa numbers of motives*, in the Grothendieck Festchrift I, Progress in Math. **86**, Birkhäuser (1990), 333-400.

[BGR] S. Bosch, U. Güntzer and R. Remmert, *Non-archimedean analysis: a systematic approach to rigid analytic geometry*, Springer Verlag, Berlin (1984).

[BCDT] C. Breuil, B. Conrad, F. Diamond and R. Taylor, *On the modularity of elliptic curves over* $\mathbb{Q}$*: wild 3-adic exercises*, J. Amer. Math. Soc. **14** (2001), 843-939.

[Bu] Séminaire de Bures, *Périodes p-adiques*, Astérisque, Société Mathématique de France **223** (1994).

[Ca1] J.W.S. Cassels, *Lectures on Elliptic Curves*, LMS Student Texts **24** (1991).

[Ca2] J.W.S. Cassels, *Arithmetic on curves of genus 1 (VIII). On the conjectures of Birch and Swinnerton-Dyer*, J. Reine Angew. Math. **217** (1965), 180-199.

[CC] F. Cherbonnier and P. Colmez, *Théorie d'Iwasawa des représentations p-adiques d'un corps local*, J. Amer. Math. Soc. **12** (1999), 241-268.

[CW] J. Coates and A. Wiles, *On p-adic L-functions and elliptic units*, J. Aust. Math. Soc. (Series A) **26** (1978), 1-25.

[Co] R. Coleman, *Division values in local fields*, Invent. Math. **53** (1979), 91-116.

[Cz] P. Colmez, *Théorie d'Iwasawa des représentations de de Rham d'un corps local*, Ann. Math. **148** (1998), 485-571.

[CSS] G. Cornell, J. Silverman, and G. Stevens, *Modular forms and Fermat's last theorem*, Proc. of a Conference held at Boston University, Springer Verlag (1997).

[Cr] J. Cremona, *Algorithms for modular elliptic curves*, Second edition, Cambridge University Press (1997).

[Db1] D. Delbourgo, *Λ-adic Euler characteristics of elliptic curves*, Documenta Math. in honour of Coates' 60th birthday (2006), 301-323.

[Db2] D. Delbourgo, *On S-units and Beilinson elements at weight one*, in preparation.

[Db3] D. Delbourgo, *Vertical heights of Heegner points and L-functions*, in preparation.

[DS] D. Delbourgo and P. Smith, *Kummer theory for big Galois representations*, Math. Proc. Camb. Phil. Soc. **142** (2007), 205-217.

[De] P. Deligne, *Formes modulaires et représentations l-adiques*, Séminaires Bourbaki, Lecture Notes in Math. **179**, Springer Verlag (1969), 139-172.

[ES] M. Eichler, *Quaternare quadratische Formen und die Riemannische Vermutung für die Kongruenzzetafunktion*, Arch. Math. **5** (1954), 355-366.

[EPW] M. Emerton, R. Pollack and T. Weston, *Variation of Iwasawa invariants in Hida families*, Invent. Math. **163** (2006), 523-580.

[Fa] G. Faltings, *Crystalline cohomology of semi-stable curves, the $\mathbb{Q}_p$-theory*, J. Alg. Geom. **6** (1997), 1-18.

[Fl] M. Flach, *A generalisation of the Cassels-Tate pairing*, J. Reine Angew. Math. **412** (1990), 113-127.

[FM] J-M. Fontaine and W. Messing, *p-adic periods and p-adic étale cohomology*, Contemporary Math. **67** (1987), 179-207.

[FPR] J-M. Fontaine and B. Perrin-Riou, *Autour des conjectures de Bloch et Kato: cohomologie galoisienne et valeurs de fonctions L*, in Motives (Seattle), Proc. Symposia in Pure Math. **55** part I (1994), 599-706.

[Fu] T. Fukaya, *Coleman power series for K_2 and p-adic zeta functions of modular forms*, Documenta Math. in honour of Kato's 50th birthday (2003), 387-442.

[Gr1] R. Greenberg, *Iwasawa theory and p-adic deformations of motives*, Proc. Symp. Pure Math. **55** part II, Amer. Math. Soc. (1994), 193-223.

[Gr2] R. Greenberg, *Iwasawa theory for elliptic curves*, in Arithmetic of Elliptic Curves – Cetraro 1997, Lecture Notes in Math. **1716** (1999), 51-144.

[GS] R. Greenberg and G. Stevens, *p-adic L-functions and p-adic periods of modular forms*, Invent. Math. **111** (1993), 401-447.

[GV] R. Greenberg and V. Vatsal, *On the Iwasawa invariants of elliptic curves*, Invent. Math. **142** (2000), 17-63.

[Ha] M. Harris, *p-adic representations arising from descent on abelian varieties*, Compositio Math. **39** no. 2 (1979), 177-245.

[Hi1] H. Hida, *Galois representations into $GL_2(\mathbb{Z}_p[[X]])$ attached to ordinary cusp forms*, Invent. Math. **85** (1986), 545-613.

[Hi2] H. Hida, *Iwasawa modules attached to congruences of cusp forms*, Ann. Sci. École Norm. Sup. (4) **19** (1986), 231-273.

[Hi3] H. Hida, *A p-adic measure attached to the zeta-functions associated with two elliptic modular forms I*, Invent. Math. **79** (1985), 159-195.

[Ho] B. Howard, *The Iwasawa theoretic Gross-Zagier theorem*, Compositio Math. **141** no. 4 (2005), 811-846.

[Hy] O. Hyodo, *A note on the p-adic étale cohomology in the semi-stable reduction case*, Invent. Math. **91** (1988), 543-557.

[Im] H. Imai, *A remark on the rational points of abelian varieties with values in cyclotomic $\mathbb{Z}_l$-extensions*, Proc. Japan Acad. Ser. A **51** (1975), 12-16.

[IS] A. Iovita and G. Stevens, *p-adic variation of p-adic periods of modular forms: the general case*, preprint (2003).

[JS] H. Jacquet and J. Shalika, *Euler products and the classification of automorphic forms I*, Amer. J. Math. **103** (1981), 499-558.

[Ja] U. Jannsen, *Continuous étale cohomology*, Mathematische Annalen **280** no. 2 (1988), 207-245.

[Ka1] K. Kato, *p-adic Hodge theory and values of zeta functions of modular forms*, Astérisque **295** ix (2004), 117-290.

[Ka2] K. Kato, *Generalized explicit reciprocity laws*, Advanced Studies in Contemp. Math. **1** (1999), 57-126.

[Ka3] K. Kato, *Euler systems, Iwasawa theory and Selmer groups*, Kodai Math. J. **22** (1999), 313-372.

[Ka4] K. Kato, *Logarithmic structures of Fontaine-Illusie*, in Algebraic Analysis, Geometry and Number Theory, John Hopkins University Press (1989), 191-224.

[KKT] K. Kato, M. Kurihara and T. Tsuji, *Local Iwasawa theory of Perrin-Riou and syntomic complexes*, preprint (1996).

[Kz] N. Katz, *p-adic L-functions for CM fields*, Invent. Math. **49** (1978), 199-297.

[Ki] K. Kitagawa, *On standard p-adic L-functions of families of elliptic cusp forms*, in p-adic Monodromy and the Birch and Swinnerton-Dyer Conjecture, Contemp. Mathematics **165**, Amer. Math. Soc. (1994), 81-110.

[Kn] A. Knapp, *Elliptic curves*, Princeton University Press **40** (1992).

[Ky] S. Kobayashi, *Iwasawa theory for elliptic curves at supersingular primes*, Invent. Math. **152** (2003), 1-36.

[Ko] N. Koblitz, *Introduction to elliptic curves and modular forms*, Graduate Texts in Mathematics, Springer Verlag **97** (1985).

[Kv] V. Kolyvagin, *Euler systems*, in the Grothendieck Festchrift II, Progress in Math. **86**, Birkhäuser (1990), 435-483.

[Mn] J. Manin, *Periods of cusp forms and p-adic Hecke series*, Math. USSR-Sbornik **21** (1973), 371-393.

[Mt] H. Matsumura, *Commutative ring theory*, Cambridge Studies in Advanced Mathematics **8**, Cambridge University Press (1989).

[Mz1] B. Mazur, *Modular curves and the Eisenstein ideal*, Institut Hautes Études Sci. Publ. Math. **47** (1977), 33-186.

[Mz2] B. Mazur, *Rational points of abelian varieties with values in towers of number fields*, Invent. Math. **18** (1972), 183-266.

[MR] B. Mazur and K. Rubin, *Kolyvagin systems*, Amer. Math. Soc. **168** no. 799 (2004).

[MSD] B. Mazur and P. Swinnerton-Dyer, *Arithmetic of Weil curves*, Invent. Math. **25** (1974), 1-61.

[MTT] B. Mazur, J. Tate and J. Teitelbaum, *On p-adic analogues of the conjectures of Birch and Swinnerton-Dyer*, Invent. Math. **84** (1986), 1-48.

[MW] B. Mazur and A. Wiles, *On p-adic analytic families of Galois representations*, Compositio Math. **59** (1986), 231-264.

[Mi] J.S. Milne, *Arithmetic duality theorems*, Perspectives in Mathematics **1**, Academic Press Incorporated, Boston (1986).

[My] T. Miyake, *Modular Forms*, Springer Verlag, Berlin-New York (1989).

[Ne1] J. Nekovář, *On p-adic height pairings*, Séminaire de Théorie des Nombres de Paris, Progr. Math. **108**, Academic Press (1993), 127-202.

[Ne2] J. Nekovář, *On the p-adic height of Heegner cycles*, Math. Annalen **302** (1995), 609-686.

[Ne3] J. Nekovář, *Selmer complexes*, Astérisque, Société Mathématique de France **310** (2006).

[NP] J. Nekovář and A. Plater, *On the parity of ranks of Selmer groups*, Asian J. Math. **4** no. 2 (2000), 437-497.

[Oc1] T. Ochiai, *A generalization of the Coleman map for Hida deformations*, J. Amer. Math. Soc. **125** (2003), 849-892.

[Oc2] T. Ochiai, *Euler system for Galois deformation*, Annales de l'Insitut Fourier **55** (2005), 113-146.

[Oh] M. Ohta, *On l-adic representations attached to automorphic forms*, in Jap. J. Math. **8** (1982), 1-47.

[PR1] B. Perrin-Riou, *Théorie d'Iwasawa et hauteurs p-adiques*, Invent. Math. **109** (1992), 137-185.

[PR2] B. Perrin-Riou, *Théorie d'Iwasawa des représentations p-adiques sur un corps local*, Invent. Math. **115** (1994), 81-149.

[PR3] B. Perrin-Riou, *Fonctions L p-adiques des représentations p-adiques*, Astérisque, Société Mathématique de France **229** (1995).

[PR4] B. Perrin-Riou, *Systèmes d'Euler p-adiques et théorie d'Iwasawa*, Annales Inst. Fourier **48** (1998), 1231-1307.

[Pl] A. Plater, *Height pairings in families of deformations*, J. Reine Angew. Math. **486** (1997), 97-127.

[Po] R. Pollack, *On the p-adic L-function of a modular form at a supersingular prime*, Duke Math. J. **118** no. 3 (2003), 523-558.

[Ro] D. Rohrlich, *On L-functions of elliptic curves and cyclotomic towers*, Invent. Math. **75** (1984), 409-423.

[Ru1] K. Rubin, *The main conjectures of Iwasawa theory for imaginary quadratic fields*, Invent. Math. **103** no. 1 (1991), 25-68.

[Ru2] K. Rubin, *Euler systems*, Ann. Math. Studies **147**, Princeton University Press (2000).

[StÉ] K. Barré-Sireix, G. Diaz, F. Gramain, and G. Philibert, *Une preuve de la conjecture Mahler-Manin*, Invent. Math. **124** (1996), 1-9.

[Sa] T. Saito, *Modular forms and p-adic Hodge theory*, Invent. Math. **129** (1997), 607-620.

[Sn1] P. Schneider, *p-adic height pairings I*, Invent. Math. **69** (1982), 401-409.

[Sn2] P. Schneider, *p-adic height pairings II*, Invent. Math. **79** (1985), 329-374.

[Sch] T Scholl, *An introduction to Kato's Euler systems*, in Galois Representations in Arithmetic Algebraic Geometry, LMS Lecture Notes **254** (1998), 379-460.

[Se] J-P. Serre, *Corps locaux*, Hermann, Paris (1968).

[Sh] G. Shimura, *Introduction to the arithmetic theory of automorphic functions*, Princeton University Press (1971).

[Si1] J. Silverman, *The arithmetic of elliptic curves*, Graduate Texts in Mathematics, Springer Verlag **106** (1986).

[Si2] J. Silverman, *Advanced topics in the arithmetic of elliptic curves*, Graduate Texts in Mathematics, Springer Verlag **151** (1994).

[ST] J. Silverman and J. Tate, *Rational points on elliptic curves*, Undergraduate Texts in Mathematics, Springer Verlag (1992).

[SU] C. Skinner and E. Urban, work in progress.

[Sm] P. Smith, *The arithmetic of Galois representations over affinoids*, PhD Thesis, University of Nottingham (2005).

[St] G. Stevens, *Stickelberger elements and the modular parametrizations of elliptic curves*, Invent. Math. **98** no. 1 (1989), 75-106.

[Ta1] J. Tate, *A review of non-archimedean elliptic functions*, Elliptic Curves, Modular Forms and Fermat's Last Theorem, International Press (1995), 162-184.

[Ta2] J. Tate, *p-divisible groups*, Proc. of a Conference on Local Fields in Driebergen, Springer Verlag (1966), 158-183.

[TW] R. Taylor and A. Wiles, *Ring theoretic properties of certain Hecke algebras*, Ann. Math. **141** (1995), 553-572.

[Ti] J. Tilouine, *Un sous-groupe p-divisible de la jacobienne de $X_1(Np^r)$ comme modules sur l'algébre de Hecke*, Bull. Soc. Math. France **115** (1987), 329-360.

[Wi1] A. Wiles, *Modular elliptic curves and Fermat's last theorem*, Ann. Math. **141** (1995), 443-551.

[Wi2] A. Wiles, *On ordinary λ-adic representations associated to modular forms*, Invent. Math. **94** (1988), 529-573.

[Ya] R. Yager, *On two-variable p-adic L-functions*, Ann. Math. **115** (1982), 411-449.

Index